Cloud Dynamics

Photograph taken by astronauts aboard the *Apollo Saturn* spacecraft during April 1970. The picture is centered over the north Pacific Ocean. The Kamchatka Peninsula of Siberia can be seen just northwest of a large oceanic extratropical cyclone. The sun's reflection off the ocean surface is seen to the south of the storm. A belt of mesoscale convective cloud systems extends east–west near the equator.

This is Volume 53 in the
INTERNATIONAL GEOPHYSICS SERIES
A series of monographs and textbooks
Edited by RENATA DMOWSKA and JAMES R. HOLTON

A complete list of the books in this series appears at the end of this volume.

Cloud Dynamics

Robert A. Houze, Jr.
DEPARTMENT OF ATMOSPHERIC SCIENCES
UNIVERSITY OF WASHINGTON
SEATTLE, WASHINGTON

ACADEMIC PRESS, INC.
A Division of Harcourt Brace & Company
San Diego   New York   Boston
London   Sydney   Tokyo   Toronto

Cover art designed by Scott Braun and K. M. Dewar, Department of Atmospheric Sciences, University of Washington.

Academic Press, Inc.
1250 Sixth Avenue, San Diego, California 92101-4311

United Kingdom Edition published by
Academic Press Limited
24–28 Oval Road, London NW1 7DX

Library of Congress Cataloging-in-Publication Data

Cloud Dynamics / Robert A. Houze, Jr.
    p.   cm. — (International geophysics ; v. 53)
   ISBN 0-12-356880-3
   1. Clouds—Dynamics. I. Title. II. Series.

  QC921.6.D95H68   1993
  551.57'6–dc20   92-33769
                                             CIP

PRINTED IN THE UNITED STATES OF AMERICA
93 94 95 96 97 98   EB   9 8 7 6 5 4 3 2 1

Contents

Part I   Fundamentals

Chapter 1   Identification of Clouds

Chapter 2   Atmospheric Dynamics

Chapter 3   Cloud Microphysics

Chapter 4   Radar Meteorology

Part II   Phenomena

Chapter 5   Shallow-Layer Clouds

Chapter 6   Nimbostratus

Chapter 7   Cumulus Dynamics

Chapter 8   Thunderstorms

Chapter 9   Mesoscale Convective Systems

Chapter 10   Clouds in Hurricanes

Chapter 11   Precipitating Clouds in Extratropical Cyclones

Chapter 12   Orographic Clouds

Preface

Clouds are a vital link in the global climate and water cycle and an integral part of weather forecasting and analysis. Serious students of the earth's atmosphere can little afford to ignore the physics of clouds and their effects on the atmosphere. Over the past twenty years, a two-quarter course of study has been developed at the University of Washington to provide the necessary background to students destined to work in some phase of atmospheric science. The course notes have evolved substantially in concert with advances in modeling, theory, and observations. Within the last few years the course notes have stabilized, and they now treat rather comprehensively and cohesively the air motions and larger-scale physics of clouds and systems of clouds. Hence, the decision was made to formalize the material into a book that could be of use to a wider community of atmospheric scientists. It should be useful to researchers, operational meteorologists, graduate students, and advanced undergraduates.

One of the most gratifying aspects of writing this book has been the help I have received from both individuals and organizations. I am particularly indebted to those who have read and commented on large portions of the text. These include Marcia Baker, Scott Braun, Randy Brown, Christopher Bretherton, Richard E. Carbone, Shuyi S. Chen, Dale R. Durran, Spiros G. Geotis, Peter V. Hobbs, James R. Holton, Brian E. Mapes, Clifford F. Mass, Arthur L. Rangno, Richard J. Reed, James Renwick, Steven Rutledge, Christoph Schär, Lloyd Shapiro, Robert Solomon, Matthias Steiner, Ming-Jen Yang, and Sandra Yuter. The detailed editing of the manuscript and darkroom photography were expertly performed by Grace C. Tutt-Gudmundson, while the graphic arts were produced by Kay M. Dewar. Deborah Houze served admirably as editorial assistant. Many authors have graciously allowed me to reproduce figures from their texts, and in many cases they provided original copies for my use. They are all acknowledged in the figure captions. A wealth of photos and satellite pictures, from which I was able to select examples for this text, were provided to me by Ernest M. Agee, Robert Breidenthal, Steven Businger, Howard B. Bluestein, Kelvin K. Droegemeier, Dale R. Durran, Ronald Holle, Joachim P. Kuettner, Arthur L. Rangno, Shenqyang Shy, Aarnout van Delden, and Morton G. Wurtele. A large part of this book reflects my own research on clouds and precipitation, which has been sponsored by the National Science Foundation, the National Oceanic and Atmospheric Administration, and the National Aeronautics and Space Administration. I began this book in 1988–1989, while I was on sabbatical leave from the University of Washington and serving as Guest Professor at the Swiss Federal Institute of Technology in Zürich, Switzerland. Without that year of focused effort, I would probably never have been able to complete the project. I am indebted to both universities for making that year possible.

Introduction

A cloud is defined as "a visible aggregate of minute particles of water or ice, or both, in the free air."[1] Clouds cover about half of the earth at any given time. Thus, from space, the earth appears as a planet semi-enshrouded by these "visible aggregates" of drops and ice crystals (see the frontispiece). The pattern of clouds seen from space fluctuates strongly as they form, dissipate, and move in conjunction with the fluid motion of the air. These masses of cloud are an integral part of weather, climate, and the global water cycle, and they have been the object of serious scientific study—as well as art, music, history, religion, poetry, proverbs, mysticism, and mythology—since at least the time of Aristotle, and probably long before.[2]

About 1940, the discipline of *cloud physics* began to emerge as an identifiable specialty in meteorology. The emphasis in these early years was on *cloud microphysics,* the study of the formation and growth of the "minute particles" making up the "visible aggregates." As early as 1957, B. J. Mason recognized that cloud microphysics was really only one branch of cloud studies, and that another branch needed to be developed. He pointed out that the microphysical processes "are largely controlled by the atmospheric motions that are manifest in clouds. These macrophysical features of cloud formation and growth, which might more properly be called a *dynamics*, provide a framework of environmental conditions confining the rates and duration of the microphysical events." In 1971, Mason reiterated "the importance of acquiring a much deeper understanding of cloud dynamics for the development of the subject [of cloud studies] as a whole."[3]

The focus of this book is on the air motions associated with clouds. In keeping with Mason's terminology, we refer to this discipline as *cloud dynamics*. The development of cloud dynamics has trailed that of cloud microphysics largely because the technologies required to observe and model cloud dynamics are so demanding. The proper documentation of cloud dynamics entails scales of air motion from <1 km to >1000 km. At the same time that the air motions are documented across these scales, the microphysical processes, occurring in the context of the dynamics, must continue to be accounted for, at least in some gross

---

[1] From the *International Cloud Atlas, Volume I* (World Meteorological Organization, 1956).

[2] According to Hobbs (1989), studies of the physics of clouds may have begun with the Chinese in 135 B.C., when it was noted that snow crystals are six-sided.

[3] From *The Physics of Clouds*, by B. J. Mason, Preface to the First Edition (1957) and Preface to the Second Edition (1971). Both prefaces are printed on pp. v–ix of the second edition.

fashion. To accomplish necessary documentation and understanding across so many spatial scales,[4] special aircraft, radars, computers, satellites, and numerical-modeling techniques must be employed.

Much of the technology needed to study cloud dynamics now exists. Consequently, the subject has reached sufficient maturity that most of the salient dynamics of all the different types of clouds in the earth's atmosphere have been identified and can be discussed as a coherent discipline together in a single volume. That is the purpose of this text.

The book is divided into two parts. Part I is called "Fundamentals" and consists of the first four chapters. It is devoted to a review of the fundamental scientific background needed to study cloud dynamics. In Chapter 1 we review the variety of clouds that exist in the atmosphere and the nomenclature used to identify them. Air motions throughout the atmosphere are governed by the laws of classical fluid dynamics, and Chapter 2 summarizes the principles of atmospheric dynamics that are essential to the study of cloud dynamics. The dynamics of the air motions are inseparable from the microphysical processes; the air motions determine whether the air is receptive to the formation and development of the microscopic particles composing the cloud, while the heat released or consumed by phase changes of water in clouds and the friction between the particles and the air constitute feedback of the cloud microphysics onto the dynamics. Since knowledge of cloud microphysics is essential to understanding cloud dynamics, Chapter 3 gives an overview of cloud microphysics. One of the primary tools for observing clouds, especially precipitating clouds, is meteorological radar, which is a remote-sensing tool with a variety of special characteristics that should be borne in mind by the student of cloud dynamics. Hence, Chapter 4 offers a summary of the subject of radar meteorology.

Part II is called "Phenomena." It treats the dynamics of each type of cloud and cloud system that occurs in the atmosphere. Each of the chapters in Part II (Chapters 5–12) examines a particular category of cloud. The discussions comprising these chapters draw upon the reservoir of observational and theoretical fundamentals provided in Part I. The reader already familiar with the fundamentals reviewed in Part I may find it possible to proceed directly to Part II. Those just being introduced to the topics in Part I may wish to supplement that material by seeking out more detailed texts, several of which will be mentioned in footnotes of Part I. For all readers (as well as the author), Part I serves as a handy reference, both while reading Part II and when studying the current scientific literature on cloud dynamics.

---

[4] Hobbs (1981) noted that the phenomena involved in cloud dynamics range from "the nucleation of cloud particles to baroclinic waves. In terms of spatial scales, 15 orders of magnitude are involved," which is the same as "the ratio of the linear dimensions of the Milky Way to those of the earth."

List of Symbols

A an area

$\mathscr{A}$ an arbitrary variable

$\mathscr{A}^{\#}$ turbulent deviation from $\mathscr{A}_r(r)$

A_c autoconversion

$\mathscr{A}_c$ value of a variable $\mathscr{A}$ in a cloud; $\mathscr{A}$ may be h, T, p, q_i, u, w, or ρ

$\mathscr{A}_e$ value of a variable $\mathscr{A}$ in the environment of a cloud; $\mathscr{A}$ may be h, T, p, q_i, u, w, or ρ

A_m effective cross-sectional area swept out by a particle of mass m

A_T total area covered by rain in a given region

A_X x-gradient of nondimensional pressure perturbation

A_Z contribution to z-gradient of nondimensional pressure perturbation by advection and turbulence

$\mathscr{A}_{BL}$ value of $\overline{\mathscr{A}}$ in a well-mixed boundary layer

$\mathscr{A}_{SFC}$ value of $\overline{\mathscr{A}}$ at the sea surface

A_o, B_o positive constants in the expression for the radiative flux divergence in an ice-cloud outflow from cumulonimbus

$\hat{A}$, $\hat{B}$, $\hat{C}$ constants

$A_E(r)$ total area covered by echoes within an annular range ring centered at range r

$\mathscr{A}_r(r)$ basic radial distribution of $\mathscr{A}^*$ across a cloud

$A(\mathfrak{R}_\tau)$ area covered by rain with rates exceeding $\mathfrak{R}_\tau$

$A(Z_e, r)$ area covered by echoes within an annular range ring centered at range r with reflectivity between Z_e and $Z_e + dZ_e$

$(\Delta \mathscr{A}_c / \Delta t)_S$ rate of change of $\mathscr{A}_c$ that would occur if a parcel were not exchanging mass with the environment

α expansion coefficient

a a positive constant relating the buoyancy generation of eddy kinetic energy and the overall generation of eddy kinetic energy

a proportionality constant relating tangential velocity to radius in the inner region of a Rankine vortex

a_H half-width of a mountain ridge

a_I constant in the expression for N_I

a_o, a_1, b_1, a_2, b_2 coefficients in the Fourier decomposition of the mean radar radial velocity

a_T autoconversion threshold

$$\hat{a} \equiv \frac{2\sigma_{LV}}{\rho_L R_v T}$$

$\bar{a}, \bar{b}, \bar{a}_1, \bar{b}_1, \bar{a}_2, \bar{b}_2$ positive constants in empirical formulas relating radar reflectivity to rainfall rate, rainwater mixing ratio, and precipitation particle fall speed

α specific volume (ρ^{-1})

α_a azimuth angle (measured clockwise from the north) toward which radar beam is pointing

α_e elevation angle of radar beam

α_i proportionality factor relating the volume of a polyhedron to that of an inscribed sphere

α_o adjustable parameter relating sensible heat fluxes at cloud base to those at cloud top

α_L^2 positive constant in the expression for $\mathcal{E}_2$

α_ε an empirical constant determined in laboratory experiments on turbulent elements

$\tilde{\alpha}$ proportionality constant in formula for autoconversion

$\hat{\alpha}, \hat{\beta}$ constants

$\mathcal{B}$ creation (destruction) of $\mathcal{K}$ by thermally direct (indirect) flow perturbations

B buoyancy

$\mathfrak{B}_i$ time rate of change of the concentration of drops per unit size interval in size range i resulting from drop breakup

$$B_T = \bar{u}\frac{\partial \bar{\theta}_a}{\partial x} + \bar{w}\frac{\partial \bar{\theta}_a}{\partial z}$$

B_X x-gradient of apparent potential temperature perturbation

BPGA   buoyancy pressure gradient acceleration

b   proportionality constant relating tangential velocity to inverse radius in the outer region of a Rankine vortex

b In Chapter 7, b represents the radius of a turbulent element; in Chapter 11, it is defined as the quantity $g\theta/\hat{\theta}$

$$\hat{b} \equiv \frac{3i_{vH}m_s M_w}{4\pi\rho_L M_s}$$

β_i proportionality factor relating the surface area of a polyhedron to that of an inscribed sphere

β_T proportionality factor in the expression for the eddy flux of cloud virtual potential temperature

$\mathcal{C}$ rate at which $\mathcal{K}$ is created by conversion from mean-flow kinetic energy, when down-gradient eddy momentum fluxes occur

C condensation of vapor

C_A empirical coefficient in the bulk aerodynamic formula for surface flux of $\mathcal{A}$

C_c condensation of cloud water

C_D empirical coefficient (drag coefficient) in the bulk aerodynamic formula for surface flux of $\mathcal{A} = m$

C_d deposition rate

$\tilde{\mathfrak{C}}_i$ time rate of change of the concentration of drops per unit size interval in size range i resulting from collection

C_R a constant depending on the characteristics of a particular set of radar equipment

C_S empirical coefficient in the bulk aerodynamic formula for surface flux of $\mathcal{A} = c_p \ln \theta_e$

CAPE convective available potential energy

CDR circular depolarization ratio

$\tilde{C}$ shape factor analogous to electrical capacitance

c one-half the horizontal convergence in a one- or two-cell vortex

c phase speed

c_o speed of light

c_p specific heat of dry air at constant pressure

c_s specific heat at constant pressure of soil

c_v specific heat of dry air at constant volume

c_w specific heat of water

$\hat{c}$ specific heat of a homogeneous fluid

D particle diameter

$\mathcal{D}$ domain of radar observations

$\mathscr{D}$ (<0) frictional dissipation of $\mathcal{K}$ by eddies or molecular friction

D_h diameter of a hailstone

$\mathcal{D}_i$ time rate of change of the concentration of drops per unit size interval in size range i resulting from vapor diffusion (condensation and/or evaporation)

D_R parameterization of the vertical pressure gradient acceleration of a rising bubble (''drag'')

D_v diffusion coefficient for water vapor in air

$\mathcal{D}_{i_c}$ time rate of change of the concentration of drops per unit size interval in size range i resulting from condensation

$\mathcal{D}_{i_e}$ time rate of change of the concentration of drops per unit size interval in size range i resulting from evaporation

$\tilde{\mathscr{D}}$ $\equiv \bar{\mathbf{v}} \cdot \bar{\tilde{\mathscr{F}}}$, a measure of turbulent dissipation of the mean flow

$\hat{D}$ viscosity

d constant proportionality factor relating the tangential velocity to the inverse of the radius of a one-cell vortex

δ_h homogeneous freezing factor

$\mathbf{E}$ electric field vector

E_c evaporation of cloud water

E_r evaporation of rainwater

$\mathscr{E}_1$ dynamic entrainment

$\mathscr{E}_2$ entrainment by lateral eddy mixing

$\mathscr{E}_3$ entrainment by the convergence of the vertical eddy flux across the top and bottom of an infinitesimal layer of cloud

e_s saturation vapor pressure over a plane surface of water

e_{si} saturation vapor pressure with respect to a plane surface of ice

$e_{si}(\infty)$ saturation vapor pressure over a plane surface of ice

ΔE net energy (Gibbs free energy) required to accomplish nucleation of a particle of water or ice bubble

Σ_c collection efficiency

Σ_{rc} collection efficiency of raindrops collecting cloud drops

$\hat{\varepsilon}$ small positive fraction parameterizing the near adjustment of frontal clouds to moist symmetric neutrality

ε_i time rate of change of the concentration of drops per unit size interval in size range i resulting from entrainment

$\mathbf{F}$ molecular friction force

F net convergence of the vertical flux of liquid water relative to the air (sedimentation of liquid water)

F_B buoyancy source for unaveraged pressure perturbation

F_D $\equiv R_v T(\infty) D_v^{-1} e_s^{-1}(\infty)$ (Chapter 3); dynamic source in diagnostic equation for unaveraged pressure perturbation (Chapters 7–9)

F_g sedimentation of snow and graupel

$\widehat{\mathfrak{F}}_i$ time rate of change of the concentration of drops per unit size interval in size range i resulting from sedimentation (or fallout)

F_r sedimentation of rainwater

$\mathscr{F}_r$ turbulence term in equation for $\bar{q}_r$

F_s sedimentation of snow

$\mathscr{F}_T$ turbulence term in equation for the total water mixing ratio

$\mathscr{F}_u$ turbulence term in the x-component of the equation of motion

$\mathscr{F}_v$ vertical flux of water vapor

$\mathscr{F}_w$ turbulence term in the vertical component of the equation of motion

$\mathscr{F}_\theta$ vertical flux of sensible heat

F_κ $\equiv L^2 \kappa_a^{-1} R_v^{-1} T^{-2}(\infty)$

F_{Di} $\equiv R_v T(\infty) D_v^{-1} e_{si}^{-1}(\infty)$

$F_{\kappa i}$ $\equiv L_s^2 \kappa_a^{-1} R_v^{-1} T^{-2}(\infty)$

$\hat{F}, \hat{G}$ conservative properties of a parcel of air

$\bar{\vec{\mathscr{F}}}$ three-dimensional convergence of the eddy flux of momentum

$\bar{\vec{\mathscr{F}}}_H$ horizontal component of the convergence of the eddy flux of momentum (turbulence term in equation for $\bar{\mathbf{v}}_H$)

$F(\mathfrak{R}_\tau)$ fraction of a rain area covered by rain by rates exceeding the threshold rain rate $\mathfrak{R}_\tau$

$\overline{F}_B$ buoyancy source for averaged pressure perturbation

$\overline{F}_D$ dynamic source for averaged pressure perturbation

$\overline{F}_M$ turbulence source for averaged pressure perturbation

Fr   Froude number

f Coriolis parameter

$\hat{f}$ fraction of a final mixture of two parcels of air constituted by one of the two initial parcels

G antenna gain

$\mathscr{G}$ generation of eddy kinetic energy

g magnitude of gravitational acceleration

g_m particle mass distribution function

Γ circulation

Γ_c circulation of a vortex

γ $\equiv -\partial \overline{T}/\partial z$, basic-state lapse rate

γ_E $\equiv \sqrt{f/2K_m}$

H a vertical distance; in Chapter 5, H represents the half-depth of an ice-cloud outflow from cumulonimbus; in Chapter 12, it represents the vertical thickness of a layer of homogeneous fluid

H_s $\equiv \hat{p}/(\hat{\rho} g)$

$\dot{H}$ heating rate

$\dot{\mathscr{H}}$ $\equiv \dfrac{1}{c_p}\left(\dfrac{\hat{p}}{p}\right)^{\kappa} \dot{H}$

$\overline{\dot{\mathscr{H}}}_I$ contribution to $\overline{D}\bar{\theta}/Dt$ by infrared radiation

$\overline{\dot{\mathscr{H}}}_L$ contribution to $\overline{D}\bar{\theta}/Dt$ by latent heating

$\overline{\dot{\mathscr{H}}}_S$ contribution to $\overline{D}\bar{\theta}/Dt$ by solar radiation

$\overline{\dot{\mathscr{H}}}_T$ contribution to $\overline{D}\bar{\theta}/Dt$ by turbulent mixing

h a vertical distance; in Chapters 2–11, it always represents a vertical depth of a layer of fluid; in some discussions it is specifically the height of the top of the planetary boundary layer; in Chapter 12 it is used to represent the height of topography

$\hbar$ moist static energy

h_a amplitude of two-dimensional topography

h_E depth of the Ekman layer

h_s amplitude of Fourier component of function describing terrain height

$\hat{h}$ $\equiv c_v T + p\alpha + Lq_{vs}$

$\hat{h}$ height of terrain

$\hat{h}_c$ a positive number dependent on Froude number and fluid thickness that must exceed the terrain height if a two-dimensional flow is not to be blocked

$\hat{h}_m$ maximum height of a mountain

η x-component of vorticity

η_r radar reflectivity

$I_1(m)$ rate of decrease of the number concentration of drops of mass m as a result of their coalescence with drops of all other sizes

$I_2(m)$ rate of generation of drops of mass m by coalescence of smaller drops

i_{vH} van't Hoff factor

i, j, k   unit vectors in the x, y, and z directions

J_v vertical flux of water vapor in soil

J_w vertical flux of liquid water in soil

K constant turbulent exchange coefficient

$\mathcal{K}$ eddy kinetic energy

K_A turbulent exchange coefficient for an arbitrary quantity $\mathcal{A}$

K_c collection of cloud water

K_H turbulent exchange coefficient for q_H

K_i turbulent exchange coefficient for water category i

K_L turbulent exchange coefficient for liquid water

K_m turbulent exchange coefficient for horizontal momentum

K_v turbulent exchange coefficient for water vapor

K_θ turbulent exchange coefficient for θ

K_ξ turbulent exchange coefficient for vorticity

$\hat{K}$ $\equiv \hat{\kappa}/\rho\hat{c}$ (Chapter 2); collection kernel for a particle of mass m collecting a particle of mass m' (Chapter 3)

$|K|^2$ a function of the complex index of refraction

k wave number in x-direction

k_B Boltzmann's constant

k_s horizontal wave number in the x-direction of a Fourier component (denoted by s) of function describing terrain height

κ R_d/c_p

κ_a thermal conductivity of air

κ_s thermal conductivity of soil

$\hat{\kappa}$ thermal conductivity

L latent heat of vaporization

$\mathcal{L}$ entrainment rate (per unit time)

L_f latent heat of fusion

L_s latent heat of sublimation

L_{af} along-front length scale

L_{cf} cross-front length scale

LDR   linear depolarization ratio

LFC   level of free convection

$\tilde{L}$ wind scale height

$\mathscr{L}$ entrainment rate (per unit time)

$\hat{\mathscr{L}}$ a constant value of $\mathscr{L}$

l wave number in y-direction

ℓ when used as a unit, ℓ stands for liter; when used as a mathematical symbol, it is the square root of the Scorer parameter

ℓ^2 Scorer parameter

ℓ_G cube root of grid volume

ℓ_L^2, ℓ_U^2 Scorer parameters for lower and upper layers, respectively

Λ entrainment rate (per unit height)

λ wavelength

λ_R Rossby radius of deformation

λ_r slope parameter of Marshall–Palmer particle size distribution

M $\equiv v + fx$, absolute momentum

$\mathscr{M}$ $\equiv f(M - M_o)$

M_g geostrophic absolute momentum

M_s molecular weight of dissolved salt

M_w molecular weight of water

M_{ev} efficiency factor for evaporation

$\mathfrak{m}$ mass of a cloud or precipitation particle

m   mass of a rising cloud element

m in some discussions, m represents a wave number in the z-direction; in other discussions it is the angular momentum about the axis of a cylindrical coordinate system

$\mathfrak{m}_k$ mass of a crystal of type k

m_l $\equiv \sqrt{\ell_L^2 - k^2}$

m_s mass of dissolved salt (Chapter 3); vertical wave number of a Fourier mode (denoted by s) of the function describing the vertical velocity of the air flowing over two-dimensional topography (Chapter 12)

$\dot{\mathfrak{m}}$ time rate of change of $\mathfrak{m}$

$\dot{\mathfrak{m}}_{col}$ time rate of change of $\mathfrak{m}$ as a result of collection

$\dot{\mathfrak{m}}_{cond}$ time rate of change of mass of a drop of mass $\mathfrak{m}$ as a result of condensation of vapor

$\dot{\mathfrak{m}}_{dif}$ time rate of change of $\mathfrak{m}$ as a result of vapor diffusion

$\dot{\mathfrak{m}}_{mel}$ time rate of change of mass of an ice particle as a result of melting

$(\Delta m)_\delta$ mass of air detrained to the environment

$(\Delta m)_\varepsilon$ mass of air entrained from environment

μ_f vertical mass flux in a jet at height z

μ_l Gibbs free energy of a liquid molecule

μ_u $\equiv \sqrt{k^2 - \ell_U^2}$

μ_v Gibbs free energy of a vapor molecule

$\hat{\mu}^2$ $\equiv -m^2$

N in discussion of dynamics, N is buoyancy frequency of the mean state; in discussions of cloud microphysics and radar meteorology, N is the particle size distribution function (number density)

$\mathcal{N}$ consumption of eddy kinetic energy

N_I number of ice nuclei per liter of air

$\mathfrak{N}_i$ time rate of change of the concentration of drops per unit size interval in size range i resulting from nucleation from the vapor phase

N_k particle size distribution function for crystal type k

N_o buoyancy frequency of reference state, except in Chapter 3, where N_o represents the intercept parameter of Marshall–Palmer particle size distribution

N_T total concentration of drops per unit volume

$\left(\dfrac{\partial N}{\partial t}\right)_{col}$ change in the number density of drops of mass m by stochastic collection

$\left(\dfrac{\partial N}{\partial t}\right)_{bre}$ net rate of production of drops of mass m by breakup

$\tilde{N}$ buoyancy frequency in pseudoheight coordinate system

n coordinate normal to a streamline

n mass of a cloud condensation nucleus

n_i number of molecules per unit volume of ice

n_l number of water molecules per unit volume of liquid

ν frequency

Ω angular speed of the earth's rotation

$\boldsymbol{\omega}$ vorticity

$\boldsymbol{\omega}_a$ $\equiv \eta\mathbf{i} + \xi\mathbf{j} + (\zeta + f)\mathbf{k}$, absolute vorticity

$\boldsymbol{\omega}_H$ horizontal vorticity

$\boldsymbol{\omega}_{ag}$ absolute geostrophic vorticity

P Ertel's potential vorticity, except in Chapter 4 where P represents a probability density function

$P_B(\mathrm{m})$ probability that a drop of mass m breaks up per unit time

P_e equivalent potential vorticity

$\mathcal{P}_g$ geostrophic potential vorticity in geostrophic space

P_g geostrophic potential vorticity in physical space

$\mathcal{P}_m$ $\equiv \mathcal{P}_{eg}$ if air is saturated, $\mathcal{P}_g$ if unsaturated

P_t transmitted power

$\mathcal{P}_{eg}$ geostrophic equivalent potential vorticity in geostrophic space

P_{eg} geostrophic equivalent potential vorticity

PRF pulse-repetition frequency

$\hat{P}$ probability that a drop of mass m will collect a drop of mass m' in time interval Δt

$\overline{P_r}$ average returned power

p pressure of air

p_c central pressure of a hurricane

$\hat{p}$ reference pressure, usually representative of conditions near the earth's surface, often taken to be 1000 mb

p_B^* pressure perturbation associated with buoyancy field

p_D^* pressure perturbation associated with wind field

Φ geopotential

$\tilde{\Phi}$ $\equiv \Phi + v_g^2/2$

ϕ latitude

ϕ' velocity potential for perturbation flow

ϕ_p phase of reflected radar waves

$\hat{\phi}$ ratio of mass of water evaporated by entrainment to the total mass of water of a rising parcel of air

Π Exner function

π $\equiv p^*/\rho_o$

ψ_s soil moisture potential

ψ, Ψ, Ψ' stream functions

$\mathbf{Q}$ Q-vector

Q_1, Q_2 components of Q-vector

Q_1' Q-vector component in geostrophic (X) space

Q_F^2 integrated effect of turbulence in a volume containing a hydraulic jump

$Q_B(\text{m}',\text{m})\, d\text{m}$ number of drops of mass m to m + dm formed by the breakup of one drop of mass m'

$\dot{Q}_c$ rate at which hailstone loses heat to air by conduction

$\dot{Q}_f$ rate at which heat is gained as a result of the riming of a hailstone

$\dot{Q}_s$ rate at which a hailstone gains heat by deposition

$\tilde{Q}$ $\equiv \overline{u}H$ in a homogeneous fluid

q_c mass of cloud liquid water per unit mass of air

q_d mass of drizzle per unit mass of air

q_g mass of graupel per unit mass of air

q_H total mass of liquid water and/or ice per unit mass of air (hydrometeor mixing ratio)

q_h mass of hail per unit mass of air

q_I mass of cloud ice per unit mass of air

q_i mass of water of type i per unit mass of air (mixing ratio of a given type of water substance)

q_L mass of liquid water per unit mass of air

$q_{\text{m}'}$ mass of liquid water contained in drops of mass m' per mass of air

q_r mass of rainwater per unit mass of air

q_s mass of snow per unit mass of air

q_s vapor-mixing ratio of soil

q_T mass of total water substance per unit mass of air

q_v mass of water vapor per unit mass of air (mixing ratio of water vapor in air)

q_{vs} saturation mixing ratio

R radius of a spherical particle (Chapter 3); radius of a cloud (Chapter 7); radius of a downdraft (Chapter 8)

$\mathcal{R}$ radiative heat flux in the vertical (positive upward)

$\mathfrak{R}$ rainfall rate

R_c critical radius of a drop

R_d gas constant for dry air

R_s radius of curvature of a streamline

R_v gas constant for a unit mass of water vapor

$\mathfrak{R}_{area}$ area-integrated rain rate

R_{ci} critical radius of an inscribed sphere used to express the volume of an ice particle

$\mathfrak{R}_\tau$ threshold rain rate

RH_a surface relative humidity in the vicinity of but not within a hurricane

Ra Rayleigh number

RH relative humidity

Ri Richardson number

$\tilde{R}$ $\equiv N\tilde{L}/U_o$

$\hat{R}$ a constant representing the e-folding radius of a turbulent jet

$\langle\mathfrak{R}\rangle_o$ average rain rate in some region

$\langle\mathfrak{R}\rangle_\tau$ average rain rate within regions where the threshold rain rate $\mathfrak{R}_\tau$ is exceeded

r radial coordinate (called range in the context of radar meteorology)

r_a radius of the outer boundary of a hurricane

r_c radius of a circle described by the intersection of the cone with a level surface at a fixed altitude above a radar

r_c radius of a jump in fluid depth at the initial time (Chapter 2); radius of the inner region of a vortex (Chapter 8)

r_{max} maximum range at which a target can be detected by a radar

ρ density

ρ_a density of the air, including gaseous components only

ρ_L density of liquid water

ρ_s density of soil

ρ_v vapor density (mass of water vapor per unit volume of air)

ρ_{vs} saturation vapor density over a plane surface of water

ρ_{vsfc} vapor density at surface of a particle

$\hat{\rho}$ density of a constant reference state representative of conditions near the earth's surface

$\mathbf{S}$ $\equiv \dfrac{\partial}{\partial z}(\bar{u}\mathbf{i} + \bar{v}\mathbf{j})$, vertical shear of the mean horizontal wind

S surface surrounding volume $\mathcal{V}$

S_c sources plus sinks of cloud water associated with ice-phase microphysical processes

S_g sources plus sinks of graupel associated with ice-phase microphysical processes

S_I sources plus sinks of cloud ice associated with ice-phase microphysical processes

S_i sources plus sinks of a particular category of water

S_r sources plus sinks of rainwater associated with ice-phase microphysical processes

S_s sources plus sinks of snow associated with ice-phase microphysical processes

S_v sources plus sinks of water vapor associated with ice-phase microphysical processes

SST sea-surface temperature

$S(V_R)$ Doppler velocity spectrum

$\mathscr{G}(\mathfrak{R}_z)$ proportionality factor relating $\langle \mathfrak{R} \rangle_o$ and $F(\mathfrak{R}_z)$

$\tilde{S}$ $\equiv e(\infty)/e_s(\infty) - 1$, ambient supersaturation with respect to a plane surface of liquid water

$\tilde{S}_i$ $\equiv e(\infty)/e_{si}(\infty) - 1$, ambient supersaturation with respect to a plane surface of ice

$\hat{S}$ saturated moist entropy

σ effective radar backscatter cross section of one particle

σ_{il} free energy of an ice–liquid interface

σ_{vl} surface energy (surface tension) of a liquid–vapor interface

T temperature

T_B temperature at the top of the boundary layer

T_G temperature at which a particular crystal habit grows

T_o temperature of outflow from hurricane eyewall

$\mathcal{T}_s$ temperature of the soil

T_s temperature at which a parcel of air at temperature T, water vapor mixing ratio q_v, and pressure p would become saturated by lowering its pressure dry adiabatically

T_v virtual temperature

T_w temperature of a drop of water

$\bar{T}_o$ mean temperature of outflow from hurricane eyewall

t time

τ_A vertical eddy flux of $\mathscr{A}$

τ_e time scale of evaporation

τ_m time scale of mixing

τ_p duration of an emitted radar pulse

τ_x vertical flux of momentum in x-direction

τ_y vertical flux of momentum in y-direction

Θ azimuth angle of a cylindrical coordinate system

θ potential temperature

θ_a apparent potential temperature perturbation

θ_e equivalent potential temperature

θ_H horizontal beamwidth angle

θ_v virtual potential temperature

θ_{ec} equivalent potential temperature at the center of a hurricane

θ_{es} saturation equivalent potential temperature

θ_{cv} cloud virtual potential temperature

$\hat{\theta}$ potential temperature of a constant reference state representative of conditions near the earth's surface

θ_V vertical beamwidth angle

$\mathscr{U}$ user-chosen weighting function that assures homogeneity of the dimensions in the integral computed in the retrieval of thermal properties of the air from Doppler radar observations and decides the relative weight to be ascribed to vertical and horizontal gradients in the integral

$U_a \equiv \dfrac{\bar{w}}{f}\dfrac{\partial v_g}{\partial Z} + u_a$

U_f speed of movement of the leading edge of a gravity current

U_o mean wind at the ground

U_{af} along-front velocity scale

U_{cf} cross-front velocity scale

U_1, U_2 horizontal velocity components of parcels in upper and lower levels of a fluid, respectively

u   radial velocity component

u individual wind component in x-direction

u_s wind speed in the layer above a lower-tropospheric layer in which wind speed increases with height

V terminal fall speed (>0)

$\mathscr{V}$ volume

V_a magnitude of the horizontal wind velocity component in the azimuthal direction from a radar

V_F ventilation factor

V_k fall speed of crystal of type k

V_p volume of a rising parcel of air

V_R radar radial velocity (component of velocity of target along the beam of a radar); special case of v

V_s wind speed at distance R_s from the center of a vortex

V_T fall speed of targets affecting radial velocity detected by Doppler radar

V_{Fc} ventilation factor for conduction

V_{Fs} ventilation factor for sublimation

V_{ice} fall speed scale of ice crystals and snow

V_{max} magnitude of the maximum unambiguous Doppler radar radial velocity

$\mathcal{V}_{res}$ radar resolution volume

$\mathbf{V}_c = u_c \mathbf{i} + v_c \mathbf{j}$ cloud or cell motion relative to the ground

$\mathbf{V}_p$ storm propagation velocity component resulting from new cell development

$\mathbf{V}_s$ velocity of a multicell thunderstorm

$\mathbf{V}_c$ velocity of an individual cell

$\mathcal{V}_\varepsilon$ volume of air entrained into a rising parcel of air

$\hat{V}$ mass-weighted fall speed

$\hat{V}_H$ mass-weighted mean-particle fall speed for ice-particle hydrometeors

$\mathbf{v}$ three-dimensional velocity of a parcel of air

v wind component in y-direction

v tangential velocity component

$\mathbf{v}_a$ $\equiv (u - u_g)\mathbf{i} + (v - v_g)\mathbf{j}$, ageostrophic wind

$\mathbf{v}_g$ $\equiv u_g \mathbf{i} + v_g \mathbf{j}$, geostrophic wind

$\mathbf{v}_H$ horizontal wind vector

v_n wind component normal to a boundary

v_t wind component tangent to a boundary

$\bar{\mathsf{v}}_{max}$ maximum mean tangential velocity in a vortex

W $\equiv \tilde{w} f / \zeta_{ag}$

$\mathcal{W}$ creation of $\mathcal{K}$ by pressure–velocity correlation

w wind component in z-direction

w_B background vertical velocity

w_e entrainment velocity

w_{eb} entrainment velocity at cloud base

w_{et} entrainment velocity at cloud top

w_l, w_u vertical velocities in lower and upper layers, respectively

$\hat{w}$ deviation from background vertical velocity w_B (Chapter 5); amplitude of sinusoidally varying vertical velocity (Chapter 12)

$\tilde{w}$ $\equiv D\mathfrak{z}/Dt$, vertical velocity in pseudoheight coordinate system

δW difference in the vertical velocities of two parcels of fluid separated by a horizontal radial distance

X $\equiv x + v_g / f$, geostrophic coordinate

χ_A constant used in the parameterization of the vertical eddy flux of $\mathcal{A}$ in a cloud

x horizontal coordinate

ξ y-component of vorticity

y horizontal coordinate

Z radar reflectivity factor, except in $\partial/\partial Z$ (see below)

Z_e equivalent radar reflectivity factor

Z_{DR} differential reflectivity

Z_{HH} radar reflectivity factor that is horizontally transmitted and horizontally received

Z_{HV} radar reflectivity factor that is horizontally transmitted and vertically received

Z_{orth} radar reflectivity factor received in circular polarization of the opposite sense as that transmitted (orthogonal component)

Z_{par} radar reflectivity factor received in circular polarization of the same sense as that transmitted (parallel component)

Z_{VV} radar reflectivity factor that is vertically transmitted and vertically received

z height

$\mathfrak{z}$ pseudoheight

z_b height of the base of an ice-cloud outflow from cumulonimbus

z_f height of top of vortex funnel

z_m height of the middle of an ice-cloud outflow from cumulonimbus

z_o height of lower tip of vortex funnel

z_s height of the top of a lower-tropospheric layer in which wind speed increases with height

z_T height at which $\theta = \bar{\theta}$

z_t height of the top of an ice-cloud outflow from cumulonimbus

ζ z-component of vorticity

ζ_g vertical component of geostrophic vorticity

ζ_{ag} vertical component of geostrophic absolute vorticity

$\dfrac{d}{dt}$ derivative operator for a variable that is a function of time only

$$\frac{D}{Dt} \equiv \frac{\partial}{\partial t} + \mathbf{v} \cdot \nabla, \text{ total derivative}$$

$$\frac{\overline{D}}{Dt} \equiv \frac{\partial}{\partial t} + \bar{u}\,\frac{\partial}{\partial x} + \bar{v}\,\frac{\partial}{\partial y} + \bar{w}\,\frac{\partial}{\partial z}$$

$$\frac{D_g}{Dt} \equiv \frac{\partial}{\partial t} + u_g\,\frac{\partial}{\partial x} + v_g\,\frac{\partial}{\partial y}$$

$$\frac{D_A}{Dt} \equiv \frac{\partial}{\partial t} + (u_g + u_a)\,\frac{\partial}{\partial x} + v_g\,\frac{\partial}{\partial y} + w\,\frac{\partial}{\partial z}$$

$$\frac{\mathfrak{D}}{\mathfrak{D}\tau} \equiv \frac{\partial}{\partial \tau} + u_g\,\frac{\partial}{\partial X} + v_g\,\frac{\partial}{\partial y}$$

$\dfrac{\partial}{\partial Z}$ height derivative in geostrophic coordinate system

$\dfrac{\partial}{\partial \tau}$ time derivative in geostrophic coordinate system

$\overline{(\ )}$ an average; in discussions of atmospheric air motions, the average is usually taken over a spatial volume or area; in discussions of radar measurements (Chapter 4), the overbar is used to indicate other types of averages (over time, over a power spectrum); in discussions of laboratory experiments (Chapter 7), it is used to indicate a time average

$\overline{\overline{(\ )}}$ average along the outer boundary of a cloud at a given height

$(\ )_o$ throughout the book, the subscript o indicates a property of a hydrostatically balanced reference state; it is also used in discussions of adjustment to geostrophic and gradient wind balance to indicate the half-amplitude of an initial discontinuity in fluid depth (Chapter 2); in discussions of Doppler radar (Chapter 4) to indicate the center of a circle centered on the radar; in discussions of the cloud-topped boundary layer (Chapter 5) and flow over topography (Chapter 12) to indicate conditions at the surface of the earth; in describing a tornado vortex (Chapter 8) to indicate the height of the lower tip of the funnel cloud; and in discussing hurricanes (Chapter 10) to indicate the temperature of the outflow from the eyewall

$(\ )_\varepsilon$ property of entrained air

$(\ )_\delta$ property of detrained air

$(\ )_g$ geostrophic value of a variable

$(\ )_b$ conditions at cloud base

$(\ )_t$ conditions at cloud top

$(\ )_h$ conditions at the top of the boundary layer

$(\ )''$ deviation from $\overline{(\ )}$

$(\ )'$ deviation from $\overline{(\ )}$ or $\langle\ \rangle$

$(\ )^*$ deviation from a hydrostatically balanced reference state

$\langle\ \rangle$ in discussions of shallow layer clouds, these brackets indicate integration with respect to height over the depth of a mixed layer (Chapter 5); in discussions of cloud modeling (Chapter 7), they indicate a horizontal average at a given height over a circular region centered on the central vertical axis of the cloud

$|_{(\ )}$ indicates evaluation at constant $(\ )$

$\nabla_p \equiv \mathbf{i}\left(\dfrac{\partial}{\partial x}\right)\Big|_{y,p,t} + \mathbf{j}\left(\dfrac{\partial}{\partial y}\right)\Big|_{x,p,t} + \mathbf{k}\left(\dfrac{\partial}{\partial p}\right)\Big|_{x,y,t}$

∇_H horizontal gradient operator

∇ three-dimensional gradient operator

Part I | Fundamentals

Chapter 1 | Identification of Clouds

"Here lions threat, there elephants will range,
And camel-necks to vapoury dragons change ..."[5]

Clouds exhibit a wide variety of sizes and shapes, which reflect variations in the dynamical processes producing clouds. In this chapter, we review the basic nomenclature for identifying clouds. This nomenclature makes no attempt to explain the observed clouds; it is based only on the sizes, heights, and physical appearance of the cloud. This purely descriptive method of cloud identification is very practical since it provides weather observers at locations all over the world a simple, direct way to report clouds without having to make a physical interpretation of what they see. For our purposes, it allows us to establish major categories of cloud type based purely on observation. The classification of cloud phenomena thus remains completely independent of our attempts to explain their structures physically and dynamically in later chapters.

Before describing the system of cloud identification, we will briefly review the nomenclature used to describe the vertical and horizontal dimensions of cloud phenomena in Sec. 1.1. Since the vertical dimensions are often described in terms of the nomenclature used to describe the vertical structure of the atmosphere, we review first the terminology used to describe the vertical layering of the atmosphere. Then, also in Sec. 1.1, we present the definitions of terms used to describe the horizontal scales of phenomena in the atmosphere. Since the horizontal sizes of clouds range from less than a kilometer to thousands of kilometers, this horizontal-scale nomenclature is quite necessary to the description and classification of clouds.

In Sec. 1.2, we examine the cloud identification scheme used by weather observers. This method has evolved over two centuries into an internationally accepted procedure. It is based solely on what clouds look like to a person observing them visually. The basic cloud types identified according to this scheme will be described with pictorial examples.

---

[5] From Goethe's *In Honour of Howard*, a poem about clouds written in recognition of Luke Howard, the early nineteenth-century British pharmacist, meteorologist, and devout Quaker who devised the naming system of clouds in use today. Goethe, who was himself interested in clouds, greatly admired Howard and was so touched upon corresponding with him that he wrote the poem (translated from German by George Soane and Sir John Bowring) from which these lines referring to the ephemeral forms of clouds are taken. Other lines from Goethe's poem appear at the headings of later chapters. For more on Howard and his relationship to Goethe, see Scott (1976).

3

The visual cloud identification scheme is inherently limited by the field of view of the observer on the earth or in an aircraft. Larger cloud patterns associated with organized storm systems can be seen only from a platform in space. In Sec. 1.3 we therefore extend the cloud identification scheme to include the systems of storm clouds that are identifiable in satellite pictures.

1.1 Atmospheric Structure and Scales

Before proceeding with our discussion of cloud types and nomenclature, we first take note of some basic terminology used in describing the structure of the atmosphere and its phenomena. The atmosphere is usually divided into several layers, based on the mean vertical profile of temperature (Fig. 1.1). Most clouds occur in the *troposphere*, which is the lowest layer. In the troposphere, which contains nearly all of the water in the atmosphere, the temperature decreases with height in the mean. The top of the troposphere, called the *tropopause*, occurs at about the 12-km level. It is lower than this over the poles and higher in equatorial regions. Above the tropopause, the mean temperature profile is first isothermal and then increases with height in the *stratosphere*. At the bottom of the troposphere, the atmosphere is affected by the presence of the earth's surface, through the transfers of heat and momentum. The layer in which this influence is felt is called the

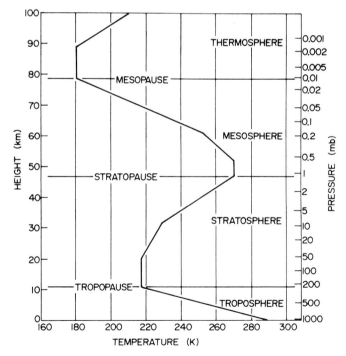

Figure 1.1 Vertical temperature profile for the U.S. Standard Atmosphere. (From Wallace and Hobbs, 1977.)

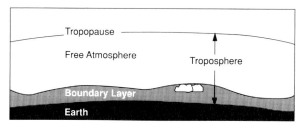

Figure 1.2   Division of the atmosphere into two layers: a boundary layer near the surface and the free atmosphere above. The top of the boundary layer is often ~1 km but may be much less, e.g., ~100 m, depending on wind and thermodynamic properties of the air near the surface. The tropopause height is ~10–12 km at high altitudes and ~14–18 km in the tropics. (From Stull, 1988. Reprinted by permission of Kluwer Academic Publishers.)

planetary *boundary layer* (Fig. 1.2). As will be discussed in Sec. 2.11.2, the depth of the planetary boundary layer is highly variable, ranging from ~10 m to 2–3 km. The lowest 10% of the planetary boundary layer is referred to as the *surface layer*. The region lying above the boundary layer is referred to as the *free atmosphere*.

The scales of air motion encountered in cloud dynamics can be divided roughly into three ranges. The *synoptic scale* encompasses phenomena exceeding about 2000 km in horizontal scale; the *mesoscale* covers phenomena between about 20 and 2000 km in scale; and the *convective scale* covers phenomena between 0.2 and 20 km.[6] These definitions are loose and somewhat overlapping. As yet, a universally accepted physically (as opposed to phenomenologically) based distinction of these scales has not been achieved.[7] The discussions and interpretations in this book are not strongly dependent on the distinctions among these three scales; however, the above ranges are useful to keep in mind as a guide.

1.2   Cloud Types Identified Visually

1.2.1   Genera, Species, and Étages

Visual observation of clouds shows that they take on several distinctive forms which are recognized and named internationally, so that weather observers can record and report the local state of the sky in a way that is readily understandable without the aid of pictures. Once every 6 hours, observers at weather stations around the world identify the amount and types of clouds present, and this information is transmitted for immediate use, as well as archived for climatological purposes. Many stations make a more limited evaluation of the state of the sky every hour.

The internationally agreed method of naming clouds serves as a convenient way to begin to organize our discussion of cloud dynamics. The categories of

---

[6] These ranges were suggested by Orlanski (1975). For an extension of Orlanski's scheme into the domain of cloud and aerosol microphysical processes, see Hobbs (1981).

[7] For a discussion of the problem of defining the mesoscale physically, see Emanuel (1986b).

clouds reported by observers are identified purely on the basis of their visual appearance. Thus, the observer is not required to make a physical interpretation, only a description. Our task then is to provide a dynamical explanation for each type of cloud. Chapters 5–8 and Chapter 12 will be devoted to the dynamics of those clouds that can be identified visually by a ground observer. Chapters 9–11 will be concerned with the dynamics of larger conglomerates of clouds, which are too extensive spatially to be identified by a ground observer and must therefore be viewed from a satellite perspective.

The method of visual identification and classification of clouds that we follow here is basically that of the World Meteorological Organization's *International Cloud Atlas*,[8] which is the guidebook for official weather observers around the world.[9] According to this scheme, cloud types are given descriptive names based on Latin root words.[10] *Cumulus* means heap or pile. *Stratus* is the past participle of the verb meaning to flatten out or cover with a layer. *Cirrus* means a lock of hair or a tuft of horsehair. *Nimbus* refers to a precipitating cloud, and *altum* is the word for height. These five Latin roots are used either separately or in combination to define 10 mutually exclusive cloud *genera*, which are organized into three groups, or *étages*, corresponding to the typical height of the base of the cloud above the local height of the earth's surface, as indicated in Table 1.1.[11] The étages overlap and their limits vary with altitude. Each genus may take on several different forms, which are designated as *species*. Species are further subdivided into *varieties*. In this book, we will refer to only a few species and varieties; however, we will consider all of the ten genera.

In addition to the genera in Table 1.1, we will consider *fog* as an eleventh cloud type. Fog is generally any cloud whose base touches the ground. It does not appear as a cloud genus in Table 1.1 because, according to the internationally specified procedures for reporting and archiving meteorological data, fog is coded by weather observers not as a cloud but rather as a "restriction to visibility." In

---

[8] There have been several editions of this atlas. The first was the *International Atlas of Clouds and Study of the Sky*, Volume I, *General Atlas*, published in 1932 by the International Commission for the Study of Clouds. The World Meteorological Organization (established in 1951) published the first edition of the *International Cloud Atlas*, Volumes I and II, in 1956. The first volume contains descriptive and explanatory text, while the second volume contains 224 illustrative plates. An abridged version combining the essential information in the two original volumes was published in 1969. A revised edition of Volume I was published under the title *Manual on the Observation of Clouds and Other Meteors* in 1975. An extensively updated version of Volume II, containing 196 plates, was published in 1987.

[9] Other useful pictorial cloud guides include: *Cloud Atlas, An Artist's View of Living Cloud* (Itoh and Ohta, 1967), *Clouds of the World: A Complete Color Encyclopedia* (Scorer, 1972), *A Field Guide to the Atmosphere* (Schaefer and Day, 1981), and *The Cloud Atlas of China* (National Meteorological Service of China, 1984), and *Spacious Skies* (Scorer and Verkaik, 1989).

[10] The Latin naming scheme was proposed in 1803 by Luke Howard. The Latin names quickly caught on and have been used in meteorological textbooks since the mid-nineteenth century.

[11] The method of dividing the atmosphere vertically into three layers (étages) where clouds form was introduced by the French naturalist Jean Babtiste Lamarck in 1802. He proposed a cloud classification with French names, which were not universally adopted. Today's cloud classification scheme is a combination of Howard's Latin names and Lamarck's organization into three étages.

Table 1.1

Genera and Étages of Clouds Identified Visually

Genus	Étage	Height of cloud base		
		Polar regions	Temperate regions	Tropical regions
Cumulus Cumulonimbus Stratus Stratocumulus Nimbostratus	Low	below 2 km	below 2 km	below 2 km
Altostratus Altocumulus	Middle	2–4 km	2–7 km	2–8 km
Cirrus Cirrostratus Cirrocumulus	High	3–8 km	5–13 km	6–18 km

this book, we depart from this convention and consider fog to be a type of cloud. It is grouped with clouds of the lowest étage, since the cloud base is at the ground.

In Secs. 1.2.2–1.2.4, we will define, describe briefly, and show pictorial examples of each of the cloud types corresponding to the 10 genera listed in Table 1.1 plus fog. We will examine these 11 cloud types in groups, according to étage, considering first the low clouds, then middle clouds, and finally high clouds. This order of discussion follows roughly the way in which a trained weather observer normally proceeds, as the observer's job is to describe the entire state of the sky, in terms of all of the clouds present. First the low clouds are identified. Then, to the extent that they are not obscured by the low clouds, the middle cloud types and amounts are determined, and finally the high clouds are evaluated, to the extent that they are not obscured by low and middle clouds.

After discussing the 11 basic cloud types and their grouping according to the three étages in Secs. 1.2.2–1.2.4, we will discuss the particular form taken by certain orographically induced clouds in Sec. 1.2.5. Many of these orographic clouds are designated as *lenticularis*,[12] which is a species of stratocumulus, altocumulus, or cirrocumulus. Lenticularis, however, is such a unique cloud form that it could be considered a genus unto itself, and we will treat it separately in this text. Chapters 5–9 are concerned with the dynamics of clouds of the 11 basic cloud types, except for the lenticularis species, which is covered in Chapter 12 as part of the separate treatment of the dynamics of orographic clouds.

1.2.2 Low Clouds

Clouds of the lowest étage include six types: the five low-cloud genera listed in Table 1.1 plus fog. These six types may be divided into two subgroups: *cumuli-*

---

[12] Meaning the shape of a lentil in Latin, or, in more modern terms, lens shaped.

form clouds (cumulus and cumulonimbus), which are composed of rapidly rising air currents that give the clouds a bubbling and towering aspect, and *stratiform* clouds (fog, stratus, stratocumulus, and nimbostratus), which are broad sheets of quiescent clouds characterized by little or no vertical movement of air. We will describe the cumuliform clouds first, then the stratiform.

1.2.2.1   Cumuliform Clouds

Cumulus clouds are "detached clouds, generally dense and with sharp outlines, developing vertically in the form of rising mounds, domes or towers, of which the bulging upper part often resembles a cauliflower. The sunlit parts of these clouds are mostly brilliant white; their base is relatively dark and nearly horizontal. Sometimes cumulus is ragged."[13] Cumulus occur in a wide range of sizes. They are less than a kilometer in horizontal and vertical extent in their early stages of development and often never become any larger (Fig. 1.3a), particularly when the individual clouds are isolated. On other occasions, when there is a tendency for the clouds to cluster, cumulus may grow to larger size (Fig. 1.3b). A large cumulus cloud (species *congestus*) consists of a heap of rapidly fluctuating bulbous towers, which give it its "cauliflower" appearance. It may have tops extending into the second étage. However, it is always considered to be a low cloud because its *base* is usually in the lowest layer.

Cumulonimbus is a "heavy and dense cloud, with a considerable vertical extent, in the form of a mountain or huge towers. At least part of its upper portion is usually smooth, or fibrous or striated, and nearly always flattened; this part often spreads out in the shape of an *anvil* or vast plume. Under the base of this cloud, which is often very dark, there are frequently low ragged clouds either merged with it or not, and precipitation sometimes in the form of virga [precipitation not reaching the ground]." The cumulonimbus is an advanced stage of cumulus development. As cumulus congestus continue to grow, they develop precipitation (hence the nimbus designation), and the top usually turns to ice. An example of cumulus congestus growing to the cumulonimbus stage is shown in Fig. 1.4a–d. In its later stages the icy structure at the top has a fibrous appearance, and high winds aloft can blow the top downwind, thus producing the anvil structure (Fig. 1.4c and d). The top of the anvil in the tallest clouds is usually very near the tropopause level; it is flattened because the rising air in the cloud cannot significantly penetrate the very stable stratosphere. As the glaciated anvil ages, large quantities of icy cloud material are injected into the upper troposphere (Fig. 1.5). From a satellite perspective, the anvil spreading downwind is the primary feature identifying a cumulonimbus (Fig. 1.6). Like cumulus congestus, the cumulonimbus is classified as a low cloud because its base is within the lowest étage. It often extends through all three layers, with its anvil occurring in the highest étage.

The anvil is not an essential feature of the cumulonimbus. In the tropics, a towering cumulus, whose top is well below the 0°C level, often produces a heavy

---

[13] Definitions of cloud genera given in quotation marks are from the *International Cloud Atlas*. (The same definitions are used in all versions of the atlas.)

shower of rain. Since it rains, the towering cumulus designation changes to cumulonimbus. At higher latitudes, an anvil shape may not be seen at the upper levels of a precipitating cumuliform cloud if the wind shear in the environment of the cloud is weak, even if the upper portion of the cloud is composed of ice particles.

1.2.2.2   Stratiform Clouds

Fog is generally any cloud whose base touches the ground. Thus, any type of cloud intersecting a hill or mountain would be reported as fog by an observer on the portion of the hill enshrouded by cloud, while an observer located below the base of the cloud would identify the cloud by one of the 10 genera listed in Table 1.1. What we will consider to be a true fog occurs as a *result* of the air being in contact with the ground and as such is not described in terms of one of the 10 genera. As will be discussed in Chapter 5, the most common and widespread types of fog form when a layer of air is in contact with a cold surface. *Radiation fog* occurs when the underlying surface has cooled by infrared radiation. It forms under very calm conditions, as the turbulence associated with any wind would destroy the fog. An example of radiation fog is shown in Fig. 1.7a. Radiation fog may be quite widespread, covering mesoscale or synoptic-scale regions of the earth (e.g., fog is seen covering the entire Central Valley of California in Fig. 1.7b). *Advection fog* forms when warm air moves over a pre-existing cold surface (Fig. 1.7c). *Steam fog* forms when cold air is over warm water and a turbulent steam rises from the water surface (Fig. 1.7d).

Stratus comprises a "generally grey cloud layer with a fairly uniform base, which may give drizzle, ice prisms, or snow grains. When the sun is visible through the cloud, its outline is clearly discernible. . . . Sometimes stratus appears in the form of ragged patches." This type of cloud is difficult to observe completely from the ground because it is often so horizontally extensive that it lies like a blanket overhead, so that it is impossible to see the top or sides of the cloud (Fig. 1.8a). When viewed from above, as in Fig. 1.8b, and the sun is not blocked by middle or high clouds, the top of the cloud may be brilliantly white, in contrast to its grayish cast when viewed from underneath. Stratus is generally not a thick cloud (≤ 1 km), as indicated by the fact that the outline of the sun is typically visible through it.

Stratocumulus refers to a "grey or whitish, or both grey and whitish, patch, sheet, or layer of cloud which almost always has dark parts, composed of tessellations, rounded masses, rolls, etc., which are nonfibrous (except for virga) and which may or may not be merged; most of the regularly arranged small elements usually have an apparent width of more than five degrees." This type of cloud is often very similar to stratus in that it is a low overhanging blanket of cloud. It is distinguishable from stratus in that it has clearly identifiable elements (Fig. 1.9a). Viewed from an aircraft, stratocumulus sometimes appears as a mosaic of clumps (Fig. 1.9b), while at other times it exhibits tessellations or rolls, called *cloud streets* (Fig. 1.10a). The cloud streets can be quite long, and the individual cloud elements in the streets can become more vigorous and take the form of small to moderate cumulus (Fig. 1.10b).

Figure 1.3

Figure 1.4

Figure 1.5

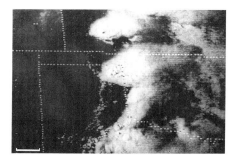

Figure 1.6

Figure 1.7   (a) Radiation fog in Woodland Park Zoo, Seattle, Washington. (Photo by Arthur L. Rangno.) (b) Satellite view of fog in California's Central Valley. (Bar ~200 km.) (c) Satellite view of advection fog along the west coast of the United States. (Bar ~500 km.) (d) Steam fog where cold air is flowing over the Gulf Stream. (Photo by Allen J. Riordan.)

Figure 1.3   (a) Cumulus humilis over a field. Platte, South Dakota. (Photo by Arthur L. Rangno.) (b) Cumulus congestus over Puget Sound, near Anacortes, Washington. (Photo by Steven Businger.)

Figure 1.4   Time sequence showing cumulus congestus developing into cumulonimbus south of Key Biscayne, Florida. (Photos by Howard B. Bluestein.)

Figure 1.5   Anvil of a cumulonimbus, as seen from Cimmaron, Colorado. The anvil is classified as cirrus spissatus cumulonimbogenitus. If the anvil were more widespread, as from a line or group of cumulonimbus clouds, it would be classified as cirrostratus cumulonimbogenitus. (Photo by Ronald L. Holle.)

Figure 1.6   Visible-wavelength satellite photograph of cumulonimbus anvils of supercell thunderstorms over Kansas, Oklahoma, and Texas. (Bar ~100 km.)

Figure 1.8   (a) Stratus cloud intersecting a hill in Orick, California. (Photo by Arthur L. Rangno.) (b) Stratus seen from Denny Mountain. Snoqualmie Pass, Washington. (Photo by Steven Businger.)

Figure 1.9   (a) Stratocumulus near Mitchell, South Dakota. (Photo by Arthur L. Rangno.) (b) Stratocumulus seen from aircraft over the Atlantic Ocean, west of the southwest coast of England, toward north. (Photo by Ronald L. Holle.)

Figure 1.10 "Cloud streets" viewed from aircraft. (a) Rows of stratocumulus as seen from a B-47 aircraft at 11 km on 10 April 1957. (Photo by Joachim P. Kuettner.) (b) Early morning cumulus cloud streets near Tampa, Florida, 12 August 1957. (Photo by Vernon G. Plank.)

Satellite pictures show that stratus and stratocumulus may cover regions ~1000 km in horizontal scale (Fig. 1.11). They also show that a field of stratus is usually not separate from stratocumulus. In the example of stratus and stratocumulus off the west coast of North America (Fig. 1.11a), the clouds nearest the coast show little texture and consist of stratus and/or fog. Farther out to sea the cloud layer turns into stratocumulus. The pattern becomes progressively more textured, breaking up into gradually larger stratocumulus clumps. In the cloud pattern off the eastern and southeastern coastlines of North America (Fig. 1.11b), the predominant cloud form is long rows of stratocumulus organized by strong cold surface winds blowing off the continent over the warm waters of the Atlantic Gulf Stream to the east and Gulf of Mexico to the southeast. The cloud streets widen farther out to sea and eventually turn into a pattern of clumpy stratocumulus. Some of the cloud elements in the cloud streets may be small to moderate cumulus or cumulonimbus rather than stratocumulus, as in Fig. 1.10b.

Nimbostratus is a "grey cloud layer, often dark, the appearance of which is rendered diffuse by more or less continuously falling rain or snow, which in most cases reaches the ground. It is thick enough throughout to blot out the sun. Low, ragged clouds frequently occur below the layer, with which they may or may not merge." The main difference between nimbostratus and stratus is that the former can be extremely deep, with a top at the tropopause level, and also that stratus never produces significant precipitation while nimbostratus usually does. Thus, deep nimbostratus clouds, like cumulonimbus, can extend through all the étages and have upper layers composed entirely of ice. It is difficult to illustrate the visual appearance of nimbostratus in a photograph. It is simply a dark, rainy cloud covering the entire sky (Fig. 1.12). Cloud base may be in either the low or middle étage. The rain area is extensive and restricts horizontal visibility. Distant objects, such as the islands in Fig. 1.12, are progressively more obscured as their range from the viewer increases.

1.2.3 Middle Clouds

There are just two genera of clouds of the middle étage (Table 1.1). *Altostratus* is a "greyish or bluish cloud sheet or layer of striated, fibrous or uniform appearance, totally or partly covering the sky and having parts thin enough to reveal the

Figure 1.11

Figure 1.12   Figure 1.13

Figure 1.14

sun at least vaguely, as through ground glass. Altostratus does not show halo[14] phenomena." Altostratus differs from stratus in that the base of the cloud is in the middle étage. The cloud base is clearly well above the mountain in the example shown in Fig. 1.13. It differs from nimbostratus in that precipitation from it does not reach the ground and it does not always obscure the sun. A *corona*, consisting of colored rings of light close to and centered on the sun or moon,[15] is sometimes seen in altostratus.

Altocumulus is a "white or grey, or both white and grey, patch, sheet or layer of cloud, generally with shading, composed of laminae, rounded masses, rolls, etc., which are sometimes partly fibrous or diffuse and which may or may not be merged; most of the regularly arranged small elements usually have an apparent width between one and five degrees." Altocumulus is usually quite thin. Its distinguishing trait is that it is composed of distinct elements. The elements may take on quite different forms, and consequently several species and varieties of altocumulus are recognized. Thin, flat layers broken up into distinct elements are called *altocumulus stratiformis* and are similar to stratocumulus except for being based in midlevels. Sometimes the elements of these clouds are clumps, either well detached from one another or like pieces of a mosaic (Fig. 1.14a). In other cases, the elements may be like long rolls[16] (Fig. 1.14b). *Altocumulus floccus* (meaning tuft of wool, fluff, or nap of a cloth, see Fig. 1.14c) and *castellanus* (meaning castle shaped, see Fig. 1.14d) are more like elevated cumulus clouds than elevated stratocumulus.

1.2.4 High Clouds

According to Table 1.1, there are three genera of high clouds: *cirrus*, *cirrostratus*, and *cirrocumulus*. Detailed descriptions of these three genera of high clouds follow.

[14] The "halo" is an optical phenomenon produced by the sun shining through a particular type of ice crystal cloud. It is associated with high clouds. See the definition of cirrostratus in the next subsection.

[15] The "corona" is of angular radius 15° or less and produced by diffraction of light by small water droplets.

[16] This type of altocumulus is sometimes called a "mackerel sky."

Figure 1.11 (a) Satellite view of stratocumulus off west coast of North America. (b) Satellite view of stratocumulus off the eastern and southeastern coast of North America. Outlines of states are superimposed on (a), as are latitude and longitude lines for every 5°. Panel (b) is roughly the same scale as (a). The width of the Florida peninsula is ~200 km.

Figure 1.12 Nimbostratus. Guemas Island, toward Orcas Island, Washington. (Photo by Steven Businger.)

Figure 1.13 Altostratus. Bodø, Norway. (Photo by Steven Businger.)

Figure 1.14 (a) Altocumulus stratiformis in the form of cellular clumps. Seattle, Washington. (b) Altocumulus stratiformis in the form of long rolls (undulatus). Seattle, Washington. (c) Altocumulus floccus. Durango, Colorado. (d) Altocumulus castellanus. Seattle, Washington. (Photos by Arthur L. Rangno.)

Cirrus consists of "detached clouds in the form of white, delicate filaments or white or mostly white patches or narrow bands. These clouds have a fibrous (hair-like) appearance, or a silky sheen, or both."

Cirrostratus is a "transparent, whitish cloud veil of fibrous (hair-like) or smooth appearance, totally or partly covering the sky, and generally producing halo[17] phenomena."

Cirrocumulus is a "thin white patch, sheet or layer of cloud without shading, composed of very small elements in the form of grains, ripples, etc., merged or separate, and more or less regularly arranged; most of the elements have an apparent width of less than one degree."

Examples of these three types of high cloud are shown in Fig. 1.15. The panels of the figure are arranged in order of stage of life cycle of the cirriform cloud. The layer of cloud in Fig. 1.15a is cirrocumulus. Individual cloud elements in the form of both grains and ripples can be seen. The cloud elements have a relatively solid, nonfibrous appearance compared to the forms of cirrus shown in Fig. 1.15b–d. This solid appearance is characteristic of cirrus elements in an early stage of development. Upward air currents in the elements are in the process of producing the ice particles, or very briefly, water droplets, which compose the elements. As the cloud ages, it takes on a progressively more diffuse and fibrous appearance.

Cirrocumulus is the high-cloud counterpart of altocumulus, and it has species and varieties similar to those of altocumulus. The example in Fig. 1.15a is *cirro-cumulus stratiformis*. Cloud elements in the form of cellular grains and undular ripples are both evident in this cloud layer. Cirrocumulus ripples are also seen in the upper left-hand portion of Fig. 1.19. The grains and ripples in these cirrocumulus patterns correspond to the cellular and roll structures seen in altocumulus stratiformis in Fig. 1.14a and b.

Cirriform clouds consist almost entirely of ice particles. Many of the individual particles become sufficiently large, and the saturation vapor pressure of ice is sufficiently low, that the particles composing the clouds evaporate slowly. The strong winds at high levels can therefore advect the particles great distances, thus giving cirrus clouds a characteristically stringy or hair-like appearance, which becomes more exaggerated as the clouds age. The clouds in Fig. 1.15b are *cirrus floccus*. Similar in appearance to the altocumulus floccus seen in Fig. 1.14c, they represent a slightly more advanced stage of cirrus development, in which solid, crisply formed elements, such as those seen in the cirrocumulus in Fig. 1.15a, have weakened and started to become diffuse and fibrous, although vestiges of their earlier clumpy structure remain evident. Slowly evaporating ice particles are beginning to be swept away from the dissipating clumps. The long strands of cirrus with a hook at their upwind ends seen in Fig. 1.15c are called *cirrus uncinus* (which is the word for hooked in Latin). They represent a more advanced stage of development than that of the cirrus floccus seen in Fig. 1.15b. In this case the cell

---

[17] The "halo" is a bright circle of angular radius 8°, 22°, or 46° surrounding the sun. It is produced by the refraction of sunlight in hexagonal prisms of ice. The 22° halo is the most common. For an explanation of the phenomenon, see Wallace and Hobbs (1977, Figs. 5.7 and 5.8).

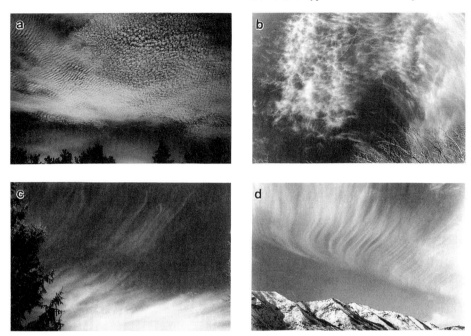

Figure 1.15 (a) Cirrocumulus stratiformis in the form of both undulatus and cells. Seattle, Washington. (b) Cirrus floccus. Durango, Colorado. (c) Cirrus uncinus. Durango, Colorado. (d) Cirrus fibratus vertebratus. Durango, Colorado. (Photos by Arthur L. Rangno.)

or clump producing the ice crystals has degenerated to a relatively weak feature, while the streamer of ice particles falling away from the cell has become quite long. This type of cirrus is sometimes called "mares' tails." Strands of cirrus with no hook at the end (Fig. 1.15d) are called *cirrus fibratus*, the most advanced stage of cirrus development. At this stage all evidence of the source of the ice particles being swept into the long streaks by the wind has disappeared. The example in Fig. 1.15d is a special case: cirrus fibratus *vertebratus*, referring to the element's arrangement into a form suggesting a fish skeleton.

Another type of cirrus is shown in Fig. 1.16. It is called *cirrus spissatus* (past participle of the Latin verb meaning to condense or make thick). It consists of dense ice cloud, which may or may not have streamers of precipitation falling from it. In Fig. 1.16, some of the ice particles generated within the thick clumps of cirrus are falling away and being swept into a curved pattern by the shearing winds at high levels. The resulting hook shape is rather like that of the cirrus uncinus seen in Fig. 1.15c. Cirrus spissatus may also be generated as the outflow or remains of a cirriform anvil of cumulonimbus (Fig. 1.5). Much of the cirrus in the atmosphere, especially in the tropics (frontispiece), arises in this manner.

When cirriform cloud occurs in a layer exhibiting no discrete grainy or undular substructure, it is called cirrostratus. It may be produced in various ways. It is sometimes produced as outflow from cumulonimbus (as in Fig. 1.5, but over a

Figure 1.16

Figure 1.17

Figure 1.18

Figure 1.19

Figure 1.20

wider area). Or it may be the upper-level leading or trailing canopy of a nimbo-stratus cloud. A common form of cirrostratus is the leading canopy of a frontal cloud system (as can be seen in satellite pictures such as those shown below in Sec. 1.3.3). A common feature of cirrostratus is the halo produced as the sun shines through the layer of ice particles (Fig. 1.17).

1.2.5 Orographic Clouds

Air moving over or around hilly or mountainous terrain often influences cloud formation. Many of the basic cloud genera and species already described can be forced, triggered, or enhanced orographically. For example, mountain ranges are typically preferred locations for fog, stratus, stratocumulus, cumulus, and cumu-lonimbus, and they affect the structure and precipitation of nimbostratus clouds associated with weather systems such as fronts. Valleys between mountains often favor fog occurrence. In addition to the modification of the cloud types that may occur anywhere, there are cloud forms that are uniquely associated with topogra-phy, and Chapter 12 is devoted to the dynamics of these truly orographic clouds. As mentioned above, the lenticularis species describes some of these cloud forms. Lenticular clouds form when air flows over a mountain. If the mountain is in the form of an isolated peak, a cap cloud may form directly on the top of the mountain (Fig. 1.18).[18] Lenticular clouds may also form downwind of the peak. An example of this phenomenon is shown in Fig. 1.19, where the lenticular cloud downwind of the peak has the shape of a horseshoe.

For reasons to be discussed in Chapter 12, the lenticular clouds in Fig. 1.18 and 1.19 are a type of *wave cloud*. If the wave clouds are associated with a long quasi-two-dimensional mountain ridge rather than an isolated peak, the wave clouds may be in the form of long cloud bands. Figure 1.20 shows wave clouds associated with the Continental Divide of the Rocky Mountains in Colorado. The main ridge

---

[18] These disk-shaped clouds over the tops of mountains appear to account for many reports of "flying saucers." The first modern report of a flying saucer (1947) was made over Mt. Rainier, Washington, where spectacular displays of lenticular clouds over the summit are sometimes seen.

Figure 1.16 Cirrus spissatus with virga (i.e., precipitation falling out of the cloud but not reaching the ground). Seattle, Washington. (Photo by Steven Businger.)

Figure 1.17 Cirrostratus with halo. Seattle, Washington. (Photo by Arthur L. Rangno.)

Figure 1.18 Cap cloud over Mount Rainier, Washington. (Photo by Peter Thomas.)

Figure 1.19 Horseshoe-shaped cloud (Turusi) in lee of Mt. Fuji, Japan. (Photo taken in 1930 by Masanao Abe.)

Figure 1.20 Looking upwind at a lee-wave cloud band (foreground) in Boulder, Colorado. Stacks of lenticular clouds give the band a bumpy appearance. In the distance is a wave cloud band capping the Continental Divide, which is the main orographic barrier. The mountains in the foreground, which appear larger in the photo, are actually smaller foothills. (Photo by Dale R. Durran.)

Figure 1.21   Looking upwind at a lenticular wave-cloud band (foreground) forming in the lee of the Continental Divide, which is in the distance, far beyond the foothill mountains seen in the foreground. Boulder, Colorado. (Photo by Dale R. Durran.)

is in the background, and a wave cloud capping the ridge can be seen in the distance. The mountains in the foreground are actually smaller foothills. Over these foothills is a wave-cloud band. Large stacks of lenticular clouds can be seen at several locations along the band. Sometimes the wave-cloud bands in the lee of a ridge can have the appearance of a smooth bar (Fig. 1.21). Often a series of lenticular clouds form in waves to the lee of a ridge or a peak. Such a series is seen looking eastward over the plane downwind of the Continental Divide in Fig. 1.22.

Sometimes violent downslope winds occur in the lee of a mountain ridge. Associated with these winds is the *rotor cloud*, which is a line of cloud that occurs immediately downwind of the ridge. The name rotor implies that the air in the cloud overturns vertically in a roll whose axis is parallel to the upwind mountain range. It is not clear that this overturning really always occurs (see Sec. 12.2.5). The primary characteristic of the clouds is that air rises abruptly in them, with the result that the air motions can be extremely turbulent. Examples of rotor clouds viewed from aircraft are shown in Figs. 1.23 and 1.24. In both examples, clearing is seen over the valley immediately downwind of the mountain range, where the strong downslope winds have suppressed all cloud formation. The blowing dust in Fig. 1.23 dramatically marks the sudden rise of the air up into the rotor cloud.

In both Figs. 1.23 and 1.24, low cloud can be seen hanging directly over the mountain ridge. This cloud, called the *Föhn*[19] *wall*, is especially clear in Fig. 1.24, where the forward edge of the cloud hangs over the edge of the mountain ridge and thins where the downslope winds begin to drop into the plain. From the plain, the Föhn wall can have the ominous appearance shown in Fig. 1.25.

---

[19] *Föhn* is the German word for a dry downslope wind. It is used throughout Austria, Switzerland, and Germany to describe the dry winds coming over the Alps from southerly directions. The origin of the word is obscure. It is sometimes associated with ancient Phoenicia (de Rudder, 1929). According to the Greek and Roman 12-part windrose, whose system of directional indications was used on German maps until the nineteenth century, a wind coming from the direction of the Phoenicians would correspond to the common direction of the Alpine *Föhn*. It is more generally accepted, however, that the word comes from the Latin word *favonius*, meaning a zephyr.

Figure 1.22 Looking downwind at a series of lenticular wave clouds in the lee of the Continental Divide. Boulder, Colorado. (Photo by Dale R. Durran.)

Figure 1.23 Turbulent rotor cloud downwind (left-hand side of the photo) of the Sierra Nevada mountain range in the Owens Valley near Bishop, California. Downslope winds gather dust on the valley floor and serve as a tracer of the air rising suddenly up into the cloud. Over the mountains themselves (upper right of photo), a portion of the *Föhn* wall cloud is seen. Photo taken by pilot Robert Symons, while flying a P-38 fighter. (Photo courtesy of Morton G. Wurtele.)

Figure 1.24 *Föhn* wall cloud (left-hand side of photo) over the Dinaric Alps and turbulent rotor cloud (right-hand side of photo) downwind of the mountains. (Photo taken from an aircraft at about 6 km by Andreas Walker.)

Figure 1.25 *Föhn* wall cloud. Boulder, Colorado. (Photo by Dale R. Durran.)

Figure   1.26   Banner   cloud   on   the
Matterhorn, Switzerland. (Photo by J. F. P.
Galvin.)

Another cloud type that sometimes occurs in the lee of an isolated sharp peak is
the *banner cloud* (Fig. 1.26). This phenomenon is also called the *smoking mountain* because the cloud emanates from the top of the mountain in a manner resembling smoke coming from a chimney.

1.3   Cloud Systems Identified by Satellite

As indicated by the foregoing discussion, cloud forms in the atmosphere can be
identified by a human observer on the ground or in an aircraft. However, what one
can see from these vantage points is limited to the immediate surroundings of the
observer. It is impossible to observe visually the full extent and structure of many
clouds. Meteorological satellites provide a viewpoint from which the broader
shapes and forms of clouds can be seen. In particular, they allow the primary
precipitation-producing clouds to be recognized and characterized. These precipitating clouds are generally mesoscale to synoptic scale in extent. We will call them
cloud *systems* because they are often composed of a mix of cloud types, which
work together to comprise a precipitating complex. The dynamics of the three
major types of precipitating cloud systems will be considered in Chapters 9–11.
Their appearance in satellite imagery is briefly described in the following three
subsections.

1.3.1   Mesoscale Convective Systems

Cumulonimbus clouds sometimes occur in organized groups, in which their anvils
merge into a single mesoscale cirriform cloud shield, which is readily identified in

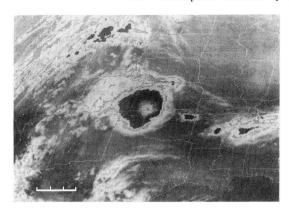

Figure 1.27   Infrared satellite view of a mesoscale convective system over Kansas, Oklahoma, and Texas. Gray shades are proportional to infrared radiative temperature at cloud top, with coldest values indicated by light shading in the interior of the cloud system. The scale of this cloud system can be compared with the size of the smaller, individual thunderstorm anvils seen over the same geographical region in Fig. 1.6. (Bar ~300 km.)

satellite pictures. This configuration is referred to as a *mesoscale convective system*. A more exact definition and more complete description of the phenomenon will be given in Chapter 9. An example of a mesoscale convective system over the United States is shown in infrared satellite imagery in Fig. 1.27. The infrared data indicate the temperature of the cloud top, confirming that the cloud shield must be cirriform and near the tropopause. The cloud shield in this case is much larger than the anvil of an individual thunderstorm (Figs. 1.4–1.6). Mesoscale convective systems are especially prevalent in the tropics, where at any given time they are scattered all along the equatorial cloudiness belt (frontispiece). This type of cloud system accounts for most tropical rainfall and a large proportion of midlatitude warm season rainfall over the continents.

1.3.2   Hurricanes

Some mesoscale convective systems in the tropics evolve into intense cyclonic storms called *hurricanes*.[20] These storms are disturbances of small synoptic scale and are recognized readily in satellite pictures by the highly circular form taken by the upper-level cloud shield, which typically surrounds a cloud-free *eye* in the center of the cyclone (Fig. 1.28). In time-lapse photography, the upper-level clouds are seen to be diverging anticyclonically outward from the center of the storm, while low-level cumulus and stratocumulus cloud elements are spiraling cyclonically inward toward the storm center. Chapter 10 investigates the dynamics of clouds associated with hurricanes.

---

[20] "Hurricane" is the name used for tropical cyclones in the Western Hemisphere. The same type of storm is known variously in the Eastern Hemisphere as "typhoon," "cyclone," and "Willy Willy."

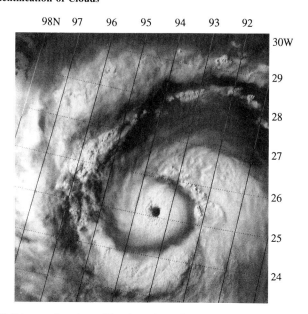

Figure 1.28   Visible wavelength satellite view of Hurricane Allen (1980). (Photo courtesy of Frank D. Marks, Jr.)

1.3.3   Extratropical Cyclones

The most significant producers of precipitation in midlatitudes are the *extratropical cyclones* that populate the midlatitude westerlies. They are always forming, moving, and dying in association with synoptic-scale waves in the westerlies. The largest and most prevalent of extratropical cyclones are *frontal cyclones*, which

Figure 1.29   Satellite view (visible channel) of a westerly moving frontal cloud system (large cloud band) over the North Pacific and a comma cloud (smaller cloud system trailing the larger one). Surface frontal positions are superimposed. The west coast of the United States and Mexico is faintly visible on the right-hand side of the picture. An approximate scale is indicated to help judge the sizes of the cloud areas. (Bar ~500 km.) (Adapted from Businger and Reed, 1989. Reproduced with permission from the American Meteorological Society.)

Figure 1.30 Infrared satellite view of a polar low off the north coast of Norway on 27 February 1987. (Bar ~200 km.)

are often, especially over water, characterized by a systematic cloud pattern aligned along *fronts*, which mark the warm edges of concentrated horizontal temperature gradients. Over land, the cloud arrangements around cyclones can be more complex.

An example of a large frontal cyclonic cloud pattern is shown in Fig. 1.29. It is marked by a long curving cloud band beginning far to the south and eventually curling in toward the center of the cyclone. This long band to the south is associated with the *cold front*. Extending eastward from the curled-up end of the cold-frontal cloud band is a wider, stubbier region of cloud associated with the *warm front*.

West of the cold-frontal cloud band in Fig. 1.29, a smaller comma-shaped cloud system is seen. It is associated with an upper-level low-pressure system (or *short wave*) that forms in the cold air behind the main frontal system. This short-wave cloud system is an example of a class of extratropical disturbances referred to as *polar lows*. Generally polar lows occur in cold air streams over oceans, and they exhibit a continuum of sizes and cloud configurations.[21] Most exhibit either comma- or spiral-shaped cloud structure in satellite pictures. The type that forms close to a large frontal cyclone, as in Fig. 1.29, is often referred to as a *comma cloud*—even though most polar lows and even the larger frontal cyclone itself all have comma-shaped cloud patterns. Some polar lows occur in cold air masses much farther away from frontal systems than the comma cloud in Fig. 1.29. These polar lows tend to be smaller in scale and sometimes develop a spiral shape and an eye reminiscent of the hurricane. An example of this type of storm is shown in Fig. 1.30. In Chapter 11 we will explore the dynamics of clouds associated with frontal cyclones and polar lows.

---

[21] See the review paper of Businger and Reed (1989) for a general discussion of the types of polar lows that occur.

Chapter 2 | Atmospheric Dynamics

"... the cloud messenger in air expires ..."[22]

Atmospheric dynamics is the application of principles of fluid dynamics to the earth's atmosphere. Cloud dynamics, more specifically, is the application of these principles to the air motions in clouds and their immediate environments. This chapter summarizes the topics from atmospheric dynamics that are most essential to the discussions of cloud dynamics in Chapters 5–12.

2.1 The Basic Equations[23]

2.1.1 Equation of Motion

The movement of air in the atmosphere is governed by *Newton's second law of motion*, which may be written in the form

$$\frac{D\mathbf{v}}{Dt} = -\frac{1}{\rho}\nabla p - f\mathbf{k} \times \mathbf{v} - g\mathbf{k} + \mathbf{F} \tag{2.1}$$

where t is time and $\mathbf{v}$ is the three-dimensional velocity of a parcel of air. ∇ is the three-dimensional gradient operator and D/Dt, called the *total derivative*, is the time derivative following a parcel of air. It is given by

$$\frac{D}{Dt} \equiv \frac{\partial}{\partial t} + \mathbf{v} \cdot \nabla \tag{2.2}$$

The parcel velocity $\mathbf{v}$ is called the *wind* velocity. If the coordinate system is one in which the horizontal directions are given by x and y and height is represented by z, then the wind velocity of the parcel is given by the three-dimensional wind vector $\mathbf{v} = u\mathbf{i} + v\mathbf{j} + w\mathbf{k}$, where $\mathbf{i}, \mathbf{j}$, and $\mathbf{k}$ are unit vectors in the x, y, and z directions and u, v, and w are the individual wind components. Other dependent variables in (2.1) are the density ρ and pressure p of the air. The terms on the right-hand side of (2.1) represent the four forces affecting the motion of a unit mass of air. They are,

---

[22] Goethe seems here to sense that clouds ultimately are controlled by the air in which they are suspended.

[23] A standard reference for large-scale dynamic meteorology is Holton (1992). Another very useful text is Gill (1982).

respectively, the pressure-gradient, Coriolis, gravitational, and frictional accelerations. The quantity g is the magnitude of the gravitational acceleration, while f is
the Coriolis parameter $2\Omega \sin \phi$, where Ω is the angular speed of the earth's
rotation and ϕ is latitude.

2.1.2   Equation of State

The thermodynamic state of dry air is well approximated by the *equation of state*
for an ideal gas, written in the form

$$p = \rho R_d T \tag{2.3}$$

where R_d is the gas constant for dry air and T is temperature. When air contains
water vapor, this equation is modified to

$$p = \rho R_d T_v \tag{2.4}$$

where T_v is the *virtual temperature*, given to a good degree of approximation by

$$T_v \approx T(1 + 0.61 q_v) \tag{2.5}$$

where q_v is the mixing ratio of water vapor in air (mass of water vapor per unit
mass of air).

2.1.3   Thermodynamic Equation

Changes of temperature of a parcel of air are governed by the *First Law of
Thermodynamics*, which for an ideal gas may be written as

$$c_v \frac{DT}{Dt} + p \frac{D\alpha}{Dt} = \dot{H} \tag{2.6}$$

where $\dot{H}$ is the heating rate, c_v is the specific heat of dry air at constant volume, T
is temperature, and α is the specific volume ρ^{-1}. The first term on the left of (2.6) is
the rate of change of internal energy of the parcel, while the second term is the
rate at which work is done by the parcel on its environment. With the aid of the
equation of state (2.4), the First Law (2.6) may be written in the form

$$c_p \frac{DT}{Dt} - \alpha \frac{Dp}{Dt} = \dot{H} \tag{2.7}$$

where c_p is the specific heat of dry air at constant pressure. It is often preferable to
write the First Law in terms of the potential temperature,

$$\theta \equiv \left(\frac{\hat{p}}{p}\right)^{\kappa} T \tag{2.8}$$

where $\kappa = R_d/c_p$, R_d is the gas constant for dry air (the difference between the
specific heats at constant pressure and volume), and $\hat{p}$ is a reference pressure,

usually assumed to be 1000 mb. In terms of θ, (2.7) takes the simple form

$$\frac{D\theta}{Dt} = \dot{\mathcal{H}}$$

(2.9)

where

$$\dot{\mathcal{H}} \equiv \frac{1}{c_p}\left(\frac{\hat{p}}{p}\right)^{\kappa}\dot{H}$$

(2.10)

In cloud dynamics, $\dot{H}$ includes the sum of heating and cooling associated with phase changes of water, radiation and molecular diffusional processes. Under adiabatic (isentropic) conditions, (2.9) reduces to

$$\frac{D\theta}{Dt} = 0$$

(2.11)

For cloud dynamics, the form (2.11) is often not useful, since the latent heat of phase change and/or radiation may render the in-cloud conditions nonisentropic. When the only phase changes are associated with condensation and evaporation of water, the First Law (2.7) can be written as

$$c_p\frac{DT}{Dt} - \alpha\frac{Dp}{Dt} = -L\frac{Dq_v}{Dt}$$

(2.12)

where contributions to the specific heat and specific volume from water vapor and condensed water have been neglected. With the aid of (2.8), (2.12) can be written as

$$\frac{D\theta}{Dt} = -\frac{L}{c_p\Pi}\frac{Dq_v}{Dt}$$

(2.13)

where q_v is the mixing ratio of water vapor (kilograms of water per kilogram of air), L is the latent heat of vaporization, and Π, called the *Exner function*, is defined as

$$\Pi = \left(\frac{p}{\hat{p}}\right)^{\kappa}$$

(2.14)

A useful quantity in this case is the *equivalent potential temperature* θ_e, defined as the potential temperature a parcel of air would have if all its water vapor was condensed and the latent heat converted into sensible heat. Integration of (2.13) shows that, as long as the air is *saturated*, the equivalent potential temperature is

$$\theta_e = \theta(T, p) \exp\left[\int_0^{q_{vs}(T,p)}\frac{L}{c_pT}dq_v\right]$$

(2.15)

where $q_{vs}(T,p)$ is the *saturation mixing ratio*, defined as the value of the mixing ratio when air at temperature T and pressure p is saturated, and the integral in

(2.15) is along a path in which the air is always maintained at saturation. Numerical evaluation of the integral shows that a good approximation to (2.15) is

$$\theta_e \approx \theta e^{Lq_{vs}(T,p)/c_pT} \tag{2.16}$$

For *unsaturated* air we define the equivalent potential temperature as

$$\theta_e \equiv \theta(T,p)e^{Lq_v/c_pT_s(T,q_v,p)} \tag{2.17}$$

where $T_s(T,q_v,p)$ is the temperature at which a parcel of air at temperature T, water vapor mixing ratio q_v, and pressure p would become saturated by lowering its pressure dry-adiabatically. As defined in (2.17), θ_e is conserved in dry-adiabatic motion, since all the variables in the expression on the right-hand side are constant in that case. Moreover, (2.15) and (2.17) are identical at and below the pressure where saturation is reached. Taking $D(2.16)/Dt$, while making the approximation that T^{-1} can be brought outside the integral, verifies that θ_e is very nearly conserved under saturated conditions. Thus, regardless of whether or not the air is saturated, we have

$$\frac{D\theta_e}{Dt} \approx 0 \tag{2.18}$$

The conservation of θ_e expressed by (2.18) is often used in place of (2.11) when condensation and evaporation are the only diabatic effects. Note also that both (2.16) and (2.17) are approximated by

$$\theta_e \approx \theta\left(1 + Lq_v/c_pT\right) \tag{2.19}$$

where it is understood that $q_v = q_{vs}$, whenever the air is saturated.

2.1.4 Mass Continuity

In addition to the equation of motion, equation of state, and the First Law of Thermodynamics, parcels of air obey two mass-continuity constraints. Overall mass conservation for an air parcel is expressed by the *continuity equation*

$$\frac{D\rho}{Dt} = -\rho\nabla \cdot \mathbf{v} \tag{2.20}$$

2.1.5 Water Continuity

Conservation of the mass of water in an air parcel is governed by a set of *water-continuity equations*

$$\frac{Dq_i}{Dt} = S_i, \quad i = 1, \ldots, n \tag{2.21}$$

where q_i is a mixing ratio of one type of water substance (mass of water per unit mass of air). In addition to the vapor mixing ratio q_v, there is a mixing ratio of total liquid water and total ice content of the air parcel. The liquid and ice content can be further subdivided into categories according to drop size in the case of liquid and according to ice particle type and size in the case of ice. Since there are so many possible categories of water substance, we indicate simply that there is some total number of categories n, including the vapor category, and that there are various sources and sinks of each type of water. The sum of the sources and sinks for a particular category of water is indicated by S_i. The water-continuity equations (2.21) are discussed further in Sec. 3.4.

2.1.6   The Full Set of Equations

Equations (2.1), (2.4), (2.9), (2.20), and (2.21) form a set of equations in $\mathbf{v}$, ρ, T, p, q_i, $\dot{\mathcal{H}}$, S_i, and $\mathbf{F}$. If $\dot{\mathcal{H}}$, S_i, and $\mathbf{F}$ can be expressed in terms of the other variables, then the set is a closed system of differential equations that can be solved for the wind, thermodynamic, and water variables as functions of x, y, z, and t.

2.2   Balanced Flow

Atmospheric motions often proceed through a series of near-equilibrium states. Although small accelerations are changing the flow, the forces in the equation of motion are so close to balanced that the instantaneous flow can largely be described as if the forces were in fact balanced. Some basic properties of the types of near force balance that occur in the context of cloud dynamics are summarized in the following subsections.

2.2.1   Quasi-Geostrophic Motion

Clouds usually involve air motions on the mesoscale and convective scale. However, it is nonetheless important to be cognizant of the dynamics of the larger-scale environment, within which the clouds are embedded. Typically, the larger-scale environment (synoptic scale and larger) is in a state where, above the friction layer, the Coriolis and horizontal pressure-gradient forces are in a state of *quasi-geostrophic* balance, especially in midlatitudes. Under this type of balance, the horizontal velocity and length scales in both the x and y directions are characteristic of the large-scale atmospheric motion. If the pressure, height, and vertical velocity scales are also characteristic of the large-scale flow, then the horizontal wind is given, to a first approximation, by its *geostrophic* value

$$\mathbf{v}_g = u_g\mathbf{i} + v_g\mathbf{j} \equiv \frac{1}{\rho f}\left(-p_y\mathbf{i} + p_x\mathbf{j}\right) \tag{2.22}$$

which is the horizontal wind for which the Coriolis force in (2.1) would be exactly balanced by the pressure-gradient force in both the x and y directions.[24] The small acceleration of the geostrophic wind components is determined by the Coriolis force acting on the *ageostrophic* part of the wind:

$$\frac{D_g \mathbf{v}_g}{Dt} = -f\mathbf{k} \times \mathbf{v}_a \tag{2.23}$$

where

$$\mathbf{v}_a \equiv \left(u - u_g\right)\mathbf{i} + \left(v - v_g\right)\mathbf{j} \tag{2.24}$$

and

$$\frac{D_g}{Dt} \equiv \frac{\partial}{\partial t} + u_g \frac{\partial}{\partial x} + v_g \frac{\partial}{\partial y} \tag{2.25}$$

and f is assumed to be constant. In general, f varies with latitude. In this book we have no need to take this variation into account. When it is necessary to consider larger scales of motion, for which the variation of f with latitude must be taken into account, an equation similar in form to (2.23) is still obtained. However, the analysis required to obtain the more general version is beyond the scope of this book.[25]

While (2.23) illustrates the physical nature of the forcing that changes the geostrophic wind components, they cannot be solved, as they stand, for the geostrophic flow, since u and v are not determined. This problem is overcome by forming a potential vorticity equation, as discussed below in Sec. 2.5.

2.2.2 Semigeostrophic Motions

In the vicinity of a front, which is one of the major producers of midlatitude clouds, the scales of motion are quite different in directions taken across and along the front. As a matter of convention, we take x and u to be in the cross-front direction and y and v to be in the along-front direction. If we then let L_{cf} and U_{cf} be the cross-front length and velocity scales, respectively, and L_{af} and U_{af} be the along-front scales, and assume that $L_{cf} \ll L_{af}$, $U_{cf} \ll U_{af}$, and $D/Dt \sim U_{cf}/L_{cf}$, then the magnitudes of the acceleration in the cross front and along-front directions, respectively, may be compared to the Coriolis acceleration as follows:

$$\left(\frac{Du}{Dt}\right)\bigg/fv \sim \left(\frac{U_{cf}^2}{U_{af}^2}\right)\left(\frac{U_{af}}{fL_{cf}}\right) \tag{2.26}$$

$$\left(\frac{Dv}{Dt}\right)\bigg/fu \sim \left(\frac{U_{af}}{fL_{cf}}\right) \tag{2.27}$$

---

[24] Independent variables used as subscripts indicate partial derivatives. Thus, in (2.22), $p_x \equiv \partial p/\partial x$. This shorthand notation will be used frequently throughout the text to indicate partial derivatives.

[25] See Holton (1992) for a full discussion of the scale analysis leading to the quasi-geostrophic equations, as well as various other topics presented in this section.

If the front is strong, accelerations in the along-front direction are large enough that $U_{af}/fL_{cf} \sim 1$ and

$$\left(\frac{Du}{Dt}\right)\bigg/fv \ll 1 \tag{2.28}$$

which implies that the v component of the wind is geostrophic. At the same time,

$$\frac{Dv}{Dt}\bigg/fu \sim 1 \tag{2.29}$$

which implies that fu is not in balance with the pressure-gradient acceleration (i.e., u is *not* geostrophic). In this case, which is referred to as *semigeostrophic*, the equation of motion in the along-front direction becomes[26]

$$\frac{D_A v_g}{Dt} = -fu_a \tag{2.30}$$

$$\frac{D_A}{Dt} \equiv \frac{\partial}{\partial t} + \left(u_g + u_a\right)\frac{\partial}{\partial x} + v_g\frac{\partial}{\partial y} + w\frac{\partial}{\partial z} \tag{2.31}$$

This form differs from the geostrophic case in that the *ageostrophic* circulation (u_a, w), which is directed transverse to the along-front flow, contributes to advection along with the geostrophic wind. The form of (2.30) and (2.31) can be simplified mathematically by a special coordinate transformation, which leads to a set of equations that are of the same form as the geostrophic equations. This coordinate transformation will be discussed in Chapter 11. As in the geostrophic case, a potential vorticity equation must be formed to predict the flow (Sec. 11.2.2).

Equations (2.30) and (2.31) are a special case of the more general *geostrophic momentum approximation*, according to which motions on a time scale greater than $1/f$ are closely approximated by the vector equation of motion

$$\frac{D\mathbf{v}_g}{Dt} = -f\mathbf{k} \times \mathbf{v}_a \tag{2.32}$$

This relation is an extension of (2.30) since the use of the full total derivative D/Dt, defined in (2.2), includes advection by the ageostrophic flow in all directions (u_a, v_a, w). Equation (2.32) is similar to the geostrophic equation of motion (2.23) except that ageostrophic advection and vertical advection of geostrophic momentum are retained.[27] This generalized form of the geostrophic momentum approximation may also be recast in terms of a coordinate transformation that leads to a

---

[26] For a thorough review of the semigeostrophic approximation in relation to fronts, see Hoskins (1982).

[27] The geostrophic momentum approximation was introduced by Eliassen (1948) and further developed by Hoskins (1975, 1982). See Bluestein (1986) for a summary of the derivation of the approximation.

set of *semigeostrophic equations*, which are analogous to the geostrophic equations.

2.2.3   Gradient-Wind Balance

When horizontal flow becomes highly circular, as in a hurricane, it is convenient to consider the air motions in a cylindrical coordinate system, with the center of the circulation as the origin. If the circulation is assumed to be axially symmetric and devoid of friction, then the horizontal components of the equation of motion (2.1) are

$$\frac{Du}{Dt} = -\frac{1}{\rho}\frac{\partial p}{\partial r} + fv + \frac{v^2}{r} \tag{2.33}$$

$$\frac{Dv}{Dt} = -fu - \frac{uv}{r} \tag{2.34}$$

where u and v represent the radial and tangential horizontal velocity components Dr/Dt and $rD\Theta/Dt$, respectively, where r is the radical coordinate and Θ is the azimuth angle of the coordinate system. The centrifugal term v^2/r in (2.33) and the term $-uv/r$ in (2.34) are apparent forces that arise from the use of the cylindrical coordinates. *Gradient-wind balance* is said to occur when the three forces, pressure-gradient, Coriolis, and centrifugal, are balanced in (2.33). Hurricanes and other strong vortices on the mesoscale to the synoptic scale tend to be in such a balance.

It is convenient to write the equations of gradient-wind balance in terms of the angular momentum m about the axis of the cylindrical coordinate system, where

$$m \equiv rv + \frac{fr^2}{2} \tag{2.35}$$

In the case of gradient-wind balance ($Du/Dt = 0$), (2.33) and (2.34) become

$$0 = -\frac{1}{\rho}\frac{\partial p}{\partial r} + \frac{m^2}{r^3} - \frac{f^2 r}{4} \tag{2.36}$$

$$\frac{Dm}{Dt} = 0 \tag{2.37}$$

The last relation expresses the conservation of angular momentum around the center of the circulation. Note that it applies whether or not the circulation is balanced, as long as it is cylindrically symmetric.

2.2.4   Hydrostatic Balance

The above discussions of geostrophic and gradient flow refer to approximate force balance in the horizontal. The atmosphere is also often in a state of force balance in the vertical. When the gravitational term in (2.1) nearly balances the vertical component of the pressure gradient acceleration so that both terms are much larger than the net vertical acceleration and the vertical component of the Coriolis force, the atmosphere is said to be in *hydrostatic balance*. This balance is expressed by

$$\frac{\partial \Phi}{\partial p} = -\alpha \qquad (2.38)$$

where Φ is the geopotential, defined as $gz +$ constant. Synoptic-scale and larger-scale circulations are typically in hydrostatic balance, as are the circulations at fronts, in hurricanes, and in a variety of other mesoscale phenomena. Under hydrostatically balanced conditions, the pressure gradient acceleration in (2.1) may be written as

$$-\frac{1}{\rho} \nabla p = -\nabla_p \Phi \qquad (2.39)$$

where

$$\nabla_p \equiv \mathbf{i} \left(\frac{\partial}{\partial x} \right) \Big|_{y,p,t} + \mathbf{j} \left(\frac{\partial}{\partial y} \right) \Big|_{x,p,t} + \mathbf{k} \left(\frac{\partial}{\partial p} \right) \Big|_{x,y,t} \qquad (2.40)$$

2.2.5   Thermal Wind

When a circulation is in both hydrostatic and geostrophic balance, the winds at different altitudes are related by the *thermal wind equation*,

$$f \frac{\partial \mathbf{v}_g}{\partial p} = \frac{\partial \alpha}{\partial y} \mathbf{i} - \frac{\partial \alpha}{\partial x} \mathbf{j} \qquad (2.41)$$

which is obtained by taking $-\mathbf{i} \cdot \partial(2.38)/\partial y + \mathbf{j} \cdot \partial(2.38)/\partial x$, substituting from the geostrophic wind equation (2.22) and making use of (2.39). When a cylindrically symmetric circulation is in both hydrostatic and gradient-wind balance, the winds (expressed by m) at different altitudes are related by the somewhat more general thermal wind equation:

$$\frac{2m}{r^3} \frac{\partial m}{\partial p} = -\frac{\partial \alpha}{\partial r} \qquad (2.42)$$

which is obtained by taking $\partial(2.38)/\partial r$, substituting from the gradient-wind equation (2.36), and making use of (2.39). From the equation of state (2.4), it is evident that at a constant pressure level, α is directly proportional to the virtual temperature. Hence, the thermal wind relations (2.41) and (2.42) imply that the vertical shear of the horizontal wind is directly proportional to the horizontal thermal gradient whenever these balanced conditions apply.

2.2.6   Cyclostrophic Balance

A cylindrical column of fluid may be rotating around an axis oriented *at any angle from the vertical*, and we may orient the axis of a cylindrical coordinate system parallel to the axis of the column. If the pressure-gradient force dominates over all other forces, the equation of motion (2.1) reduces to

$$\frac{D\mathbf{v}}{Dt} = -\frac{1}{\rho}\nabla p \tag{2.43}$$

If the flow is assumed to be axisymmetric, the component accelerations in the radial and azimuthal directions are

$$\frac{Du}{Dt} = -\frac{1}{\rho}\frac{\partial p}{\partial r} + \frac{v^2}{r} \tag{2.44}$$

$$\frac{Dv}{Dt} = -\frac{uv}{r} \tag{2.45}$$

where, as in the case of gradient flow, u and v represent the radial and tangential velocity components of motion Dr/Dt and $rD\Theta/Dt$, respectively. However, in this case, these components are not necessarily horizontal. They lie in the plane normal to the central axis of the coordinate system, whatever its orientation. In the case in which the radial forces are in balance ($Du/Dt = 0$), (2.44) reduces to a simple balance between the pressure gradient and centrifugal acceleration:

$$\frac{v^2}{r} = \frac{1}{\rho}\frac{\partial p}{\partial r} \tag{2.46}$$

From (2.46), it is evident that the pressure decreases toward the center of the vortex, regardless of the sense of the rotation, and that the strength of the pressure gradient at a given radius is proportional to the square of the tangential velocity. It is also evident that this balance cannot hold at the center of the vortex, where the pressure gradient would be infinite. Cyclostrophic flows occur at various locations in thunderstorms and near the eyes of hurricanes.

2.3   Anelastic and Boussinesq Approximations

Since the large-scale environment is usually in hydrostatic balance, it is convenient to write the equation of motion (2.1) in terms of the deviations of pressure and density from a hydrostatically balanced reference state, whose properties vary only with height. If we denote this reference state by subscript o and deviation from the reference state by an asterisk, then (2.1) is closely approximated by

$$\frac{D\mathbf{v}}{Dt} = -\frac{1}{\rho_o}\nabla p^* - f\mathbf{k}\times\mathbf{v} + B\mathbf{k} + \mathbf{F} \tag{2.47}$$

where B is the *buoyancy*, defined as

$$B \equiv -g \frac{\rho^*}{\rho_o} \tag{2.48}$$

In applying the equation of motion (2.47) to clouds, hydrometeors must be considered. Since liquid and ice particles in the air quickly achieve their terminal fall velocities, the frictional drag of the air on the particles can be regarded as being in balance with the downward force of gravity acting on the particles. Thus, according to Newton's third law of motion, the drag of the particles on the air is $-gq_H$, where q_H is the mixing ratio of hydrometeors in the air (total mass of liquid water and/or ice per unit mass of air). One way to incorporate this acceleration into (2.47) is simply to add this additional acceleration to the right-hand side. An equivalent way to incorporate this effect, which does not require modifying any of the above equations, is to consider the density ρ to be broadly enough defined to include the mass of the hydrometeors as well as the air itself. If we let ρ_a be the density of the air, then

$$\rho = \rho_a\left(1 + q_H\right) \tag{2.49}$$

With this definition of ρ, the hydrometeor weighting automatically becomes a negative contribution to the buoyancy. Substitution of (2.49) and the equation of state (2.4) into (2.48) leads to

$$B \approx g\left(\frac{T^*}{T_o} - \frac{p^*}{p_o} + 0.61 q_v^* - q_H \right) \tag{2.50}$$

The hydrometeor drag appears as the last term, and the other three terms indicate clearly the separate contributions of temperature, water vapor, and pressure perturbation to the buoyancy. The base-state atmosphere is assumed to contain no hydrometeors; therefore, the hydrometeor mixing ratio in (2.50) is the total hydrometeor content of the air and the asterisk is appended to $q_{H'}$.

The buoyancy may be alternatively written in terms of the *virtual potential temperature* θ_v, which is defined as the potential temperature that dry air would have if its pressure and density were equal to those of a given sample of moist air. It is approximated by

$$\theta_v \approx \theta\left(1 + 0.61 q_v\right) \tag{2.51}$$

Substituting this expression into (2.50), we obtain

$$B \approx g\left[\frac{\theta_v^*}{\theta_{v_o}} + (\kappa - 1)\frac{p^*}{p_o} - q_H \right] \tag{2.52}$$

If we make use of the Exner function defined by (2.14), we may rewrite the equation of motion as

$$\frac{D\mathbf{v}}{Dt} = -c_p\theta_{v_o}\ \nabla\Pi^* - f\mathbf{k} \times \mathbf{v} + g\left(\frac{\theta_v^*}{\theta_{v_o}} - q_H\right)\mathbf{k} + \mathbf{F} \qquad (2.53)$$

In this form the pressure perturbation term disappears from the buoyancy force. For this and a variety of other reasons, a system of thermodynamic variables based on θ_v and Π instead of T and p is often used in cloud dynamics and other branches of mesoscale meteorology. For most of this book we will use T and p.

The continuity equation (2.20) can be written to a high degree of approximation as

$$\nabla \cdot \rho_o\mathbf{v} = 0 \qquad (2.54)$$

This form retains the essential density variation with height. The absence of the time derivative of density eliminates sound waves from the equations, but these waves are usually of no interest in cloud dynamics. The use of (2.47) and (2.54) in place of (2.1) and (2.20) is usually referred to as the *anelastic approximation*. Actually, in its purest form, the anelastic approximation requires that the reference state be isentropic as well as hydrostatic. In practice, the reference state is often taken to be nonisentropic. For example, the assumed reference state might be that of the environment observed by a radiosonde. When the nonisentropic form is used, some caution is required because the full set of equations does not conserve energy exactly. On the other hand, the atmospheric basic state is usually substantially nonisentropic, and the errors involved in assuming a nonisentropic basic state do not appear to be especially large.[28]

If the vertical extent of the air motions is confined to a shallow layer, ρ_o can be replaced by a constant in (2.47) and (2.54). The latter then takes the incompressible form

$$\nabla \cdot \mathbf{v} = 0 \qquad (2.55)$$

The use of (2.47) and (2.55) in place of (2.1) and (2.20) is usually referred to as the *Boussinesq approximation*. For many applications in cloud dynamics, the Boussinesq equations are either sufficient or easily generalized [through the use of (2.54) in place of (2.55)] to include compressibility effects without changing the physical interpretation that can be gleaned from the simpler form. We will frequently use the Boussinesq approximation in this text.

---

[28] For further discussion of these problems, see Ogura and Phillips (1962), who provide a rigorous derivation of the anelastic equations; Wilhelmson and Ogura (1972), who introduced a commonly used nonisentropic form and asserted that the nonisentropic form is accurate to within 10% in the context of a numerical cloud model; and Durran (1989), who has reexamined all the previous forms of the anelastic approximation and has suggested further improvements to it.

2.4   Vorticity

An intrinsic characteristic of a fluid is its tendency to rotate about a local axis. The air motions associated with clouds are particularly prone to rotation, and an appreciation of fluid rotation is most helpful to understanding the dynamics of clouds. The local measure of the rotation is the *vorticity*, which is defined as the curl of the wind velocity:

$$\boldsymbol{\omega} \equiv \nabla \times \mathbf{v} = \left(w_y - v_z \right)\mathbf{i} + \left(u_z - w_x \right)\mathbf{j} + \left(v_x - u_y \right)\mathbf{k} \equiv \eta\mathbf{i} + \xi\mathbf{j} + \zeta\mathbf{k} \qquad (2.56)$$

Equations governing time changes of the components of this vector are found by taking the curl of the Boussinesq equation of motion with frictional forces neglected and making use of (2.55). The resulting equations are

$$\frac{D\eta}{Dt} = B_y + \eta u_x + \xi u_y + (\zeta + f)u_z \qquad (2.57)$$

$$\frac{D\xi}{Dt} = -B_x + \xi v_y + (\zeta + f)v_z + \eta v_x \qquad (2.58)$$

$$\frac{D(f + \zeta)}{Dt} = (f + \zeta)w_z + \left(\xi w_y + \eta w_x \right) \qquad (2.59)$$

where the terms B_x and B_y comprise the baroclinic generation of horizontal vorticity. The quantity $(\zeta + f)$ is the sum of the vertical component f of the earth's vorticity and the vertical component ζ of the vorticity of the wind. When the vertical component of the earth's vorticity is added to the vorticity of the wind, $\boldsymbol{\omega}$, we obtain

$$\boldsymbol{\omega}_a \equiv \eta\mathbf{i} + \xi\mathbf{j} + (\zeta + f)\mathbf{k} \qquad (2.60)$$

which is called the *absolute vorticity*, since it incorporates the rotating motion of the rotating coordinate system. The terms ξv_y, ηu_x, and $(f + \zeta)w_z$ in (2.57)–(2.59) are vortex-stretching terms; they express the concentration of vorticity that occurs when vortex tubes are elongated. The last two terms in each equation [i.e., $(\zeta + f)v_z + \eta v_x$, $\xi u_y + (\zeta + f)u_z$, and $(\xi w_y + \eta w_x)$] are tilting terms; they express the conversion of vorticity parallel to one coordinate axis to vorticity around another axis as the orientations of vortex tubes change.

Often in cloud dynamics, the scales of phenomena are sufficiently small that the Coriolis effect is negligible, and f disappears from (2.57)–(2.59). We also frequently encounter flows which are quasi-two-dimensional (e.g., fronts and squall lines). When the flow is two-dimensional in the x–z plane and f is negligible, the horizontal vorticity-component equation (2.58) simplifies to

$$\xi_t = -B_x - u\xi_x - w\xi_z \quad \text{or} \quad \frac{D\xi}{Dt} = -B_x \qquad (2.61)$$

Thus, in this case, vorticity is generated only baroclinically and is redistributed by advection in the x–z plane.

2.5   Potential Vorticity

An important property of the atmosphere is *Ertel's potential vorticity*, defined as

$$P \equiv \frac{\boldsymbol{\omega}_a \cdot \nabla\theta}{\rho} \tag{2.62}$$

For a dry-adiabatic, inviscid flow, this quantity is conserved:

$$DP/Dt = 0 \tag{2.63}$$

Under semigeostrophic conditions, P is approximated by

$$P_g \equiv \frac{\boldsymbol{\omega}_{ag} \cdot \nabla\theta}{\rho} \tag{2.64}$$

where

$$\boldsymbol{\omega}_{ag} \equiv -\frac{\partial v_g}{\partial z}\mathbf{i} + \frac{\partial u_g}{\partial z}\mathbf{j} + \left(\zeta_g + f\right)\mathbf{k} \tag{2.65}$$

and the subscript g indicates geostrophic values of the variables. In the case of two-dimensional flow in the x–z plane (i.e., $\partial/\partial y \equiv 0$), (2.64) becomes

$$P_g = \rho^{-1}\left[-\frac{\partial v_g}{\partial z}\frac{\partial\theta}{\partial x} + \left(\zeta_g + f\right)\frac{\partial\theta}{\partial z}\right] \tag{2.66}$$

A useful characteristic of the potential vorticity is that the terms on the right-hand sides of (2.64) and (2.66) depend entirely on the geopotential Φ. This fact becomes evident by noting first that, from the hydrostatic relationship (2.38) and the definition of the specific volume α as ρ^{-1}, we have

$$\rho^{-1} = -\frac{\partial\Phi}{\partial p} \tag{2.67}$$

Second, from the definition of the geostrophic wind (2.22) and the expression for the horizontal pressure gradient on an isobaric surface under hydrostatic conditions (2.39), the geostrophic wind and vertical vorticity in (2.65) can be expressed as

$$u_g = -\frac{1}{f}\Phi_y, \quad v_g = \frac{1}{f}\Phi_x \tag{2.68}$$

and

$$\zeta_g = \frac{1}{f}\nabla_p^2\Phi \tag{2.69}$$

Third, from the definition of θ in (2.8), the definition of α as ρ^{-1}, the equation of state (2.4), and the hydrostatic relationship (2.38), we have

$$\theta \equiv -\frac{\hat{p}^{\kappa}}{R_d p^{\kappa-1}}\left(\frac{\partial\Phi}{\partial p}\right) \tag{2.70}$$

Substitution of (2.67)–(2.70) into (2.64) and (2.66) demonstrates that the geostrophic potential vorticity depends on only one dependent variable (Φ) in a pressure-coordinate system. Conservation of P, expressed by (2.63), then implies a prediction of Φ and [according to (2.68)] the geostrophic wind. This behavior is an extremely useful and important characteristic of a geostrophically balanced flow.

Under saturated conditions, θ is not conserved, and hence neither is P. A quantity analogous to P, which we call the *equivalent potential vorticity*, is defined as

$$P_e \equiv \frac{\boldsymbol{\omega}_a \cdot \nabla\theta_e}{\rho} \tag{2.71}$$

Because of (2.18), P_e behaves in an analogous fashion to P when the air is saturated; thus, P_e is conserved under saturated conditions. However, P_e is not exactly conserved under unsaturated conditions. The *geostrophic equivalent potential vorticity* P_{eg} is obtained by replacing θ with θ_e in (2.64).

2.6   Perturbation Form of the Equations

For a wide variety of analytical purposes, it is useful to consider air motions in terms of deviations from an average over some arbitrary spatial volume of air (e.g., a grid volume in a numerical model). Any variable $\mathscr{A}$ is then expressed as

$$\mathscr{A} = \overline{\mathscr{A}} + \mathscr{A}' \tag{2.72}$$

where the overbar represents the average value and the prime the deviation. When variables are decomposed in this way, the basic equations split into two sets: the *mean-variable equations*, which predict the behavior of the mean variables, and the *perturbation equations*, which predict the departures from the mean state. In the following subsections, we will write out the mean-variable and perturbation equation forms of the anelastic equations. Since the Boussinesq equations are a simplification of the anelastic form, the results for the anelastic case will also indicate the mean-variable and perturbation equations in the Boussinesq case. In addition to the basic equations for all the mean and perturbation variables, the form of the kinetic energy equation for the eddy motions will be indicated.[29]

---

[29] For further discussion of the averaging techniques used in this section, see Chapter 2 of Stull (1988).

2.6.1 Equation of State and Continuity Equation

Since the equation of state (2.4) and the continuity equation for anelastic flow (2.54) contain no time derivatives, they take on rather simple forms when split into mean-variable and perturbation forms. In the atmosphere, perturbations of density and virtual temperature are always small compared to their mean values. Hence, if the variables p, ρ, and T_v are written as in (2.72), and (2.4) is then averaged, we obtain the approximate equation of state for the mean variables,

$$\bar{p} \approx \bar{\rho} R_d \bar{T}_v \tag{2.73}$$

If this relation is subtracted from (2.4), we obtain the equation of state for the perturbation variables,

$$\frac{p'}{\bar{p}} \approx \frac{T_v'}{\bar{T}_v} + \frac{\rho'}{\bar{\rho}} \tag{2.74}$$

If $\mathbf{v}$ is written as in (2.72), and (2.54) is averaged, we obtain the mean-variable continuity equation,

$$\nabla \cdot \rho_o \bar{\mathbf{v}} = 0 \tag{2.75}$$

To obtain this relation, it must be assumed that the derivative of a spatial average is equal to the spatial average of the spatial derivative. The assumption is valid as long as the deviations from the average occur on a considerably smaller scale than the region over which the average is taken. This assumption is sometimes called *scale separation*. If this relation is subtracted from (2.54), we obtain the continuity equation for the perturbation wind,

$$\nabla \cdot \rho_o \mathbf{v}' = 0 \tag{2.76}$$

2.6.2 Thermodynamic and Water-Continuity Equations

If the First Law of Thermodynamics (2.9) is combined with the continuity equation (2.54), we obtain

$$\frac{\partial \theta}{\partial t} + \rho_o^{-1} \nabla \cdot \rho_o \theta \mathbf{v} = \dot{\mathcal{H}} \tag{2.77}$$

This form of the equation is referred to as the *flux form*, since $\rho_o \theta \mathbf{v}$ is the three-dimensional potential temperature flux. If the variables θ, $\dot{\mathcal{H}}$, and $\mathbf{v}$ are written as in (2.72), both sides of (2.77) are averaged, and we again make the scale separation assumption, then the mean-variable thermodynamic equation is found to be

$$\frac{\overline{D}\bar{\theta}}{Dt} = \overline{\dot{\mathcal{H}}} - \rho_o^{-1} \nabla \cdot \rho_o \overline{\mathbf{v}'\theta'} \tag{2.78}$$

where

$$\frac{\overline{D}}{Dt} \equiv \frac{\partial}{\partial t} + \bar{u}\frac{\partial}{\partial x} + \bar{v}\frac{\partial}{\partial y} + \bar{w}\frac{\partial}{\partial z} \qquad (2.79)$$

Since products of the dependent variables θ and $\mathbf{v}$ appear in (2.77), positive or negative covariances of the variables (i.e., $\overline{\mathbf{v}'\theta'}$) may contribute to changes of $\bar{\theta}$ following a parcel of air. The covariance is proportional to the *eddy flux* of potential temperature $\rho_o\overline{\mathbf{v}'\theta'}$. When the averaged thermodynamic equation (2.78) is used, the convergence of the eddy flux becomes an additional diabatic effect.

If (2.78) is subtracted from (2.77) and (2.75) is invoked, we obtain the perturbation thermodynamic equation,

$$\frac{\overline{D}\theta'}{Dt} = \left(\dot{\mathcal{H}}\right)' - \rho_o^{-1}\left[\nabla\cdot\rho_o\bar{\theta}\mathbf{v}' + \nabla\cdot\rho_o\theta'\mathbf{v}' - \nabla\cdot\rho_o\overline{\theta'\mathbf{v}'}\right] \qquad (2.80)$$

If the perturbations remain small enough, the last two terms can be neglected. In that case, the equation is said to be *linearized* about a state of mean motion $\bar{\mathbf{v}}$ and mean potential temperature $\bar{\theta}$.

The water-continuity equations (2.21) have the same form as the thermodynamic equation (2.9), where q_i replaces θ and S_i replaces $\dot{\mathcal{H}}$. Hence, the mean-variable and perturbation forms of the water-continuity equations (2.21) are, by analogy to (2.78) and (2.80),

$$\frac{\overline{D}\bar{q}_i}{Dt} = \bar{S}_i - \rho_o^{-1}\nabla\cdot\rho_o\overline{\mathbf{v}'q_i'}, \quad i = 1, \ldots, n \qquad (2.81)$$

and

$$\frac{\overline{D}q_i'}{Dt} = S_i' - \rho_o^{-1}\left[\nabla\cdot\rho_o\bar{q}_i\,\mathbf{v}' + \nabla\cdot\rho_o q_i'\,\mathbf{v}' - \nabla\cdot\rho_o\overline{q_i'\mathbf{v}'}\right], \quad i = 1, \ldots, n \qquad (2.82)$$

where, as in (2.21), $i = 1, \ldots, n$ represent the various categories of water substance being considered.

2.6.3   Equation of Motion

Now consider the equation of motion for the Boussinesq flow (2.47). If $\mathbf{v}$, p^*, and B and $\mathbf{F}$ are written as in (2.72) and substituted in (2.47), then averaging (2.47) and making the scale separation assumption yields the mean-variable form of the equation of motion,

$$\frac{\overline{D}\bar{\mathbf{v}}}{Dt} = -\frac{1}{\rho_o}\nabla\overline{p}^* - f\mathbf{k}\times\bar{\mathbf{v}} + \bar{B}\mathbf{k} + \bar{\mathbf{F}} + \bar{\vec{\mathcal{F}}} \qquad (2.83)$$

where

$$\bar{\vec{\mathcal{F}}} \equiv -\rho_o^{-1}\left[\left(\nabla\cdot\rho_o\overline{u'\mathbf{v}'}\right)\mathbf{i} + \left(\nabla\cdot\rho_o\overline{v'\mathbf{v}'}\right)\mathbf{j} + \left(\nabla\cdot\rho_o\overline{w'\mathbf{v}'}\right)\mathbf{k}\right] \qquad (2.84)$$

Equation (2.83) is the momentum analog of (2.78). The term $\bar{\vec{\mathcal{F}}}$ is the three-dimensional convergence of the eddy flux of momentum. It may be regarded as the frictional force associated with eddy motions of the air, while $\overline{\mathbf{F}}$ is the smaller-scale molecular friction force. Subtracting (2.83) from (2.47) leads to an equation for the velocity perturbation:

$$\frac{\overline{D}\mathbf{v}'}{Dt} = -(\mathbf{v}' \cdot \nabla)\overline{\mathbf{v}} - (\mathbf{v}' \cdot \nabla)\mathbf{v}' - \frac{1}{\rho_o}\nabla(p^*)' - f\mathbf{k} \times \mathbf{v}' + B'\mathbf{k} + \mathbf{F}' - \overline{\overline{\mathscr{F}}} \quad (2.85)$$

If the perturbations remain small enough and molecular friction is negligible, the last two terms in (2.85) can be neglected, and the linearized form of the equation is obtained.

We now have obtained the complete set of equations for anelastic flow decomposed into the mean-variable equations (2.73), (2.75), (2.78), (2.81), and (2.83) and the perturbation equations (2.74), (2.76), (2.80), (2.82), and (2.85). In the case of Boussinesq flow, the continuity equation (2.55) is used instead of (2.54). Exactly the same equations as (2.73)–(2.85) are then obtained, except that the density term ρ_o disappears from (2.75)–(2.78), (2.80)–(2.82), and (2.84).

2.6.4 Eddy Kinetic Energy Equation

Taking $\mathbf{v}' \cdot$ (2.85) and then averaging both sides of the equation leads to the *eddy kinetic energy equation*

$$\frac{\overline{D}\mathscr{K}}{Dt} = \mathscr{C} + \mathscr{B} + \mathscr{W} + \mathscr{D} \quad (2.86)$$

where

$$\mathscr{K} \equiv \tfrac{1}{2}\left(\overline{\mathbf{v}' \cdot \mathbf{v}'}\right) \quad (2.87)$$

$$\mathscr{C} \equiv -\overline{u'\mathbf{v}'} \cdot \nabla \overline{u} - \overline{v'\mathbf{v}'} \cdot \nabla \overline{v} - \overline{w'\mathbf{v}'} \cdot \nabla \overline{w} \quad (2.88)$$

$$\mathscr{B} \equiv \overline{w'B'} \quad (2.89)$$

$$\mathscr{W} \equiv -\rho_o^{-1}\overline{\mathbf{v}' \cdot \nabla\left(p^*\right)'} \quad (2.90)$$

$$\mathscr{D} \equiv \overline{\mathbf{v}' \cdot \mathbf{F}'} - \overline{\mathbf{v}' \cdot \overline{\overline{\mathscr{F}}}} - \overline{\mathbf{v}' \cdot (\mathbf{v}' \cdot \nabla)\mathbf{v}'}$$

$$= \overline{\mathbf{v}' \cdot \mathbf{F}'} - \overline{\mathbf{v}' \cdot \overline{\overline{\mathscr{F}}}} - \overline{\mathbf{v}' \cdot \nabla\mathscr{K}} \quad (2.91)$$

$\mathscr{K}$ is the eddy kinetic energy, and $\mathscr{C}$ is the rate at which $\mathscr{K}$ is created by conversion from mean-flow kinetic energy, when down-gradient eddy momentum fluxes occur. $\mathscr{B}$ is the creation (destruction) of $\mathscr{K}$ by thermally direct (indirect) flow perturbations, which are those eddies for which buoyancy is positively (negatively) correlated with vertical velocity [e.g., eddies characterized by warm (cold) air rising and cold (warm) air sinking]. $\mathscr{W}$ is the creation of $\mathscr{K}$ by pressure–velocity correlation, which physically is the rate at which work is done by the force of the gradient of perturbation pressure acting on parcels over distances given by the perturbation velocity. $\mathscr{D}$ (a negative quantity) is frictional dissipation of $\mathscr{K}$ by eddies or molecular friction.

2.7 Oscillations and Waves

2.7.1 Buoyancy Oscillations

If we take the reference state to be the mean large-scale environment, as would be measured by radiosonde, and assume that the potential temperature in this environment increases with height ($\partial \theta_o / \partial z > 0$), then a parcel of air displaced upward or downward dry adiabatically in this environment will experience a restoring force; upward-displaced parcels become negatively buoyant, while downward-displaced parcels become positively buoyant. This situation suggests that a stable atmosphere is conducive to parcels oscillating vertically about their equilibrium altitude. To illustrate this oscillatory tendency mathematically, we can make use of *parcel theory*, which is the name given to the analysis of parcels whose pressure does not deviate from that of the large-scale environment [i.e., $p^* = 0$ in (2.47)]. As we will see in Chapter 7, the assumption of zero pressure perturbation is often not a good assumption, as parcels of air with nonzero buoyancy must have an associated pressure perturbation field in order for mass continuity to be preserved. However, parcel theory nonetheless helps develop physical intuition regarding the behavior of parcels accelerated and decelerated by buoyancy forces.

In this spirit, we consider a dry, frictionless Boussinesq flow in which $p^* = 0$ for all parcels of air. The buoyancy (2.52) then simplifies to

$$B = g \frac{\theta^*}{\theta_o} \tag{2.92}$$

while the vertical component of the equation of motion (2.47) and the First Law of Thermodynamics (2.11) reduce to

$$\frac{Dw}{Dt} = B \tag{2.93}$$

and

$$\frac{DB}{Dt} \approx -wN_o^2 \tag{2.94}$$

where

$$N_o^2 \equiv \frac{g}{\theta_o} \frac{\partial \theta_o}{\partial z} \tag{2.95}$$

Combining (2.93) and (2.94), we obtain

$$\frac{D^2 w}{Dt^2} + wN_o^2 = 0 \tag{2.96}$$

which is the equation for an undamped harmonic oscillation, with frequency $\sqrt{N_o^2}$. The vertical velocity therefore has the form

$$w = \hat{w} e^{i\sqrt{N_o^2}\, t} \tag{2.97}$$

where $\hat{w}$ is the complex amplitude.[30] The frequency of the oscillation N_o is called the *buoyancy frequency*. The subscript o is used here to emphasize that this value of the frequency refers to the reference state and to distinguish this value from the buoyancy frequency of the mean state, which is written without subscript and defined as

$$N^2 \equiv \frac{g}{\bar{\theta}} \frac{\partial \bar{\theta}}{\partial z} \qquad (2.98)$$

Usually, this distinction is academic, as we will typically consider the reference state (always indicated by subscript o) and the mean state variables (indicated by an overbar) to be equal, in which case $N^2 = N_o^2$. The important point is that the buoyancy frequency is the primary indicator of the stability of the atmosphere to vertical displacements. If $N_o^2 > 0$, parcels undergoing vertical motion oscillate sinusoidally. The case in which $N_o^2 < 0$ will be discussed in Sec. 2.9.1.

2.7.2 Gravity Waves

The presence of buoyancy (i.e., the effect of gravity acting on density anomalies) as a restoring force in a fluid leads to the occurrence of wave motions. An idealized case, for which these wave motions are readily identified, is one in which a layer of fluid of uniform density ρ_1 lies below another layer of uniform but lower density ρ_2. Both layers are assumed to be in hydrostatic balance. The mean depth of the lower layer is $\bar{h}$. When the depth is perturbed by a small amount h', the total depth is given by $h = \bar{h} + h'$. If, for simplicity, we assume that there is no horizontal pressure gradient in the upper layer,[31] then in the lower layer the horizontal pressure gradient in the x-direction is $-g(\delta\rho/\rho_1)h_x'$, where $\delta\rho \equiv \rho_1 - \rho_2$. For simplicity, assume that there are no variations in the y-direction, that the Coriolis effect is nil, and that the fluid is inviscid. Then the linearized perturbation forms of the Boussinesq equation of motion and continuity equation (Sec. 2.3) for a basic state of constant mean motion only in the x-direction are

$$u_t' + \bar{u}u_x' = -g\frac{\delta\rho}{\rho_1}h_x' \qquad (2.99)$$

and

$$u_x' + w_z' = 0 \qquad (2.100)$$

The vertical velocity at h is $w(h) = h_t + uh_x$, and, if the lower layer is bounded below by a rigid flat surface, $w(0) = 0$. Then (2.100) may be integrated from $z = 0$

---

[30] When solutions to wave equations are given in complex form in this text, it is to be understood that only the real part of the solution has physical significance.

[31] This is the case if the upper layer is infinitely deep.

to h to obtain

$$h'_t + \bar{u}h'_x + u'_x \bar{h} = 0 \tag{2.101}$$

Equations (2.99) and (2.101) combine to yield the equation

$$\left(\frac{\partial}{\partial t} + \bar{u}\frac{\partial}{\partial x}\right)^2 h' - \frac{g\bar{h}\,\delta\rho}{\rho_1} h'_{xx} = 0 \tag{2.102}$$

which has solutions of the form

$$h' \propto e^{ik(x-ct)} \tag{2.103}$$

where

$$c = \bar{u} \pm \left(\frac{g\bar{h}\,\delta\rho}{\rho_1}\right)^{1/2} \tag{2.104}$$

Thus, the perturbation in the depth of the lower layer of fluid propagates in the form of a wave of wave number k moving at phase speed c, as the buoyancy force successively restores the depth of the fluid to its mean height at one location and the excess mass is transferred to the adjacent location in x, where a new perturbation is created upon which the buoyancy must act. In the case where the mean flow is zero ($\bar{u} = 0$) and the upper layer consists of air while the lower layer is water ($\delta\rho \approx \rho_1$), $c = \pm\sqrt{g\bar{h}}$. This speed is that of gravity waves in water of depth $\bar{h}$. It is referred to as the *shallow-water wave speed*, and (2.99) and (2.101) are referred to as the *linearized shallow-water equations*.

The shallow-water prototype is a useful analog for certain types of atmospheric motions. However, it is a highly restrictive one since shallow-water waves propagate only horizontally. As the atmosphere is continuously stratified, gravity waves can propagate vertically as well as horizontally. This fact can be deduced from the two-dimensional linearized perturbation form of the horizontal vorticity equation (2.61). If we continue to consider the case of constant mean motion in the x-direction only and to disregard friction and diabatic heating, this equation is

$$\left(\frac{\partial}{\partial t} + \bar{u}\frac{\partial}{\partial x}\right)(u'_z - w'_x) + B'_x = 0 \tag{2.105}$$

while the incompressible (constant density) version of the perturbation thermodynamic equation (2.80), after multiplying it by g/θ_o, substituting from (2.98), and ignoring the pressure perturbation, water vapor, and hydrometeor contributions to B in (2.52), becomes

$$B'_t + \bar{u}B'_x = -w'\bar{B}_z = -w'N^2 \tag{2.106}$$

With the aid of (2.106) and (2.100), u' and B' can be eliminated from (2.105) to obtain

$$\left(\frac{\partial}{\partial t} + \bar{u}\frac{\partial}{\partial x}\right)^2 (w'_{xx} + w'_{zz}) + N^2 w'_{xx} = 0 \tag{2.107}$$

which has solutions of the form

$$w' \propto e^{i(kx+mz-vt)} \qquad (2.108)$$

where the frequency is given by the dispersion relationship

$$v = \bar{u}k \pm Nk\big/\left(k^2 + m^2\right)^{1/2} \qquad (2.109)$$

These solutions represent waves called *internal gravity waves*, which propagate with a vertical as well as a horizontal component of phase velocity. The factor multiplying N in (2.109) is the cosine of the angle of the phase line from the vertical. If the fluid is in hydrostatic balance, $m^2 \gg k^2$, then the phase speed $c = v/k$ implied by (2.109) is $c = \bar{u} \pm N/m$, which is equivalent to (2.104) for the case of the two homogeneous fluid layers. It can be shown further[32] that internal gravity waves transport energy in the direction of the *group velocity*, defined as the vector $(\partial v/\partial k, \partial v/\partial m)$. This vector is parallel to the phase lines.

2.7.3 Inertial Oscillations

The Coriolis force acts as a horizontally directed restoring force for small perturbations from a geostrophically balanced basic state in the same way that the buoyancy force acts as a vertically directed restoring force for small perturbations from a hydrostatically balanced basic state. To illustrate this fact, let the mean-state geostrophic wind be entirely in the y-direction, with a value $\bar{v} = v_o$, which depends only on x while $\bar{u} = \bar{w} = 0$. We again make use of parcel theory by setting $p^* = 0$ in the dry, inviscid, Boussinesq equations, and we assume that the flow is two-dimensional, with no variation in the y-direction. The *absolute momentum* is defined as

$$M \equiv v + fx \qquad (2.110)$$

The x- and y-components of (2.47) may then be written as

$$\frac{Du}{Dt} = f\left(M - M_o\right) \qquad (2.111)$$

and

$$\frac{DM}{Dt} = 0 \qquad (2.112)$$

where M_o is the basic-state absolute momentum $v_o + fx$. The absolute momentum in a two-dimensional geostrophic flow thus behaves similarly to the angular momentum m in gradient flow [recall (2.37)]. Equations (2.111) and (2.112) may be

---

[32] See Durran (1990) or Holton (1992).

written alternatively as

$$\frac{Du}{Dt} = \mathcal{M}$$

(2.113)

and

$$\frac{D\mathcal{M}}{Dt} = -uf\frac{\partial M_o}{\partial x}$$

(2.114)

where

$$\mathcal{M} \equiv f(M - M_o)$$

(2.115)

Equations (2.113) and (2.114) are analogous to (2.93) and (2.94). Hence, u is governed by the harmonic-oscillator equation

$$\frac{D^2 u}{Dt^2} + uf\frac{\partial M_o}{\partial x} = 0$$

(2.116)

which is an analog of (2.96). It has solutions of the form

$$u' \propto e^{i v t}, \quad v = \pm\sqrt{f \partial M_o / \partial x} = \pm\sqrt{f\left(\partial v_o / \partial x + f\right)}$$

(2.117)

Thus, the Coriolis force and horizontal shear of the basic-state absolute momentum determine the frequency of the oscillation, just as gravity and the vertical gradient of basic-state potential temperature determine the frequency of buoyancy oscillations according to (2.97). The oscillations expressed by (2.117) are called *inertial oscillations*. From the quantity in parentheses in (2.117), it is evident that the horizontal shear of the absolute momentum of the basic-state flow is equivalent to the absolute vorticity $(\zeta_o + f)$ of the basic state.

2.7.4 Inertio-Gravity Waves

Since the Coriolis force gives rise to horizontal oscillations while the buoyancy force gives rise to vertical oscillations, it is not surprising that propagating waves exist for which the Coriolis and buoyancy forces act jointly to provide the net restoring force. These waves, called *inertio-gravity waves*, are solutions to the three-dimensional, dry adiabatic, inviscid Boussinesq equations linearized about a motionless, hydrostatic basic state. Under these assumptions, the x- and y-component equations of motion from (2.85) are

$$u'_t = -\pi'_x + fv', \quad v'_t = -\pi'_y - fu'$$

(2.118)

where $\pi \equiv p^*/\rho_o$. The vertical equation of motion

$$w'_t = -\pi'_z + B'$$

(2.119)

and the thermodynamic equation

$$B'_t = -w'N^2$$

(2.120)

may be combined to yield the wave equation

$$w''_{tt} + \pi'_{zt} + N^2 w' = 0 \tag{2.121}$$

The continuity equation is

$$u'_x + v'_y + w'_z = 0 \tag{2.122}$$

It can be verified by substitution that (2.118), (2.121), and (2.122) have solutions of the form

$$u', v', w', \pi' \propto e^{i(kx+ly+mz-vt)} \tag{2.123}$$

where

$$v^2 = N^2 \left(\frac{k^2 + l^2}{k^2 + l^2 + m^2} \right) + f^2 \left(\frac{m^2}{k^2 + l^2 + m^2} \right) \tag{2.124}$$

The factor multiplying N^2 is the cosine squared of the angle of the phase lines from the vertical, while the multiplier of f^2 is the sine squared of that angle. In the case of hydrostatic motions (i.e., $m^2 \gg k^2 + l^2$), (2.124) reduces to

$$v^2 = N^2 \left(\frac{k^2 + l^2}{m^2} \right) + f^2 \tag{2.125}$$

2.8 Adjustment to Geostrophic and Gradient Balance

As noted in Sec. 2.2.1, the large-scale environment in which cloud systems are embedded is typically in a state of near-geostrophic balance. Any disruption of this balance (for example by a cloud system) can lead to the excitation of inertio-gravity waves, which act to readjust the pressure field to geostrophic balance. The physical mechanism of the adjustment is well illustrated by the shallow-water versions of (2.118), which are

$$u'_t = -gh'_x + fv' \tag{2.126}$$

and

$$v'_t = -gh'_y - fu' \tag{2.127}$$

The continuity equation, obtained by integrating (1.22) over the depth of the water and applying boundary conditions similar to those leading to (2.101), is

$$h'_t + \left(u'_x + v'_y \right) \bar{h} = 0 \tag{2.128}$$

The divergence equation obtained by taking $\partial(2.126)/\partial x + \partial(2.127)/\partial y$ is

$$h'_{tt} - g\bar{h} \left(h'_{xx} + h'_{yy} \right) + f\bar{h}\zeta' = 0 \tag{2.129}$$

When there is no Coriolis effect ($f = 0$), this equation reduces to a two-dimensional shallow-water equation [cf. (2.102)] with gravity-wave solutions of the form

$$h' \propto e^{i(kx+ly-vt)} \tag{2.130}$$

with $\nu^2 = g\bar{h}(k^2 + l^2)$. When $f \neq 0$, the height–perturbation and perturbation vorticity fields are related by (2.129). Under geostrophic conditions, the time derivatives in (2.126)–(2.129) disappear, and (2.129) simply states that the vorticity field is in geostrophic balance. When the initial state is not in geostrophic balance (e.g., Fig. 2.1), (2.129) describes the approach to geostrophic balance. However, to accomplish this description a further relation between h' and ζ' is needed. For this, we turn to the equation for the vertical component of vorticity obtained by taking $\partial(2.127)/\partial x - \partial(2.126)/\partial y$. The result is

$$\zeta_t' = -f\left(u_x' + v_y'\right) \tag{2.131}$$

which is a linearized perturbation form of (2.59). This equation may be combined with the continuity equation to obtain the conservation property

$$\frac{\partial}{\partial t}\left(\frac{\zeta'}{f} - \frac{h'}{\bar{h}}\right) = 0 \tag{2.132}$$

The quantity in parentheses is a linearized form of the *shallow-water potential vorticity*. It retains its initial value everywhere in the spatial domain. Because of this behavior, the final velocity field, after adjustment to geostrophic balance, can be found from (2.129). If at $t = 0$ the fluid is motionless ($u' = v' = 0$) but its depth h' is distorted into the step-function configuration illustrated in Fig. 2.1, then, according to (2.132), the potential vorticity at a later time t is

$$\frac{\zeta'}{f} - \frac{h'}{\bar{h}} = \frac{h_o}{\bar{h}}\,\mathrm{sgn}(x) \tag{2.133}$$

where $h_o \geq 0$ and $\mathrm{sgn}(x)$ indicates "sign of x." Substituting the value of ζ' given by this relation into the last term of (2.129) yields

$$h_{tt}' - g\bar{h}\left(h_{xx}' + h_{yy}'\right) + f^2 h' = -f^2 h_o\,\mathrm{sgn}(x) \tag{2.134}$$

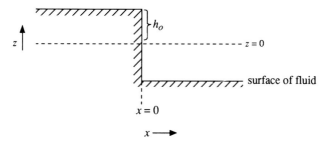

Figure 2.1   Surface of shallow-water fluid distorted into a stair-step configuration. Notation is that used in the text. The infinite horizontal pressure gradient, which must apply across the discontinuity, is incompatible with a state of hydrostatic balance.

If the surface is undisturbed ($h_o = 0$), (2.134) has shallow-water, inertio-gravity wave solutions of the form (2.130) with

$$\nu^2 = f^2 + g\bar{h}\left(k^2 + l^2\right) \tag{2.135}$$

which is the shallow-water version of (2.125). For a finite disturbance of the surface ($h_o > 0$), the final steady-state form of (2.134) is

$$-\frac{d^2 h'}{dx^2} + \frac{h'}{\lambda_R^2} = -\frac{h_o}{\lambda_R^2}\,\mathrm{sgn}(x) \tag{2.136}$$

where we have introduced the *Rossby radius of deformation*,[33] defined as

$$\lambda_R \equiv \sqrt{\frac{g\bar{h}}{f^2}} \tag{2.137}$$

The y-derivatives disappear from (2.136) because the perturbation is independent of y. The solution of (2.136) is

$$h' = h_o \begin{cases} -1 + \exp\left(-x/\lambda_R\right), & x > 0 \\ 1 - \exp\left(+x/\lambda_R\right), & x < 0 \end{cases} \tag{2.138}$$

Substitution of (2.138) into (2.126) and (2.127) show that the final perturbation wind field is geostrophic (and nondivergent) with values

$$u' = 0, \quad v' = -\frac{gh_o}{f\lambda_R}\exp\left(-|x|/\lambda_R\right) \tag{2.139}$$

This flow is illustrated in Fig. 2.2. It has adjusted to a point where the Coriolis force is just balanced by the pressure gradient force, and no further. If there were no Coriolis effect, then geostrophic balance would never be achieved and gravity waves would carry off all the potential energy of the initial perturbation. The final pressure gradient would be flat, and the final values of u' and v' would be zero. However, with the Coriolis effect present, the initial pressure gradient (manifested by the step-like jump in the depth of the fluid) would not be wiped out altogether by the gravity waves but rather would be spread over a finite distance of e-folding width $2\lambda_R$. This more gentle gradient of h' is in geostrophic balance. Thus, some of the potential energy of the initial disturbance is retained as geostrophic perturbation kinetic energy (since $v' \neq 0$ in the final steady state). It turns out that in this example one-third of the potential energy of the initial disturbance goes into the steady geostrophic flow, while the remaining two-thirds is carried off by the inertio-gravity waves.

A process analogous to geostrophic adjustment occurs in a strong circular flow, such as a hurricane. If we again assume shallow-water flow but assume the basic state is an axially symmetric circular vortex in gradient balance, with a mean

---

[33] Named for the Swedish-American meteorologist Carl-Gustav Rossby (1898–1957). Many of the basic principles of modern atmospheric dynamics were first elucidated by him.

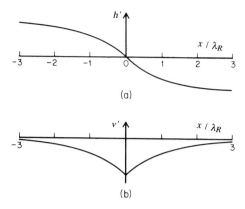

Figure 2.2 The geostrophic equilibrium solution corresponding to adjustment from an initial state that is one of rest but has uniform infinitesimal surface elevation $-h_o$ for $x > 0$ and elevation h_o for $x < 0$. (a) The equilibrium surface level h', which tends toward the initial level as $x \rightarrow \pm\infty$. The unit of distance in the figure is the Rossby radius. (b) The corresponding equilibrium velocity distribution. (From Gill, 1982.)

azimuthal velocity $\bar{v}$ that is in gradient balance and solid body rotation ($\bar{v} \propto r$), then the linearized equations for the radial and azimuthal velocity perturbations become

$$u'_t = -gh'_r + v'\left(f + \bar{\zeta}\right) \tag{2.140}$$

$$v'_t = -u'\left(f + \bar{\zeta}\right) \tag{2.141}$$

In writing (2.140), we have made use of the fact that the proportionality constant between $\bar{v}$ and r is $\bar{\zeta}/2$ in the case of solid-body rotation (Sec. 8.5.2). The perturbation continuity equation is

$$\frac{\partial h'}{\partial t} = -\frac{\bar{h}}{r}\frac{\partial(u'r)}{\partial r} \tag{2.142}$$

while the shallow-water potential vorticity equation is

$$\frac{\partial}{\partial t}\left[\frac{\zeta'}{f} - \frac{\left(f + \bar{\zeta}\right)h'}{f\bar{h}}\right] = 0 \tag{2.143}$$

Applying similar arguments to (2.140)–(2.143) as were applied to their counterparts above, one finds the equation corresponding to (2.136) to be

$$-\frac{1}{r}\frac{d}{dr}\left(r\frac{dh'}{dr}\right) + \frac{h'}{\lambda_R'^2} = -\frac{h_o\,\mathrm{sgn}(r - r_c)}{\lambda_R'^2} \tag{2.144}$$

where r_c is the radius of a jump in h' at the initial time, similar to that at $x = 0$ in Fig. 2.1, and

$$\lambda'_R \equiv \sqrt{\frac{g\bar{h}}{\left(\bar{\zeta} + f\right)^2}} \tag{2.145}$$

which is the Rossby radius of deformation for the specified circular flow. The Rossby radius in this case is influenced by the relative vorticity of the vortex as well as by the earth's vorticity f. In a strong vortex (like a hurricane), the relative vorticity dominates. When the radius of curvature of the basic flow vortex is infinite (i.e., when the flow is straight) λ'_R reduces to λ_R.

2.9 Instabilities

2.9.1 Buoyant, Inertial, and Symmetric Instabilities

In Secs. 2.7 and 2.8 we have summarized some of the types of atmospheric motions relevant to cloud dynamics. However, we have considered only those situations for which $\partial\bar{\theta}/\partial z > 0$, where $\bar{\theta}$ is the potential temperature of a hydrostatically balanced mean state, and $\partial\bar{M}/\partial x > 0$, where $\bar{M}$ is the absolute momentum of a two-dimensional geostrophically balanced mean state. Referring to (2.97) and (2.117), we see that these are the cases for which, according to parcel theory, the buoyancy and Coriolis restoring forces produce stable sinusoidal oscillations about a parcel's initial equilibrium position. The oscillations are vertical in the case of (2.97) and horizontal in the case of (2.117). Now we investigate briefly the situations in which $\partial\bar{\theta}/\partial z < 0$ and $\partial\bar{M}/\partial x < 0$. According to (2.97) and (2.117), the solutions to the basic equations no longer oscillate stably, but may grow exponentially. The environments in these situations are said to be *unstable*. Negative thermal stratification ($\partial\bar{\theta}/\partial z < 0$) is called *buoyant instability*. Negative horizontal shear ($\partial\bar{M}/\partial x < 0$) is called *inertial instability*.

These two types of instability can be inferred heuristically by making the parcel-theory assumption ($p^* = 0$) and using the fact that θ is conserved under adiabatic conditions [according to (2.9)] and M is conserved under two-dimensional, inviscid conditions [according to (2.112)]. If a parcel of dry air is displaced upward in a buoyantly unstable environment, it immediately becomes buoyant and accelerates upward, according to (2.93). If a parcel of air is displaced in the positive x-direction in an inertially unstable environment, it immediately obtains an excess of absolute momentum ($M - \bar{M} > 0$) and is accelerated in the positive x-direction, according to (2.111). In both cases, the displacement leads to the parcel being accelerated in the direction of the displacement (i.e., away from the equilibrium position).

Actually, the atmosphere is rarely unstable to dry-adiabatic displacements. Pure buoyant instability is generally eliminated as rapidly as it builds up, since it is released by any perturbation, no matter how small. However, as a result of the conservation of θ_e expressed by (2.18), there exists an analog to buoyant instability that is relevant to motions within clouds. If a parcel of air in a saturated environment (mean state) is displaced vertically upward, the parcel, which conserves its value of θ_e, will become positively buoyant if $\partial\bar{\theta}_e/\partial z < 0$. A saturated environment is thus buoyantly unstable in the same way that a dry atmosphere is unstable when $\partial\bar{\theta}/\partial z < 0$. However, this type of instability is also quickly eliminated and also rather rare.

Since the atmosphere is usually moist (i.e., contains water vapor) but *unsaturated*, a situation called *conditional instability* can arise. A necessary condition for this instability to exist is for the temperature in the undisturbed atmosphere to decrease with height less rapidly than a dry parcel of air, conserving its θ but more rapidly than that of a saturated parcel conserving its θ_e. In such an atmosphere, a mass of moist but unsaturated air lifted under parcel-theory conditions above its saturation level may eventually become warmer than the undisturbed environment. Nearly all convective clouds (cumulus and cumulonimbus) form in conditionally unstable environments and acquire their buoyancy in this manner.

Another way to state the necessary condition for conditional instability is that we must have $\partial\bar{\theta}_{es}/\partial z < 0$, where θ_{es} is the *saturation equivalent potential temperature* defined by

$$\theta_{es} \equiv \theta(T, p)e^{Lq_{vs}(T, p)/c_p T} \tag{2.146}$$

That is, the environment is conditionally unstable if the equivalent potential temperature computed as if the environment were saturated decreases with height.

If a whole layer of air, which is initially moist but unsaturated and characterized by $\partial\bar{\theta}_e/\partial z < 0$, is brought to saturation (e.g., by lifting), the whole layer becomes buoyantly unstable. Such a layer of air is said to be *potentially unstable*. Severe thunderstorm occurrences are often preceded by a period in which a layer of the atmosphere is potentially unstable. The storms occur when the potentially unstable layer is lifted to saturation.[34]

In the atmosphere, buoyancy and Coriolis forces act simultaneously. If the large-scale mean state is both geostrophically and hydrostatically balanced and if there is no friction or heating, motions are two-dimensional in the $x–z$ plane, and we continue to make the assumption that $p^* = 0$ on a parcel, then both (2.93) and (2.111) apply. An absolute momentum excess or deficit of a parcel displaced from equilibrium leads to an acceleration in the horizontal, while a deficit or excess in potential temperature leads to an acceleration in the vertical. These accelerations may be in either the stable or unstable sense. It is possible for the atmosphere to be stable for purely vertical or purely horizontal displacements but nonetheless unstable to sloping or slantwise displacements. Such an environment is said to be *symmetrically unstable* and can lead to strong slantwise air motions.

A symmetrically unstable two-dimensional mean state is illustrated in Fig. 2.3. The isolines of $\bar{\theta}$ and $\bar{M}$ both slope in the $x–z$ plane, but both $\partial\bar{\theta}/\partial z > 0$ and

---

[34] For a more complete discussion of conditional and potential instability, see Wallace and Hobbs (1977).

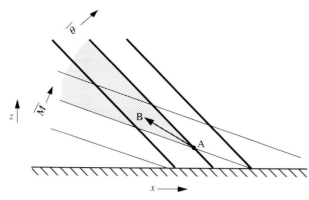

Figure 2.3 A symmetrically unstable two-dimensional mean state. The isolines of $\bar{\theta}$ and $\bar{M}$ in the x–z plane slope such that $\partial\bar{\theta}/\partial z > 0$ and $\partial\bar{M}/\partial x > 0$. A parcel lifted from A to B has a higher $\bar{\theta}$ and a lower $\bar{M}$ than its environment.

$\partial\bar{M}/\partial x > 0$. Thus, this mean state is both buoyantly and inertially stable. A parcel of air displaced vertically upward from point A acquires negative buoyancy and is accelerated back downward, while, analogously, a parcel of air displaced horizontally in the negative x-direction from point A acquires a positive value of $M - \bar{M}$ and is accelerated in the positive x-direction back toward A. However, a parcel of air displaced upward along a slantwise path anywhere within the shaded wedge acquires both a positive buoyancy and a negative value of $M - \bar{M}$. Thus, the parcel is accelerated farther away from its equilibrium position A.

The condition that characterizes the environment in Fig. 2.3 as symmetrically unstable is that the slope of the $\bar{M}$ surfaces must be less than the slope of the $\bar{\theta}$ surfaces. This difference of slope can be expressed in three different but equivalent ways. First, the lines must cross such that

$$\frac{\partial\bar{\theta}}{\partial z}\bigg|_{\bar{M}} < 0 \tag{2.147}$$

where the notation $|_{()}$ means at constant $()$. Thus, the environment must be buoyantly unstable on a surface of constant $\bar{M}$. Alternatively this condition may be stated

$$\frac{\partial\bar{M}}{\partial x}\bigg|_{\bar{\theta}} < 0 \tag{2.148}$$

which indicates that the environment is inertially unstable on a surface of constant $\bar{\theta}$. Finally, we note that for the slope of the $\bar{M}$ surface (a negative quantity in the case of Fig. 2.3) to greater than the slope of the $\bar{\theta}$ surface (also negative), we must have

$$\frac{\partial\bar{M}}{\partial x}\frac{\partial\bar{\theta}}{\partial z} - \frac{\partial\bar{M}}{\partial z}\frac{\partial\bar{\theta}}{\partial x} < 0 \tag{2.149}$$

However, from (2.66), this is equivalent to saying that the two-dimensional geostrophic mean state satisfies

$$P_g < 0 \tag{2.150}$$

(i.e., the mean state must have negative potential vorticity). Note that since P is conserved in adiabatic frictionless flow, fluid parcels cannot rid themselves of the negative potential vorticity in the absence of heating and/or turbulent mixing.

If the air is saturated and parcels undergo adiabatic motion except for liquid–vapor phase changes, so that (2.18) applies rather than (2.11), then the concept of symmetric instability carries over if $\bar{\theta}$ is replaced with $\overline{\theta_e}$ in (2.147)–(2.149), and P_g is replaced with the geostrophic equivalent potential vorticity P_{eg} in (2.150). If the air is unsaturated, replacement of $\bar{\theta}$ with $\overline{\theta_e}$ in (2.147)–(2.149), may be referred to as *potential symmetric instability*, which is analogous to the potential instability described above; in other words, a layer in which $\overline{\theta_e}$ decreases with height on a surface of constant $\bar{M}$ would be potentially unstable for slantwise parcel displacement, if the air were brought to saturation.

If the air is moist, but unsaturated, a condition called *conditional symmetric instability* can also be identified, which is analogous to the conditional instability described above. Conditional symmetric instability is said to exist when the temperature lapse rate of a moist but unsaturated environment is conditionally unstable *on a surface of constant $\bar{M}$* (i.e., the slope of the $\bar{M}$ surfaces must be less than the slope of the $\bar{\theta}_{es}$ surfaces). This condition is stated by replacing $\bar{\theta}$ by $\bar{\theta}_{es}$ in (2.147)–(2.149). To illustrate, let us imagine the lines of $\bar{\theta}$ in Fig. 2.3 to be replaced by lines of $\bar{\theta}_{es}$. If the parcel at point A has already been lifted above its saturation level and conserves θ_e as it is displaced slantwise farther upward into the region of the shaded wedge, then it will be positively buoyant at the same time that it is subjected to inertial acceleration.

In a circular two-dimensional flow under gradient balance (Sec. 2.2.3), the conserved momentum variable is the angular momentum m rather than the absolute momentum M [cf. Eqs. (2.37) and (2.112)]. Hence, by analogy to (2.147) and (2.148), the condition for symmetric instability in a two-dimensional gradient flow is

$$\left.\frac{\partial \bar{\theta}}{\partial z}\right|_{\bar{m}} < 0, \quad \left.\frac{\partial \bar{m}}{\partial r}\right|_{\bar{\theta}} < 0 \tag{2.151}$$

Conditions for potential and conditional symmetric instability cast in terms of m follow by arguments analogous to those above.

2.9.2 Kelvin–Helmholtz Instability

In the previous section, we have seen that the vertical gradient of potential temperature and the horizontal shear of the horizontal wind component are sources of flow instability. Another source is the vertical shear of the horizontal wind. This

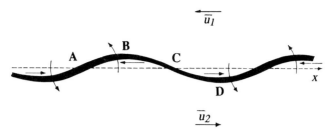

Figure 2.4 Conceptual illustration of the growth of a sinusoidal disturbance to an initially uniform sheet of vorticity (dashed) located between layers of fluid moving at different horizontal velocities ($\bar{u}_1$ and $\bar{u}_2$). The local strength density of the sheet is represented by its thickness. The arrows indicate the direction of the self-induced movement of the vorticity in the sheet and show both the accumulation of vorticity at points like A and the general rotation about points like A, which together lead to exponential growth of the disturbance. (From Batchelor, 1967.)

type of instability, referred to as *Kelvin–Helmholtz instability*,[35] can be visualized by the heuristic argument summarized in Fig. 2.4. Two fluids, one beneath the other, are moving parallel to the x-axis, but at different speeds. When the system is undisturbed, the interface of the two fluids is horizontal. The interface is regarded as a thin layer of strong vorticity, or "vortex sheet," which may be thought of as consisting of a layer of small discrete vortices, all rotating in the same direction. The up and down motions of the vortices cancel, while the horizontal components reinforce and account for the net velocity difference from top to bottom of the vortex sheet. If the sheet is perturbed sinusoidally (as shown), the vortex elements interact to produce velocity perturbations along the sheet. The vortex elements at A and C produce leftward motion in the region under wave crest B and rightward motion above trough D. These motions advect the vortex sheet in the direction of exaggerating the perturbation at A and weakening the one at C. They concentrate vorticity at A and lessen its intensity at C. The stronger counterclockwise rotation at A lifts the crest and lowers the trough, and the fluid-motion perturbations are amplified.

Kelvin–Helmholtz instability is related quantitatively to the differences of velocity and density between the two fluids. The velocity difference is a measure of the intensity of the vorticity, which is locally intensified by perturbations according to Fig. 2.4. The greater the velocity difference, the stronger the instability. The density difference is important because, the more the density of the lower fluid exceeds that of the upper fluid, the greater the buoyant stability, and thus the more the buckling of the interface will be suppressed by the buoyancy restoring force. From Fig. 2.4, positive buoyancy exists above the troughs and negative buoyancy below each crest. The vorticity associated with the velocity difference across the interface must be strong enough for perturbations in the interface to overcome these restoring forces.

[35] Named after Lord Kelvin and Herman von Helmholtz, who first developed the mathematics of this type of instability in the late nineteenth century. Further discussions of the topic can be found in Lamb (1932), Turner (1973), Batchelor (1967), Lilly (1986), and Acheson (1990).

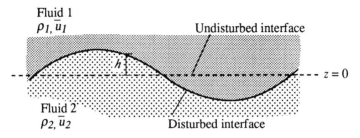

Figure 2.5 Notation used to describe a sinusoidal disturbance to an initially uniform vorticity sheet (dashed) located between layers of fluid of different densities (ρ_1 and ρ_2) moving at different horizontal velocities ($\bar{u}_1$ and $\bar{u}_2$).

The classical mathematical expression of the instability condition,[36] which includes the relative amounts of velocity shear and density difference across the interface, can be obtained by considering a highly idealized case in which the two fluids on either side of the interface are homogeneous and all vorticity in the system is contained in an infinitesimally thin interface. We will first review the analysis of this classical case, which indicates how waves form on the interface and results in the instability conditions given in (2.163) and (2.164) below. Then we will derive the instability condition (2.170), which applies to a continuously stratified fluid.

The classical case is obtained by considering the situation depicted in Fig. 2.5. The upper fluid is indicated by subscript 1 and the lower fluid by 2. The mean velocities and densities are $\bar{u}_1, \bar{u}_2$ and ρ_1, ρ_2, respectively. The pressures are $p_1 = \bar{p}(z) + p_1'(x,z,t)$ and $p_2 = \bar{p}(z) + p_2'(x,z,t)$ where

$$\frac{\partial \bar{p}}{\partial z} = \begin{cases} -\rho_1 g, & z > 0 \\ -\rho_2 g, & z < 0 \end{cases} \tag{2.152}$$

The undisturbed height of the interface is taken to be $z = 0$, while the height of the disturbed interface relative to the undisturbed height is represented by h. We consider the range of x for which $h < 0$ and integrate the vertical equation of motion from $z = -\infty$ to ∞. In carrying out the integration, we have three versions of the linearized vertical equation of motion to consider:

$$\left(\frac{\partial}{\partial t} + \bar{u}_2 \frac{\partial}{\partial x}\right) w_2' = -\frac{1}{\rho_2} \frac{\partial p_2'}{\partial z}, \quad \text{for } z < 0 \text{ below the interface} \tag{2.153}$$

$$\left(\frac{\partial}{\partial t} + \bar{u}_1 \frac{\partial}{\partial x}\right) w_1' = -\frac{1}{\rho_1} \frac{\partial p_1'}{\partial z} + \left(\frac{\rho_2 - \rho_1}{\rho_1}\right) g, \quad \text{for } z < 0 \text{ above the interface} \tag{2.154}$$

$$\left(\frac{\partial}{\partial t} + \bar{u}_1 \frac{\partial}{\partial x}\right) w_1' = -\frac{1}{\rho_1} \frac{\partial p_1'}{\partial z}, \quad \text{for } z > 0 \text{ above the interface} \tag{2.155}$$

[36] See Lamb (1932).

Equation (2.154) contains the positive buoyancy, which attempts to restore the perturbation to its rest state. We represent the vertical velocity perturbation in terms of a velocity potential ϕ', which is defined such that

$$w'_1 = -\frac{\partial \phi'_1}{\partial z}, \quad w'_2 = -\frac{\partial \phi'_2}{\partial z} \tag{2.156}$$

The use of the velocity potential implies that the flow is irrotational (i.e., has no vorticity) in both the upper and lower layers, which is consistent with the assumption that all of the vorticity is concentrated in the infinitesimally thin interface. We also make use of the fact that at the interface we must have

$$p'_1(h) = p'_2(h) \tag{2.157}$$

Integration of (2.153)–(2.155) with substitution from (2.156) and (2.157) yields

$$\rho_1 \left(\frac{\partial}{\partial t} + \bar{u}_1 \frac{\partial}{\partial x} \right) \phi'_1 + (\rho_2 - \rho_1) g h = \rho_2 \left(\frac{\partial}{\partial t} + \bar{u}_2 \frac{\partial}{\partial x} \right) \phi'_2 \tag{2.158}$$

A point on the material interface surface must obey the following two relations:

$$-\frac{\partial \phi'_1}{\partial z} = \frac{dh}{dt} \approx \left(\frac{\partial}{\partial t} + \bar{u}_1 \frac{\partial}{\partial x} \right) h \tag{2.159}$$

$$-\frac{\partial \phi'_2}{\partial z} = \frac{dh}{dt} \approx \left(\frac{\partial}{\partial t} + \bar{u}_2 \frac{\partial}{\partial x} \right) h \tag{2.160}$$

Equations (2.158)–(2.160) have wave solutions of the form

$$h \propto e^{i(kx-vt)}, \quad \phi'_1 \propto e^{i(kx-vt)} e^{kz}, \quad \phi'_2 \propto e^{i(kx-vt)} e^{-kz}, \quad \text{at } z = h \tag{2.161}$$

where k is a positive number and

$$\frac{v}{k} = \frac{\rho_2 \bar{u}_2 + \rho_1 \bar{u}_1}{\rho_2 + \rho_1} \pm \sqrt{\frac{-\rho_1 \rho_2 (\bar{u}_2 - \bar{u}_1)^2}{(\rho_2 + \rho_1)^2} + \frac{g(\rho_2 - \rho_1)}{k(\rho_2 + \rho_1)}} \tag{2.162}$$

The coefficients of z in (2.161) are obtained by assuming that the magnitude of the vertical velocity at the peak of the disturbance pictured in Fig. 2.5 is equal in magnitude and opposite in sign to the vertical velocity at the trough, by applying the continuity equation (2.100), and by eliminating physically unrealistic cases that have no possible stable solutions. The physically unrealistic cases would arise if the coefficients of z in (2.161) were allowed to have the same sign. An expression would be obtained in place of (2.162) that would have a number under the radical that was always negative. Thus, the solutions would always be exponentially growing. The solutions given by (2.161) and (2.162) also have unstable (exponentially growing) solutions when the expression under the radical in (2.162) is negative. However, these unstable solutions occur only when

$$k > \frac{g\left(\rho_2^2 - \rho_1^2\right)}{\rho_1\rho_2\left(\bar{u}_2 - \bar{u}_1\right)^2} \tag{2.163}$$

If $\rho_2 = \rho_1 + \Delta\rho$, where $\Delta\rho$ is small compared to ρ_1, then (2.163) becomes

$$k > \frac{2g\rho^{-1}\Delta\rho}{\left(\Delta\bar{u}\right)^2} \tag{2.164}$$

where $\Delta\bar{u} \equiv \bar{u}_2 - \bar{u}_1$. It is thus apparent that there are more unstable modes the larger the wind difference (i.e., the greater the vorticity available to the perturbations) and the smaller the buoyancy (i.e., the weaker the restoring force).

The infinitesimally thin interface is rather idealized. However, an instability condition similar to (2.164) can be derived for a continuously stratified, incompressible (or Boussinesq) fluid by considering two parcels of fluid, each of unit mass, initially separated by a small distance δz in a two-dimensional fluid (no variation in the y-direction). The lower parcel is denser by an amount $\delta\rho = -(\partial\bar{\rho}/\partial z)\delta z$, where $\bar{\rho}$ is the mean state density of the fluid. If the parcels switch positions in the vertical, the increase of potential energy per unit volume of fluid is

$$g \, \delta\rho \, \delta z \tag{2.165}$$

The only source of energy for the exchange is the shear of the mean wind $\partial\bar{u}/\partial z$. Let U_1 and U_2 be the initial velocities of the upper and lower parcels, respectively. Since momentum must be conserved in the exchange, the final velocities can be represented by $U_1 + \Delta$ and $U_2 - \Delta$. The change of kinetic energy per unit mass that occurs in the exchange is

$$\left(U_1 - U_2\right)\Delta + \Delta^2 \tag{2.166}$$

which is maximum for

$$\Delta = \frac{U_2 - U_1}{2} \tag{2.167}$$

Substitution of $U_2 = U$ and $U_1 = U + \delta U$ as well as (2.167) into (2.166) shows that the maximum kinetic energy that can be lost in the exchange, per unit volume of fluid, is approximately

$$\frac{\bar{\rho}(\delta U)^2}{4} \tag{2.168}$$

From (2.165) and (2.168), it is evident that for energy to be released in the exchange (in the form of an instability), we must have

$$g \, \delta\rho \, \delta z < \tfrac{1}{4} \, \bar{\rho}(\delta U)^2 \tag{2.169}$$

Or, since $\delta U = (\partial\bar{u}/\partial z)\delta z$, this may be restated as

$$\text{Ri} \equiv \frac{-(g/\bar{\rho})\left(\partial\bar{\rho}/\partial z\right)}{\left(\partial\bar{u}/\partial z\right)^2} < \frac{1}{4} \tag{2.170}$$

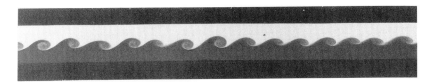

Figure 2.6 A laboratory example of waves resulting from Kelvin–Helmholtz instability. (From Thorpe, 1971. Reprinted with permission from Cambridge University Press.)

where Ri is called the *Richardson number*. Thus, again, we see that the vertical shear of the horizontal motion must be strong enough that the vorticity will be able to overcome the static stability.

An example of waves resulting from Kelvin–Helmholtz instability are shown in Fig. 2.6. A long covered trough was filled with a lower layer of salt water and an upper layer of fresh water. As the trough was tilted, salt water flowed downward, while fresh water flowed upward. Strong shear was thus produced at an interface. Waves formed on the interface, at first sinusoidal, then curling over and breaking, leading to turbulent mixing. The development of Kelvin–Helmholtz waves in time is further illustrated by Fig. 2.7. An example of Kelvin–Helmholtz waves on the top edge of a small cloud is shown in Fig. 2.8. It is not uncommon to observe Kelvin–Helmholtz waves on the borders of clouds, where there is often a discontinuity in density and wind.

2.9.3 Rayleigh–Bénard Instability

Another kind of instability that is relevant to cloud dynamics arises when a thin layer of fluid is subjected to heat fluxes at the top and/or bottom of the layer. The motion that results from this type of instability is referred to as *Rayleigh–Bénard convection*, and it is manifest in certain types of stratocumulus, altocumulus, and cirrocumulus, which exhibit substructure in the form of rolls and cells (e.g., Figs. 1.11b, 1.14a and b, 1.15a). An instability theory to explain the phenomenon was provided by Rayleigh.[37] His theory is based on the Boussinesq equations (Sec. 2.3) applied to an incompressible fluid that expands as its temperature is increased. For such a fluid the buoyancy is given by

$$B = g\alpha T' \qquad (2.171)$$

where α is the expansion coefficient, defined such that $\rho'/\rho_o = -\alpha T'$. Friction and heat conduction are parameterized in terms of a constant viscosity $\hat{D}$ and thermal conductivity $\hat{\kappa}$. The equations are linearized about a state of zero mean motion

[37] This type of convection was described by Thomson (1881), who observed a pattern of cells of overturning fluid in a barrel of warm soapy water behind an inn used for cleaning glasses. The top surface of the water was evaporating into the cool air. Bénard (1901) devised a laboratory experiment to reproduce the cells and pointed out the analogy between the laboratory phenomenon and "mackerel cloud," a name sometimes used to describe altocumulus rolls. Lord Rayleigh (1916) proposed the instability theory summarized here to explain the cells.

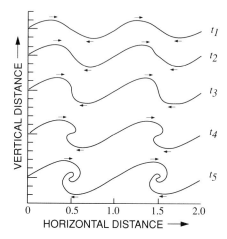

VERTICAL DISTANCE →

HORIZONTAL DISTANCE →

Figure 2.7 Development of Kelvin–Helmholtz waves in time. The shape of the disturbed fluid interface is shown at a succession of times, t_1–t_5. (From Rosenhead, 1931.)

and horizontally uniform temperature. Perturbations are then governed by the equation of motion,

$$\frac{\partial \mathbf{v}'}{\partial t} = -\frac{1}{\rho_o}\nabla p' + g\alpha T'\mathbf{k} + \hat{D}\nabla^2\mathbf{v}' \tag{2.172}$$

the continuity equation,

$$\nabla \cdot \mathbf{v}' = 0 \tag{2.173}$$

and the thermodynamic equation

$$\frac{\partial T'}{\partial t} - w\gamma = \hat{K}\nabla^2 T' \tag{2.174}$$

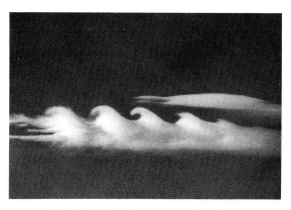

Figure 2.8 An example of Kelvin–Helmholtz waves on the edge of a small fragment of stratus cloud. (Photo by Brooks E. Martner, NOAA Wave Propagation Laboratory.)

where $\hat{K} \equiv \hat{\kappa}/\rho\hat{c}$, $\hat{c}$ is the specific heat of the homogeneous fluid, and γ is the basic-state lapse rate ($\equiv -\partial\bar{T}/\partial z$) maintained by heating below and/or cooling above. Substitution of solutions of the form[38]

$$w', T' \propto \sin mz\, e^{i(kx+ly)} e^{vt} \tag{2.175}$$

$$u', v', p' \propto \cos mz\, e^{i(kx+ly)} e^{vt} \tag{2.176}$$

into (2.172)–(2.174) leads to the dispersion relation

$$v^2\left(k^2 + l^2 + m^2\right) + v\left(\hat{K} + \hat{D}\right)\left(k^2 + l^2 + m^2\right)^2$$
$$+ \hat{D}\hat{K}\left(k^2 + l^2 + m^2\right)^3 - \gamma g a\left(k^2 + l^2\right) = 0 \tag{2.177}$$

Solving this quadratic equation leads further to the conclusion that unstable solutions (v positive and real) occur when

$$-\gamma g a\left(k^2 + l^2\right) + \hat{D}\hat{K}\left(k^2 + l^2 + m^2\right)^3 < 0 \tag{2.178}$$

If there is no friction ($\hat{D} = 0$), this relation reduces to simply

$$\gamma > 0 \tag{2.179}$$

That is, the lapse rate must be positive to get unstable growth. If both friction and conduction are finite, then terms in (2.178) can be rearranged to

$$\text{Ra} > \frac{\left(k^2 + l^2 + m^2\right)^3 h^4}{k^2 + l^2} \tag{2.180}$$

where Ra is the Rayleigh number, defined as

$$\text{Ra} \equiv \frac{h^4 \gamma g a}{\hat{D}\hat{K}} \tag{2.181}$$

and h is the depth of the fluid. Ra contains all of the prescribed characteristics of the fluid. If it is assumed that the vertical wave number is related to h by

$$m = \frac{\pi}{h} \tag{2.182}$$

then it is clear that (i) instability can occur for any number of combinations of k and l, including both cells ($k = l$) and rolls ($k = 0, l \neq 0$); and (ii) the value which Ra must exceed, according to (2.180), is a function of horizontal wave number $\sqrt{k^2 + l^2}$, as shown in Fig. 2.9. In order to be unstable at all, the Rayleigh number characterizing the fluid must exceed the minimum value:

$$(\text{Ra})_c = \frac{27\pi^4}{4} \tag{2.183}$$

---

[38] The vertical form assumed by Rayleigh is actually rather unrealistic when the fluid has a rigid bottom or top, since horizontal velocity perturbations are not required to disappear at $z = 0$ or D. With the requirement of zero horizontal velocity at the top and bottom, the analysis becomes more difficult.

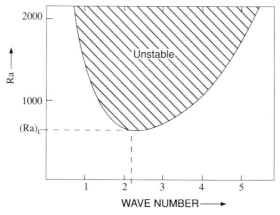

Figure 2.9 Value which Ra must exceed for Rayleigh–Bénard instability. $(Ra)_c$ is called the critical Rayleigh number.

which is found by differentiating the right-hand side of (2.180) with respect to $(k^2 + l^2)$. The most unstable solution for the special case of $\hat{K} = \hat{D}$ is obtained by differentiating (2.177) with respect to $(k^2 + l^2 + m^2)$ [or $(k^2 + l^2)$] and setting $dv/d(k^2 + l^2 + m^2) = 0$ (to obtain the condition of maximum v). A second equation in v is thus obtained, which may be combined with (2.177) to eliminate v and obtain

$$\frac{m^2}{k^2 + l^2 + m^2} = 1 - \frac{Ra}{h^4}\left(\frac{m}{k^2 + l^2 + m^2}\right)^4 \tag{2.184}$$

which is the relation among the wave numbers when v has its maximum value. If we again assume that $m = \pi/h$ and that we have square cells such that $k = l = 2\pi/S$, where S is the spacing of the cells, then (2.184) implies an expected ratio of S to h. Under laboratory conditions this ratio is about 3 : 1. No simple relation like (2.184) is obtained for the more general case of $\hat{K} \neq \hat{D}$.

2.10   Representation of Eddy Fluxes

We have seen that deviations from mean motions on the scale of interest (defined by the overbar) can be important if the deviations are systematically correlated, thus constituting fluxes [recall eddy flux terms in (2.78), (2.81), and (2.83)–(2.84)]. This is usually the case where clouds are involved and in the planetary boundary layer (to be discussed in Sec. 2.11 below). An important problem thus arises because the eddy fluxes $\rho\overline{v'\mathscr{A}'}$, where $\mathscr{A}$ represents any of the variables θ, π, q_i, u, v, or w, become additional variables in the equations for the mean variables [e.g., $\rho_o\overline{v'\theta'}$ in (2.78) and $\rho_o\overline{v'u'}$ in (2.83)]. There must, therefore, be some rational way to represent these fluxes in order for the system of equations governing atmospheric motions to remain a closed set. Representation of the fluxes for purposes of closure is difficult. The method adopted for this purpose depends, first, on the scale of the region represented by the average variables. This scale, in turn, determines the scale of motions represented as eddy fluxes. When the averages represent very large scales of motion (e.g., the synoptic scale relevant to weather forecasting and analysis), entire clouds and cloud systems are part of the perturbation motions. When the averaging is done on smaller scales (e.g., mesoscale or convective scale), the mean variables explicitly describe the general flow in and around a cloud, while the eddy fluxes represent turbulence on only those scales smaller than basic mesoscale or convective-scale motions within the cloud. These eddy fluxes are always important in cloud dynamics, even in the most apparently quiescent clouds such as fog and stratus.

There are several approaches to representing eddy fluxes in the mean-variable equations. Three of these approaches are summarized briefly below.[39]

2.10.1 K-Theory

If turbulent motions are sufficiently small, the turbulent fluxes of a quantity $\mathscr{A}$ can be considered as analogous to molecular diffusion, which proceeds down-gradient at a rate proportional to the spatial gradient of $\bar{\mathscr{A}}$; that is,

$$\rho \overline{\mathbf{v}'\mathscr{A}'} = -K_A \rho \nabla \bar{\mathscr{A}} \qquad (2.185)$$

In the case of atmospheric fluid motions, the exchange coefficient K_A is presumed to be proportional to the product of eddy size and velocity. This coefficient depends strongly on wind shear and thermodynamic stratification (i.e., on the Richardson number Ri). This dependence is understandable, since we have seen in (2.170) that any layer of fluid is unstable and subject to small-scale waves and turbulence when Ri $< 1/4$. Conversely, such breakdown of the fluid motion is highly suppressed when Ri is sufficiently large. Near the ground, K_A becomes a function of other variables such as height above ground and roughness of the underlying surface. When K_A is expressed quantitatively in terms of all these parameters, and they are known, then the basic equations for the mean variables are said to be closed. This approach is called K-theory or first-order closure.

For many atmospheric situations, including the planetary boundary layer, the vertical fluxes are predominant. Some of the most frequently referred to are

$$\left.\begin{array}{l}\text{Momentum flux}\\\text{in } x\text{ - direction}\end{array}\right\} \quad \tau_x \equiv \rho\overline{u'w'} = -\rho K_m \frac{\partial \bar{u}}{\partial z} \qquad (2.186)$$

$$\left.\begin{array}{l}\text{Momentum flux}\\\text{in } y\text{ - direction}\end{array}\right\} \quad \tau_y \equiv \rho\overline{v'w'} = -\rho K_m \frac{\partial \bar{v}}{\partial z} \qquad (2.187)$$

$$\left.\begin{array}{l}\text{Sensible heat}\\\text{flux}\end{array}\right\} \quad \mathscr{F}_\theta \equiv \rho\, c_p \overline{w'\theta'} = -c_p \rho K_\theta \frac{\partial \bar{\theta}}{\partial z} \qquad (2.188)$$

$$\left.\begin{array}{l}\text{Water vapor}\\\text{flux}\end{array}\right\} \quad \mathscr{F}_v \equiv \rho\overline{w'q_v'} = -\rho K_v \frac{\partial \bar{q_v}}{\partial z} \qquad (2.189)$$

[39] For more extended discussions of the representation of eddy fluxes, the books by Panofsky and Dutton (1984), Arya (1988), and Sorbjan (1989) are recommended, especially Chapter 6 of Sorbjan.

2.10.2   Higher-Order Closure

A more detailed approach, called *higher-order closure*, is sometimes used to close the equations when turbulent fluxes are important. Equations for covariances (i.e., $\overline{w'\theta'}$, $\overline{w'u'}$, etc.) are derived from the basic equations by procedures similar to that used to obtain the turbulent kinetic energy equation (2.86) (which is the equation for the variance of the velocity). The derived covariance equations predict the components of any flux $\rho\overline{\mathbf{v}'\mathscr{A}'}$. However, these new equations involve triple-product terms [analogous to $\mathscr{D}$ in the turbulent kinetic energy equation (2.86)], which must in turn be dealt with, and the approach is generally very complicated.

2.10.3   Large-Eddy Simulation

A third approach to achieving closure of the equations, when turbulence is a factor, is called *large-eddy simulation*. The largest eddies, which account for the largest portion of the turbulent fluxes, are resolved by the grid of a numerical model. Smaller-scale eddies are parameterized, for example by K-theory.

2.11   The Planetary Boundary Layer

All of the low-cloud types (fog, stratus, stratocumulus, cumulus, and cumulonimbus) either occur within or interact strongly with the *planetary boundary layer (PBL)*, which is the layer of the atmosphere in which the air motion is strongly influenced by interaction with the surface of the earth. The interaction occurs in the form of vertical exchanges of momentum, heat, moisture, and mass. These exchanges are largely effected by atmospheric turbulence. Since the wind velocity must vanish at the earth's surface, the PBL is always characterized by shear production of turbulent kinetic energy [$\mathscr{C}$ in (2.86)]. Buoyant production may either enhance or reduce the amount of turbulence [$\mathscr{B}$ in (2.86)]. However, there is always at least a small amount of mixing present, and the mixing, whether small or large, can have a substantial effect on low-cloud development. The clouds, in turn, can have important feedbacks on the PBL through condensation, evaporation, radiation, downdrafts, and precipitation.

2.11.1   The Ekman Layer

The balanced flow regimes, quasi-geostrophic, semigeostrophic, and gradient flow, discussed in Sec. 2.2 all apply in the *free atmosphere*, which is the part of the atmosphere lying above the PBL. The forces associated with molecular friction and turbulence [terms $\bar{\mathbf{F}}$ and $\bar{\mathscr{F}}$ in (2.83)] were ignored to obtain these balanced flows. Turbulence is indeed negligible in large-scale flow in the free atmosphere (although it is *not* negligible in the mesoscale and convective-scale motions within clouds in the free atmosphere). However, in the PBL, the turbulence is not negli-

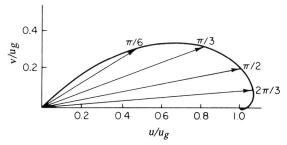

Figure 2.10 Hodograph of the Ekman spiral solution. Points marked on the curve are values of $\gamma_E z$, which is a nondimensional measure of height. (From Holton, 1992.)

gible. In fact, the PBL can be thought of as a layer in which the turbulence force is comparable in magnitude to the pressure-gradient and Coriolis forces. Let us assume that these three forces are in balance in the horizontal components of the Boussinesq version of the equation of motion (2.83). Under the Boussinesq assumption the density terms ρ_o disappear from (2.84). If we take the horizontal average of the equations, assume that the horizontal fluxes are small compared to vertical fluxes, and express the eddy flux of momentum in terms of K-theory, with a constant value of K_m, we obtain

$$K_m \bar{u}_{zz} + f\left(\bar{v} - v_g\right) = 0 \tag{2.190}$$

$$K_m \bar{v}_{zz} - f\left(\bar{u} - u_g\right) = 0 \tag{2.191}$$

It is further assumed that the geostrophic wind components are constant, that $\bar{u} = \bar{v} = 0$ at $z = 0$, and that $\bar{u} \rightarrow u_g$, $\bar{v} \rightarrow v_g$ as $z \rightarrow \infty$. For convenience, the coordinate axes are oriented such that $v_g = 0$. Then (2.190) and (2.191) have the solutions

$$\bar{u} = u_g\left(1 - e^{-\gamma_E z} \cos \gamma_E z\right), \quad \bar{v} = u_g e^{-\gamma_E z} \sin \gamma_E z \tag{2.192}$$

where $\gamma_E \equiv \sqrt{f/2K_m}$. The vector wind composed of these components describes a hodograph in the shape of a spiral (Fig. 2.10), and these solutions are referred to as the *Ekman spiral*.[40] The depth of the Ekman layer is

$$h_E \equiv \pi/\gamma_E = \pi\sqrt{2K_m/f} \tag{2.193}$$

which is the height at which the wind is parallel to and approximately equal to the geostrophic value.

Ideal Ekman layers are not usually observed in the PBL. There are several reasons for this fact. First, K_m is not a constant. Second, the K-theory approximation is often not accurate for momentum fluxes. Third, the Ekman solution is actually unstable. If the solutions in (2.192) are assumed to be the mean state, the

---

[40] Named after the Swedish oceanographer V. W. Ekman, who first discussed this spiral to explain the change of current with depth in the ocean boundary layer (Ekman, 1902). See Sverdrup *et al.* (1942) for further discussion.

time-dependent perturbation equations have unstable solutions, which grow to finite size and alter the mean state.[41]

2.11.2   Boundary-Layer Stability

Although the Ekman layer is an idealization, it provides a useful way to associate the depth of the PBL h_E with the degree of mixing in the layer. It is evident from (2.193) that the important factors in determining this depth are the mixing coefficient for momentum flux K_m and the Coriolis parameter f. Values of $K_m = 5$ m^2 s^{-1} and $f = 10^{-4}$ s^{-1} imply $h_E \approx 1$ km, which is a fairly common value for the depth of the PBL. However, this depth can vary greatly, primarily as a result of the variation of K_m with the Richardson number (Ri).

When the Richardson number is sufficiently high, the buoyant production $\mathscr{B}$ in the turbulent kinetic energy equation (2.86) constitutes a large negative effect. The shear production of turbulence $\mathscr{C}$ is thus counteracted by $\mathscr{B}$, as a result of the buoyancy restoring force suppressing the shear-induced motion perturbations. Under these conditions, the value of K_m may be greatly lowered. In extreme cases, such as where strong radiational cooling of the ground is occurring, the PBL top may drop to as little as 10 m or so. Fog and stratus clouds are often associated with such stable boundary-layer conditions.

In contrast, when the PBL stratification is tending toward unstable, Ri is low, and the buoyant production $\mathscr{B}$ of turbulent kinetic energy reinforces or completely dominates the shear production $\mathscr{C}$. In this case, K_m may be greatly increased and the turbulent layer can become very deep, say 2–3 km. In these cases, the boundary-layer depth may be controlled by conditions in the free atmosphere above. For example, large-scale subsidence aloft may produce a stable layer in the low troposphere, through which the turbulent eddies generated in the PBL cannot penetrate. This way of establishing the boundary-layer depth is particularly important in the tropics, where f is small and (2.193) loses its meaning.

Unstable boundary layers capped by large-scale subsidence are often marked by stratus or stratocumulus at the top of the mixed layer, where the buoyant, rising eddies ascend above their condensation level. Once formed, the cloud layer produces radiative feedback to the PBL structure. Clouds in both stable and unstable boundary layers are discussed in more detail in Chapter 5.

2.11.3   The Surface Layer

The amount of turbulence in the PBL, of course, generally decreases with height. In the lowest 10% of the PBL, however, it is a reasonable and useful approximation to consider the turbulent fluxes to be nearly constant with height. As we will see in Chapter 5, this view turns out to be helpful in analyzing fog, which occurs next to the ground. One of the useful simplifications of the near constancy of the wind stress vector (τ_x, τ_y) is that the wind *direction* does not vary with height.

---

[41] See the reviews of Brown (1980, 1983) for further discussion of this topic.

Chapter Three | Cloud Microphysics

"... fleecy piles dissolved in dew drops ..."[42]

As noted in the Introduction, cloud physics consists of two branches: cloud microphysics and cloud dynamics. While the topic of this book is the latter, it is impossible to divorce a discussion of the dynamics from a knowledge of the microphysics. Just as the discussions in Chapters 5–12 assume a certain level of background knowledge of atmospheric dynamics, so do they assume some background in cloud microphysics. To provide this background, the present chapter summarizes the aspects of cloud microphysics that are crucial to the discussions of later chapters. First, we describe some of the basic microphysical processes that are involved in the formation, growth, shrinkage, breakup, and fallout of cloud and precipitation particles.[43] In Sec. 3.1, we describe the microphysics of *warm clouds,* where the temperature is everywhere above 0°C. Section 3.2 extends the review of microphysical processes to *cold clouds,* in which the temperature drops below 0°C and both ice and liquid particles may exist. After this review of the individual microphysical processes that may occur in clouds, we consider in Secs. 3.3–3.6 how these microphysical processes occur simultaneously in a real cloud and how they may be linked to the cloud dynamics through a set of water-continuity equations.

3.1 Microphysics of Warm Clouds

3.1.1 Nucleation of Drops

The particles in a cloud form by a process referred to as *nucleation,* in which water molecules change from a less ordered to a more ordered state. For example, vapor molecules in the air may come together by chance collisions to form a liquid-phase drop. To see how this process takes place, consider the conditions required for the formation of a drop of pure water from vapor. This case is called *homogeneous nucleation* to distinguish it from the case of *heterogeneous nucleation,* which refers to the collection of molecules onto a foreign substance. If the

---

[42] Goethe realizes that the clouds are composed of microscopic particles.

[43] These microphysical processes are described in more detail in basic texts on cloud microphysics, such as Fletcher (1966), Mason (1971), Pruppacher and Klett (1978), and Rogers and Yau (1989). The physics of ice is presented comprehensively by Hobbs (1974).

embryonic drop of pure water has radius R, then the net energy required to accomplish its nucleation is

$$\Delta E = 4\pi R^2 \sigma_{vl} - \frac{4}{3}\pi R^3 n_l \left(\mu_v - \mu_l \right)$$ (3.1)

The first term on the right is the work required to create a surface of vapor–liquid interface around the drop. The factor σ_{vl} is the work required to create a unit area of the interface. It is called the *surface energy* or *surface tension*. The second term on the right of (3.1) is the energy change associated with the vapor molecules going into the liquid phase. It is expressed as the change in the Gibbs free energy of the system. The Gibbs free energy of a single vapor molecule is μ_v, while that of a liquid molecule is μ_l, and the factor n_l is the number of water molecules per unit volume in the drop. If the work required to create the surface exceeds the change in Gibbs free energy ($\Delta E > 0$), the embryonic drop formed by chance aggregation of molecules has no chance of surviving and immediately evaporates. If, on the other hand, the work required to create the surface is less than the change in Gibbs free energy ($\Delta E < 0$), then the drop survives and is said to have been nucleated.

It can be shown[44] that

$$\mu_v - \mu_l = k_B T \ln \frac{e}{e_s}$$ (3.2)

where k_B is Boltzmann's constant, e is the vapor pressure, and e_s is the saturation vapor pressure over a plane surface of water. Substituting this expression into (3.1), seeking the condition for which the work required to change the drop's surface is exactly matched by the change in Gibbs free energy ($\Delta E = 0$), and rearranging terms, we obtain an expression for the critical radius R_c at which this equilibrium condition holds. This expression is

$$R_c = \frac{2\sigma_{vl}}{n_l k_B T \ln\left(e/e_s \right)}$$ (3.3)

and is referred to as *Kelvin's formula*.[45] This radius is evidently crucially dependent on the *relative humidity* (defined as $e/e_s \times 100\%$). Air is said to be *saturated* whenever the relative humidity is 100% ($e/e_s = 1$). However, it is clear from (3.3) that it is impossible for a cloud droplet to form under saturated conditions since $R_c \to \infty$ as $e/e_s \to 1$. Rather, the air must be supersaturated ($e/e_s > 1$) for R_c to be positive. The greater the *supersaturation* [defined, in percent, as $[(e/e_s - 1) \times 100\%]$, the smaller the size of the drop that must be exceeded by the initial chance collection of molecules.

It should be noted that R_c is also a function of temperature. Not only does T appear in the denominator of (3.3) explicitly, but σ_{vl} and e_s are functions of T. However, at atmospheric temperatures, the dependence of R_c on temperature is comparatively weak. In view of the primary dependence of R_c on ambient humid-

---

[44] See problem 2.19 of Wallace and Hobbs (1977).
[45] Named after Lord Kelvin, who first derived it.

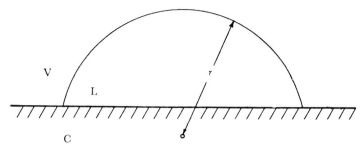

Figure 3.1 A spherical-cap embryo of liquid (L) in contact with its vapor (V) and a nucleating surface (C). (From Fletcher, 1966. Reprinted with permission from Cambridge University Press.)

ity, it is not surprising that the rate of nucleation of drops exceeding the critical size R_c is a strong function of the degree of supersaturation. The rate at which the vapor molecules collide to form aggregates of various sizes can be computed using principles of statistical quantum mechanics applied to an ideal gas whose molecules are in a state of random motion.[46] This rate of formation of drops exceeding the critical size is the *nucleation rate*. It is found to increase from undetectably small values to extremely large values over a very narrow range of e/e_s. The value of e/e_s at which this rise occurs is in the range of 4–5. Thus, the air must be supersaturated by 300–400% for a drop of pure water to be nucleated homogeneously. Since supersaturation in the atmosphere seldom exceeds 1%, one concludes that homogeneous nucleation of water drops plays no role in natural clouds. However, the physics of the process are nonetheless relevant, as will become evident below.

Heterogeneous nucleation is the process whereby cloud drops actually form. The atmosphere is filled with small *aerosol particles,* and molecules of vapor may collect onto the surface of aerosol particles as illustrated ideally in Fig. 3.1. If the surface tension between the water and the nucleating surface is sufficiently low, the nucleus is said to be *wettable,* and the water may form a spherical cap on the surface of the particle. A particle onto which the molecules collect in this manner is referred to as a *cloud condensation nucleus (CCN).*

If a CCN is insoluble in water, the physics governing the survival of an embryonic cloud droplet are the same as in the case of homogeneous nucleation. It can be shown that Eq. (3.3) still applies, but R_c has the more general interpretation in that it refers to the critical radius *of curvature* of the embryonic drop. Since the radius of curvature of the droplet forming on a particle is greater than what it would be if the same number of molecules were to aggregate in the absence of the particle (Fig. 3.1), the aggregation of the vapor molecules has a greater chance of producing a drop exceeding the critical radius. If the aggregated water molecules form a film of liquid completely surrounding a particle, then a complete droplet is formed whose radius is larger than it would be in the absence of the nucleus. Clearly, the larger such a nucleus is, the more likely is the survival of a drop

---

[46] See Chapter 2 of Fletcher (1966).

formed by a film around it. For this reason, the larger the aerosol particle, the more likely it is to be a site for drop formation in a natural cloud.

If the cloud condensation nucleus happens to be composed of a material that is soluble in water, the efficacy of the nucleation process is further enhanced. Since the saturation vapor pressure over the liquid solution is generally lower than that over a surface of pure water, e/e_s is increased. According to (3.3), the critical radius is then reduced, and nucleation is easier to achieve at the ambient vapor pressure.

There are generally more than enough wettable aerosol particles in the air to accommodate the formation of all cloud droplets. However, the physics of the nucleation process just described indicate that the first droplets in a cloud will tend to form around the largest and most soluble CCN. The sizes and compositions of the aerosol particles in a sample of air thus have a profound effect on the size distribution of particles nucleated in a cloud.

3.1.2 Condensation and Evaporation

Once formed, water drops may continue to grow as vapor diffuses toward them. This process is called *condensation*. The reverse process, drops decreasing in size as vapor diffuses away from them, is called *evaporation*. Particle growth by condensation and evaporation may be represented quantitatively by assuming that the flux of water vapor molecules through air is proportional to the gradient of the concentration of vapor molecules.[47] In this case, the vapor density ρ_v (defined as the mass of vapor per unit volume of air) is governed by the diffusion equation

$$\frac{\partial \rho_v}{\partial t} = \nabla \cdot \left(D_v \nabla \rho_v \right) = D_v \nabla^2 \rho_v \tag{3.4}$$

where $D_v \Delta \rho_v$ is the flux of water vapor by molecular diffusion and D_v is the diffusion coefficient (assumed constant) for water vapor in air. The concentration of vapor around a spherical pure-water drop of radius R is assumed to be symmetric about a point located at the center of the drop, and the diffusion is assumed to be in a steady state. Under these assumptions, ρ_v depends only on radial distance r from the center of the drop, and (3.4) reduces to

$$\nabla^2 \rho_v(r) = \frac{1}{r^2} \frac{\partial}{\partial r} \left(r^2 \frac{\partial \rho_v}{\partial r} \right) = 0 \tag{3.5}$$

The vapor density at the surface is $\rho_v(R)$. As $r \to \infty$, the vapor density approaches the ambient or free-air value $\rho_v(\infty)$. The solution to (3.5) satisfying these boundary conditions is

$$\rho_v(r) = \rho_v(\infty) - \frac{R}{r} \left[\rho_v(\infty) - \rho_v(R) \right] \tag{3.6}$$

---

[47] This assumption is called Fick's first law of diffusion.

If the drop has mass m, the flux of molecules causes its mass to increase or decrease at a rate given by

$$\dot{m}_{dif} = 4\pi R^2 D_v \frac{d\rho_v}{dr}\bigg|_R \tag{3.7}$$

where $D_v d\rho_v/dr|_R$ is the flux of vapor in the radial direction across a spherical surface of radius R. Substitution of (3.6) into (3.7) yields

$$\dot{m}_{dif} = 4\pi R D_v [\rho_v(\infty) - \rho_v(R)] \tag{3.8}$$

Since m $\propto R^3$, there are two unknowns in (3.8), $\rho_v(R)$ and either m or R. Conditions in the environment ($r = \infty$) are assumed to be known. To obtain a solution for m or R, other relationships are needed. First, a heat-balance equation is introduced. In the condensation of water vapor on a drop, latent heat is released at a rate $L\dot{m}_{dif}$, where L is the latent heat of vaporization. Assuming that heat is conducted away from the drop as rapidly as it is being released, we have by analogy to (3.8)

$$L\dot{m}_{dif} = 4\pi\kappa_a R[T(R) - T(\infty)] \tag{3.9}$$

where κ_a is the thermal conductivity of air and T is temperature.

The equation of state for an ideal gas applied to water vapor under saturated conditions over a plane surface of pure water is

$$e_s = \rho_{vs} R_v T \tag{3.10}$$

where R_v is the gas constant for a unit mass of water vapor, and e_s and ρ_{vs} are the saturation vapor pressure and density over a planar surface of water. Since e_s depends only on temperature,[48] it is evident from (3.10) that ρ_{vs} is a known function of T. If it is then assumed that the vapor density at the drop's surface is given by the saturation vapor density, we may write

$$\rho_v(R) = \rho_{vs}[T(R)] \tag{3.11}$$

and (3.8), (3.9), and (3.11) can be solved numerically for $\dot{m}_{dif}$, $T(R)$, and $\rho_v(R)$. These equations can, moreover, be combined analytically for the special case of a drop growing or evaporating in a saturated environment (i.e., for the case in which $e(\infty) = e_s[T(\infty)]$). In this special case, use is made of the Clausius–Clapeyron equation:[49]

$$\frac{1}{e_s}\frac{de_s}{dT} \cong \frac{L}{R_v T^2} \tag{3.12}$$

Combination of (3.10) and (3.12) yields

$$\frac{d\rho_{vs}}{\rho_{vs}} = \frac{L}{R_v}\frac{dT}{T^2} - \frac{dT}{T} \tag{3.13}$$

---

[48] See pp. 72–73 of Wallace and Hobbs (1977).
[49] See p. 95 of Wallace and Hobbs (1977).

Then (3.8), (3.9), (3.11), and (3.13) may be combined[50] under saturated environmental conditions to obtain

$$\dot{m}_{dif} = \frac{4\pi R\tilde{S}}{F_\kappa + F_D} \tag{3.14}$$

where $\tilde{S}$ depends on the humidity of the environment, F_κ on the heat conductivity, and F_D on the vapor diffusivity. More specifically, $\tilde{S}$ is the ambient supersaturation (expressed as a fraction):

$$\tilde{S} \equiv \frac{e(\infty)}{e_s(\infty)} - 1 \tag{3.15}$$

The other factors are given by

$$F_\kappa \equiv \frac{L^2}{\kappa_a R_v T^2(\infty)} \tag{3.16}$$

and

$$F_D \equiv \frac{R_v T(\infty)}{D_v e_s(\infty)} \tag{3.17}$$

From (3.14)–(3.17), it is evident that the diffusional growth rate of a drop depends on the temperature and humidity of the environment and on the radius of the drop.

The relation (3.11) used in deriving (3.14) assumes that saturation at the drop's surface may be approximated as if it obtained over a plane surface of water (i.e., that the growing drop were large enough for the curvature of the drop's surface to have negligible influence upon the equilibrium vapor pressure). The drop has also been assumed to be sufficiently dilute with respect to dissolved nuclei or other impurities that the drop may be regarded as being composed of pure water. For very small drops, however, curvature and solution effects must be included. If a drop is growing on a water-soluble nucleus, $\rho_v(R)$ becomes

$$\rho_v(R) = \rho_{vs}[T(R)]\left(1 + \frac{\hat{a}}{R} - \frac{\hat{b}}{R^3}\right) \tag{3.18}$$

where the term $\hat{a}/R$ represents the effect of drop curvature on the equilibrium vapor pressure above the drop. The factor $\hat{a}$ is given by

$$\hat{a} = \frac{2\sigma_{vl}}{\rho_L R_v T} \tag{3.19}$$

where σ_{vl} is the surface tension of liquid–vapor interface and ρ_L is the density of liquid water. The term $\hat{b}/R^3$ represents the effect of salt dissolved in the drop on

---

[50] See pp. 99–102 of Rogers and Yau (1989) for details of the derivation.

the equilibrium vapor pressure above the drop. The factor $\hat{b}$ is given by

$$\hat{b} = \frac{3i_{vH}m_s M_w}{4\pi\rho_L M_s} \tag{3.20}$$

where i_{vH} is the van't Hoff factor,[51] m_s and M_s are the mass and molecular weight of the dissolved salt, respectively, and M_w is the molecular weight of water.

Replacing (3.11) with (3.18) leads, following steps similar to those leading to (3.14), to the equation

$$\dot{m}_{dif} = \frac{4\pi R}{F_\kappa + F_D}\left(\tilde{S} - \frac{\hat{a}}{R} + \frac{\hat{b}}{R^3}\right) \tag{3.21}$$

which applies when the air is saturated. When the air is unsaturated, (3.8), (3.9), and (3.18) must be solved numerically to obtain $\dot{m}_{dif}$ for the evaporation rate of the drop.

When drops are falling relative to the surrounding air, the diffusion of vapor and heat is altered. To account for this process, the right-hand sides of (3.8) and (3.9) may be multiplied by a ventilation factor V_F. In this case, (3.14) and (3.21), the growth/evaporation rate under saturated conditions become

$$\dot{m}_{dif} = \frac{4\pi R V_F \tilde{S}}{F_\kappa + F_D} \tag{3.22}$$

and

$$\dot{m}_{dif} = \frac{4\pi R V_F}{F_\kappa + F_D}\left(\tilde{S} - \frac{\hat{a}}{R} + \frac{\hat{b}}{R^3}\right) \tag{3.23}$$

respectively.[52]

3.1.3 Fall Speeds of Drops

Growing cloud droplets are subject to downward gravitational force. This force can lead to their fallout as precipitation particles. The gravitational force on a drop is, however, largely offset by the frictional resistance of the air. As a particle is accelerated downward by gravity, its motion is increasingly retarded by the growing frictional force. Its final speed is called the *terminal fall speed V*. For drops of water in air, V is a function of the drop radius R. Generally V is negligible until the drops reach a radius of about 0.1 mm. This is usually considered to be the threshold size separating *cloud droplets*, which are suspended in the air indefinitely, from falling *precipitation* drops. The smallest precipitation drops (taken by con-

---

[51] This factor is equal to the number of ions into which each molecule of salt dissociates. See p. 162 of Wallace and Hobbs (1977).

[52] See pp. 440–463 of Pruppacher and Klett (1978).

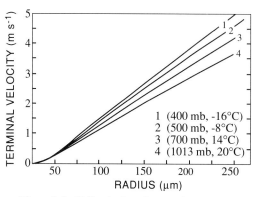

Figure 3.2 Fall velocity of water drops $<500 \ \mu m$ in radius for various atmospheric conditions. (From Beard and Pruppacher, 1969. Reprinted with permission from the American Meteorological Society.)

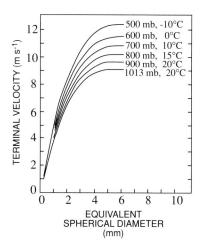

Figure 3.3 Fall velocity of water drops $>500 \ \mu m$ in radius. (From Beard, 1976. Reprinted with permission from the American Meteorological Society.)

vention[53] to be those 0.1–0.25 mm in radius) are called *drizzle*. Drops >0.25 mm in radius are called *rain*. Drizzle and raindrops have terminal fall speeds that increase with increasing drop radius. We will represent this function as $V(R)$. For drops $<500 \ \mu m$ in radius, V increases approximately linearly with increasing drop radius (Fig. 3.2). For larger drops, $V(R)$ increases at a lower rate (Fig. 3.3), becoming a constant at a radius of about 3 mm. This asymptotic behavior is associated with the fact that a drop becomes increasingly flattened, into the shape of a horizontally oriented disc, at larger sizes (see Fig. 4.2).

3.1.4   Coalescence

3.1.4.1   Continuous Collection

Cloud drop growth by coalescence with other drops can be envisioned in terms of a drop of mass m falling through a cloud of particles of mass m'. The water contained in the particles of mass m' is assumed to be distributed uniformly through the cloud with liquid water content $\rho q_{m'}$ (g m^{-3}), where $q_{m'}$ is the cloud water mixing ratio (mass of cloud water per mass of air). As it falls, the particle of mass m is assumed to increase in mass continually at a rate given by the *continuous collection equation*,

$$\dot{m}_{col} = A_m |V(m) - V(m')| \rho q_{m'} \Sigma_c(m, m') \tag{3.24}$$

[53] See the *Glossary of Meteorology* (Huschke, 1959).

where V represents the fall speed of the drops of masses m and m' (Figs. 3.2 and 3.3), ρ is the density of the air, Σ_c (m, m') is the *collection efficiency*, and A_m is the effective cross-sectional area swept out by a particle of mass m. The absolute value notation is used in (3.24) since it is only the relative motion of the particles that matters for collectional growth. For the case of a large drop collecting smaller drops, the absolute value symbol is redundant since the fall velocity of the larger drop always exceeds that of the smaller drops. However, (3.24) may also be used to calculate the increase of mass of a smaller drop coalescing with larger drops. If the absolute value were not used in that case, negative growth would be calculated. Moreover, as will be seen below, (3.24) is applied also to cold clouds where in some special cases (e.g., an ice particle collecting water drops) the fall velocity of the larger particle may not be the greater of the two.

For the purpose of calculating collectional growth, water drops are usually assumed to be spherical. In that case, the factor A_m in (3.24) is given by

$$A_m = \pi(R + R')^2 \tag{3.25}$$

where R and R' are the radii of drops of mass m and m', respectively. This area is based on the sum of the drop radii since any drop centered within a distance $R + R'$ of the center of the drop of radius R can be intercepted by that drop.

The collection efficiency Σ_c(m, m') is the efficiency with which a drop intercepts and unites with the drops it overtakes. It is the product of a collision efficiency and a coalescence efficiency. The collision efficiency (Fig. 3.4) is determined primarily by the relative airflow around the falling drop. Smaller particles may be carried out of the path of a larger particle (efficiency <1), or small particles not in the direct path of a large particle may collide with the large particle if they are pulled into its wake (efficiency >1). The coalescence efficiency expresses the

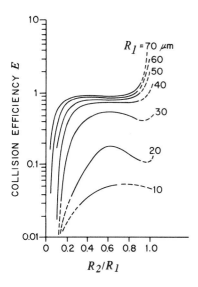

Figure 3.4 Collision efficiency for collector drops of radius R_1 with droplets of radius R_2. The dashed portions of the curve represent regions of doubtful accuracy. (From Wallace and Hobbs, 1977.)

fact that a collision between two drops does not guarantee coalescence; the drops may bounce off each other or remain united only temporarily. Under most conditions, coalescence efficiency is high, especially if the droplets are electrically charged or if an electric field is present. The electrical conditions are often met in clouds, and little else is known about the coalescence efficiency. Hence, the most common practice in theoretical or modeling studies is to assume a coalescence efficiency of unity. The collection efficiency then reduces to the collision efficiency.

A more general version of (3.24) may be written for the case in which a particle of mass m is falling relative to a population of particles of varying size. For that case, the generalized continuous collection equation is

$$\dot{m}_{col} = \int_0^\infty A_m |V(m) - V(m')| \, m' N(m') \Sigma_c(m, m') \, dm' \tag{3.26}$$

where $N(m') \, dm'$ is the number of particles per unit volume of air in the size range m' to $m' + dm'$.

3.1.4.2 *Stochastic Collection*

Cloud drop growth by coalescence is actually not a continuous process, as assumed in (3.24), but rather proceeds in a discrete, stepwise, probabilistic manner. In a time interval Δt, drops of a given initial size do not grow uniformly. Some may undergo more than the average number of collisions and thus grow faster than others. Consequently, a drop size distribution develops.

The probabilistic nature of collection may be accounted for by considering the size distribution $N(m,t)$, where $N(m,t) \, dm$ is the number of particles per unit volume of air in mass range m to $m + dm$ at time t. The change in $N(m,t)$ with time is computed as follows. The rate at which the space within which a particle of mass m' is located is swept out by a particle of mass m is given by the *collection kernel,* defined as

$$\hat{K}(m, m') \equiv A_m |V(m) - V(m')| \Sigma_c(m, m') \tag{3.27}$$

The probability that a particular drop of mass m will collect a drop of mass m' in time interval Δt is

$$\hat{P} \equiv N(m',t) \, dm' \, \hat{K} \Delta t \tag{3.28}$$

where it is assumed that Δt is small enough that the probability of more than one collection in this time is negligible. Making use of (3.27) and (3.28), we note that the mean number of drops of mass m that will collect drops of mass m' at time Δt is

$$\hat{P} N(m,t) \, dm = \hat{K}(m,m') N(m',t) N(m,t) \, dm \, dm' \, \Delta t \tag{3.29}$$

Rearranging this expression we obtain

$$\frac{\hat{P} N(m,t)}{\Delta t} = \hat{K}(m,m') N(m',t) N(m,t) \, dm' \tag{3.30}$$

which expresses the rate at which the number of drops of mass m is reduced as a result of coalescence with drops of mass m' per unit volume of air. It follows that the rate of decrease of the number concentration of drops of mass m as a result of their coalescence with drops of all other sizes is given by the integral

$$I_1(m) = \int_0^\infty \hat{K}(m,m')N(m',t)N(m,t)\,dm' \tag{3.31}$$

By reasoning similar to that given above we may express the rate of generation of drops of mass m by coalescence of smaller drops as

$$I_2(m) = \frac{1}{2}\int_0^m \hat{K}(m-m',m')N(m-m',t)N(m',t)\,dm' \tag{3.32}$$

where the factor of 1/2 is included to avoid counting each collision twice. The net rate of change in the number density of drops of mass m is obtained by subtracting (3.32) from (3.31) and may be written as

$$\left(\frac{\partial N(m,t)}{\partial t}\right)_{col} = I_2(m) - I_1(m) \tag{3.33}$$

This result is referred to as the *stochastic collection equation*.

Computations may be made with (3.33) starting with some arbitrary initial drop size distribution $N(m,0)$. The result obtained by integrating (3.33) over time yields the drop size distribution altered by the stochastic collection process. In addition to the initial distribution, one must also assume reasonable values of the collection efficiencies and fall velocities appearing in (3.31) and (3.32). For realistic conditions, it is generally found that a large portion of the liquid water accumulates in the tail of the distribution. An example of such a calculation is shown in Fig. 3.5.

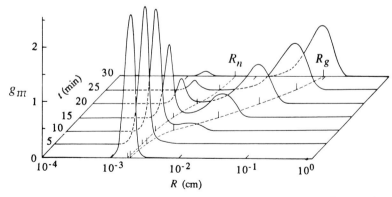

Figure 3.5 Example of the evolution of a drop size distribution as a result of stochastic collection. g_m is the mass distribution function; R is the drop radius. The two dashed lines show the radii (R_n and R_g) corresponding to the means of the number and mass distributions, respectively. (From Berry and Reinhardt, 1973. Reprinted with permission from the American Meteorological Society.)

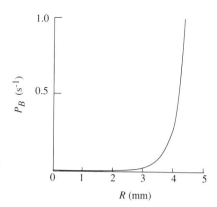

Figure 3.6 The probability $P_B(\mathrm{m})$ that a drop of radius R breaks up per unit time. Based on empirical formula of Srivastava (1971).

The drop size distribution at successive times is plotted as mass distribution $g_\mathrm{m} \equiv \mathrm{m}N(\mathrm{m})$, rather than number distribution $N(\mathrm{m})$, so that the area under each curve is proportional to the total liquid water content in the distribution. The mass distribution is plotted versus the radius of a drop of mass m on a logarithmic scale. This plotting convention emphasizes the result that a large portion of the liquid water becomes concentrated in the large drops as time progresses. The two peaks in the mass distribution after 30 min correspond to the amount of water contained in cloud droplets (radii $\sim 10^{-3}$ cm) and raindrops (radii $\sim 10^{-1}$ cm). The two dashed lines following the centers of the two peaks correspond to the means of the number and mass concentrations. The mean of the number distribution follows the cloud droplet peak. This result illustrates that the cloud droplets are far more numerous than the raindrops but that the latter nonetheless contain a large part of the liquid water after half an hour of stochastic collection. Stochastic collection can thus quickly convert cloud water to rainwater.

3.1.5 Breakup of Drops

When raindrops achieve a certain size, they become unstable and break up into smaller drops. Breakup has been studied in the laboratory, and empirical functions based on the experimental data are used to describe breakup quantitatively.[54] One empirical function is the probability $P_B(\mathrm{m})$ that a drop of mass m breaks up per unit time. It is nearly zero for drops less than about 3.5 mm and increases exponentially with size for radii greater than this value (Fig. 3.6). The function shown in the plot is

$$P_B(\mathrm{m}) = 2.94 \times 10^{-7} \exp(3.4R) \tag{3.34}$$

where R is the radius in millimeters of a drop of mass m and $P_B(\mathrm{m})$ is in s^{-1}. A second empirical function is $Q_B(\mathrm{m}',\mathrm{m})$, which is defined such that $Q_B(\mathrm{m}',\mathrm{m})\, d\mathrm{m}$ is the number of drops of mass m to $\mathrm{m} + d\mathrm{m}$ formed by the breakup of one drop of

---

[54] The formulation of breakup presented in this subsection was developed by Srivastava (1971).

mass m'. $Q_B(m',m)$ is approximately exponential. It is given by

$$Q_B(m', m) = 0.1R'^3 \exp(-15.6R) \tag{3.35}$$

where the radii are in cm. The empirical functions $P_B(m)$ and $Q_B(m',m)$ can be used to determine the net effect of breakup on the drop size distribution $N(m,t)$. The net rate of production of drops of mass m by breakup implied by these functions is

$$\left(\frac{\partial N(m,t)}{\partial t}\right)_{bre} = -N(m,t)P_B(m) + \int_m^\infty N(m',t)Q_B(m',m)P_B(m')\,dm' \tag{3.36}$$

3.2   Microphysics of Cold Clouds

3.2.1   Homogeneous Nucleation of Ice Particles

Ice particles in clouds may be nucleated from either the liquid or vapor phase. Homogeneous nucleation of ice from the liquid phase is analogous to nucleation of drops from the vapor phase. An embryonic ice particle can be considered a polyhedron of volume $\alpha_i 4\pi R^3/3$ and surface area $\beta_i 4\pi R^2$, where R is the radius of a sphere that can just be contained within the polyhedron, and α_i and β_i are both greater than unity but approach unity as the polyhedron tends toward a spherical shape. By reasoning analogous to that leading to (3.3), the expression for the critical radius R_{ci} of the inscribed sphere is

$$R_{ci} = \frac{2\beta_i \sigma_{il}}{\alpha_i\, n_i k_B T \ln\left(e_s/e_{si}\right)} \tag{3.37}$$

where σ_{il} is the free energy of an ice–liquid interface, n_i is the number of molecules per unit volume of ice, and e_{si} is the saturation vapor pressure with respect to a plane surface of ice. The saturation vapor pressures of liquid and ice in the denominator and the free energy in the numerator are all functions of temperature. The critical radius is thus a function of temperature.

   Theoretical and empirical results indicate that homogeneous nucleation of liquid water occurs at temperatures lower than about -35 to $-40°C$, depending somewhat on the size of the drops being subjected to the low temperature.[55] This threshold lies within the range of temperatures in natural clouds, which may have cloud-top temperatures below $-80°C$. It is therefore possible, in a natural cloud, to have unfrozen liquid (i.e., *supercooled*) drops in the temperature range of $0°C$ to about $-40°C$. However, wherever the temperature in the cloud is below about $-40°C$, any liquid drops that happen to be present freeze spontaneously by homogeneous nucleation. This conclusion is consistent with the fact that at tempera-

---

[55] Larger drops freeze homogeneously at slightly higher temperatures than smaller ones (Rogers and Yau, 1989, p. 151).

tures below −40°C atmospheric clouds are always composed entirely of ice, in which case they are said to be *glaciated*.

In principle, an ice particle may be nucleated directly from the vapor phase in the same manner as a drop. The critical size for homogeneous nucleation of an ice particle directly from the vapor phase is given by an expression similar in form to (3.3). In this case, the critical size depends strongly on both temperature and ambient humidity. Theoretical estimates of the rate at which molecules in the vapor phase aggregate to form ice particles of critical size indicate, however, that nucleation occurs only at temperatures below −65°C and at supersaturations ~1000%. Such high supersaturations do not occur in the atmosphere. Since liquid drops would nucleate from the vapor phase before these supersaturations were reached, and since the liquid drops thus formed would freeze homogeneously below −40°C, it is concluded that homogeneous nucleation of ice directly from the vapor phase never occurs in natural clouds.

3.2.2   Heterogeneous Nucleation of Ice Particles

From observations of the particles in clouds it is readily determined that ice crystals form at temperatures between 0°C and −40°C. Since homogeneous nucleation does not occur in this temperature range, the crystals must form by a heterogeneous process. As in the case of heterogeneous nucleation of liquid drops, the foreign surface on which an ice particle nucleates reduces the critical size that must be attained by chance aggregation of molecules. However, in the case of drops nucleating from the vapor phase, the atmosphere has no shortage of wettable nuclei. In contrast, ice crystals do not form readily on many of the particles found in air. The principal difficulty with the heterogeneous nucleation of the ice is that the molecules of the solid phase are arranged in a highly ordered *crystal lattice*. To allow the formation of an interfacial surface between the ice embryo and the foreign substance, the latter should have a lattice structure similar to that of ice. Figure 3.7 illustrates schematically an ice embryo which has formed on a crystalline substrate with a crystal lattice different from that of the ice. There are two ways in which the embryo could form. Either the ice could retain its normal lattice dimensions right to the interface, with dislocations in the sheets of

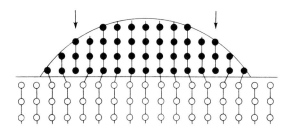

Figure 3.7   Schematic illustration of an ice embryo growing upon a crystalline substrate with a slight misfit. Dislocations of the interface are indicated by arrows. (From Fletcher, 1966. Reprinted with permission from Cambridge University Press.)

molecules, or the ice lattice could deform elastically to join the lattice of the substrate. The effect of dislocations is to increase the surface tension of the ice–substrate interface. The effect of elastic strain is to raise the free energy of the ice molecules. Both of these effects lower the ice-nucleating efficiency of a substance. These effects, moreover, are temperature dependent in the sense that the higher the temperature, the more the surface tension and elastic strain are increased.

There are several modes of action by which an ice nucleus can trigger the formation of an ice crystal. An ice nucleus contained within a supercooled drop may initiate heterogeneous freezing when the temperature of the drop is lowered to the value at which the nucleus can be activated. There are two possibilities in this case. If the cloud condensation nucleus on which the drop forms is the ice nucleus, the process is called *condensation nucleation*. If the nucleation is caused by any other nucleus suspended in supercooled water, the process is referred to as *immersion freezing*. Drops may also be frozen if an ice nucleus in the air comes into contact with the drop; this process is called *contact nucleation*. Finally, the ice may be formed on a nucleus directly from the vapor phase, in which case the process is called *deposition nucleation*.

From the above considerations, it is evident that the probability of ice particle nucleation should increase with decreasing temperature and that substances possessing a crystal lattice structure similar to that of ice should be the most likely to serve as a nucleating surface. In this respect, ice itself provides the best nucleating surface; whenever a supercooled drop at any temperature $\leq 0°C$ comes into contact with a surface of ice it immediately freezes. Other than ice, the natural substances possessing a crystal lattice structure most similar to that of ice appear to be certain clay minerals found in many soil types and bacteria in decayed plant leaves. They may nucleate ice at temperatures as high as $-4°C$ but appear to occur in low concentrations in the atmosphere. Most ice particle nucleation in clouds occurs at temperatures lower than this. In general, particles in the air on which ice crystals are able to form are called *ice nuclei*. Measurements can be made to indicate how many ice nuclei can be activated by lowering the temperature of a sample of air in an expansion chamber.[56] Generally, these measurements do not distinguish among condensation, immersion, contact, or deposition nucleation, nor do they indicate the composition of the nuclei. They also do not indicate the effect of varying the humidity. However, extensive measurements of this type indicate that the average number of ice nuclei N_I per liter of air generally increases exponentially with decreasing temperature according to the empirical formula

$$\ln N_I = a_I \left(253° \, K - T \right) \tag{3.38}$$

where a_I varies with location but has values in the range of 0.3–0.8. Note that according to this relationship, there is only about one ice nucleus per liter at $-20°C$. For a value of $a_I = 0.6$, the concentration increases by approximately a factor of ten for every $4°C$ of temperature decrease.

---

[56] See p. 184 of Wallace and Hobbs (1977) for a description of the technique.

3.2.3 Deposition and Sublimation

Growth of an ice particle by diffusion of ambient vapor toward the particle is called *deposition*. The loss of mass of an ice particle by diffusion of vapor from its surface into the environment is called *sublimation*. These processes are the ice-phase analogs of condensation and evaporation. However, since ice particles take on a variety of shapes, the spherical geometry assumed in evaluating the growth and evaporation of drops by vapor diffusion (Sec. 3.1.2) may not always be assumed in calculations of the change of mass of ice particles. Diffusion of vapor toward or away from nonspherical ice particles is accounted for by replacing R in (3.8), and thus in (3.14) and (3.22), by a shape factor $\tilde{C}$, which is analogous to electrical capacitance.[57] Thus, the analog to (3.8) is

$$\dot{m}_{\text{dif}} = 4\pi\tilde{C}D_v\left[\rho_v(\infty) - \rho_{v\text{sfc}}\right] \tag{3.39}$$

where $\rho_{v\text{sfc}}$ is the vapor density at the particle's surface. It follows that the analogs to (3.14) and (3.22) are

$$\dot{m}_{\text{dif}} = \frac{4\pi\,\tilde{C}\tilde{S}_i}{F_{\kappa i} + F_{Di}} \tag{3.40}$$

and

$$\dot{m}_{\text{dif}} = \frac{4\pi\tilde{C}V_F\tilde{S}_i}{F_{\kappa i} + F_{Di}} \tag{3.41}$$

respectively. $\tilde{S}_i$, $F_{\kappa i}$, and F_{Di} are the same as $\tilde{S}$, F_{κ}, and F_D in (3.15)–(3.17) except that L is replaced by the latent heat of sublimation L_s in (3.16), and $e_s(\infty)$ is replaced by the saturation vapor pressure over a plane surface of ice $e_{si}(\infty)$ in (3.15) and (3.17). The relations (3.40) and (3.41), like (3.14) and (3.22), apply only when the air is saturated (in this case with respect to ice). As in the case of drops, $\dot{m}_{\text{dif}}$ must be obtained numerically if the air is unsaturated.

The shape, or *habit*, adopted by an ice crystal growing by vapor diffusion is a sensitive function of the temperature T and supersaturation $\tilde{S}_i$ of the air.[58] These growth modes are known from observations in the laboratory and in clouds themselves. The basic crystal habits exhibit a hexagonal face. Let a crystal be imagined to have an axis normal to its hexagonal face. If this axis is long compared to the width of the hexagonal face, it is said to be *prismlike*. If this axis is short compared to the width of the hexagonal face, the crystal is said to be *platelike*. The basic crystal habits are illustrated schematically in Fig. 3.8. The habits change back and forth between prismlike and platelike as the ambient temperature changes (Table 3.1). The effect of increasing the ambient supersaturation is to increase the surface-to-volume ratio of the crystal. The additional surface area gives the increased

---

[57] The analogy between the vapor field around an ice crystal and the field of electrostatic potential around a conductor of the same size and shape was first applied by Houghton (1950). See Hobbs (1974) for further notes on the origin of the analogy.

[58] See Chapters 8 and 10 of Hobbs (1974).

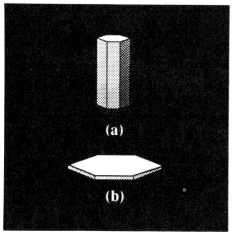

Figure 3.8 Schematic representation of the main shapes of ice crystals: (a) columnar, or prismlike; (b) plate; (c) dendrite. (Adapted from Rogers and Yau, 1989.)

ambient vapor more space on which to deposit. The multiarmed, fernlike crystals that appear at temperatures of −12 to −16°C have six main arms and several secondary branches (Fig. 3.8c). They may be thought of as hexagonal plates with sections deleted to increase the surface-to-volume ratio of the crystal. They occur in the temperature range where the difference between the saturation vapor pressure over water (an approximation to the actual vapor pressure in many cold clouds) and the saturation vapor pressure over ice (an approximation to the condition at the surface of the crystal) is greatest.

3.2.4 Aggregation and Riming

If ice particles collect other ice particles, the process is called *aggregation*. If ice particles collect liquid drops, which freeze on contact, the process is called *riming*. The continuous collection equation (3.24) may be used to describe the growth of ice particles by aggregation or riming.

Table 3.1

Variations in the Basic Habits of Ice Crystals with Temperature

Temperature (°C)	Basic habit	Types of crystal at slight water supersaturation
0 to −4	Platelike	Thin hexagonal plates
−4 to −10	Prismlike	Needles (−4 to −6°C)
		Hollow columns (−5 to −10°C)
−10 to −22	Platelike	Sector plates (−10 to −12°C)
		Dendrites (−12 to −16°C)
		Sector plates (−16 to −22°C)
−22 to −50	Prismlike	Hollow columns

Source: Wallace and Hobbs (1977).

Aggregation depends strongly on temperature. The probability of adhesion of colliding ice particles becomes much greater when the temperature increases to above $-5°C$, at which the surfaces of ice crystals become sticky. Another factor affecting aggregation is crystal type. Intricate crystals, such as dendrites, become aggregated when their branches become entwined. These facts are known from laboratory experiments and observations of natural snow. The sizes of collected snow aggregates are shown as a function of the temperature at which they were observed in Fig. 3.9. The sizes increase sharply at temperatures above $-5°C$, while aggregation does not appear to exist below $-20°C$. A secondary maximum occurs between -10 and $-16°C$, where the arms of the dendritic crystals growing at these temperatures apparently become entangled. In correspondence to these observations, the collection efficiency for aggregation is often assumed to be an exponentially increasing function of temperature in calculations using (3.24) or (3.26).

The collection efficiency for riming is not well known theoretically or empirically, but it is generally thought to be quite high and often assumed to be unity in calculations using (3.24) or (3.26). If the ice particle is viewed as the collector and the liquid drops as the collected particles in (3.24), the degree of riming that is achievable is determined primarily by the mixing ratio of the liquid water ($q_{in'}$). Lightly to moderately rimed crystals retain vestiges of the original crystal habit of the collector (Fig. 3.10a–d). Under heavy riming the identity of the collector

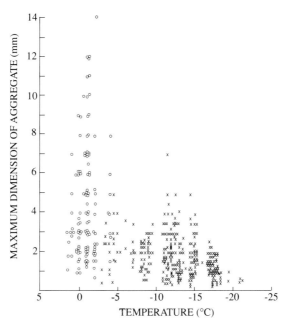

Figure 3.9 Maximum dimensions of natural aggregates of ice crystals as a function of the temperature of the air where they were collected. × indicates crystals collected from an aircraft. Circles represent crystals collected on the ground. (From Hobbs, 1973b. Reprinted with permission from Oxford University Press.)

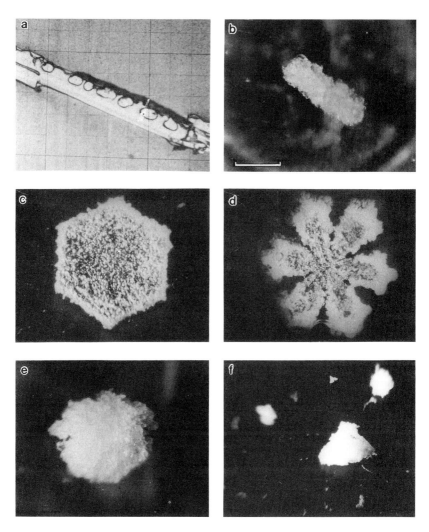

Figure 3.10 (a) A lightly rimed needle; (b) densely rimed column; (c) densely rimed plate; (d) densely rimed stellar; (e) lump graupel; (f) cone graupel. (From Wallace and Hobbs, 1977.)

becomes lost, and the particle is referred to as *graupel,* which may be in the form of lumps or cones (Fig. 3.10e and f).

3.2.5 Hail

Extreme riming produces *hailstones*. These particles are commonly 1 cm in diameter but have been observed to be as large as 10–15 cm. They are produced as graupel or frozen raindrops collect supercooled cloud droplets. So much liquid water is accreted in this fashion that the latent heat of fusion released when the collected water freezes significantly affects the temperature of the hailstone. The hailstone may be several degrees warmer than its environment. This temperature difference has to be taken into account in calculating the growth of hail particles, which is determined by considering the heat balance of the hailstone.

The rate at which heat is gained as a result of the riming of a hailstone of mass m is

$$\dot{Q}_f = \dot{m}_{col}\left\{L_f - c_w\left[T(R) - T_w\right]\right\} \tag{3.42}$$

The factor $\dot{m}_{col}$ is the rate of increase of the mass of the hailstone as a result of collecting liquid water. It is given by (3.26). The hailstone is assumed to be spherical with radius R. L_f is the latent heat of fusion released as the droplets freeze on contact with the hailstone. The second term in the curly brackets is the heat per unit mass gained as the collected water drops of temperature T_w come into temperature equilibrium with the hailstone. The factor c_w is the specific heat of water. If the air surrounding the particle is subsaturated, the temperature T_w is approximated by the wet-bulb temperature of the air, which is the equilibrium temperature above a surface of water undergoing evaporation at a given air pressure.[59] This temperature may be several degrees less than the actual air temperature when the humidity of the air is very low. If the air surrounding the particle is saturated $T_w = T(\infty)$.

The rate at which the hailstone gains heat by deposition (or loses heat by sublimation) is obtained from a modified form of (3.8)

$$\dot{Q}_s = 4\pi RD_v\left[\rho_v(\infty) - \rho_v(R)\right]V_{Fs}L_s \tag{3.43}$$

where V_{Fs} is a ventilation factor for sublimation, and L_s is the latent heat of sublimation.

The rate at which heat is lost to the air by conduction is obtained from a modified version of (3.9), which may be written as

$$\dot{Q}_c = 4\pi R\kappa_a\left[T(R) - T(\infty)\right]V_{Fc} \tag{3.44}$$

where V_{Fc} is a ventilation factor for conduction.

In equilibrium we have

$$\dot{Q}_f + \dot{Q}_s = \dot{Q}_c \tag{3.45}$$

which upon substitution from (3.42)–(3.44) may be solved for the hailstone equilibrium temperature as a function of size. As long as this temperature remains below 0°C, the surface of the hailstone remains dry, and its development is called *dry growth*. The diffusion of heat away from the hailstone, however, is generally too slow to keep up with the release of heat associated with the riming (depositional growth is much less than the riming). Therefore, if a hailstone remains in a supercooled cloud long enough, its equilibrium temperature can rise to 0°C. At this temperature, the collected supercooled droplets no longer freeze spontaneously upon contact with the hailstone. Some of the collected water may then be

[59] See pp. 75–76 of Wallace and Hobbs (1977).

lost to the warm hailstone by shedding. However, a considerable portion of the collected water becomes incorporated into a water–ice mesh forming what is called *spongy hail*. This process is called *wet growth*. During its lifetime, a hailstone may grow alternately by the dry and wet processes as it passes through air of varying temperature. When hailstones are sliced open, they often exhibit a layered structure, which is evidence of these alternating growth modes.

3.2.6 Ice Enhancement

When the concentrations of ice particles are measured in natural clouds, it is often found that there are far more ice particles present than can be accounted for by the typical concentrations of ice nuclei activated by lowering the temperature of air in expansion chambers.[60] Figure 3.11 compares some measurements of ice particle concentrations in cumuliform clouds with the concentration of ice nuclei expected from expansion-chamber measurements to be active at the cloud-top temperature. The latter concentration is calculated from (3.38). The cloud-top temperature is the lowest temperature anywhere in the cloud and hence provides an estimate of the maximum possible ice nucleus concentration in the cloud according to (3.38). The actual particle concentration is seen typically to exceed the maximum possible nucleus concentration by one or more orders of magnitude.

These high concentrations are found in many cold clouds. They do not, however, occur in uniform spatial and temporal patterns within a cloud. A common characteristic is that they occur in older rather than newly formed portions of clouds, and they are found in association with supercooled cloud droplets. They are most likely to be found when the size distribution of the droplets is broad, with largest drops exceeding about 20 μm in diameter. The high concentrations develop initially near the tops of clouds, and the high concentrations may develop suddenly (e.g., the concentration may rise from 1 to 1000 ℓ^{-1} in less than 10 min).

The microphysical process, or processes, by which the concentrations of ice particles become so highly enhanced relative to the number of nuclei which would appear to be active according to (3.38) are not certain. Some hypotheses that have been suggested are[61]

(i) *Fragmentation of ice crystals.* Delicate crystals may break into pieces as a result of collisions and/or thermal shock.

(ii) *Ice splinter production in riming.* It has been found in laboratory experiments that when supercooled droplets >23 μm in diameter collide with an ice surface at a speed of $\geq$1.4 m s^{-1} at temperatures of -3 to $-8°$C, small ice splinters are produced.[62]

---

[60] For a more complete account of the observations and hypotheses regarding the occurrence of high ice particle concentrations in clouds, see Hobbs and Rangno (1985, 1990) and Rangno and Hobbs (1988, 1991).

[61] For a more full discussion and many references to the literature, see the Hobbs and Rangno papers mentioned in the previous footnote.

[62] The laboratory experiments were performed by Hallett and Mossop (1974). This ice-enhancement process is often called the Hallett–Mossop mechanism.

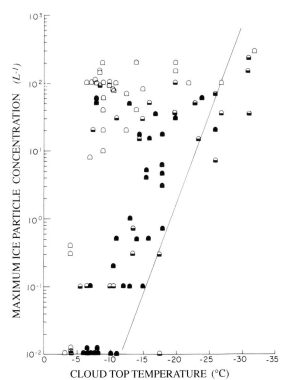

Figure 3.11 Maximum ice particle concentrations observed in mature and aging maritime (open humps), continental (closed humps), and transitional (half-open humps) cumuliform clouds. The line represents the concentrations expected at the cloud-top temperature. (From Hobbs and Rangno, 1985. Reprinted with permission from the American Meteorological Society.)

(iii) *Contact nucleation.* It is thought that when certain aerosol particles come into contact with supercooled droplets they can cause nucleation at higher temperatures than they would through other forms of nucleation.

(iv) *Condensation or deposition nucleation.* There is evidence that the ice-nucleating activity of atmospheric aerosol particles by either condensation or deposition nucleation is greatly increased when the ambient supersaturation rises above 1% with respect to water. Ice nucleus counters whose data lead to the expression in (3.38) are usually operated near water saturation. A pocket of high supersaturation in a cloud might be favorable for the sudden appearance of a large number of ice particles.

The latter two mechanisms, (iii) and (iv), do not require the pre-existence of ice particles and may thus help to account for the sudden appearance of high concentrations in cloudy air at relatively high temperatures.

3.2.7 Fall Speeds of Ice Particles

The fall speeds of ice particles encompass a wide range. Observations show that these speeds depend on particle type, size, and degree of riming and that the more heavily rimed a particle, the more its fall speed depends on its size. Individual snow crystals (lower curves of Fig. 3.12) and unrimed to moderately rimed aggre-

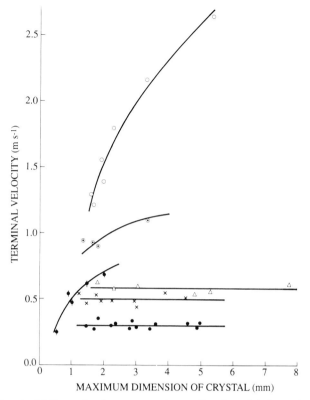

Figure 3.12   Terminal fall speeds of snow crystals as a function of their maximum dimensions. Open circle (uppermost curve) indicates graupel fall speeds. Other curves are for rimed crystals (dot in circle), needles (filled circle with slash), spatial dendrites (triangle), powder snow (×), and dendrites (filled circle). (From Nakaya and Terada, 1935.)

gates of crystals (Fig. 3.13) drift downward at speeds of 0.3–1.5 m s⁻¹, with the aggregates showing a tendency to increase slightly in fall speed as they approach 12 mm in dimension. Graupel fall speeds increase sharply from 1 to 3 m s⁻¹ over a narrow size range of 1–3 mm (upper curve of Fig. 3.12 and Fig. 3.14). Empirical formulas for fall speeds of snow and graupel for the data set represented in Fig. 3.12 and Fig. 3.14 are listed in Table 3.2. These formulas apply near the earth's surface. They do not take into account the fact that the fall speed also depends on the density of the air through which the particles are falling. The dependence on air density has been determined experimentally and theoretically.[63]

Hailstone fall velocities are an order of magnitude larger than those for snow and graupel. At a pressure of 800 mb and a temperature of 0°C, they obey the empirical formula[64]

$$V\left(\text{m s}^{-1}\right) \approx 9D_h^{0.8} \tag{3.46}$$

[63] See, for example, Böhm (1989).
[64] From Pruppacher and Klett (1978, p. 345), based on data of Auer (1972).

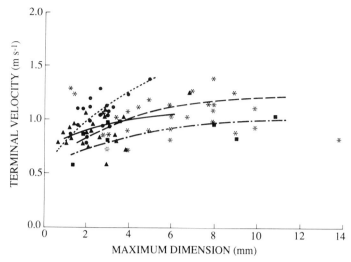

Figure 3.13 Terminal velocity and maximum dimension measurements for unrimed to moderately rimed aggregates. Combinations of: sideplanes (a type of branched crystal), bullets and columns (circles, dotted curve); sideplanes (triangles, solid curve); radiating assemblages of dendrites (asterisks, dashed curve); dendrites (squares, dash–dot curve). (From Hobbs, 1974, based on data from Locatelli and Hobbs, 1974. Reprinted with permission from Oxford University Press.)

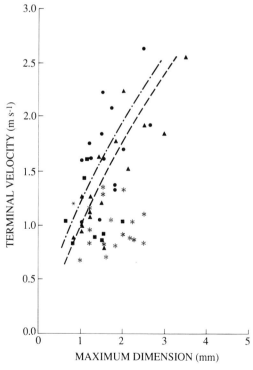

Figure 3.14 Terminal velocity and maximum dimension measurements for graupel and graupel-like snow. Cone-shaped graupel (circles, dash–dot curve); hexagonal graupel (triangles, dashed curve); graupel-like snow of hexagonal type (asterisks). (From Hobbs, 1974, based on data from Locatelli and Hobbs, 1974. Reprinted with permission from Oxford University Press.)

Table 3.2

Empirical Relationships between the Terminal Velocities v (in m s^{-1}) of Solid Precipitation Particles and Their Maximum Dimension D_m (in mm)

	Type of particle[a]	v–D_m relationship
Graupel and graupel-like snow	Cone-shaped graupel	$v = 1 \cdot 2 D_m^{0.65}$
	Hexagonal graupel	$v = 1 \cdot 1 D_m^{0.57}$
	Graupel-like snow of lump type	$v = 1 \cdot 1 D_m^{0.28}$
	Graupel-like snow of hexagonal type	$v = 0 \cdot 86 D_m^{0.25}$
	Combination of sideplanes, plates, bullets, and columns	$v = 0 \cdot 69 D_m^{0.41}$
Unrimed aggregates	Sideplanes	$v = 0 \cdot 82 D_m^{0.12}$
	Radiating assemblages of dendrites or dendrites	$v = 0 \cdot 8 D_m^{0.16}$
Densely rimed aggregates	Radiating assemblages of dendrites or dendrites	$v = 0 \cdot 79 D_m^{0.27}$
Densely rimed columns		$v = 1 \cdot 1 D_m^{0.56}$

Source: Locatelli and Hobbs (1974).
[a] Based on Magono and Lee's (1966) classification.

where D_h is the diameter of the hailstone in cm. This formula was obtained for hailstones in the size range 0.1–8 cm. Over this range, the fall speeds indicated by (3.46) are roughly 10–50 m s^{-1}. These large values imply that updrafts of comparable magnitude must exist in the cloud to support the hailstones long enough for them to grow. Hence, hail is found only in very intense thunderstorms, of the types considered in Chapters 8 and 9.

3.2.8   Melting

Ice particles can change into liquid water when they come into contact with air or water that is above 0°C. A quantitative expression for the rate of melting of an ice particle of mass m can be obtained by assuming heat balance for the particle during the melting. According to (3.45), this balance may then be written as

$$-L_f \dot{m}_{\text{mel}} = 4\pi R \kappa_a [T(\infty) - 273 \text{ K}] V_{Fc} + \dot{m}_{\text{col}} c_w (T_w - 273 \text{ K}) + \dot{Q}_s \quad (3.47)$$

where $\dot{m}_{\text{mel}}$ is the rate of change of the mass of the ice particle as a result of melting. The first term on the right is the diffusion of heat toward the particle from the air in the environment. The second term is the rate at which heat is transferred to the ice particle from water drops of temperature T_w that are collected by the melting particle. The third term, $\dot{Q}_s$, is the gain or loss of heat by vapor diffusion [given by (3.43) in the case of a spherical hailstone]. If both the air and drop temperatures exceed 273 K, then both the first and second term in (3.47) contribute to the melting.

3.3   Types of Microphysical Processes and Categories of Water Substance in Clouds

From the foregoing review of cloud microphysics (Secs. 3.1 and 3.2), it is evident that water substance can take on a wide variety of forms in a cloud and that these forms develop under the influence of *seven basic types of microphysical processes*:

1. Nucleation of particles
2. Vapor diffusion
3. Collection
4. Breakup of drops
5. Fallout
6. Ice enhancement
7. Melting

These individual processes may sometimes be isolated for study in numerical models or in the laboratory. However, in a natural cloud several or all of these processes occur simultaneously, as the entire ensemble of particles comprising the cloud forms, grows, and dies out. Thus, the various forms of water and ice particles coexist and interact within the overall cloud ensemble. It is the behavior of the overall ensemble that is of primary interest in cloud dynamics, and it is generally unnecessary to keep track of every particle in the cloud in order to describe the cloud's gross behavior. At the same time, it is also impossible to ignore the microphysical processes and accurately represent the cloud's overall behavior. To retain the essentials of the microphysical behavior, it is convenient to group the various forms of water substance in a cloud into several broad *categories of water substance*:

- *Water vapor* is in the gaseous phase.
- *Cloud liquid water* is in the form of small suspended liquid-phase droplets (i.e., drops that are too small to have any appreciable terminal fall speed constitute cloud liquid water and therefore are generally carried along by the air in which they are suspended).
- *Precipitation liquid water* is in the form of liquid-phase drops that are large enough to have an appreciable fall speed toward the earth. This water may be subdivided into drizzle (drops 0.1–0.25 mm in radius) and rain (drops >0.25 mm in radius), as defined in Sec. 3.1.3.
- *Cloud ice* is composed of particles that have little or no appreciable fall speed. These particles may be in the form of pristine crystals nucleated directly from the vapor or water phase, or they may be tiny particles of ice produced in some form of ice enhancement process.
- *Precipitation ice* refers to ice particles that have become large and heavy enough to have a terminal fall speed ~ 0.3 m s^{-1} or more. These particles may be pristine crystals, larger fragments of particles, rimed particles, aggregates, graupel, or hail. To simplify the description of the gross behavior

of a cloud these particle types are sometimes grouped into categories according to their density or fall speed. Such groupings are arbitrary; however, a commonly used scheme (employed in discussions below) is to divide these particles into *snow*, which has lower density and falls at speeds of ~0.3–1.5 m s^{-1} (see Figs. 3.12–3.13); *graupel*, which falls at speeds of ~1–3 m s^{-1} (see Figs. 3.12 and 3.14); and *hail*, which falls at speeds of ~10–50 m s^{-1} [see Eq. (3.46)].

According to these categories of particles, the water substance in a sample of air may be represented by eight mixing ratios:

$$q_v \equiv \text{mass of } \textit{water vapor}/\text{mass of air}$$
$$q_c \equiv \text{mass of } \textit{cloud liquid water}/\text{mass of air}$$
$$q_d \equiv \text{mass of } \textit{drizzle}/\text{mass of air}$$
$$q_r \equiv \text{mass of } \textit{rainwater}/\text{mass of air}$$
$$q_I \equiv \text{mass of } \textit{cloud ice}/\text{mass of air} \qquad (3.48)$$
$$q_s \equiv \text{mass of } \textit{snow}/\text{mass of air}$$
$$q_g \equiv \text{mass of } \textit{graupel}/\text{mass of air}$$
$$q_h \equiv \text{mass of } \textit{hail}/\text{mass of air}$$

The evolution of a cloud can be characterized in terms of fields of these mixing ratios of water substance. The various categories are interactive. For example, drops grow at the expense of vapor during nucleation and condensation, precipitation ice grows at the expense of cloud and precipitation liquid water during riming, rainwater is produced at the expense of precipitation ice during melting, and so on. The many conversions of water substance from one form to another in a given sample of air are illustrated in Fig. 3.15 for a cloud whose water substance is divided into water vapor, cloud liquid water, rainwater, cloud ice, snow, and graupel. Also indicated are the possible gains or losses by precipitation fallout. The six categories in Fig. 3.15 are a subset of the eight listed above, drizzle and hail being omitted. This six-category scheme is frequently used in cloud dynamics. Sometimes hail rather than graupel is used as the sixth category. To be perfectly general, all the categories should be included. It is easy to see from Fig. 3.15 that the number of interactions to be considered would increase greatly by expanding to a seven- or eight-category scheme. It is for this reason that the number of categories is often limited to six or less.

3.4 Water-Continuity Equations

In cloud dynamics one is concerned with the overall development of a cloud, in which any combination of the categories of water substance defined above and any combination of the microphysical processes linking the various categories (as shown in Fig. 3.15) may be present simultaneously in the context of a particular set of air motions. It is therefore necessary to have a way to keep track of all of the

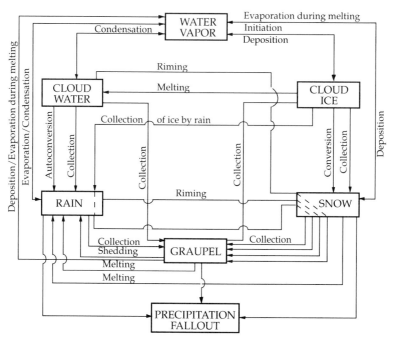

Figure 3.15 Conversions of water substance from one form to another in a bulk water-continuity model in which there are six categories of water substance. Dashed lines indicate various interactions leading to production of graupel: collection of cloud ice by rain freezes the ice by contact nucleation and produces either snow or graupel; or riming of snow by collection of either cloud water or raindrops can also lead to production of graupel. In the model, the mass of ice produced by these two processes passes temporarily through the snow category. (Adapted from Rutledge and Hobbs, 1984. Reproduced with permission from the American Meteorological Society.)

processes in some systematic way. For this purpose, the water-continuity equations (2.21) are used. They allow one to account numerically for the amount of water contained in the form of vapor and in the form of particles of different types and sizes throughout a cloud, as it evolves. This system of equations is referred to as a *water-continuity model*. In (2.21), each category of water is assigned a mixing ratio q_i, where the subscript i refers to a particular category of water and the mixing ratio is defined as the mass of water of category i per unit mass of air. The total water substance in a parcel of air is then given by the sum of the water contained in each of the categories:

$$q_T = \sum_{i=1}^{n} q_i \tag{3.49}$$

where n is the total number of categories into which the total water q_T in the air parcel has been divided.

There is no limit to the number of categories into which the total water in a parcel of air can be divided. To begin with, we can divide the water into the eight categories listed in (3.48). Each of these categories, however, can be further subdivided. The cloud liquid water, drizzle, and rainwater categories can be subdivided according to drop size. Each drop size category can be further subdivided according to such factors as the type of nucleus on which the drops formed, chemical composition of the drops, etc. The snow category can be subdivided by type of particle (columns, plates, aggregates, ice splinters, etc.), and these particle types can be further subdivided according to particle size, density, or other factors. The graupel may be subdivided by size and shape (some graupel particles are cone shaped while some are lumpy), and hail may be subdivided by particle size, whether it is spongy or hard, and other factors. Obviously, calculations can become highly complex if all of these subdivisions are employed. Generally the strategy in cloud dynamics is to identify and use only the categories and subcategories of water substance that are essential to keep account of in the particular problem being considered. Hence, practically every water-continuity model employed is tailored to the problem at hand.

Once the n categories to be used in a particular water-continuity model have been established, the source terms on the right-hand side of (2.21) must be formulated to allow for all the possible interactions among the different categories of water (e.g., the interactions shown for the six-category model in Fig. 3.15). The source terms are formulated in terms of the seven basic cloud microphysical mechanisms mentioned earlier (nucleation, vapor diffusion, collection, particle breakup, fallout, ice enhancement, and melting). Two general strategies have been employed to formulate the source terms. In *bulk* models, the liquid and ice water mixing ratios are grouped into categories according to particle type only. In *explicit* models, the hydrometeors are subdivided according to size within each particle-type grouping. The following sections summarize the salient features of these two types of models.

3.5 Explicit Water-Continuity Models

3.5.1 General

Hydrometeors may be grouped into categories according to particle type as in (3.48). In an explicit water-continuity model, one or more of these categories are subdivided according to particle size. Since the mass of water contained in particles of different sizes is calculated, the size distributions of particles are able to evolve naturally in each air parcel associated with the cloud. The only disadvantage of the explicit model is computational. A large number of size categories ($\sim 10-100$) and associated interactions have to be included to represent the size distribution of the particles of a given category of water substance accurately. From a physical standpoint, the explicit method is the more direct approach. The microphysical principles reviewed in previous sections can be applied directly to the calculation of the size distributions within a given category of water substance.

3.5.2   Explicit Modeling of Warm Clouds

3.5.2.1   *Drops Subdivided by Size*

We consider first clouds without ice. Therefore, none of the ice categories in (3.48) are relevant; $q_i = q_s = q_g = q_h = 0$. The cloud and precipitation liquid water categories are combined into a single total liquid water mixing ratio $q_L \equiv q_c + q_d + q_r$. The total liquid water content is thus viewed as a continuous spectrum of drops, ranging from small, freshly nucleated cloud droplets to large raindrops. The size distribution is represented by $N(m)$, where $N(m)\,dm$ is the number of particles per unit volume of air in mass range m to $m + dm$, in which case

$$q_L = \frac{1}{\rho} \int_0^\infty mN(m)\,dm \tag{3.50}$$

For computational purposes the size distribution is approximated by k discrete mass categories so that q_L is approximated by

$$q_L \approx \frac{1}{\rho} \sum_{i=1}^{k} m_i N_i (\Delta m)_i \tag{3.51}$$

The water continuity in the cloud may then be expressed by $k + 1$ equations[65]

$$\frac{DN_i}{Dt} = -N_i \nabla \cdot \mathbf{v} + \mathfrak{N}_i + \mathfrak{D}_i + \mathfrak{C}_i + \mathfrak{B}_i + \mathfrak{F}_i, \quad i = 1, \ldots, k \tag{3.52}$$

and

$$\frac{Dq_v}{Dt} = -\frac{1}{\rho} \sum_{i=1}^{k} m_i \left(\mathfrak{N}_i + \mathfrak{D}_i \right)(\Delta m)_i \tag{3.53}$$

The first term on the right-hand side of (3.52) represents changes in N_i associated with air motions. It expresses the decrease (increase) in concentration associated with the expansion (contraction) of the volume of air within which the population of drops is located. The remainder of the terms in (3.52) represent the microphysical processes affecting the drop size distribution. They express the changes in the concentration of drops in size range i resulting from nucleation from the vapor phase $\mathfrak{N}_i$, vapor diffusion (condensation or evaporation) $\mathfrak{D}_i$, collection $\mathfrak{C}_i$, drop breakup $\mathfrak{B}_i$, and sedimentation (or fallout) $\mathfrak{F}_i$. These microphysical processes correspond to the first five of the seven basic microphysical mechanisms listed in Sec. 3.3. The remaining two on that list, ice enhancement and melting, do not apply in a warm cloud.

The nucleation term $\mathfrak{N}_i$ in (3.52) is calculated by first making assumptions about the characteristics of the condensation nuclei present in the air. Then the appro-

---

[65] This type of water-continuity model was used in early studies by Takeda (1971), Ogura and Takahashi (1973), and Soong (1974).

priate concentration of drops is nucleated according to the supersaturation of the air and the relations governing nucleation (Sec. 3.1.1). The diffusion term $\mathfrak{D}_i$ represents the rate of change in the number concentration of drops resulting from differing diffusional growth or evaporation rates of drops in adjacent size categories. This rate of change is expressed by

$$\mathfrak{D}_i = -\frac{\partial}{\partial m}\left[\dot{m}_{\text{dif}}N\right]\Big|_{m=m_i} \tag{3.54}$$

where $\dot{m}_{\text{dif}}$ is the diffusional growth (or evaporation) rate of an individual drop of mass m. If the air is saturated with respect to liquid water, $\dot{m}_{\text{dif}}$ is given by (3.22). If not, it is obtained by solving (3.8), (3.9), and (3.11) numerically. The term $\mathfrak{D}_i$ may be thought of as the convergence of number concentration of drops as a result of the growth rate $\dot{m}_{\text{dif}}$ varying with drop size. The collection term $\mathfrak{C}_i$ in (3.52) is given by the stochastic collection equation (3.33), which may be written for the i-th drop size category as

$$\mathfrak{C}_i = \left(\frac{\partial N(m_i,t)}{\partial t}\right)_{\text{col}} = I_2(m_i) - I_1(m_i) \tag{3.55}$$

The rate of production of drops of mass m_i by breakup $\mathfrak{B}_i$ can be expressed according to (3.36) as

$$\mathfrak{B}_i = \left(\frac{\partial N(m_i,t)}{\partial t}\right)_{\text{bre}}$$
$$= -N(m_i,t)P_B(m_i) + \int_{m_i}^{\infty} N(m',t)Q_B(m',m_i)P_B(m')dm' \tag{3.56}$$

Finally, the sedimentation term $\mathfrak{F}_i$ in (3.52) represents the local accumulation of drops of mass m_i at a point as a result of their fall speeds. It is given by

$$\mathfrak{F}_i = \frac{\partial}{\partial z}(N_iV_i) \tag{3.57}$$

where V_i is the terminal velocity (defined to be positive downward) of a drop of mass m_i.

3.5.2.2 *Variable Drop Composition*

During the early stages of drop growth, the condensation nucleus around which the drop formed may become dissolved and affect the rate of growth of the drop by vapor diffusion. To account for this effect, we can further subdivide the drop size distribution according to nuclear mass n, with $N(m,n)\,dm\,dn$ representing the number of particles of mass m to $m + dm$ and nuclear mass n to $n + dn$ per unit volume of air.[66] The distribution could be further subdivided if nuclei of different

---

[66] Subdivision of the drop size spectrum according to the size of condensation nuclei has been used by Silverman (1970), Tag *et al.* (1970), Árnason and Greenfield (1972), Clark (1973), and Silverman and Glass (1973). Further details of the technique can be found in these articles.

types of dissolved substances were considered. For the purpose of illustration, we will consider the nuclei all to consist of the same substance but to be of different sizes. Each of the k discrete categories of drop size can then be subdivided into l discrete categories of nucleus size. The subdivided water-continuity equations are then

$$\frac{DN_{ij}}{Dt} = -N_{ij}\nabla \cdot \mathbf{v} + \mathfrak{N}_{ij} + \mathfrak{D}_{ij} + \mathfrak{C}_{ij} + \mathfrak{B}_{ij} + \mathfrak{F}_{ij} \tag{3.58}$$

where $i = 1, \ldots, k, j = 1, \ldots, l$, and

$$\frac{Dq_v}{Dt} = -\frac{1}{\rho}\sum_{j=1}^{l}(\Delta \mathfrak{n})_j \sum_{i=1}^{k} \mathfrak{m}_i\left(\mathfrak{N}_{ij} + \mathfrak{D}_{ij}\right)(\Delta \mathfrak{m})_i \tag{3.59}$$

The change in the concentration N_{ij} of drops in size category i and nucleus category j as a result of nucleation is represented by $\mathfrak{N}_{ij}$ and calculated by postulating a spectrum of nuclei to be present in the air initially and activating them in accordance with the supersaturation of the air and the relations governing nucleation (Sec. 3.1.1). The change in the concentration N_{ij} of drops in size category i and nucleus category j as a result of vapor diffusion $\mathfrak{D}_{ij}$ is given by

$$\mathfrak{D}_{ij} = -\frac{\partial}{\partial \mathfrak{m}}\left[\dot{\mathfrak{m}}_{\text{dif}}N_{ij}\right]\bigg|_{\mathfrak{m}=\mathfrak{m}_i, \mathfrak{n}=\mathfrak{n}_j} \tag{3.60}$$

which is similar in form to (3.54). However, the drop growth rate $\dot{\mathfrak{m}}_{\text{dif}}$ in this case is given, under saturated conditions, by (3.23), which takes into account solution effects in depositional growth. If the air is unsaturated, it is obtained by numerical solution of (3.8), (3.9), and (3.18). The collection term $\mathfrak{C}_{ij}$ is conceptually more difficult when the nuclear composition of drops is accounted for. Appropriate assumptions must be made concerning the coalescence or noncoalescence of the nuclei of coalescing drops. If the nuclei are assumed to coalesce whenever two parent drops coalesce, the nuclear mass $\mathfrak{n}$ of the newly formed drop is determined by adding the $\mathfrak{n}$'s of the coalescing drops.[67] In this case, the stochastic collection equation is

$$\mathfrak{C}_{ij} = \frac{1}{2}\int_0^{\mathfrak{m}_i}\int_0^{\mathfrak{n}_i}\hat{K}\left(\mathfrak{m}_i - \mathfrak{m}', \mathfrak{m}'\right)N\left(\mathfrak{m}_i - \mathfrak{m}', \mathfrak{n}_j - \mathfrak{n}', t\right)N(\mathfrak{m}', \mathfrak{n}', t)\, d\mathfrak{n}'\, d\mathfrak{m}'$$

$$-\int_0^{\infty}\int_0^{\infty}\hat{K}\left(\mathfrak{m}_i, \mathfrak{m}'\right)N\left(\mathfrak{m}_i, \mathfrak{n}_j, t\right)N(\mathfrak{m}', \mathfrak{n}', t)\, d\mathfrak{n}'\, d\mathfrak{m}' \tag{3.61}$$

where the first term on the right-hand side is the rate of formation of drops of mass $\mathfrak{m}_i$ and nuclear mass $\mathfrak{n}_j$ by coalescences of drops with masses smaller than $\mathfrak{m}_i$ and $\mathfrak{n}_j$. The second term is the rate of removal by combinations of drops mass $\mathfrak{m}_i$ and

---

[67] Suggested by Silverman (1970).

nuclear mass n_j with other drops. Drop breakup $\mathfrak{B}_{ij}$ is also conceptually difficult. One must assume something about what happens to the nucleus during the breakup process in order to calculate this term, and there is no standard way to do this.[68] Finally, the sedimentation $\mathfrak{F}_{ij}$ is expressed by

$$\mathfrak{F}_{ij} = \frac{\partial}{\partial z}\left(N_{ij}V_i\right) \tag{3.62}$$

which is similar to (3.57).

3.5.3   Explicit Modeling of Cold Clouds

In an explicit water-continuity model of cold clouds one must subdivide the ice categories listed in (3.48) as well as the liquid categories. This case is particularly complex since the liquid-phase drops in each size category potentially may interact with ice particles of every size and type. A staggering multiplicity of interactions is possible. Nonetheless, explicit water-continuity models have been developed that treat a substantial subset of the possible microphysical interactions that can occur in a mixed-phase cloud. Figure 3.16 indicates the categories of hydrometeors included in one example of such a model.[69] The categories of particle types in this model parallel those of the six-category model illustrated in Fig. 3.15, except that cloud ice has been subdivided into pristine "ice crystals" and "ice splinters," the latter being the product of a postulated ice-enhancement mechanism. The other categories, "cloud droplets," "drops," "snowflakes," and "graupel," correspond to the categories cloud liquid water, rainwater, snow, and graupel in Fig. 3.15. The ice crystals are subdivided into categories of crystal habit (plates, columns, and dendrites), where the crystal habit is determined by temperature. The graupel particles are subdivided according to particle density, ranging from 0.1 to 0.8 $\times$ 10^3 kg m^{-3}. Size distributions are computed for each particle category or subcategory. Stochastic collection concepts are used in the generation of drops from cloud droplets, snowflakes (aggregates) from ice crystals, and graupel from the riming of ice crystals. Graupel is also allowed to form when drops freeze.

3.6   Bulk Water-Continuity Models

As noted in Sec. 3.5.1, the disadvantage of the explicit water-continuity models is that one has to keep track of so many individual categories of water. The basic idea of *bulk water-continuity models* is to assume as few categories of water as possible in order to minimize the number of equations and calculations in the

---

[68] One attempt to include this effect was made by Silverman and Glass (1973).

[69] This example is from Scott and Hobbs (1977). Another cold-cloud explicit water-continuity model was developed by Hall (1980).

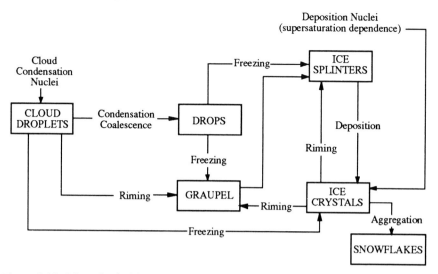

Figure 3.16 Microphysical interactions among different categories of water substance in an explicit model of a mixed-phase cloud. (From Scott and Hobbs, 1977. Reprinted with permission from the American Meteorological Society.)

water-continuity model. To accomplish this simplification, the shapes and size distributions of particles must be assumed and the basic microphysical processes must be parameterized. This method is used extensively in cloud dynamics and the following subsections outline its essential features.

3.6.1 Bulk Modeling of Warm Clouds

The simplest type of cloud is a warm, nonprecipitating cloud. The minimum number of categories that describe it is two: vapor, represented by q_v, and cloud liquid water, represented by q_c. The total water-substance mixing ratio, $q_T = q_v + q_c$, is conserved, and the water-continuity model (2.21) consists simply of the two equations

$$\frac{Dq_v}{Dt} = -C \tag{3.63}$$

$$\frac{Dq_c}{Dt} = C \tag{3.64}$$

where C represents the condensation of vapor when $C > 0$ and evaporation when $C < 0$.

For a warm precipitating cloud, rain is included as an additional category of water substance (drizzle is ignored). The water-continuity model (2.21) then consists of three equations:

$$\frac{Dq_v}{Dt} = -C_c + E_c + E_r \tag{3.65}$$

$$\frac{Dq_c}{Dt} = C_c - E_c - A_c - K_c \tag{3.66}$$

$$\frac{Dq_r}{Dt} = A_c + K_c - E_r + F_r \tag{3.67}$$

where q_r is the mixing ratio of rainwater, as defined in (3.48), C_c is the *condensation of cloud water*, E_c is the *evaporation of cloud water*, and E_r is the *evaporation of rainwater*. A_c is the *autoconversion*, which is the rate at which cloud water content decreases as particles grow to precipitation size by coalescence and/or vapor diffusion. K_c is the *collection of cloud water*, which is the rate at which the precipitation content increases as a result of the large falling drops intercepting and collecting small cloud droplets lying in their paths. F_r is the *sedimentation* of the raindrops in the air parcel; it is the net convergence of the vertical flux of rainwater relative to the air. All of the terms on the right of (3.65)–(3.67) are defined to be positive quantities, except for F_r, which is positive or negative depending on whether more rain is falling into or out of the air parcel. According to this model, cloud water q_c first appears by condensation C_c. Vapor is not condensed directly onto raindrops. Once sufficient cloud water has been produced, microphysical processes can then lead to the autoconversion (A_c) of some of the cloud water to rain. After autoconversion has begun to act, the amount of precipitation can then increase further through either A_c or K_c, or both. Once sufficient rainwater q_r has been produced, the additional microphysical processes E_r and F_r can become active.[70]

To calculate the microphysical sources and sinks of rainwater, A_c, K_c, E_r, and F_r, in terms of the mixing ratios q_v, q_c, and q_r, several key assumptions are made about the raindrops. First, the terminal fall speeds (defined to be positive in the downward direction) of individual raindrops are related to drop diameter D such that

$$V = V(D) > 0 \tag{3.68}$$

where $V(D)$ is given by an empirical curve like that in Fig. 3.3. Second, it is assumed that all the rain in a parcel of air falls with the mass-weighted fall velocity

$$\hat{V} \equiv \frac{\int_0^\infty V(D)\mathrm{m}(D)N(D)\,dD}{\int_0^\infty \mathrm{m}(D)N(D)\,dD} \tag{3.69}$$

---

[70] The bulk warm-cloud water-continuity model, consisting of the three categories vapor, cloud water, and rain interrelated by autoconversion and collection, was proposed by Kessler (1969). Virtually all bulk water-continuity models used today are a direct outgrowth of the concepts introduced in Kessler's seminal monograph.

where $m(D)$ is the mass of a drop of diameter D, and $N(D) \, dD$ is the number of drops per unit volume of air with diameter D to $D + dD$. Third, the precipitation particles are assumed to be exponentially distributed in size:

$$N = N_o \exp(-\lambda_r D) \tag{3.70}$$

where N_o is an empirically determined constant. This distribution is called the *Marshall–Palmer distribution*.[71] A typical value of N_o in rain is $\sim 8 \times 10^6 \text{ m}^{-4}$. The quantity λ_r is a function of the total rainwater mixing ratio q_r. Its value is determined by inverting the integral:

$$q_r = \rho^{-1} \int_0^\infty m(D) N_o \exp(-\lambda_r D) \, dD \tag{3.71}$$

The above assumptions allow the microphysical source/sink terms K_c, E_r, and F_r to be calculated. According to the continuous collection equation (3.24), the rate of increase of the mass m of an individual raindrop of diameter D is given by

$$\dot{m}_{col} = \frac{\pi D^2}{4} V(D) \rho q_c \Sigma_{rc} \tag{3.72}$$

where Σ_{rc} is the collection efficiency of raindrops collecting cloud drops. The variables Σ_{rc} and $V(D)$ are assumed to be known empirically. The fall velocity of the cloud liquid water is assumed to be zero, in accordance with the definitions given in Sec. 3.3. Σ_{rc} is usually taken to be ~ 1, which is an approximation that is representative of the majority of rainfall situations (Sec. 2.5.1). The depletion of cloud water by collection K_c by all of the raindrops in a parcel of air is computed from the integral

$$K_c = \rho^{-1} \int_0^\infty \dot{m}_{col} N(D) \, dD \tag{3.73}$$

where $N(D)$ is assumed to have the exponential form (3.70). The bulk rate of evaporation of rainwater mass from all raindrops E_r is computed in an analogous manner from the integral

$$E_r = \rho^{-1} \int_0^\infty \dot{m}_{dif} N(D) \, dD \tag{3.74}$$

where $\dot{m}_{dif}$ is the rate of evaporation by diffusion of vapor mass away from a single raindrop of diameter D falling through unsaturated air. A relationship for $\dot{m}_{dif}$ as a function of D must be obtained by numerical solution of (3.8), (3.9), and (3.11). It is not given in general by (3.22), which applies only when the air is saturated (i.e., in cloud). Often the rain is falling below cloud base, where the air is quite unsaturated.

---

[71] After Marshall and Palmer (1948), who analyzed images of a large sample of raindrops collected on treated filter paper.

The mass-weighted fall speed of the rain, given by (3.69), can be used to calculate the sedimentation of raindrops in the air parcel according to

$$F_r = \frac{\partial}{\partial z}\left(\hat{V}q_r\right) \tag{3.75}$$

The autoconversion rate A_c is usually assumed to be proportional to the amount by which the cloud liquid water mixing ratio exceeds a selected threshold; that is,

$$A_c = \tilde{\alpha}\left(q_c - a_T\right) \tag{3.76}$$

where a_T is the autoconversion threshold (often assumed arbitrarily to be 1 g kg^{-1}) and $\tilde{\alpha}$ is a positive constant when $q_c > a_T$ and 0 otherwise.[72] Thus, whenever cloud water exceeds the threshold amount, it is converted to rainwater at an exponential rate.

3.6.2   Bulk Modeling of Cold Clouds

The bulk water-continuity model described in the previous subsection can be extended to cold clouds by adding the categories of cloud ice, snow, and graupel, represented by the mixing ratios q_I, q_s, and q_g defined in (3.48).[73] We thus obtain the six-category water-continuity scheme illustrated in Fig. 3.15. The water-continuity equations (2.21) may then be written as

$$\frac{Dq_v}{Dt} = \left(-C_c + E_c + E_r\right)\delta_4 + S_v \tag{3.77}$$

$$\frac{Dq_c}{Dt} = \left(C_c - E_c - A_c - K_c\right)\delta_4 + S_c \tag{3.78}$$

$$\frac{Dq_r}{Dt} = \left(A_c + K_c - E_r + F_r\right)\delta_4 + S_r \tag{3.79}$$

$$\frac{Dq_I}{Dt} = S_I \tag{3.80}$$

$$\frac{Dq_s}{Dt} = F_s + S_s \tag{3.81}$$

$$\frac{Dq_g}{Dt} = F_g + S_g \tag{3.82}$$

---

[72] This autoconversion formulation was postulated intuitively by Kessler (1969) as part of his basic warm-cloud bulk parameterization scheme. Other autoconversion formulas have been developed from the physics of droplet coalescence (Cotton, 1972; Berry and Reinhardt, 1973). Cotton (1972), however, showed that Kessler's simple formula works about as well as the other formulas.

[73] This extension of the bulk water-continuity method to cold clouds was developed by Lin et al. (1983).

where the S terms represent all of the sources and sinks associated with ice-phase microphysical processes, except for the sedimentation of snow and graupel, which are represented by terms F_s and F_g, respectively. The term δ_4 is defined as

$$\delta_4 = \begin{cases} 0 & \text{if } T < -40°C \\ 1 & \text{otherwise} \end{cases} \tag{3.83}$$

Thus, it is assumed that, if the air temperature drops below $-40°C$, all super-cooled water freezes by homogeneous nucleation (Sec. 3.2.1) and hence all the terms in the liquid-water part of the model are set to zero.

The terms on the right in (3.77)–(3.82) include all of the possible interactions among the six categories of water, as illustrated in Fig. 3.15. Among these interactions are several bulk collection terms of the form (3.73). These represent graupel collecting cloud water and rain water, snow collecting cloud ice, etc. There are also several evaporation terms of the form (3.74). These include the sublimation and depositional growth of snow, graupel, and cloud ice. In addition, there are melting terms representing the increase of rainwater mixing ratio as a result of the melting of snow and graupel. The process of shedding liquid water collected by but not frozen to the surface of graupel or hail particles is also included. There are also three-way interactions that can occur, such as rain collecting cloud ice to produce graupel or hail.

To obtain mathematical expressions for the F_g, F_s, and S terms in (3.77)–(3.82), the same types of basic assumptions are made about the precipitating ice particles as were made for raindrops in the warm-cloud scheme. Crude assumptions are made regarding the collection efficiencies of ice particles, since very little is known about them. For riming (i.e., ice particles collecting liquid particles), the collection efficiency is usually assumed to be ~ 1. The collection efficiencies of ice particles collecting other ice particles is sometimes assumed to be a function of temperature that drops off exponentially from a value ~ 1 at $0°C$ to zero at lower temperatures. This assumption mirrors the observation of more frequent aggregation of falling particles as they near the melting level (Fig. 3.9). The precipitation particles are assumed to be exponentially distributed, as in (3.70), but with different values of N_o. For example, N_o might be assumed to be $\sim 8 \times 10^6$–2×10^7 m^{-4} for snow,[74] $\sim 4 \times 10^6$ m^{-4} for graupel,[75] and $\sim 3 \times 10^4$ m^{-4} for hail.[76] The fall speeds of snow and high-density ice particles are assumed to be known empirically as functions of particle diameter, as in (3.68), and the precipitation in a parcel of air is assumed to fall with the mass-weighted fall velocity, similar to that expressed by (3.69).[77]

---

[74] This value exhibits a temperature dependence. See Houze *et al.* (1979).

[75] See Rutledge and Hobbs (1983, 1984).

[76] See Lin *et al.* (1983).

[77] See Lin *et al.* (1983) for further details of how the cold-cloud bulk parameterization terms may be formulated. Rutledge and Hobbs (1983) give a concise summary of the technique.

Chapter 4 | Radar Meteorology

"That which no hand can reach, no hand can clasp..."[78]

Clouds occur primarily on the mesoscale and convective scale, which are among the most difficult size ranges for which to obtain atmospheric data. Mesoscale and convective phenomena (as defined in Sec. 1.1) tend to be too small to be resolved by synoptic surface and upper-air networks and too large to be easily observed locally. One effective instrument for obtaining observations in cloud systems on these scales is radar, which is especially suited to detect the precipitation falling from clouds. It has the capability, moreover, to map the precipitation with meso-scale coverage and convective-scale resolution. Because of this unique capability, it is one of the most important instruments for understanding cloud systems. In addition to the precipitation, which is the primary field observed, some meteoro-logical radars have the capability to receive signals from smaller cloud particles and from turbulent clear air in the planetary boundary layer. Meteorological ra-dars have been installed on land, ships, and aircraft to obtain key observations. Soon they will also be on spacecraft. Their use is so widespread that some knowl-edge of them is essential to a study of cloud dynamics. It is especially important to gain this knowledge since the parameters measured by meteorological radars provide a somewhat indirect indication of the information we most need in cloud physics and dynamics. The field of *radar meteorology* has evolved around the development of techniques for deriving meteorologically useful information from these measurements. This chapter summarizes the main techniques of radar mete-orology that have been developed for studies of precipitating clouds.[79]

4.1 General Characteristics of Meteorological Radars

A radar alternately switches between emitting and receiving pulses of microwave radiation with a common antenna (Fig. 4.1). The antenna is used to focus the radiation into a narrow beam, so that the transmitted signals travel outward in a

[78] In this line Goethe expresses the difficulty and frustration of observing clouds and thus seems to presage the development of remote-sensing techniques, like radar, to reach finally "that which no hand can clasp."

[79] For more extensive treatments of radar meteorology, see the books by Battan (1973), Doviak and Zrnić (1984), Rinehart (1991), and the compendium *Radar in Meteorology* edited by Atlas (1990). For shorter treatments, see Rogers and Yau (1989) and Burgess and Ray (1986). For more technical aspects, see Skolnik (1980).

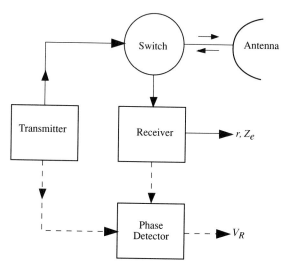

Figure 4.1 Simplified diagram of a weather radar. Dashed lines connect parts of the system included in a Doppler radar. The non-Doppler part provides measurements of target range r and equivalent radar reflectivity factor Z_e. The Doppler components provide the radial velocity of the target V_R.

specific direction. The original purpose was to detect and locate military targets.[80] The received signals are reflected from targets lying in the path of the beam, and the distance, or range r, of the target from the radar can be determined accurately from the time between the transmitted and received signals. With appropriate data processing equipment, the received signals can be further interpreted in terms of physical quantities relevant to precipitation physics and cloud dynamics. The basic parameters of the returned signal that are measured include (1) the power received, from which the reflectivity of the target is derived, (2) the Doppler shift in frequency, from which the target velocity is determined, and (3) the polarization of the signal, from which information on target shape and/or orientation can be derived. The interpretation of these quantities requires some understanding of how radars work.

As indicated in Fig. 4.1, the information in the received and processed signals is typically presented to the user of the radar data in three quantities: r, Z_e, and V_R. The range r has already been defined. The quantity Z_e represents the reflectivity, which will be defined more precisely in Sec. 4.2. V_R denotes the velocity component of the target along the beam of the radar; it is called the *radial velocity* since it indicates motion along a radius extending outward in some direction from the radar antenna. Polarization information is obtained from reflectivity measure-

---

[80] Radar was developed in Britain and the United States during World War II as an air defense system, for which the accurate determination of the range of an aircraft from the radar site was the primary measurement to be made. The word itself is a palindromic acronym meaning *RA*dio *D*etection *A*nd *R*anging. For more on the history of radar, see Page (1962).

ments made by transmitting and receiving at various polarizations. In Secs. 4.2–4.4, we will examine the physical basis of the parameters Z_e and V_R and how they may be used in the analysis of precipitating clouds.

Most meteorological radars measure only the range and reflectivity. They are called *noncoherent* or *conventional* radars. A radar that is instrumented to determine the velocity of the target is called a *Doppler radar*. A radar that provides information on the polarization of the signal is called a *polarimetric* or *multiple-polarization* radar. These radars generally operate at wavelengths $\lambda = 1$–30 cm, with 3, 5, and 10 cm being the most commonly used. Radiation at these wavelengths is scattered primarily by precipitation-sized particles, and the main use of meteorological radars is for mapping and analyzing the precipitation field and its associated physics and dynamics. Useful meteorological radar signals can also be obtained at these wavelengths from strong turbulent fluctuations in clear air, chaff particles intentionally introduced into the air, and insects. These data help determine the structure of the atmosphere surrounding and just preceding the occurrence of precipitating cloud systems.

Radar signals emitted at the shorter end of the 1–30-cm wavelength range are sensitive to more weakly reflecting targets, and they do not require a large-diameter antenna to focus the beam; however, they are subject to attenuation of the signal in heavy rain. The 10-cm wavelength (S-band) is the shortest at which attenuation by rain is essentially eliminated. However, the antenna required to focus the 10-cm wavelength beam is very large (a parabolic dish about 8 m in diameter is required for a 1° beamwidth), and the radar equipment is therefore bulky and expensive. Nonetheless, primary land-based radar facilities are typically 10 cm in wavelength. Shipborne and airborne radars, on the other hand, tend to be 5 cm (C-band) or 3 cm (X-band) systems because of space limitations. Although satellite-borne meteorological radars have yet to be employed, they are envisioned as having a wavelength of about 2 cm (K_u band)—also because of size limitations.[81] One must be cautious about the effects of attenuation when using the data from these shorter-wavelength systems. However, when proper care is taken, the data are useful.

Most meteorological radars have scannable antennas, which allow the radar beam to be pointed in a specified direction. By continuously scanning in azimuth and elevation, radars can obtain data in three spatial dimensions with high time resolution throughout a volume of space surrounding the radar. Depending on the size of the sector to be scanned, a set of three-dimensional data can be obtained about every 2–10 min, although more rapidly scanning antennas will be used on a few radars in the near future.[82] Most S-, C-, and X-band weather radars can detect precipitation at horizontal ranges up to 200–400 km. However, quantitative measurements can usually be obtained only within about 100–200 km of the radar site. Nonetheless, this capability allows precipitation to be mapped over mesoscale regions and data are obtained with high resolution. Each radar pulse extends over

[81] See *TRMM: A Satellite Mission to Measure Tropical Rainfall*, edited by Simpson (1988).
[82] For example, see Joss and Collier (1991).

a very small (~ 100 m) increment of range. The more restrictive limit on spatial resolution is the beamwidth (angular distance between half-power points), which for most precipitation radars is between 1 and $2°$.[83] As a result of the beam resolution limits, data are obtained with resolution of less than a kilometer close to the radar and a few kilometers at the outer range limits.

Meteorological radars with very short wavelength (bands centered around 1 and 8 mm) are useful for detecting cloud particles. However, these bands are too highly attenuated by rain to be useful in heavy precipitation. Lidar uses laser light (at $\lambda \sim 0.3-10$ μm) for observing atmospheric aerosol particles and cloud structure.[84] Radars with wavelengths in the UHF (75 cm) and VHF (6 m) bands are useful for clear air sounding of the atmosphere; these radars are referred to as *profilers*.[85] The discussions in Secs. 4.2 and 4.3 apply mainly to centimetric wavelength radars, which detect precipitation particles. In general, these microwave radars are the most widely used for precipitation physics and cloud dynamics. However, many of the Doppler radar techniques discussed in Sec. 4.4 are applicable to radars at any wavelength.

4.2   Reflectivity Measurements

4.2.1   Obtaining Reflectivity from Returned Power

Precipitation particles (rain, snow, graupel, and hail) detected by meteorological radar are called distributed targets, since many scattering elements are simultaneously illuminated by the radar beam. The volume containing the illuminated particles is called the *resolution volume* of the radar. This volume is determined by the beamwidth and duration of the transmitted pulse of radiation. Since the distributed targets are generally moving relative to each other, as a result of differing fall speeds and sheared and/or turbulent winds, the power (as well as phase) returned from the resolution volume centered at a range r fluctuates in time. The instantaneous power returned depends upon the arrangement of scatter-

[83] Usually, the horizontal and vertical beamwidths are the same. However, some airborne systems require a restricted antenna dimension and hence broader beamwidth in the vertical. For example, the U.S. National Oceanographic and Atmospheric Administration WP-3D research aircraft have a C-band radar with a $1.5°$ horizontal beamwidth and $4°$ vertical beamwidth. The antenna is mounted below the fuselage.

[84] Lidar is an acronym for *Li*ght *I*ntensity *D*etection *A*nd *R*anging. In a lidar, the returned laser light is collected in a telescope and focused on a photomultiplier detector, after which it is amplified, digitized, and recorded. The lidar is used essentially like a radar and is often called a "laser radar." Its beam can be used to scan a volume of the atmosphere, and it can be instrumented to make Doppler and multiple-polarization measurements. See p. 412 of Stull (1988) for a brief nontechnical discussion and some examples of the types of measurements that can be made.

[85] These long-wavelength radars do not usually scan but rather operate with several fixed beams: for example, with one beam vertical and two $\sim 15°$ off vertical. Their utility derives from their ability to receive reflections from refractive index variations of the air as well as from particles suspended in the atmosphere (see Röttger and Larsen, 1990 for a review of this type of radar).

ers. However, a time average over ~0.01–0.1 s averages out the fluctuations, and the average returned power $\overline{P}_r$ is given by

$$\overline{P}_r = \frac{P_t G^2 \lambda^2 \theta_H \theta_V \tau_p c_o \eta_r}{512(2 \ln 2)\pi^2 r^2} \qquad (4.1)$$

where P_t is the transmitted power, G is the antenna gain (a dimensionless number allowing for the focusing effect of the antenna), τ_p is the duration of the emitted radar pulse, c_o is the speed of light, θ_H and θ_V are the horizontal and vertical beamwidth angles (i.e., the angles between power points expressed in radians), respectively, and η_r is called the *radar reflectivity* per unit volume of air. It is the effective scattering cross section of a unit volume of air, and it is what is deduced from the measurement of average returned power, since all the other terms on the right-hand side of (4.1) are known. The radar equation (4.1) is obtained from geometrical considerations and by assuming that there is no attenuation by intervening targets and that the resolution volume is uniformly filled with targets. There are no losses of signal in the radar system itself, and the antenna gain is uniform across the beamwidth.[86] The term $2 \ln 2$ is a factor that adjusts the half-power beamwidths to an effective beamwidth, which has a constant distribution of power. This adjustment is necessary since the antenna gain in (4.1) is assumed to be constant across the beam.

The radar reflectivity η_r is the sum of the scattering cross sections of the individual scatterers within the resolution volume. It is given by the summation $\Sigma \sigma$, where σ is the effective backscatter cross section of one particle within the resolution volume and the summation is over all the particles in the volume. The backscatter cross section is a factor introduced to account for the fact that the target may not scatter radiation isotropically. It is the cross-sectional area an isotropic scatterer (which scatters all impinging radiation equally in all directions) would need in order to produce the scattering by the actual target. For a single spherical target whose diameter D is small compared to the radar wavelength, Rayleigh[87] scattering theory applies, in which case

$$\sigma = \pi^5 |K|^2 D^6 \lambda^{-4} \qquad (4.2)$$

where $|K|^2$ is a function of the complex index of refraction of the substance of which the scatterer is composed.[88]

Precipitation particles are usually small enough that the particle diameter $D < 0.1\lambda$.[89] Consequently, Rayleigh scattering theory can be used to approximate the characteristics of signals reflected back to the radar. Substitution of $\Sigma \sigma$ for η_r in (4.1), with σ given by (4.2), and some algebraic rearrangement of terms show that

---

[86] See Chapter 4 of Battan (1973) or Chapters 4 and 5 of Rinehart (1991).

[87] The same Lord Rayleigh who derived the theory of convection described in Sec. 2.9.3.

[88] See Chapter 4 of Battan (1973) or Chapters 4 and 5 of Rinehart (1991) for further discussion.

[89] This condition holds for drizzle, as well as for most rain, snow, and graupel, but not for hailstones, which are ~1–15 cm in diameter, and often not for very large raindrops or aggregated snowflakes, especially at shorter wavelengths.

the returned power $\overline{P_r}$ from scatterers of known $|K|^2$ located within a resolution volume $\mathcal{V}_{\rm res}$ centered at a range r from a radar yields the *radar reflectivity factor*

$$Z \equiv \frac{1}{\mathcal{V}_{\rm res}} \sum D^6 = \frac{r^2 \overline{P_r} C_R}{|K|^2} \tag{4.3}$$

where

$$C_R = \frac{64 \lambda^2 r^2}{P_t G^2 \pi^2 \mathcal{V}_{\rm res}} \tag{4.4}$$

and

$$\mathcal{V}_{\rm res} = \pi \theta_H \theta_V \left(\frac{r}{2}\right)^2 \frac{c_o \tau_p}{2} \tag{4.5}$$

The term $\sum D^6$ in (4.3) is the sum of the sixth powers of the diameters of all of the scatterers in the volume $\mathcal{V}_{\rm res}$. Note that since $\mathcal{V}_{\rm res}$ is proportional to r^2, C_R is a constant depending only on the characteristics of the particular set of radar equipment being used.

Usually there is no way to be certain of the value of $|K|^2$. The scatterers could be composed of liquid, ice, melting ice, insects, turbulent eddies, or chaff. The convention therefore is to use the measured $\overline{P_r}$ and r and (4.3) to calculate the *equivalent radar reflectivity factor*,

$$Z_e \equiv \frac{r^2 \overline{P_r} C_R}{0.93} \tag{4.6}$$

where 0.93 is the value of $|K|^2$ for liquid water. Z_e thus would be the value that the reflectivity factor of the particles producing the returned power $\overline{P_r}$ detected at range r would have if they were composed purely of liquid water. The value of $|K|^2$ for ice is usually set to 0.197. If it were known that the reflectors were actually ice particles, then the true reflectivity factor could be obtained simply by multiplying Z_e by the ratio (0.93)/(0.197). However, since the composition of the scatterers is not normally known with certainty, radar data are usually expressed as Z_e, in units of $mm^6 \, m^{-3}$, which refer to particle size to the sixth power per unit volume of air. Typically, the numbers are given in decibel units:

$$dBZ_e \equiv 10 \log_{10} Z_e \tag{4.7}$$

Typical values of $dB[Z_e(mm^6 \, m^{-3})]$ are ~ -30 to 0 in marginally detectable precipitation; ~ 0–10 in drizzle, very light rain, or light snow; ~ 10–30 in moderate rain and heavier snow; ~ 30–45 in melting snow; ~ 30–60 in moderate to heavy rain; and ~ 60–70 or more in hail.

4.2.2   Relating Reflectivity to Precipitation

One of the most important uses of meteorological radar is to estimate the precipitation content of the air, the precipitation rate, and the fall speed of the precipitation at the earth's surface. In this section, we will outline the techniques used to make these estimates from measurements of radar reflectivity.

4.2.2.1   Particle Size Method

Since the radar reflectivity factor Z is related to particle size, there is a physical basis for a quantitative relationship between the precipitation content of the air and the received radar echo intensity. According to the definition (4.3), the radar reflectivity factor is the sixth moment of the particle size distribution. If the number of particles in a radar-sampled volume is very large, a continuous version of (4.3) can be used:

$$Z = \int_0^\infty D^6 N(D)\, dD \tag{4.8}$$

where $N(D)$ is the particle size distribution function defined such that $N(D)dD$ is the number of particles of diameter D to $D + dD$ per unit volume of air. If the scatterers are liquid water, $Z = Z_e$. The particle size distribution also determines the mixing ratio of rainfall,

$$q_r = \frac{\pi \rho_L}{6\rho} \int_0^\infty D^3 N(D) dD \tag{4.9}$$

where ρ_L is the density of liquid water, ρ is the density of air, and the rainfall rate $\Re$ is given by

$$\Re = \frac{\pi \rho_L}{6} \int_0^\infty V(D) D^3 N(D)\, dD \tag{4.10}$$

where $V(D)$ is the fall velocity related empirically or theoretically to drop of diameter D [Eq.(3.68)]. It is evident from (4.8)–(4.10) that measurements of the drop size distribution $N(D)$ can be used to determine a set of values $(Z, q_r, \Re)$. For rainfall of a given type in a given climatological setting, the curves of $\log Z$ vs. $\log \Re$ and $\log Z$ vs. $\log \rho q_r$ are usually linear and therefore yield empirical relationships of the form

$$Z = \tilde{a}\Re^{\tilde{b}} \tag{4.11}$$

$$Z = \tilde{a}_1 \left(\rho q_r\right)^{\tilde{b}_1} \tag{4.12}$$

where $\tilde{a}$, $\tilde{b}$, $\tilde{a}_1$, and $\tilde{b}_1$ are positive constants derived from the slopes and intercepts of the log–log plots. From (4.9) and (4.10), it is evident that the ratio $\Re/(\rho q_r)$ is the mass-weighted particle fall speed $\hat{V}$ defined by (3.69). The relationships (4.11) and (4.12) imply that this ratio is a function of Z of the form

$$\hat{V} = \tilde{a}_2 Z^{\tilde{b}_2} \tag{4.13}$$

where $\tilde{a}_2$ and $\tilde{b}_2$ are constants. Sometimes a factor accounting for the variable density of air through which the particle is falling is included in the fall speed formula.[90] The relationships (4.11)–(4.13) are used for estimating rainfall rate, rain mixing ratio, and rain fall speed from radar reflectivity measurements. Similar relationships can be obtained for snow. The values of the constants should be determined from particle size measurements made within the particular type of precipitation being studied.

To be able to use a Z–$\Re$ relation like (4.11), the equivalent radar reflectivity factor Z_e obtained from the observed radar data must first be converted to an appropriate value of Z. Several types of sampling problems typically make this conversion difficult: (i) The composition of the particles (liquid or ice) must be assumed. This problem is hard to overcome when the beam is partially filled with liquid and ice particles, contains melting ice, or is partially filled by ground targets. (ii) Because of both the curvature of the earth and the elevation angle of the beam, the center of the radar beam is typically found at higher altitude the greater the range. The radar thus indicates a value of Z for a radar resolution volume $\mathcal{V}_{\text{res}}$ located some distance above the earth's surface, whereas the drop size distribution measurements are usually made at a point on the surface. The particle size distribution may undergo significant evolution as a result of diffusional and collectional growth, evaporation, breakup, or sedimentation, which may occur in the population of raindrops between the time the drops are in the radar resolution volume and when they finally reach the earth's surface [recall Eq. (3.52)]. This problem is greatly exacerbated if the radar beam at low elevation angles is blocked by an intervening mountain. In such a case, the precipitation at lower altitudes behind the mountain cannot be observed, and a vertical profile of reflectivity must be assumed for the precipitation below the lowest elevation angle that is not blocked.[91] (iii) According to (4.6), Z_e is computed from the returned power $\overline{P_r}$ under the assumption that the distributed targets completely fill the resolution volume $\mathcal{V}_{\text{res}}$ at a given range. The returned power may therefore be underestimated if the resolution volume is not filled completely. This problem becomes more likely the greater the range since the beamwidth angles in (4.5) are fixed. These sampling problems lead to an uncertainty of about a factor of 2 (or more in the case of topographic shielding) in estimating rain from Z–$\Re$ relations based on drop size distribution measurements. Some cancellation of errors can be obtained by integrating radar measurements over long times or large areas.[92] The accuracy of Z–q_r and Z–$\hat{V}$ relations is less known but probably similar.

4.2.2.2   *Rain Gauge Method*

Although the particle size methodology described above for determining precipitation rates from radar data is elegant in that it is developed around the physical relationship between particle size distribution and radar reflectivity, it is limited by the uncertainties mentioned at the end of the preceding subsection.

---

[90] See Foote and DuToit (1969) or Beard (1985).

[91] The assumed profile could be based on a climatology of the vertical profile of reflectivity. For a discussion of the problem of shielding by mountains, see Joss and Waldvogel (1990).

[92] For further details see Joss and Waldvogel (1990) or Austin (1987).

To avoid these problems, one can directly relate surface rain gauge and radar-measured values of Z_e.

Rain gauge data can be used to obtain the probability density function $P(\mathfrak{R})$, which is defined such that $P(\mathfrak{R})\,d\mathfrak{R}$ is the probability that if rain is falling, its rate (in mm h^{-1}) is between $\mathfrak{R}$ and $\mathfrak{R} + d\mathfrak{R}$. For the same population of rain-producing clouds, radar data can be used to determine another probability density function $P(Z_e)$, which is defined such that $P(Z_e)\,dZ_e$ is the probability that if rain (and hence radar echo) is present at a given range from the radar, the equivalent radar reflectivity is between Z_e and $Z_e + dZ_e$. The probability density function for the equivalent radar reflectivity can be determined for an interval of range centered at r. Let the sum of the area covered by echo within this annular region be $A_E(r)$. Let the area covered by those echoes with reflectivity between Z_e and $Z_e + dZ_e$ be $A(Z_e,r)$. Then the probability density function $P(Z_e)$ for the region centered at r is given by

$$P(Z_e)\big|_r \, dZ_e = \frac{A(Z_e,r)}{A_E(r)} \tag{4.14}$$

It can be shown[93] that since Z_e and $\mathfrak{R}$ are functionally related, the correct transformation of one into the other will produce equal probabilities, such that

$$P(\mathfrak{R}_i)\,d\mathfrak{R} = P(Z_{ei})\,dZ_e \tag{4.15}$$

where the pairs $(Z_{ei},\mathfrak{R}_i)$ define the Z_{ei}–$\mathfrak{R}_i$ relationship. It follows also that

$$\int_{\mathfrak{R}_i}^{\infty} P(\mathfrak{R})\,d\mathfrak{R} = \int_{Z_{ei}}^{\infty} P(Z_e)\,dZ_e \tag{4.16}$$

The expression in (4.14) can be substituted in the right-hand integral. If rain gauges located within the annular zone centered at range r from the radar are used to derive $P(\mathfrak{R})\big|_r$, which is the probability density function for $\mathfrak{R}$ that applies specifically in this zone, then this empirical probability density can be substituted for the integrand on the left-hand side of (4.16). The Z_e–$\mathfrak{R}$ relationship is then obtained by finding the pairs of lower limits $(Z_{ei},\mathfrak{R}_i)$ that make the integrals equal. Thus, the empirical probability density functions for Z_e and $\mathfrak{R}$ can be used to obtain a Z_e–$\mathfrak{R}$ relationship that applies at a given range from the radar. By repeating the process for various ranges, Z_e–$\mathfrak{R}$ relationships can be found that apply throughout the field of view of the radar. This technique thus allows one to relate Z_e and $\mathfrak{R}$ without having to account for radar-beam geometry or differences between Z_e and Z.

4.2.3 Estimating Areal Precipitation from Radar Data

One important use of radar data is to estimate rainfall over an area covered by radar observations. A straightforward way to make this estimate is to apply a

---

[93] See Calheiros and Zawadzki (1987).

Z_e–$\Re$ relationship to all the low-altitude reflectivity data obtained over a region in the x–y plane to estimate the rainfall rate as a function $\Re(x,y)$. Then the area-integrated rain rate $\Re_{\mathrm{area}}$ (mass of water per unit time) is obtained by integrating as follows:

$$\Re_{\mathrm{area}} = \iint\limits_{A_T} \Re(x,y)\,dx\,dy \qquad (4.17)$$

where A_T is the total area covered by rain. Although up to now this approach has been the standard, it depends on having accurate rain rates everywhere within A_T.

An alternative method for obtaining the area-wide rain rate, which is outlined below,[94] allows an estimate of this rate to be made by using radar data to identify the area covered by rain and using a statistical knowledge of the rainfall character-istics to estimate the amount of precipitation falling within the area. This tech-nique can be useful in the case of observations with a satellite-borne radar, which operates at a wavelength of ~2 cm. Because of attenuation at this wavelength and other difficulties of making spaceborne radar observations, this type of radar may accurately estimate only the lower rain rates. The rain rates at every point within the area, especially where the rain is heavy, may be difficult to estimate. How-ever, the measurement of the lower rates will allow the areas in which the rain exceeds a given threshold to be outlined accurately. Airborne radar measure-ments also tend to describe best the areas covered by rain but do not usually accurately describe the rain rate at all the interior points of the area. The following technique can be useful in estimating the areal rain rates in these cases.

Let $A(\Re_\tau)$ be the area covered by rain with rates exceeding a given threshold rain rate $\Re_\tau$. The ratio of $A(\Re_\tau)$ to the total area covered by rain is then

$$F(\Re_\tau) \equiv \frac{A(\Re_\tau)}{A_T} = \int_{\Re_\tau}^{\infty} P(\Re)\,d\Re \qquad (4.18)$$

where, as in Sec. 4.2.2.2, $P(\Re)$ is the probability density function for rainfall rate. $F(\Re_\tau)$ is thus the fraction of the total area covered by rain with rates exceeding the threshold rain rate $\Re_\tau$. The term on the right-hand side of (4.18) expresses the fact that the ratio of areas is equivalent to the cumulative probability of rain with rates $>\Re_\tau$. The average rain rate within regions where the threshold is exceeded is

$$\langle \Re \rangle_\tau = \int_{\Re_\tau}^{\infty} \Re P(\Re)\,d\Re \bigg/ \int_{\Re_\tau}^{\infty} P(\Re)\,d\Re \qquad (4.19)$$

The overall average rain rate is then

$$\langle \Re \rangle_o = \int_0^{\infty} \Re P(\Re)\,d\Re \qquad (4.20)$$

---

[94] For further details, see Doneaud *et al.* (1981, 1984), Atlas *et al.* (1990), and Rosenfeld *et al.* (1990).

Multiplying the right-hand sides of (4.18) and (4.19), and substituting from (4.20), we obtain

$$F(\mathfrak{R}_\tau)\langle\mathfrak{R}\rangle_\tau = \langle\mathfrak{R}\rangle_o\left[\int_{\mathfrak{R}_\tau}^\infty \mathfrak{R}P(\mathfrak{R})d\mathfrak{R}\bigg/\int_0^\infty \mathfrak{R}P(\mathfrak{R})d\mathfrak{R}\right] \tag{4.21}$$

Both of the terms in brackets and $\langle\mathfrak{R}\rangle_\tau$ are determined by the empirical probability density function $P(\mathfrak{R})$. It follows that the ratio of $\langle\mathfrak{R}\rangle_\tau$ to the term in brackets is also determined solely by $P(\mathfrak{R})$. If we call this ratio $\mathcal{S}(\mathfrak{R}_\tau)$, then the area-wide average rain rate is given by

$$\langle\mathfrak{R}\rangle_o = F(\mathfrak{R}_\tau)\mathcal{S}(\mathfrak{R}_\tau) \tag{4.22}$$

Thus, the area-wide average rain rate can be determined by using radar to measure the fractional area covered by rates above the threshold $\mathfrak{R}_\tau$, provided the probability density function $P(\mathfrak{R})$ [and hence the factor $\mathcal{S}(\mathfrak{R}_\tau)$] is known empirically from other sources of information, such as rain gauge records. The area-integrated rain rate $\mathfrak{R}_{\text{area}}$ (mass of water per unit time) is obtained by multiplying $\langle\mathfrak{R}\rangle_o$ by the total area covered by rain:

$$\mathfrak{R}_{\text{area}} = \langle\mathfrak{R}\rangle_o A_T \tag{4.23}$$

The limitation of this method of estimating the areal rainfall is that the probability density function $P(\mathfrak{R})$ must be determined for the particular climatological setting. It might have more value in determining averages of large populations of storms than the rainfall in individual cases. However, it may sometimes (e.g., with satellite or airborne radar) be the only method available.

4.3 Polarization Data

To interpret the returned radar signal, atmospheric scientists have four fundamental properties of electromagnetic waves at their disposal: amplitude, phase, frequency, and polarization. For precipitation, the polarization characteristics are related to the mean values and distributions of size, shape, and spatial orientation of the particles filling the radar resolution volume, their thermodynamic phase state, and fall behavior. To take advantage of polarization properties, polarimetric radars are instrumented to transmit and receive radiation polarized in several specified orientations.[95] *Circularly polarized polarimetric radars* transmit an electric field vector whose direction normal to the path of the radiation rotates with time. *Linearly polarized polarimetric radars* transmit and receive pulses of radiation polarized in two orthogonal directions.

---

[95] For further discussions of multiple polarization in meteorological radar, see Bringi *et al.* (1986a,b), Wakimoto and Bringi (1988), Bringi and Hendry (1990), Jameson and Johnson (1990), and Herzegh and Jameson (1992).

A parameter that can be derived from the data of a circularly polarized radar is the *circular depolarization ratio (CDR)*. It is defined as

$$\text{CDR} \equiv 10 \log_{10}\left(\frac{Z_{\text{par}}}{Z_{\text{orth}}}\right) \tag{4.24}$$

where Z_{par} is the reflectivity factor received in circular polarization of the same sense as that transmitted (called the parallel component) and Z_{orth} is that received in the opposite (or orthogonal) polarization. If the transmitted pulse is right-hand circularly (RHC) polarized, the ratio in (4.24) would be the ratio of the right-hand to the left-hand (counterclockwise) polarized received reflectivities. CDR indicates the shapes of the particles producing the signal. When circularly polarized waves are scattered by a spherical particle, the backscattered signal is also circularly polarized but in the opposite sense, because the direction of propagation is reversed. In this case, CDR $= -\infty$. A typical value of CDR measured in rain is ~ -20 to -35. On the other hand, if the target is infinitely long and thin, the parallel and orthogonal components are equal in magnitude, and CDR $= 0$. The value of the CDR, on a scale of $-\infty$ to 0, thus indicates the relative sphericity of the particles producing the radar echo. It may therefore be regarded as an indicator of the shapes of the detected precipitation particles. However, as will be noted below, the rather small index of refraction of ice reduces the magnitude of the expected shape effects.

Linearly polarized radars usually transmit and receive radiation in the horizontal and vertical planes, although, in principle, the two orthogonal planes of polarization could be at any chosen orientation. The motivation for choosing horizontal and vertical planes is that raindrops tend to become oblate as they increase in size,[96] and they fall with the major axis oriented rather horizontally (Fig. 4.2). Two useful parameters that can be obtained from linearly polarized radars that operate in the horizontal and vertical are the *differential reflectivity* Z_{DR} and the *linear depolarization ratio* LDR. They are defined as

$$Z_{\text{DR}} \equiv 10 \log_{10}\left(\frac{Z_{\text{HH}}}{Z_{\text{VV}}}\right) \tag{4.25}$$

and

$$\text{LDR} \equiv 10 \log_{10}\left(\frac{Z_{\text{HV}}}{Z_{\text{HH}}}\right) \tag{4.26}$$

where Z_{HH}, Z_{VV}, and Z_{HV} are the horizontally transmitted/horizontally received, vertically transmitted/vertically received, and horizontally transmitted/vertically received reflectivity factors, respectively.

Since raindrops tend to be oblate and horizontally oriented, positive values of Z_{DR} are produced, generally in the range 0–5, as shown in Fig. 4.2. The larger drops produce the larger values since the degree of oblateness increases with drop size. Regions of ice particles tend to produce values of Z_{DR} near zero. These

---

[96] The oblateness of raindrops is discussed by Pruppacher and Klett (1978).

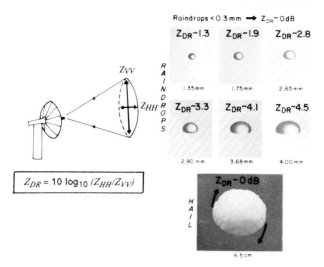

Figure 4.2 Typical Z_{DR} values for raindrops of various sizes and hail. Black arrows on the hail particle represent tumbling motions. (Adapted from Wakimoto and Bringi, 1988. Reproduced with permission from the American Meteorological Society.)

values are low partly because ice particles do not generally exhibit a preferred orientation as they fall. Ice crystals and snowflakes may be flat; however, they oscillate or tumble as they fall, so that their axes are rather randomly oriented. Hail particles also tend to tumble or spin as they fall, as indicated in Fig. 4.2. Thus, regions of snow or hail particles exhibit no preferred orientation and hence have near-zero values of Z_{DR}. Echoes produced by rain can therefore often be distinguished from those produced by ice particles by whether or not they have a finite value of Z_{DR}.

Another reason the Z_{DR} of signals returned from ice particles is generally low is related to the nature of the interaction that occurs when electromagnetic radiation impinges on a particle of low complex index of refraction. An extreme aspect ratio is required to produce a measurable Z_{DR} (similar to CDR) in this case. This condition is eased when the ice particles are wet and thus have a larger complex index of refraction. Consequently, large and wet ice particles tend to produce a noticeable signal, if the particles have a preferred orientation. These conditions prevail in the melting layer, which produces a strong signal. Thus, differential reflectivity Z_{DR} can be used to distinguish between ice and water, to delineate the melting layer, and to indicate the mean drop size in the rain.

According to (4.26), the linear depolarization ratio LDR is a measure of how much of the transmitted signal is depolarized. In general, precipitation particles depolarize propagating electromagnetic waves only if they are aspherical and if they have no axis of symmetry parallel to the direction of the wave polarization. Therefore a spherical particle illuminated by horizontally transmitted radiation produces an echo pulse that is purely horizontally polarized. All of the vertical components of electric field oscillation excited within the sphere by the horizontally polarized wave cancel by symmetry (i.e., no net depolarization occurs). Hence, only horizontal components are returned to the radar, and LDR $= -\infty$. If a

particle is elongated and not oriented either purely horizontally or purely vertically, then the oscillations parallel to the long axis of the particle predominate, and
the returned signal has both a horizontal and a vertical component (i.e., part of the
transmitted wave has been depolarized). Thus, LDR $> -\infty$. The more strongly the
transmitted radar waves are depolarized by precipitation particles, the stronger
the backscattered cross-polar signal will be and thus the larger (i.e., the less
negative) the LDR becomes. Since the cross-polar signal is generally smaller than
the copolar signal, LDR never exhibits positive values. Regions of snow characterized by flat ice particles of varying orientation tend collectively to produce a
measurable value of LDR. However, the small index of refraction of ice might
suppress the magnitude of the expected shape effects. The LDR values are dramatically enhanced (less negative) for wet ice particles because of the increased
complex index of refraction. LDR has typical values of -10 to -20 for melting ice
to -25 to -30 for ice particles or rain. Thus, the linear depolarization ratio LDR is
an excellent indicator of the melting layer and also provides some information
about the shape and fall behavior of the precipitation particles.

4.4   Doppler Velocity Measurements

The information derived from reflectivity and polarimetric data, as described in
Secs. 4.2 and 4.3, is related primarily to the physics of the precipitation producing
the signals returned to the radar. The same signals can be further processed to
obtain information on the motions of the precipitation particles and of the air in
which they are located. This processing takes advantage of the Doppler-shifted
frequency of the echoes. The kinematic information derived in this manner allows
the precipitation measurements to be viewed in the context of the circulation of air
through the storms producing the precipitation. Since radars operating at 3–10-cm
wavelengths are sometimes sufficiently sensitive to receive echoes from clear air,
mainly from turbulent fluctuations of the index of refraction of the air and/or from
insects, aspects of the air motions in the environments of storms can also be
detected. The clear air echoes are returned primarily from the boundary layer, and
they can be useful for mapping the flow of boundary-layer air into and out of the
storm. They are also useful for analyzing the boundary-layer air motions that exist
before and after precipitating cloud systems occur.[97] Most of the techniques described below in Secs. 4.4.1–4.4.7 are applicable to the processing of both clear air
echoes and echoes from precipitation particles.

4.4.1   Radial Velocity

If a target detected by a radar is moving, the phase ϕ_p of the reflected waves
changes with time at a rate

$$\frac{d\phi_p}{dt} = \frac{4\pi V_R}{\lambda} \tag{4.27}$$

---

[97] See Gossard (1990) for a review of radar observation of the boundary layer.

where the radial velocity V_R is the component of the velocity of the target parallel to the beam of the radar. Thus, V_R can be determined from $d\phi_p/dt$. This principle is the same as that of a police radar used for detecting the speed of an automobile, except that the backscattered radiation comes from the population of targets (usually precipitation particles) in the radar resolution volume, rather than from a single target, and these scatterers are all moving at somewhat different velocities.[98] For one pulse of emitted and received radiation, the radar detects the net amplitude ($|\mathbf{E}|$) and phase (ϕ_p) of the returned electric field vector $\mathbf{E}$, which is the vector sum of all the electric field vectors returned by individual scatterers. Each successive pulse received provides a new estimate of the net $\mathbf{E}$. Vector subtraction of the net $\mathbf{E}$ vectors of two successive pulses yields an estimate of the net phase difference $\Delta\phi_p$ between the two pulses. Vector averaging of the net $\mathbf{E}$ vectors of the two successive pulses yields the vector average $\bar{\mathbf{E}}$, whose amplitude is $|\bar{\mathbf{E}}|$. A new vector given by ($|\bar{\mathbf{E}}|$, $\Delta\phi_p$) may then be formed for each pulse pair. The vector average of ($|\bar{\mathbf{E}}|$, $\Delta\phi_p$) from a number of successive pulse pairs can be calculated to improve the estimate of this vector. The phase of the vector that results from the average, which we will call $\overline{\Delta\phi_p}$, is an improved estimate of the net phase change. The estimate $\overline{\Delta\phi_p}$ is substituted in (4.27) to obtain an estimate of the *mean radial velocity* $\overline{V_R}$.[99]

The amplitude of the vector that results from the average of ($|\bar{\mathbf{E}}|$, $\Delta\phi_p$) over several pairs of pulses is related to the average power $\overline{P_r}$ returned by the scatterers in the resolution volume, since it can be shown that the amplitude of the returned electric field vector is proportional to $\sqrt{\overline{P_r}}$.[100] Since the mean radial velocity $\overline{V_R}$ estimated by the above-described vector averaging is weighted by $|\bar{\mathbf{E}}|$ it may be regarded as the mean of the *Doppler velocity spectrum*, which is the distribution of power returned from the sampling volume expressed as a function of radial velocity. This spectrum may be denoted $S(V_R)$, where $S(V_R)\,dV_R$ is the amount of returned power accounted for by targets with radial velocities between V_R and $V_R + dV_R$. It is related to $\overline{P_r}$ by

$$\overline{P_r} = \int_{-\infty}^{\infty} S(V_R)\,dV_R \tag{4.28}$$

The mean radial velocity $\overline{V_R}$, defined as the mean (first moment) of this distribution, may be written as

$$\overline{V_R} \equiv \frac{\int_{-\infty}^{\infty} V_R S(V_R)\,dV_R}{\int_{-\infty}^{\infty} S(V_R)\,dV_R} \tag{4.29}$$

---

[98] Actually, a police radar is a much simpler device than a meteorological radar since it measures only the speed and not the location of the target. However, the speed of the target is determined by the rate of phase angle change in either case.

[99] For further discussion of Doppler radar velocity estimation, see Sirmans and Bumgarner (1975).

[100] See Doviak and Zrnic (1984, pp. 34–38).

The Doppler velocity spectrum $S(V_R)$ contains several types of useful information. For example, when the antenna is pointing horizontally, a sharply bimodal spectrum of radial velocity can indicate the presence of a tornado (see Sec. 8.8) lying in the beam of the radar. Or if the antenna is pointing vertically, the Doppler spectrum can be related to the particle size spectrum of falling precipitation (see Sec. 4.4.3). The variance (i.e., the second moment) of the Doppler spectrum contains information related to the shear and turbulence of the air motions producing the Doppler velocities.

Although the radial velocity spectrum contains some useful information, most work with Doppler radar data considers only the power-weighted mean radial velocity $\overline{V_R}$. This means target velocity can be related to air motions and particle fall speeds through the geometry of the pointing angle of the antenna. If we assume that the scatterers detected by the radar move horizontally with the wind and vertically as a combination of vertical air motion and (in the case of precipitation particles) fall speed V_T, then

$$\overline{V_R} = \left(u \sin \alpha_a + v \cos \alpha_a\right) \cos \alpha_e + \left(w - V_T\right) \sin \alpha_e \qquad (4.30)$$

where α_a is the azimuth angle (measured clockwise from the north) toward which the radar beam is pointing, and α_e is the elevation angle. The angles are always known, and they can be varied by pointing or scanning the antenna in various directions. Ways can be devised to direct the beams of one or more radars such that information about the wind components (u,v,w) and the fall speed V_T can be derived from the measurement of $\overline{V_R}$. The main strategies used are summarized in Secs. 4.4.3–4.4.6.

4.4.2   Velocity and Range Folding

Since the radar emits pulses of radiation at a set frequency (called the pulse-repetition frequency, or PRF), the radial velocity of a target $\overline{V_R}$ can be determined by comparing the phases of the signals reflected back to the antenna from two successive pulses. The radar-detected difference in phase angle $\Delta\phi_p$ between successive pulses is a number between $-\pi$ and $+\pi$. Since the pulses are separated by a finite interval of time Δt, we could in principle determine $\overline{V_R}$ from (4.27). However, since the phase is not monitored continuously during the interval Δt, there is no way to be certain that the true phase change is not the detected difference plus some integral multiple of $\pm 2\pi$. Hence, the detected radial velocity is unambiguous only if the actual velocities of the targets are never so large that they produce a true phase change outside the range of $\pm \pi$, that is, if they move less than a quarter of a radar wavelength in the interval between two successive pulses [recall (4.27)]. The unambiguous velocity range is then

$$\left|V_R\right| \le \frac{\text{PRF} \cdot \lambda}{4} \equiv V_{\max} \qquad (4.31)$$

If the true radial velocity lies outside this range (or Nyquist interval), it is said to be *folded*. It is often possible to correct for folded data by adding or subtracting the correct number of Nyquist intervals ($2V_{max}$) to the detected radial velocity. This correction (called *unfolding*) is possible because the primary factor contributing to the magnitude of V_R is the horizontal wind velocity and (i) there is usually some independent knowledge of the prevailing wind that can be used to indicate the correct Nyquist interval for a given situation, and (ii) since gradients of the wind are continuous, the correct Nyquist interval can usually be assumed to be the one which removes discontinuities in the field of V_R.

The procedure of unfolding can be tedious and undesirable, and it would appear that it could be avoided simply by setting the PRF to a sufficiently high value. However, a high PRF creates another type of folding. The maximum range at which a target can be detected before the next pulse is emitted is

$$r_{max} = \frac{c_o}{2 \cdot \text{PRF}} \tag{4.32}$$

where c_o is the speed of light. The factor 2 appears because the pulse must travel out to r_{max} and back before the next pulse is emitted. Ideally, the PRF should be set to a low enough value that r_{max} exceeds the maximum range at which echoes can be detected by the radar equipment. Otherwise a second pulse will be emitted before the preceding pulse returns to the antenna from the distant target. The radar will then automatically position the echo of a distant target as though it were a reflection from the second pulse. The echo will thus be placed too close to the radar by an amount equal to r_{max}. This type of aliasing is called *range folding*, and the misplaced echoes are called *second-trip echoes*.

It is thus desirable to have the PRF set to as low a value as possible to avoid range folding and to as high a value as possible to avoid velocity folding. Usually a compromise is made, according to which the PRF is set to an intermediate value for which moderate amounts of both types of folding occur. Consequently, an important step in Doppler radar analysis is data editing, in which folded data are eliminated or corrected.[101]

4.4.3 Vertical Incidence Observations

One way to greatly simplify the geometry in (4.30) is to hold the antenna in a vertically pointing position. Then

$$\overline{V_R} = w - V_T \tag{4.33}$$

If the antenna is held in this position, a time series of $(w - V_T)$ as a function of height is obtained. In stratiform precipitation, where $|w| \ll |V_T|$ (see Chapter 6), these data become a time series of precipitation fall speed as a function of height. In convective precipitation, where $w \sim V_T$, the time–height series of $\overline{V_R}$ observed at vertical incidence can be converted to a time–height plot of vertical air velocity

---

[101] Some Doppler radars have the capability to operate at more than one PRF, either continuously or by alternating rapidly between low and high PRF. These capabilities allow some of the ambiguities of range and velocity aliasing to be removed from the radar data.

w, if an independent estimate of the particle fall speed can be obtained to substitute into (4.33).

In the stratiform case, where $|w| \ll |V_T|$, it is also possible to use vertical incidence Doppler data to study the size spectrum of raindrops. In this case, $V_R \approx V_T$. Since V_T is a function only of the drop diameter, the radial velocity spectrum $S(V_R)$ can be converted to a spectrum $S(D)$ of the total returned power accounted for by drops in the size range D to $D + dD$. Since the returned power is proportional to the sum of the sixth power of the diameter of all the particles scattering the radiation, $S(D)$ is proportional to $D^6 N(D)$, and the measured velocity spectrum evidently can be inverted to obtain the drop-size distribution $N(D)$. This procedure is made difficult, however, by the fact that the inversion requires division by D^6. Small errors in the measurements at small drop sizes can therefore be exaggerated greatly by division by a very small number. In addition, turbulent air motions, which do not affect the mean velocity of the air, can broaden the Doppler velocity spectrum in a way that is unrelated to particle size. This inversion method is even more difficult to apply to ice-particle spectra, since the fall velocity of snow depends not only on particle size but also on particle shape and density. In principle, the inversion of the Doppler spectrum could be applied to convective precipitation if the air motion could be provided. However, the turbulence effects on the spectrum become overwhelming.

4.4.4　Range–Height Data

Another way to simplify the geometry in (4.30) is to consider data at a single azimuth angle α_a while the elevation angle α_e is varied by scanning through a range of 0° to about 20°. At these quasi-horizontal angles, the second term in (4.30) is small, and

$$\overline{V_R} \approx V_a \cos \alpha_e \qquad (4.34)$$

where V_a is the magnitude of the horizontal wind velocity component in the azimuthal direction. This method can be used to study quasi–two-dimensional phenomena, such as fronts, hurricanes, or squall lines.[102] If the azimuth angle α_a is directed normal to the echo line, then the velocity component V_a determined from (4.34) is the cross-line component of the airflow. If the airflow has little variation in the along-line direction, then the cross-line derivative of V_a can be computed and integrated vertically to obtain the vertical component of the flow from the two-dimensional form of the anelastic mass continuity equation (2.54).

4.4.5　Velocity–Azimuth Display Method

If the region surrounding a radar is characterized by winds that vary approximately linearly (in the horizontal) across the region, one may obtain a consider-

---

[102] As discussed in Chapter 9, a squall line is a type of mesoscale convective system that exhibits a line of cumulonimbus, which can sometimes be approximated as two-dimensional.

able amount of information about the horizontal wind, its divergence and deformation, and the vertical air motion by scanning the antenna through a full revolu-
tion of azimuth, $\alpha_a = 0$ to $360°$, while holding the elevation angle α_e fixed, that is,
by obtaining measurements on the surface of an inverted cone centered on the
radar and extending up through the region of echo. Echoes with winds varying
linearly across the region might be encountered either in clear-air echo or in a
region of stratiform precipitation.

The data collected on the surface of the inverted cone can be analyzed by a
technique called the *velocity–azimuth display (VAD)* method.[103] This method considers a circle of radius r_c described by the intersection of the cone with a level
surface at a fixed altitude above the radar. The wind components u and v are
assumed to vary linearly in x and y across the region of the circle and to be
constant in time over the observation period. V_T is assumed to be a constant over
the region of the circle. The wind components on the circle are then

$$u = u_o + \left(\partial u/\partial x\right)_o r_c \sin \alpha_a + \left(\partial u/\partial y\right)_o r_c \cos \alpha_a \tag{4.35}$$

$$v = v_o + \left(\partial v/\partial x\right)_o r_c \sin \alpha_a + \left(\partial v/\partial y\right)_o r_c \cos \alpha_a \tag{4.36}$$

where the subscript o indicates evaluation at the center of the circle ($r_c = 0$) of
radar observations at a given height. Equation (4.30) may be rewritten for the
measurements on a circle surrounding the radar, with the aid of trigonometric
identities.[104]

$$\overline{V_R} = a_o + a_1 \sin \alpha_a + b_1 \cos \alpha_a + a_2 \sin 2\alpha_a + b_2 \cos 2\alpha_a \tag{4.37}$$

where

$$a_o = \left[\left(\partial u/\partial x\right)_o + \left(\partial v/\partial y\right)_o\right]\frac{r_c \cos \alpha_e}{2} + \left(w - V_T\right)\sin \alpha_e \tag{4.38}$$

$$a_1 = u_o \cos \alpha_e \tag{4.39}$$

$$b_1 = v_o \cos \alpha_e \tag{4.40}$$

$$a_2 = \left[\left(\partial u/\partial y\right)_o + \left(\partial v/\partial x\right)_o\right]\frac{r_c \cos \alpha_e}{2} \tag{4.41}$$

and

$$b_2 = -\left[\left(\partial u/\partial x\right)_o - \left(\partial v/\partial y\right)_o\right]\frac{r_c \cos \alpha_e}{2} \tag{4.42}$$

The wind field in the neighborhood of a point where the wind varies linearly in
the horizontal can be decomposed into components called translation, divergence,
rotation, stretching deformation, and shearing deformation.[105] If the radar location

---

[103] The basic method was demonstrated by Browning and Wexler (1968). Refined versions of the
technique have been developed by Srivastava *et al.* (1986) and Matejka and Srivastava (1991).

[104] Specifically, $\sin^2 x = (1 - \cos 2x)/2$, $\cos^2 x = (1 + \cos 2x)/2$, and $\sin x \cos x = (\sin 2x)/2$.

[105] See Haltiner and Martin (1957, pp. 292–293) or Saucier (1955, pp. 316–319).

is taken as the central point, all of these components except rotation can be deduced from the coefficients in (4.37). These coefficients have the form of Fourier coefficients. Therefore, the radar measurements of the velocity $\overline{V_R}$ on a circle surrounding the radar provide the left-hand side of (4.37) as a function of α_a, and the coefficients can be determined by standard methods of harmonic decomposition.

The rotational component of the wind cannot be determined from the coefficients because it depends on the wind tangential to a circle surrounding the point, and the radar measurement of $\overline{V_R}$ is made up of velocity components along the beam, which are always perpendicular to the circle surrounding the radar. The radar data therefore contain no information on rotation.

The translational component of the wind field is the horizontal wind vector at the center of the circle (u_o, v_o). It is determined from the coefficients a_1 and b_1, since the azimuth angle is given. By analyzing the data obtained on circles over a range of heights, the vertical profile of the wind velocity is obtained. Since the azimuth at which the horizontal velocity component along the beam of the radar is zero is orthogonal to the vector wind direction, the shape of the zero radial velocity contour in a polar coordinate display of the radial velocity data on a conical surface indicates the sense of the wind shear. As shown in Fig. 4.3, an S-shaped contour indicates veering wind (i.e., a wind whose direction is changing in the clockwise sense—northerly to easterly to southerly to westerly), while a backward S indicates backing wind (direction changing in the counterclockwise sense). This method is used to determine wind profiles in the clear air boundary layer and in stratiform precipitation with radars of centimetric wavelengths. A similar procedure is used to derive winds from UHF and VHF profilers.

The term in brackets in (4.41) is the shearing deformation of the wind field centered at $\mathfrak{r}_c = 0$, while that in (4.42) is the stretching deformation. They are determined from the coefficients a_2 and b_2, respectively. As we will see in Chapter 11, the deformation components of the wind field play an important role in frontogenesis, and these properties of the wind field are therefore particularly useful in analysis of frontal precipitation systems.[106]

The term in brackets in (4.38) is the divergence of the wind field. It is not, however, straightforward to determine from the value of the coefficient a_o, since this coefficient depends on two unknowns, the divergence and $(w - V_T)$. Therefore, it is necessary to measure $\overline{V_R}$ around at least two circles at the same altitude. From the two estimates of a_o, both the divergence and $(w - V_T)$ can be determined. It is better to obtain a_o around several circles and determine the divergence and $(w - V_T)$ that best fit all of the data. Once the divergence is obtained, it can be substituted into the anelastic continuity equation (2.54), and the vertical air velocity w can be obtained by integrating vertically, if a boundary condition (e.g., zero vertical velocity at echo top) can be reasonably assumed.

VAD analysis is one of the best ways to obtain the vertical air motion in stratiform precipitation. It is also one of the best ways to estimate particle fall

---

[106] For an example of this use of radar data, see Carbone *et al.* (1990).

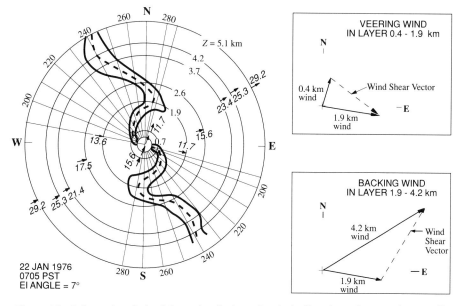

Figure 4.3 Information derived from the display of a single Doppler radar scanning stratiform precipitation at a location in Washington State. The antenna rotated through a full 360° in azimuth while pointing at an elevation angle of 7°, so the concentric circles on the radar screen are found at increasing altitude. Nearly horizontally uniform echo covered the screen. The zero radial velocity contour is dashed. The solid contours either side of the zero contour are for ±0.5 m s⁻¹. Wind directions are obtained from the zero-velocity contour by subtracting 90° from the azimuth on the northwest side or adding 90° to the azimuth on the southeast side. The two direction scales make this adjustment. Upwind and downwind speeds read from the complete velocity display are shown on the height contours. The hodographs are derived from the data on the screen. They show veering winds at lower levels and backing winds above. (Adapted from Baynton *et al.*, 1977. Reproduced with permission from the American Meteorological Society.)

speed V_T. As noted in the discussion of vertical incidence measurements in Sec. 4.4.3, stratiform precipitation is characterized by vertical air motions which are small in magnitude compared to the precipitation fall speed. Therefore, the value of $(w - V_T)$ determined by the above procedures is approximately $-V_T$. This assumption need not be made if the vertical air motion is successfully derived by mass continuity. In that case, the fall speed can be obtained by subtracting $(w - V_T)$ from w.

4.4.6 Multiple-Doppler Synthesis

When radar beams can be directed at the same target volume from two different directions, two measurements of $\overline{V_R}$ can be obtained. Two such measurements can be obtained by two radars separated by some distance and pointing simultaneously at the same point in space. They can also be obtained by a single radar aboard a rapidly moving platform, such as an aircraft, and pointing a beam at the target from first one position of the platform and then a short time later from another position. These methods of obtaining dual-Doppler radar data are probably the most common; however, many other strategies can be imagined. Whatever the strategy used to obtain them, the data derived from the same target by two beams originating from different locations provide two relations of the form (4.30). Each beam corresponds to a different azimuth–elevation pair (α_a, α_e) and provides a different value of $\overline{V_R}$. However, four variables are sought for each point observed by the two beams (u, v, w, and V_T). Some assumption must be made to obtain V_T. Since the reflectivity is measured at each point at which $\overline{V_R}$ is measured, the typical procedure is to invoke a reflectivity fall speed relation of the form (4.13). It remains to determine the vertical air motion w. For this purpose, the anelastic continuity equation (2.54) is employed. Since this equation relates the wind components u, v, and w to each other, it provides the final physical relationship required to close the problem. However, since the continuity equation is a differential equation relating the horizontal derivatives of u and v to the vertical derivative of w, some further problems arise. First, the continuity equation must be integrated vertically, and appropriate boundary conditions must be supplied at the top and/or bottom of the volume containing the radar echo. Since echoes do not always extend to cloud top or to the ground, where w may readily be assumed zero, the choice of a boundary condition is sometimes difficult. Second, a basic-state density profile $\rho_o(z)$ must be provided. Usually, it can be based on nearby sounding data or simply on climatology. The effect of this exponentially varying density weighting in the anelastic continuity equation is that the equation must typically be integrated downward to avoid the rapid accumulation of errors resulting from the large air density at low levels. Third, since the horizontal derivatives of the velocity are required to apply the continuity equation, three-dimensional fields of u and v are required. A commonly followed procedure is to make a first guess of w at each data point to solve (4.30) for an initial estimate of u and v. Then, these values are used in the continuity equation to make a second estimate of w to substitute in (4.30), and so on. Iterations are performed until the fields of u, v, and w converge. When they do, the three-dimensional wind field within the region of radar echo is said to be *synthesized*.

The iteration just described is not always necessary. Two Doppler radars can be located some distance apart and coordinated in their scanning such that they simultaneously scan the same tilted planes intersecting the baseline running between them. The continuity equation can be transformed to a cylindrical coordinate system centered on the baseline. If again some assumption is made to obtain the fall speed, this form of the continuity equation can be solved for the velocity orthogonal to the planes scanned given the observed radial velocities in scanned planes. The velocities obtained can then be transformed geometrically to u, v, and w. This technique is called *coplane scanning*.[107]

Uncertainties in the determination of the wind field by multiple-Doppler radar can be reduced if a third radar beam is directed at all the targets in a given region. The additional information from the third radar can be used in two ways. In

---

[107] Coplane scanning was devised by Lhermitte (1970).

principle, it can be used to provide a third equation of the form (4.30) at each data point and thus eliminate the use of the continuity equation. However, it is desirable to retain the use of the continuity equation so that the inferred winds will satisfy mass continuity. If the continuity equation is retained, the problem can be solved by using statistical (variational) techniques to find the fields of u, v, and w that satisfy mass continuity and give the best fit to all the observed radial velocities. According to this procedure, there is no limit to the number of radar observations (either from additional radars or by the radars increasing their scanning rate) that can be incorporated into the synthesized wind field to reduce its uncertainty.[108] Despite its difficulties, multiple-Doppler synthesis is by far the most important and widely used method for determining the air motions within precipitating cloud systems.[109]

4.4.7   Retrieval of Thermodynamic and Microphysical Variables

Once a field of u, v, and w has been constructed from multiple-Doppler radar data, the equations of motion, the First Law of Thermodynamics, and water-continuity equations can be employed to diagnose (or *retrieve*) the fields of thermodynamic and microphysical variables that are consistent with the observed velocity fields. The basic premise of this retrieval methodology is that by detecting the velocity fields with high resolution in space and time, the radar measures the horizontal and vertical components of the acceleration of the wind. The basic equations can then be solved for the buoyancy and pressure gradient fields that produce the accelerations.

There are various approaches and strategies for carrying out a retrieval. To illustrate the basic idea, we consider a simplified case in which the storm being observed is steady-state in time, two-dimensional in x and z, and the microphysics can be described by the bulk warm-cloud parameterization discussed in Sec. 3.6.1.[110] Molecular friction is ignored. Under these conditions, the horizontal component of the anelastic equation of motion (2.53) can be written in mean-variable form (see Sec. 2.6) as

$$\frac{\partial \overline{\Pi^*}}{\partial x} = A_X \tag{4.43}$$

where

$$A_X \equiv -\frac{1}{c_p \theta_o} \left(\overline{u}\, \frac{\partial \overline{u}}{\partial x} + \overline{w}\, \frac{\partial \overline{u}}{\partial z} - \overline{\mathscr{F}_u} \right) \tag{4.44}$$

---

[108] Further discussion of the variational approach to multiple-Doppler radar synthesis can be found in the paper by Kessinger *et al.* (1987).

[109] For further discussion of multiple-Doppler synthesis techniques, see Ray (1990).

[110] This example is based on a study by Hauser *et al.* (1988). Their paper cites other key papers on the general technique of retrieval. Important among these are Gal-Chen (1978), to whom the concept is usually attributed, Hane *et al.* (1981), and Roux (1985).

and θ_{v_o} in (2.53) has been approximated by θ_o in the denominator of (4.44). Except for new terms defined here, the notation used in the equations of this section is the same as that used in Chapters 2 and 3. As in Sec. 2.3, the asterisk indicates a perturbation from the hydrostatic base state used in the anelastic form of the dynamical equations, while the base state is indicated by the subscript o. Π is the nondimensional pressure, or Exner function, defined by (2.14). The symbol $\mathcal{F}_u$ represents the turbulence term in the x-component of the equation of motion [recall (2.83) and (2.84)].

If accurate Doppler synthesis of the wind field provides the horizontal component of acceleration A_X, then the horizontal gradient of pressure perturbation is determined, according to (4.43). The vertical component of the equation of motion (2.53) can be written

$$\frac{\partial \overline{\Pi}^*}{\partial z} = A_Z + \frac{g\overline{\theta}_a}{c_p\theta_o^2} \tag{4.45}$$

where

$$A_Z \equiv -\frac{1}{c_p\theta_o}\left(\overline{u}\,\frac{\partial \overline{w}}{\partial x} + \overline{w}\,\frac{\partial \overline{w}}{\partial z} - \overline{\mathcal{F}_w}\right) \tag{4.46}$$

and

$$\theta_a \equiv \theta^* + \theta_o\left(0.61q_v^* - q_r - q_c\right) \tag{4.47}$$

The quantity θ_a represents the buoyancy in (4.45); it is the apparent potential temperature perturbation that corresponds to the buoyancy term in (2.53). $\mathcal{F}_w$ is the turbulent mixing term in the vertical component of the equation of motion. The horizontal vorticity equation [similar to (2.61)] obtained by taking $\partial(4.43)/\partial z - \partial(4.45)/\partial x$ provides an expression for the horizontal gradient of $\overline{\theta}_a$,

$$\frac{\partial \overline{\theta}_a}{\partial x} = B_X \tag{4.48}$$

where

$$B_X \equiv \frac{c_p\theta_o^2}{g}\left(\frac{\partial A_X}{\partial z} - \frac{\partial A_Z}{\partial x}\right) \tag{4.49}$$

Thus, Doppler measurements of the acceleration components imply the horizontal gradient of buoyancy according to (4.48) as well as the horizontal gradient of pressure perturbation according to (4.43). The radar observations do not as readily indicate the vertical gradient of Π^* since the buoyancy appears in the vertical equation of motion (4.45).

Since the buoyancy, as expressed in the third term on the right in (2.53), is a function of the thermodynamic properties and water content of the air, appeal is made to the First Law of Thermodynamics and water-continuity equations. Under the assumed conditions, the First Law (2.13) can be written in mean-variable form as

$$\bar{u}\frac{\partial\bar{\theta}}{\partial x} + \bar{w}\frac{\partial\bar{\theta}}{\partial z} = -\frac{L}{c_p\Pi_o}\left(\bar{u}\frac{\partial\bar{q}_v}{\partial x} + \bar{w}\frac{\partial\bar{q}_v}{\partial z} - \mathscr{F}_v\right) + \mathscr{F}_\theta \qquad (4.50)$$

where $\mathscr{F}_v$ and $\mathscr{F}_\theta$ represent the turbulent mixing effects on $\bar{q}_v$ and $\bar{\theta}$. The advection of $\bar{\theta}_a$ [defined by (4.47)] can be written with the aid of (4.50) as

$$\bar{u}\frac{\partial\bar{\theta}_a}{\partial x} + \bar{w}\frac{\partial\bar{\theta}_a}{\partial z} = B_T \qquad (4.51)$$

where

$$B_T \equiv -\frac{L}{c_p\Pi_o}\left(\bar{u}\frac{\partial\bar{q}_v}{\partial x} + \bar{w}\frac{\partial\bar{q}_v}{\partial z} - \mathscr{F}_v\right) + \mathscr{F}_\theta - \bar{w}\frac{\partial\theta_o}{\partial z}\left(1 - 0.61\overline{q_v^*} + \bar{q}_r + \bar{q}_c\right)$$

$$+ \theta_o\left(\bar{u}\frac{\partial}{\partial x} + \bar{w}\frac{\partial}{\partial z}\right)\left(0.61\overline{q_v^*} - \bar{q}_r - \bar{q}_c\right) \qquad (4.52)$$

According to the bulk warm-cloud water-continuity model (Sec. 3.6.1), the rain-water content of the air is governed by (3.67), which under the present assumptions may be written as

$$\bar{u}\frac{\partial\bar{q}_r}{\partial x} + \bar{w}\frac{\partial\bar{q}_r}{\partial z} - \frac{\partial}{\partial z}\left(\hat{V}q_r\right) - \mathscr{F}_r = A_c + K_c - E_r \qquad (4.53)$$

where $\mathscr{F}_r$ is the turbulent mixing effect on $\bar{q}_r$. Addition of (3.65)–(3.67) implies that continuity of total water substance is governed by the continuity equation

$$\bar{u}\frac{\partial\bar{q}_T}{\partial x} + \bar{w}\frac{\partial\bar{q}_T}{\partial z} - \frac{\partial}{\partial z}\left(\hat{V}\bar{q}_r\right) - \mathscr{F}_T = 0 \qquad (4.54)$$

where $\mathscr{F}_T$ is the effect of turbulent mixing on the total water mixing ratio, defined as

$$q_T \equiv q_c + q_r + q_v \qquad (4.55)$$

Remembering that, in addition to (4.55), $A_c + K_c - E_r = f(\bar{q}_r,\bar{q}_c,\bar{q}_v)$, $\hat{V} = f(\bar{q}_r)$, $q_v^* = \bar{q}_v - q_{v_o}$, and $\theta^* = \bar{\theta} - \theta_o$, and that A_X and A_Z are measured by radar, we see that (4.43), (4.45), (4.48), (4.51), (4.53), and (4.54) comprise a set of equations in $\bar{\Pi}^*$, $\bar{\theta}$, $\bar{q}_r$, $\bar{q}_c$, and $\bar{q}_v$, provided that the turbulent mixing terms $\mathscr{F}_u$, $\mathscr{F}_w$, $\mathscr{F}_\theta$, $\mathscr{F}_r$, and $\mathscr{F}_T$ can be parameterized appropriately. There is not really an extra equation since the information in (4.48) is redundant with that in (4.43) and (4.45). The relation is nevertheless valuable because in the solution of the set of equations it is helpful to have the horizontal gradient of θ_a expressed directly in terms of the observed quantities A_X and A_Z.

This set of relationships among the thermodynamic and microphysical variables is a very awkward set of differential equations. One way it has been solved is by iteratively taking first-guess values of $\bar{q}_r$, $\bar{q}_c$, and $\bar{q}_v$ as given; determining $\overline{\Pi^*}$ and $\bar{\theta}$ from (4.43), (4.45), (4.48), and (4.51); and then taking $\overline{\Pi^*}$ and $\bar{\theta}$ as given and determining $\bar{q}_r$, $\bar{q}_c$, and $\bar{q}_v$ from (4.53), (4.54), and the assumption that

$$
\begin{cases}
\bar{q}_c = \bar{q}_T - \bar{q}_r - \bar{q}_{vs} \ \text{ and } \ \bar{q}_v = \bar{q}_{vs}, \ \text{ if } \bar{q}_T - \bar{q}_r - \bar{q}_{vs} > 0 \\
\bar{q}_c = 0 \ \text{ and } \ \bar{q}_v = \bar{q}_T - \bar{q}_r \ \text{ if } \bar{q}_T - \bar{q}_r - \bar{q}_{vs} < 0
\end{cases}
\tag{4.56}
$$

where q_{vs} is the saturation mixing ratio. These alternate solutions are repeated until the five variables stabilize in value.

The values of $\overline{\Pi^*}$ and $\bar{\theta}$ that are most consistent with the wind observations are deduced by retrieving the fields of $\bar{\theta}_a$ and $\overline{\Pi^*}$ that best fit the observational information contained in the terms A_X, A_Z, B_X, and B_T. Since the water fields are regarded as given for this step of the procedure, θ is readily determined from $\bar{\theta}_a$, according to (4.47). Note also that $\bar{\theta}_a$ and $\overline{\Pi^*}$ could be determined even if nothing were known about the water fields, and this more limited thermodynamic retrieval is often the only part of the retrieval that is possible to carry out. However, in this partial retrieval, $\bar{\theta}_a$ cannot be decomposed to determine $\bar{\theta}$ or $\bar{\theta}_v$.

To retrieve the buoyancy ($\bar{\theta}_a$) and pressure perturbation ($\overline{\Pi^*}$) fields, the first step is to seek a fit of the $\bar{\theta}_a$ field to the data, in the least-squares sense, by minimizing the integral

$$
I_\theta = \iint_{\mathcal{D}} \left\{ \left(\frac{\partial \bar{\theta}_a}{\partial x} - B_X \right)^2 + \left[\frac{w}{\mathcal{U}} \frac{\partial \bar{\theta}_a}{\partial z} - \frac{1}{\mathcal{U}} (B_T - \bar{u} B_X) \right]^2 \right\} dx \, dz
\tag{4.57}
$$

where $\mathcal{D}$ refers to the domain of the radar observations. As a horizontal boundary condition, it is assumed that the first squared term in the integral is zero on the horizontal borders of the radar-echo region. Minimization of the first term in the integral assures that the horizontal gradient of the retrieved $\bar{\theta}_a$ field is a least-squares fit to the observed B_X data [recall (4.48)] and is thus simultaneously consistent with the horizontal and vertical equations of motion at each level. Minimization of the second term assures the field of $\bar{\theta}_a$ is vertically consistent by requiring that the vertical gradient of $\bar{\theta}_a$ be as consistent as possible with the wind observations and retrieved microphysical variables contained in B_X and B_T [recall (4.51)] and thus simultaneously with the equations of motion and thermodynamic equations. The velocity $\mathcal{U}$ is a user-chosen weighting function that assures homogeneity of the dimensions in the integral and decides the relative weight to be ascribed to the vertical and horizontal gradients in fitting $\bar{\theta}_a$ to the observations.

Once the buoyancy field ($\bar{\theta}_a$) is determined, a fit of the pressure perturbation ($\overline{\Pi^*}$) field to the data can be obtained by minimizing the integral

$$
I_p = \iint_{\mathcal{D}} \left\{ \left(\frac{\partial \overline{\Pi^*}}{\partial x} - A_X \right)^2 + \left[\frac{\partial \overline{\Pi^*}}{\partial z} - \left(A_Z + \frac{g \bar{\theta}_a}{c_p \theta_o^2} \right) \right]^2 \right\} dx \, dz
\tag{4.58}
$$

Horizontal boundary conditions are applied in the same manner as for (4.57). Minimization of the first term in (4.58) assures that the horizontal gradient of the retrieved $\overline{\Pi^*}$ field best fits the observed A_X values [recall (4.43)] and is thus consistent with the horizontal equation of motion at each level. Minimization of the second term assures the fields are vertically consistent by requiring that the vertical gradient of $\overline{\Pi^*}$ be as consistent as possible with the wind observations contained in A_Z [recall (4.45)] and thus with the vertical as well as the horizontal equations of motion, as well as with the retrieved thermodynamic field.

The overall accuracy of thermodynamic and microphysical fields derived by the retrieval technique is difficult to determine. The confirmatory data are hard to obtain by aircraft or other direct means. When the retrieval technique is applied to artificial fields of wind and reflectivity produced by numerical cloud models,[111] the retrieved fields compare well with the model fields for steady-state, horizontally uniform, and non-turbulent regions of convective systems. When the precipitation is varying rapidly in time, the technique described above must be extended to include time derivatives, which are very difficult to measure. In the future, more rapidly scanning radars may provide the time resolution necessary to estimate these terms.

---

[111] For examples of applying the retrieval technique to model output, see Hane *et al.* (1981) and Sun and Houze (1992).

Part II | Phenomena

Chapter 5 | Shallow-Layer Clouds

"The cold mist hangs like a stretch'd canopy"[112]

In the remainder of this book, we will examine the dynamics of specific types of clouds. We begin in this chapter with clouds confined to shallow layers of air in which the rate of cooling resulting in cloud formation is rather slight. These clouds include fog, stratus, stratocumulus, altostratus, altocumulus, cirrus, cirrostratus, and cirrocumulus. The only type of layer cloud we do not consider in this chapter is nimbostratus, which is very deep and produces substantial rain and snow. Chapter 6 is devoted to the subject of nimbostratus. The clouds we consider in this chapter are generally nonprecipitating, though they may occasionally produce drizzle. Although there are certain important exceptions, water contents in shallow layer clouds are mostly <1 g kg^{-1}, while mean vertical motions are generally $\sim$1–10 cm s^{-1}. These values are very small compared to the convective clouds we will consider in Chapters 7–9. The cloud layers are generally $\sim$1 km or less in vertical extent, although altostratus may sometimes be several kilometers in depth. Because the mean vertical air motions in shallow layer clouds are small, the condensation cannot be accounted for by upward air motion alone. Other physical mechanisms are also important, especially radiation and eddy mixing. In this chapter, we will see how the vertical air motion, radiation, and turbulent mixing all interact to give shallow layer clouds their particular character.

Shallow-layer clouds can occur quite locally; however, they can also be very widespread, covering mesoscale or even synoptic-scale regions. Fog, for example, can cover an area of the size of the entire central valley of California (Fig. 1.7b). Stratus occurs in huge persistent sheets covering the Arctic Ocean. Stratocumulus often covers the wide subtropical oceanic regions dominated by the circulations of the semipermanent subtropical anticyclones off the west coasts of continents (Fig. 1.11a). It also typically covers large midlatitude oceanic regions off the east coasts of continents, when cold air flows out over the ocean behind strong fronts (Fig. 1.11b). Layers of shallow midlevel cloud (altostratus and altocumulus) can also be quite widespread; they are among the most ubiquitous of cloud types, providing vivid optical effects (e.g., the corona, mock sun, sun dog) as well as frequent illustrations of atmospheric fluid dynamical behavior (rolls and cellular structures of various types). Upper-level layer clouds also occur in large sheets covering great expanses of the earth, as can be seen by viewing routine

---

[112] Goethe's words to describe stratus.

satellite pictures, which show wide expanses of high-cloud tops in both the tropics and midlatitudes (e.g., frontispiece and Figs. 1.27–1.30). The propensity of layer clouds at low, middle, and upper levels to cover great areas has a major impact on the climate of the earth through the absorptive and scattering effects of these cloud layers in the earth's radiation balance.

In examining the dynamics of layer clouds, we will begin in Sec. 5.1 with the case of fog and stratus occurring under highly stable conditions, governed by the mechanics of a stable boundary layer cooled from below (Sec. 2.11.2). We will see that even when air is generally calm and stable, mixing by eddy motions is quite important to the development and maintenance of fog and stratus. We will further see that, as fog thickens, differential radiative heating can make the upper layer of the fog less stable. Vertical mixing is thereby enhanced, and the cooling is shifted aloft to deepen the fog layer or change the cloud from fog (in contact with the ground) to elevated stratus. Once a layer of stratus has formed it produces greenhouse heating effects below its base. As examples of this type of fog and stratus formation, we will examine briefly the life cycle of a nocturnal radiation fog over a soil surface and the formation of widespread and persistent Arctic stratus, which occurs as warm air from the south moves over the melting pack ice of the Arctic Basin during summer.

After considering fog and stratus under essentially stable conditions, we proceed in Sec. 5.2 to the case of stratus and stratocumulus over a warm ocean, where the boundary layer is so strongly heated from below that it is unstable and convective and thus tends to be well mixed, governed by the dynamics of an unstable boundary layer. Wind shear can provide a second source of mixing in these more active boundary layers. In contrast to the stable case, in which the cloud first appears as fog at the bottom of the boundary layer, the cloud stratum appears at the top of the boundary layer when the strong convective mixing begins to have upward plumes reaching above the lifting condensation level. The convective mixed layer continues to deepen as the plumes of convection push farther upward and environmental air from above the boundary layer continues to be mixed downward across the top of the mixed layer. The cloud layer evolves into a more or less continuous layer of cloud, which no longer depends so much on the surface heating as on the differential radiative heating between the top and bottom of the cloud layer to drive its turbulence. Mesoscale organization of the active eddy motions resulting from various combinations of differential heating and wind shear in the boundary layer can give the mixed-layer cloud sheet a spatial texture. The cloud then takes on a discrete stratocumulus structure, in which the cloud elements consist sometimes of rolls and sometimes of cells. When the mixing is especially vigorous the elements of the cloud layer are more of the form of small cumulus than stratocumulus. Sometimes they even achieve the size of small cumulonimbus.

After considering the physics and dynamics of boundary-layer fog and stratus and cloud-topped mixed layers, we will proceed to Secs. 5.3 and 5.4, where we will examine shallow-layer clouds that occur well above the boundary layer (i.e., altostratus, altocumulus, cirrus, cirrostratus, and cirrocumulus). These cloud forms are governed by essentially the same physical processes as stratus and stratocumulus, except that the layers in which the clouds occur are not bounded below by an interface at which strong forcing (either heating or cooling) occurs. Eddy mixing and differential radiative heating and cooling again are important, and organized patterns of cells and rolls again appear in these cloud types in evidence of mesoscale organization of the convection confined to the cloud layers aloft. We will consider the high-altitude cirrus clouds in Sec. 5.3 and then finally the middle-level altocumulus and altostratus in Sec. 5.4. It is convenient to proceed in this order (considering the high clouds before the midlevel clouds) since more is known about cirrus than of middle clouds and since our approach to examining the middle clouds will be in part to consider their similarities to and differences from cirrus.

5.1 Fog and Stratus in a Boundary Layer Cooled from Below

5.1.1 General Considerations

The cause of fog under stable conditions is the cooling of moist air near the earth's surface, by either radiation or conduction.[113] When the air is cooled sufficiently, small drops or ice crystals form, constituting the fog. This apparently simple mechanism is made highly complex both by particle microphysics and by the macroscale processes that spatially redistribute the cooling and the drops in the air. As already noted, turbulent air motions are important in this respect. Also critical are the thermodynamic characteristics of the underlying surface, the sedimentation of the drops or crystals, and the spectrum of nuclei on which the particles form.

In this section, we examine the dynamics of fog and stratus formation over a surface that is substantially colder than the overlying air. The governing dynamics of the air are thus those of the stable boundary layer, in which turbulence is highly suppressed. Probably the most stable situation is that of a radiation fog, in which not only is the lapse rate very stable but the mean wind is close to zero, or nearly so. Thus, the sources of turbulent kinetic energy are nearly zero. However, even in this situation, turbulence is not wholly absent. In fact, enough remains to be quite important in the development of fog.

To obtain a feeling for the importance of the turbulent mixing in stable boundary-layer fog, we will examine briefly in Sec. 5.1.2 the problem of artificial modification of fog, which is concerned with artificially changing the microphysical characteristics of the fog to obtain a temporarily local increase in visibility. The role of turbulence is so profound that it acts to refill gaps in the fog within a few

---

[113] "Steam fog" ("arctic sea-smoke") is not produced by radiation or conduction but rather occurs when parcels of air of vastly different temperatures mix and attain a state of supersaturation. This phenomenon, however, occurs only in an *unstable* boundary layer, where cold air moves over a surface of warm water.

minutes to an hour of when they are created. After this brief look at the role of turbulence, we will consider the problem of forecasting the formation, persistence, and dissipation of fog. The forecasting problem is concerned with the grosser aspects of the fog, such as its depth and total water content, on time scales of an hour to days. In Sec. 5.1.3 we consider the forecasting of fog formation over underlying soil undergoing nocturnal radiative cooling and its dissipation as the sun rises and solar insolation increases. Then in Sec. 5.1.4 we examine arctic fog and stratus formation, which illustrates what happens if the diurnal variations are removed and the fog layer evolves into steady-state fog or stratus under the conditions of a constant lower-boundary temperature and constant weak solar heating.

5.1.2   Turbulent Mixing in Fog

The strong role of turbulence in fog is illustrated by the problem of attempting to modify the local visibility in a warm fog consisting entirely of small nonsupercooled drops.[114] In this problem, the thermodynamics and dynamics are greatly simplified. The thermodynamic equation is completely ignored and replaced by the simple assumption that the mean temperature is constant over the time scale of interest (~1/2 h). The effect of turbulence is represented by a constant mixing coefficient ($K = 4$ m^2 s^{-1}). Except for the turbulence, the air is assumed to be almost calm. Thus, the equation of motion is replaced by assuming all the mean wind components are constant and so small that advection terms are negligible in all the equations. With these assumptions, the only predictive equations are the water-continuity equations, which must be written in a form that allows the effects of artificial seeding to be represented.

The water-continuity equations are formulated in very explicit form, taking into account both nucleus size and drop size [as in Eqs. (3.58) and (3.59)]. We consider an example in which a hypothetical pre-existing fog occurs in a 100-m-deep, 300-m-wide region. The pre-existing natural fog drops are assumed to contain condensation nuclei small enough that their composition is irrelevant. The liquid water content of the initial fog is 0.3 g m^{-3}. The temperature is assumed to be 10°C. The assumed size spectrum is such that the horizontal visibility is initially 85 m. It is assumed that over a volume 50 m wide by 40 m deep an aircraft distributes 0.35 g m^{-3} of droplets containing very large NaCl nuclei. According to the diffusional

---

[114] Here, we take as an example the study of Silverman (1970), who was looking for a technique for artificially clearing fogs from airport runways. In a paper presented at a conference on weather modification, he reported on efforts to find a theoretical basis for such a technology. A somewhat similar effort was reported by Tag *et al.* (1970) at the same conference. This type of research was abandoned shortly thereafter and never formally published. Perhaps the explanation for the abrupt termination of work on warm-fog seeding is the preliminary result that the technique would not have been effective since a hole created in a warm fog by artificial seeding would be filled up as a result of the natural turbulence almost as fast as it could be created. The termination of the research would thus appear to be a further effect of the turbulence. It should be noted that this discouraging result applies only to warm fogs. As will be pointed out later in this section, seeding of cold (i.e., supercooled) fogs turns out to be more effective because it is aided by the glaciation process.

growth equation (3.23), these seed particles, because of size and solution effects, grow at a lower relative humidity than do the smaller pure water drops comprising the fog. The latter require supersaturation for growth. As the solution drops grow, they deplete the ambient humidity, reducing it to subsaturation. They continue to grow and fall out, while the small fog drops in their vicinity evaporate because of the lowered humidity. Thus, a hole is produced in the fog.

The mean drop size distribution N_{ij} at a point in the domain of the fog evolves according to

$$\frac{\partial \bar{N}_{ij}}{\partial t} = \mathfrak{D}_{ij} + \tilde{\mathfrak{F}}_{ij} + K\nabla^2 \bar{N}_{ij} \tag{5.1}$$

which is the mean-variable form of (3.58) under the assumed conditions, with i representing drop size and j nucleus size. The effect of turbulent mixing, which must appear in the mean-variable equation (recall Sec. 2.6) is expressed here in terms of K-theory by $K\nabla^2 \bar{N}_{ij}$. For the small time and space scales in this problem the horizontal eddy mixing is just as important as vertical mixing, and the mixing coefficient K is applied equally to eddy mixing in both the horizontal and vertical. The other terms in (5.1) represent the effects of vapor diffusion $\mathfrak{D}_{ij}$ and fallout $\tilde{\mathfrak{F}}_{ij}$ on the drop size distribution. The vapor diffusion term $\mathfrak{D}_{ij}$ is given by (3.60) with $\dot{m}_{\mathrm{dif}}$ given by (3.23). To obtain (5.1), some of the terms in the original version of (3.58) are taken to be zero in accordance with the conditions in which the warm fog seeding is imagined to take place: Because of the shallowness of the layer of fog, coalescence $\mathfrak{C}_{ij}$ is ineffective and, owing to the smallness of fog drops, breakup $\mathfrak{B}_{ij}$ is unimportant (Fig. 3.6). Nucleation of new drops $\mathfrak{N}_{ij}$ is set to zero since the introduction of the seeding material maintains the relative humidity at or below 100%. The divergence on the right-hand side of (3.58) and the advection terms on the left disappear because the mean wind is zero. The water–vapor mixing ratio q_v is calculated according to (3.59).

The terms on the right-hand side of (5.1) then represent the physical essence of the problem, which is an interplay of vapor diffusion $\mathfrak{D}_{ij}$, fallout $\tilde{\mathfrak{F}}_{ij}$, and turbulent mixing $K\nabla^2 \bar{N}_{ij}$. The large solution drops grow by diffusion at the expense of the smaller natural fog drops and fallout, thereby increasing the visibility through the fog by decreasing the liquid water content and the number concentration of small drops. This idea would lead to the successful clearing of fog if it were not for the turbulent mixing represented by $K\nabla^2 \bar{N}_{ij}$, which acts quickly to fill the hole in the fog created by the seeding.

Calculations illustrate the strong interplay of the vapor diffusion, fallout, and turbulence (Fig. 5.1). One minute after the seeding, a maximum of liquid water content aloft contains the seeding material and the natural fog droplets. As a result of fallout, this maximum migrates quickly downward, reaching the surface by 4 min after seeding, leaving a region of low liquid water content and high visibility aloft. There the small, natural fog drops evaporate and their moisture deposits onto the large solution drops, which fall out because of their large size. The hole is most clear at 8–12 min after the seeding. By 18 min the turbulence has already reduced the visibility again and by 30 min is not much better than it was initially,

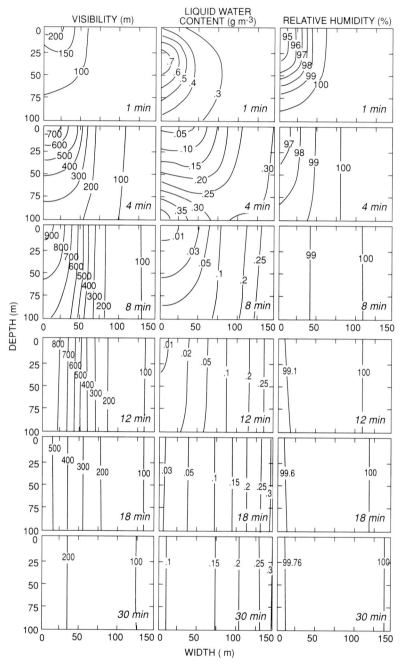

Figure 5.1 Calculated visibility, liquid water content, and relative humidity in a warm fog at a sequence of times after it is seeded. Initial values of these parameters were 85 m, 0.3 g m^{-3}, and 100%, respectively. (From Silverman, 1970. Reprinted with permission from the American Meteorological Society.)

and the liquid water content aloft is back up to 0.1 g m^{-3}. These calculations thus both illustrate the important influence of turbulence in a fog and indicate why warm-fog seeding is not very effective.

Seeding of supercooled or "cold" fogs is found to be more effective. The idea is to glaciate the fog by introducing dry ice or some heterogeneous freezing nuclei to convert some of the drops to ice particles, which grow rapidly at the expense of the drops because the ice saturation vapor pressure is lower than that of liquid water.[115] The lowered concentration of small drops again means improved visibility. In this case though, once a region of the fog becomes glaciated, it remains clear for a longer time, since any liquid drops mixed into the glaciated region either quickly freeze through contact with the ice particles or evaporate as the ice particles, growing by vapor diffusion, lower the vapor content of the air to a state of subsaturation with respect to liquid water. Eventually, though, the ice particles, like the large solution drops considered above, settle out, and the hole can then fill back in by turbulent mixing of small drops into the clear region, just as in the warm-fog case.

5.1.3   Radiation Fog

The time scale involved in the fog modification problem discussed in the preceding section is ~1/2 h. Another important problem is the forecasting of fog on a time scale of 12–24 h. As in other types of weather forecasting, a numerical prediction of fog is based on an appropriate form of the fluid dynamical equations, in which the conservation equations for momentum (Newton's second law of motion) and heat (First Law of Thermodynamics) play the central role. We will consider here the forecasting of radiation fog, which is the simplest form of fog. In this case, the momentum conservation is simplified considerably by the fact that the wind is, to a first approximation, calm. Advection fogs involve stronger winds and thereby the effects of advection as well as radiation on the temperature of the air. However, the wind is never completely calm, even in radiation fog, and the slight turbulence present is very important.

The physical parameters of the problem are illustrated schematically in Fig. 5.2. They include the radiative cooling of the earth's surface and the eddy flux of sensible heat $\mathscr{F}_\theta$ [defined by the left-hand equality in (2.188)], which is predominantly downward owing to the strong surface cooling. Since the boundary layer is very stable as a result of the cooling, the buoyancy generation term $\mathscr{B}$ in the turbulent kinetic energy (2.86) acts to suppress turbulence. The only way that turbulent kinetic energy can be generated is through term $\mathscr{C}$, which represents the conversion of mean-flow kinetic energy to eddy kinetic energy by means of the eddy momentum fluxes. Therefore, the mean flow, represented by the wind profile

---

[115] The same process apparently produces holes in clouds when they are seeded naturally by a streamer of ice particles falling from a higher cloud or by aircraft-produced ice particles. For an amusing discussion of these phenomena, see "Holes in clouds: A case of scientific amnesia" by Hobbs (1985).

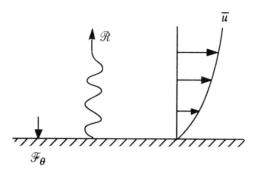

Figure 5.2 Physical parameters of the physics of radiation fog: the net upward radiative heat flux $\mathcal{R}$, which is dominated by infrared and is strongly upward at the ground; the eddy flux of sensible heat $\mathcal{F}_\theta$, which is predominantly downward owing to the strong surface cooling; and the profile of mean wind $\bar{u}$.

in Fig. 5.2, must be specified or predicted in the fog forecasting problem to determine the amount of turbulent mixing that occurs.

The magnitude of the wind at and just above the top of the stable boundary layer may be quite small (e.g., 2–3 m s^{-1}). Nevertheless, it is strong enough to generate turbulent mixing of a magnitude sufficient to be of importance to fog formation, evolution, and dissipation. The form of the momentum equation appropriate for predicting the wind in the planetary boundary layer in a situation of radiation fog is based on the mean-variable equation of motion (2.83). It is assumed that the mean vertical motion is zero. Hence, only the horizontal component of (2.83) is predictive. We further assume that the flow is Boussinesq [i.e., the density term ρ_o disappears from (2.84)], that the mean horizontal wind is so weak that horizontal advection is negligible, and that the horizontal pressure gradient is constant throughout the boundary layer. Then, the horizontal component of the equation of motion (2.83) becomes

$$\frac{\partial \bar{\mathbf{v}}_H}{\partial t} = -f\mathbf{k} \times \left(\bar{\mathbf{v}}_H - \mathbf{v}_g \right) + \vec{\mathcal{F}}_H \tag{5.2}$$

where the horizontal pressure gradient has been expressed in terms of its corresponding geostrophic wind $\mathbf{v}_g$ [recall (2.22)], and the subscript H indicates a horizontal component of a vector. As indicated in Sec. 2.10, $\vec{\mathcal{F}}_H$ can be formulated with varying degrees of sophistication. Here we use the K-theory formulation, which serves to illustrate the physics of fog formation with a minimum of mathematical complexity, and we assume that on the time scales of interest to forecasting ($\sim$10 h), the vertical mixing dominates over horizontal mixing. Then, according to (2.84), (2.186), and (2.187), the frictional stresses are given by

$$\vec{\mathcal{F}}_H = -\frac{\partial}{\partial z}\left(-K_m \frac{\partial \bar{\mathbf{v}}_H}{\partial z} \right) \tag{5.3}$$

The mixing coefficient K_m is inversely related to the Richardson number, Ri [defined in (2.170)]. Radiative cooling near the ground produces a temperature inversion and correspondingly high values of Ri, and mixing is confined to a shallow surface layer. After a fog layer becomes established in an inversion layer, radiation drives the lapse rate back toward a less stable configuration, Ri decreases, and the mixing coefficient is increased to allow more mixing.

While radiative cooling of the ground ultimately drives the radiation fog's formation, it is the delicate *im*balance of the radiation, turbulent mixing of the heat, and phase changes of water that determine the thermodynamic history of the fog, including its formation, evolution, and dissipation. It is important, therefore, to have an accurate and appropriate form of the thermodynamic equation to predict the radiation fog. The essential processes are incorporated in the mean-variable form of the First Law of Thermodynamics (2.78). If we assume for simplicity that the fog consists entirely of liquid drops (i.e., no ice), then with all the assumptions we have made about the air motions, (2.78) may be written as

$$\frac{\partial \bar{\theta}}{\partial t} = \frac{LC}{c_p \bar{\Pi}} - \frac{\partial \mathcal{R}}{\partial z} - \frac{1}{\rho_o c_p} \frac{\partial \mathcal{F}_\theta}{\partial z} \tag{5.4}$$

where the latent heating term has been written as in (2.13), with C standing for the net condensation (in units of kg of water per kg of air per second), and $\mathcal{R}$ is the net radiative heat flux (positive upward). The density term in (2.78) does not appear because we are using the Boussinesq version of the relation. Since we are expressing all of the turbulent fluxes by K-theory, $\mathcal{F}_\theta$ is given by the right-hand side of (2.188). Thus, the eddy flux convergence in (5.4) varies with the mixing coefficient K_θ according to

$$\frac{\partial \mathcal{F}_\theta}{\partial z} = c_p \frac{\partial}{\partial z} \left(-K_\theta \frac{\partial \bar{\theta}}{\partial z} \right) \tag{5.5}$$

In order to calculate the heat sources associated with condensation and radiation correctly, one must keep track of the amounts of water in the form of vapor and liquid. Hence, an appropriate set of the mean-variable water-continuity equations (2.81) is required. Under the present assumptions, this set may be written as

$$\frac{\partial \bar{q}_v}{\partial t} = -C - \frac{\partial}{\partial z} \left(-K_v \frac{\partial \bar{q}_v}{\partial z} \right) \tag{5.6}$$

$$\frac{\partial \bar{q}_i}{\partial t} = S_i - \frac{\partial}{\partial z} \left(-K_i \frac{\partial \bar{q}_i}{\partial z} \right), \quad i = 1, \dots, k \tag{5.7}$$

where k is the total number of size categories of water drops considered. The eddy flux of water vapor in (5.6) is expressed by K-theory according to (2.189), and the eddy flux of hydrometeors is expressed by an analogous term in (5.7). S_i represents the microphysical sources and sinks of hydrometeors.

To illustrate the forecasting of the timing of the formation, evolution, and dissipation of radiation fog on a time scale ~ 10 h, we do not try to account for the distribution of water among different sizes and types of particles. We will be satisfied with a knowledge of the bulk liquid water mixing ratio on a time scale ~ 10 h. For these purposes, the equations for predicting the hydrometeor mixing ratios (5.7) can be simplified to the point of considering only one category, namely the total liquid water, represented by mixing ratio q_L. In this case,

$$\bar{q}_i \equiv \bar{q}_L \tag{5.8}$$

The major assumption required to make this simplification is that the rate of fallout, which depends on drop size, can be parameterized. This is accomplished by assuming that the convergence of fallout of water can be expressed as

$$F = \frac{\partial}{\partial z}\left(\hat{V}\bar{q}_L\right) \tag{5.9}$$

where $\hat{V}$ is the mass-weighted average particle fall speed, as defined in (3.69). The expression in (5.9) is like that in (3.75), except that the falling drops are not considered to be large raindrops, but rather slowly settling cloud or drizzle (Sec. 3.1.3) particles. The size spectrum $N(D)$ of such drops can be measured and substituted into (3.69) to obtain an empirical value of $\hat{V}$. Similarly the measured $N(D)$ may be substituted into an integral like that on the right-hand side of (4.9) to obtain a value of q_L. Repeated measurements of $N(D)$ can be used to obtain a correlation between $\hat{V}$ and q_L. A commonly used formula obtained in this way[116] is

$$\hat{V} = \tilde{a}_3 q_L \tag{5.10}$$

where $\tilde{a}_3 = 6.25$, q_L is in g kg^{-1}, and $\hat{V}$ is in cm s^{-1}.

Since we have only one category of condensed water q_L, the hydrometeor equation (5.7) reduces to a single equation in which the microphysical source terms are C and F. Thus, with substitution from (5.9) and (5.10), (5.7) becomes

$$\frac{\partial \bar{q}_L}{\partial t} = C + \frac{\partial}{\partial z}\left(\tilde{a}_3 \bar{q}_L^2\right) - \frac{\partial}{\partial z}\left(K_L \frac{\partial \bar{q}_L}{\partial z}\right) \tag{5.11}$$

where K_L is the mixing coefficient for liquid water.

Calculating the behavior of fog on the 10-h time scale is critically dependent on accurate calculation of the surface temperature and humidity. We must have

$$\bar{T} = \mathcal{T}_s \quad \text{and} \quad \bar{q}_v = \mathfrak{q}_s \quad \text{at } z = 0 \tag{5.12}$$

where $\mathcal{T}_s$ and $\mathfrak{q}_s$ are the temperature and vapor-mixing ratio of the soil at the earth–air interface, respectively. A traditional but somewhat oversimplified way to treat an underlying soil is by invoking the simple diffusion equation

$$\rho_s c_s \frac{\partial \mathcal{T}_s}{\partial t} = -\frac{\partial}{\partial z}\left(\kappa_s \frac{\partial \mathcal{T}_s}{\partial z}\right) \tag{5.13}$$

where ρ_s, c_s, and κ_s are the density, specific heat at constant pressure, and thermal conductivity of the soil, respectively. At the interface, the following

[116] Suggested by Brown and Roach (1976).

balance is assumed:

$$0 = \rho c_p \overline{\Pi} \mathcal{R} - \rho c_p K_\theta \frac{\partial \overline{\theta}}{\partial z} + \kappa_s \frac{\partial \mathcal{T}_s}{\partial z} - \rho L K_v \frac{\partial \overline{q}_v}{\partial z} \tag{5.14}$$

The humidity at the soil surface is estimated according to a parameterization such as the following:[117]

$$\overline{q}_v = M_{ev}\,\overline{q}_{vs} + \left(1 - M_{ev}\right)\overline{q}_v \Big|_{(z\, =\, \varepsilon)} \tag{5.15}$$

where q_{vs} is the saturation mixing ratio and M_{ev}, the "efficiency factor for evaporation," is a specified parameter, which ranges from 0 for a dried-out soil to 1 for conditions of saturated soil or dew, and ε is some small distance (e.g., 1 cm).

The physical balance assumed by (5.14) is that the diffusion of sensible heat through the soil exactly balances the net radiative flux plus turbulent eddy fluxes of sensible and latent heat through the air. The flux of moisture within the soil is ignored. To include this flux, a more explicit treatment of the soil moisture and thermodynamics is required. Such a treatment can be implemented, but it is very complex. It gives results that differ moderately in magnitude from the simpler model. In the more explicit case, fluxes of both vapor and liquid water within the soil are calculated along with the thermal diffusion. A soil "system" is envisioned as consisting of four elements: a "soil matrix," dry air, water vapor, and liquid water.[118] The "porosity" of the soil is assumed, that is, the fraction of a given volume of the soil system occupied by the soil matrix. The First Law of Thermodynamics applied to the soil system, together with expressions of mass continuity for each of the four elements of the soil system, leads to coupled prognostic equations for the temperature and liquid water content of the soil. These equations are complex and not sufficiently instructive to repeat here. The interface conditions in the case of the explicit soil system are

$$0 = \rho c_p \overline{\Pi} \mathcal{R} - \rho c_p K_\theta \frac{\partial \overline{\theta}}{\partial z} + \kappa_s \frac{\partial \mathcal{T}_s}{\partial z} + \left(-\rho K_v \frac{\partial \overline{q}_v}{\partial z} - J_v\right)\left(L_v - \psi_s\right) \tag{5.16}$$

and

$$0 = -J_v - J_w - \rho K_v \frac{\partial \overline{q}_v}{\partial z} \tag{5.17}$$

where J_v and J_w are the fluxes of water vapor and liquid water in the soil, respectively, and ψ_s is the "soil moisture potential."[119] Equation (5.16) states that the diffusion of sensible heat *and* latent heat through the soil exactly balances the net radiative flux plus turbulent eddy fluxes of sensible and latent heat through the air.

---

[117] See Atwater (1972) as cited in Sievers *et al.* (1983).

[118] See Sievers *et al.* (1983), Forkel *et al.* (1984), and Welch *et al.* (1986) for detailed discussions of this problem.

[119] The soil moisture potential is defined as the difference between the chemical potential of the soil water and free water. It arises because absorption and surface tension (van der Waals forces) are not negligible in the soil water. See Sievers *et al.* (1983).

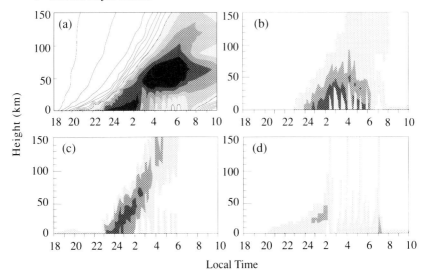

Figure 5.3   Calculated radiation fog development as a function of local time. (a) Temperature contours are at 0.48°C intervals. The colder air is indicated by shading. The lightest shading begins at a threshold of 8.7°C. The darkest shading has a threshold of 6.3°C. (b) Liquid water content (threshold values of 0, 0.15, 0.25 and 0.35 g m^{-3}). In some places the gradient is too tight for all the contours to be seen. (c) Diabatic heating rate (thresholds -1, -2, and -3°C h^{-1}). (d) Mixing coefficient (thresholds 0.001 and 1 m^2 s^{-1}). (Adapted from Welch *et al.*, 1986. Reproduced with permission from the American Meteorological Society.)

Because the latent heat transfer in the soil is now included, the additional equation (5.17) is required, which states that the diffusion of water vapor and liquid water in the soil is exactly balanced at the interface by the turbulent eddy flux of vapor through the atmosphere.

Dew forms whenever the soil temperature is less than the dew point. The downward flux of liquid from the atmosphere and the upward flux of vapor from the soil contribute to the mass of dew.

The usefulness of the thermodynamic equation (5.4) depends critically on its being coupled to an appropriate radiative transfer model. Since the radiative fluxes are affected by the aerosol particles as well as the drops and gaseous constituents, it is important to know the concentration and nature of the dry aerosol particles. Therefore, one or more additional predictive equations may be required to calculate the concentration of aerosol particles.

An example of radiation fog formation, evolution, and dissipation calculated by a scheme of the type described above is shown in Fig. 5.3. In this calculation,[120] the initial local time (LST) is 1800. The initial temperature at the surface is 14°C,

[120] This model calculation was made by Welch *et al.* (1986). In addition to the simulation results summarized here, they present observational information confirming the behavior of the model and discuss the model's sensitivity to the turbulence parameterization. Their paper should be consulted for further details about the initial and boundary conditions and how turbulent mixing coefficients are calculated, radiative transfer is formulated, aerosol concentration is estimated, and soil parameters are assigned.

decreasing to 12°C at a height of 0.1 m. The atmosphere is assumed to be isothermal at 12°C up to a height of 0.5 km, decreasing to 11°C at 1 km and to −3°C at 3 km. Initial soil temperature is assumed to increase to 15.5°C at a depth of −5 cm and then to decrease gradually to 15°C at a depth of −1 m. Relative humidity at the initial time is assumed to be 80% up to a height of 0.5 km, then to decrease linearly to 60% at 3 km. The wind at the top of the domain (3-km altitude) is assumed to be geostrophic and 2 m s^{-1}. Atmospheric aerosol particles activated as a function of relative humidity to form embryo droplets, and appropriate schemes for computing solar and infrared radiation are employed. Explicit equations are used for predicting the soil temperature and liquid water content down to a depth of 1 m below the ground.

From 1800 to 2000 LST in the example, the radiative cooling is seen to produce a temperature inversion rapidly. Thus, the height of the boundary layer drops quickly to near the earth's surface. The geostrophic wind layer, lying just above the boundary layer, thus drops almost to the ground. After 2000 LST, the rate of surface cooling decreases considerably, but the inversion continues to increase in intensity. It begins to rise above the surface as turbulent mixing becomes stronger (as indicated by the increase in exchange coefficient shown in Fig. 5.3d). The increased mixing is related to the intensifying shear across the inversion layer. The mixing in the stable layer shifts some of the surface cooling to higher levels, just below the inversion. By 2300 LST cooling below the inversion layer is sufficiently strong that the fog begins to form (as indicated by the appearance of liquid water in Fig. 5.3b). At this time, the strong effect of the fog on the radiation and thus on the generation of turbulence via cooling aloft is seen as the inversion layer suddenly jumps ~10 m upward (Fig. 5.3a). From 0000 to 0200 LST, the inversion layer continues to rise and the strongest cooling begins to occur at the top of the fog layer, rather than at the surface (Fig. 5.3c). As the inversion layer continues to rise, the turbulence continues to increase as the eddies are freer to move about in the deepening boundary layer. The turbulence continues to transfer heat downward, and by just after 0200 LST the minimum temperature lifts off the ground to a position just below the inversion layer bounding the top of the fog. This point in time marks the beginning of a series of oscillations in the fog, which are probably the model's attempt to produce a convective mixed layer in response to the radiative cooling at the top of the fog layer. The net effect of the mixing is to bring liquid water up to higher levels, and the fog layer deepens through the night. Just after 0600 LST, the upper portion of the fog decouples from the lower portion and the oscillations cease. The upper part thus becomes a stratus layer, and only a weak layer of fog is left below. As we will see in the next subsection, it is typical for this layering to occur in fog and stratus in the stable boundary layer. Greenhouse warming below the base of the upper stratus is apparently too great to allow condensation in the layer between the stratus and the weaker fog below. At sunrise, the fog layer dissipates to a thin ground fog, which, after 0750 LST, somewhat surprisingly increases in depth, as dew and soil moisture are evaporated. After 1030, as the insolation intensifies, only high relative humidity and haze persist.

5.1.4   Arctic Stratus

In the simulation of radiation fog development illustrated in Fig. 5.3, the fog deepens as a result of the turbulent mixing, and the radiative cooling at the top of the fog layer encourages stronger mixing, which further deepens the layer until finally the main cloud layer separates from the lower layer of weak fog and becomes a layer of stratus aloft. But soon after this occurs the whole process is terminated by the rising of the sun, rapid solar heating of the boundary layer, and dissolution of the cloud. Had there been no diurnal cycle and had the surface layer been maintained at a constant temperature, the stratus layer and the lower fog layer might have reached an equilibrium and persisted indefinitely. This latter, hypothetical case is similar to what actually occurs over the Arctic Basin during the summer months, when the melting pack ice comprises a widespread uniform lower boundary (of temperature 0°C) over which warm air from the south moves. As the warmer air is chilled, condensation can occur in the lower boundary layer. Stratus cloud evolves, which is never dissolved because the sun remains at a constant position, very low in the sky. As a result of this process, the Arctic Basin is noted for its ever-present widespread summertime stratus. During the summer, the monthly mean cloud amount is over 80%, and low clouds are the dominant type (Fig. 5.4).

The formation of arctic stratus has been simulated numerically from the same basic equations [(5.2)–(5.7)] used to describe the nighttime radiation fog in Sec. 5.1.3.[121] Somewhat different parameterizations for the radiation, turbulence, and condensation were used in the equations. The basic idea of the calculation is the same, however, except that in this case there is a nonzero basic-state (geostrophic) horizontal flow $\mathbf{v}_g$, in which the air is envisaged to be moving in the x-direction over a lower surface of ice. The basic current is assumed to be constant with respect to height in the layer of interest. The form of (5.2)–(5.7) then remains unchanged, as long as the time derivative is simply considered to be calculated in a coordinate system moving with the basic current. The equations are integrated in time as before, but under rather simpler conditions. There is no need for a complicated treatment of the lower boundary, which is considered to be melting ice. The temperature is simply maintained at 0°C at the surface. There is also no diurnal variation of solar radiation; the sun is held at a fixed position low in the sky (zenith angle 74°). The air moving over the surface is initially 4°C at the surface and has a potentially stable lapse rate characterized by $\partial\theta_e/\partial z = 1°C\ km^{-1}$ and 90% relative humidity. To correspond to conditions over the polar region, a mean downward vertical velocity (of the order of a few tenths of a millimeter per second) is assumed, and a vertical advection term is added to the thermodynamic equation. It is found that this term is negligible compared to the radiative and turbulent terms, but its inclusion is nonetheless illustrative of how stratiform clouds need not depend on lifting for their existence. Indeed, they can thrive in an environment of mean *downward* motion.

---

[121] For the detailed account of how the equations are set up, see the original paper by Herman and Goody (1976), on which this discussion is based.

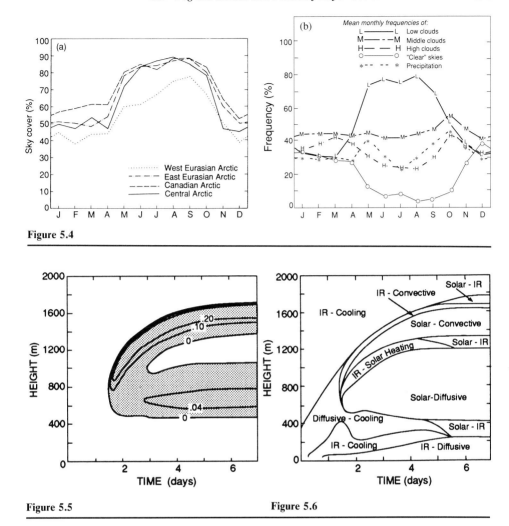

Figure 5.4

Figure 5.5   **Figure 5.6**

Figure 5.4   Monthly mean cloud conditions over the Arctic Basin. (a) Total cloud amount. (b) Frequency of various types of cloud conditions and precipitation. (From Huschke, 1969.)

Figure 5.5   Calculated development of arctic stratus. Liquid water mixing ratio as a function of height and time in isopleths of g kg^{-1}. (From Herman and Goody, 1976. Reprinted with permission from the American Meteorological Society.)

Figure 5.6   The radiative and turbulent processes producing the structure seen in Fig. 5.5. (From Herman and Goody, 1976. Reprinted with permission from the American Meteorological Society.)

Results of the calculations are shown in Fig. 5.5. The stratus, which develops and reaches a steady state after about a week, has a decidedly layered structure, characterized by a more dense upper layer and a more tenuous lower layer. The radiative and turbulent processes producing this structure are illustrated in Fig. 5.6. The terms "diffusive" and "convective" in this diagram refer to the divergence of the turbulent flux under stable and unstable conditions, respectively. As the warm air moves over the ice, the boundary layer is rapidly cooled by contact with the surface, and turbulent mixing rapidly spreads the cooling upward. In slightly more than a day's time, the diffusive cooling together with infrared cooling leads to condensation. Once the cloud forms, the radiative regime is greatly altered by the absorptive properties of the droplets. The upper, more dense cloud layer becomes unstable by long-wave loss from the top of the cloud, and a radiative–convective balance is established in the upper cloud layer. Convective warming is approximately equal to radiative cooling at the top, while the convective cooling balances the heating by solar absorption in the interior. The upper cloud retains this character as the steady-state configuration is approached toward the end of the week. The lower, more tenuous cloud layer is characterized by a general balance between the solar heating penetrating into it from above and the diffusive cooling from below. The clear layer separating the upper and lower clouds is in a state of purely radiative equilibrium. Intense heating produced by a greenhouse mechanism precludes condensation in this layer.[122] Another such radiative equilibrium layer exists just below the base of the lower cloud.

The fact that no fog occurs just above the surface during the first hour is a feature of the turbulence parameterization used. When a different, weaker eddy mixing was adopted, the result in Fig. 5.7 was obtained. In this case, fog formed first at the surface and then lifted, as in the case of radiation fog (Fig. 5.3). In this case the steady state is reached and maintained since there is no diurnal cycle to terminate the cloud by the sun rising higher in the sky. When no turbulent fluxes were allowed at all, the lower cloud layer never left the surface, but the cloud nonetheless split into two layers after 3.5 days. These results suggest that in the Arctic, generally persistent layered stratus should prevail, with fog occasionally appearing when the mixing is especially weak. This inference from the calculations is borne out by experience. For example, observers aboard a drifting ice station in the Arctic Basin during the International Geophysical Year observed sky coverage to be over eight-tenths on average throughout the summer months of 1957 and 1958 (Fig. 5.8). These clouds have been described by Professor N. Untersteiner, who was present on the ice station for 366 days, as almost continuously present low stratus, of a "boring" character, interrupted only occasionally by breaks in the cloud cover and occasional periods of fog.

5.2   Stratus, Stratocumulus, and Small Cumulus in a Boundary Layer Heated from Below

5.2.1   General Considerations

Satellite pictures show that great sheets of stratus and stratocumulus are almost always present over the oceans at subtropical latitudes to the west of continents in

---

[122] See Appendix B of Herman and Goody (1976) for a theoretical treatment of the greenhouse mechanism.

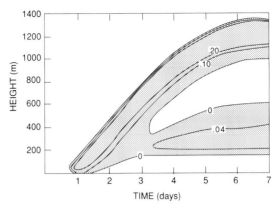

Figure 5.7 Calculated development of arctic stratus. Same as Fig. 5.5 except a weaker diffusivity is assumed. (From Herman and Goody, 1976. Reprinted with permission from the American Meteorological Society.)

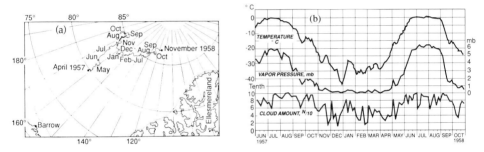

Figure 5.8 Observations taken aboard a drifting ice station in the Arctic Basin during the International Geophysical Year. (a) Location of the station. (b) Observations of temperature (11-day running means), vapor pressure (5-day means), and cloudiness (5-day means). (From Untersteiner, 1961.)

association with subtropical anticyclones (Fig. 1.11a). In middle and high latitudes, large regions of stratocumulus occur where cold air streams offshore across the coastlines of cold continents or ice sheets and forms a layer of low cloud as the cold air suddenly comes in contact with the warm underlying ocean (Fig. 1.11b). These two situations are both examples of a cloud-topped boundary layer heated from below. This type of shallow-layer cloud formation stands in stark contrast to the radiation fog and arctic stratus considered in Sec. 5.1, which form in a boundary layer cooled from below. As the boundary layer is heated from below, it can become conditionally unstable, and sometimes the clouds take the form of a broad field of small cumulus clouds rather than stratus or stratocumulus. In the present discussion, we do not make a strong distinction regarding whether the shallow layer of cloud is entirely stratus, stratocumulus, or cumuliform, as the form of a particular cloud layer is probably a matter of the relative strength of the heating and vigor of the mixing in the particular boundary layer in which the layer of cloud forms. The salient dynamics of the boundary layer leading to the formation and maintenance of the cloud layer have much in common from case to case.

The cloud-topped boundary layer is such a widespread phenomenon over the oceans that it is of great importance to the global climate. The maps shown in Fig. 5.9 illustrate this point. In summer, the maxima of low-cloud amount in the Atlantic and Pacific Oceans near 30°N and 10°S to the west of the continents show the major areas of low cloud layers associated with the subtropical anticyclones. The maxima on the east coasts of North America and Asia and south of Greenland in the boreal winter and north of Antarctica in the austral winter indicate the regions of oceanic stratocumulus forming in response to outbreaks of cold air from

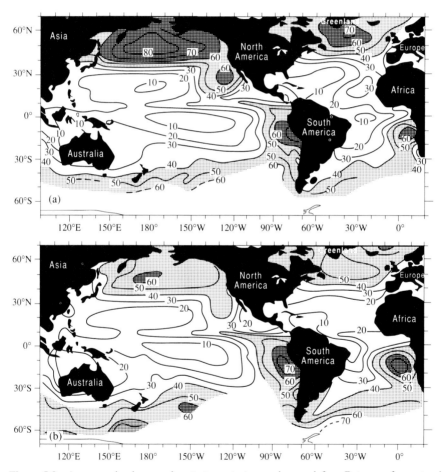

Figure 5.9 Average cloud cover by stratus, stratocumulus, and fog. Data are from standard surface observations and are expressed in percent of sky covered. The years 1952 to 1981 are included in the data set. (a) June, July, August. (b) September, October, November. (Analysis by C. Leovy of data published in the atlas of Warren *et al.*, 1988.)

the continents. The extent of these regions of oceanic low clouds is so great that it has been estimated that "a mere 4% increase in the area of the globe covered by low-level stratus clouds would be sufficient to offset the 2–3 K predicted rise in global temperature due to a doubling of CO_2."[123]

5.2.2 Cloud-Topped Mixed Layer

5.2.2.1 Conceptual Models

A conceptual model of the development of a cloud-topped mixed layer is given in Fig. 5.10. We will refer to this illustration throughout the following discussion. Panel (a) of the figure represents the boundary-layer structure before any cloud appears. The boundary layer is envisioned to be characterized by a population of buoyant plumes rising from the warm sea surface. It becomes well mixed as a result of the vigorous upward fluxes of sensible and latent heat. Unlike a radiation fog layer, in which buoyancy is suppressed by thermodynamic stability, turbulent kinetic energy is generated strongly by the buoyancy [term $\mathcal{B}$ in (2.86)], as well as by wind shear. The mixing maintains a layer of constant mean equivalent potential temperature $\bar{\theta}_e$ in a layer of depth h. This *mixed layer* is bounded above by a layer of distinctly different $\bar{\theta}_e$, usually higher. The $\bar{\theta}_e$ of the layer aloft is typically maintained by large-scale subsidence in the lower troposphere. In the case of subtropical and tropical oceanic stratus and stratocumulus, the subsidence is associated with the descending branch of the Hadley cell, concentrated in the eastern sectors of the subtropical oceanic anticyclones. In the case of cloud layers in polar airstreams off continents in winter, the subsidence occurs in the cold-air mass behind a front. The change of $\bar{\theta}_e$ across h is represented by $\Delta\theta_e$.[124] The higher $\bar{\theta}_e$ air in the upper layer is entrained by turbulence into the lower layer, thus diluting the boundary layer but at the same time increasing its depth h.

As the depth of the boundary layer increases, the tops of the turbulent elements rise above the lifting condensation level, and small cloud elements form (Fig. 5.10b). As the cloud elements gradually thicken to form a more continuous sheet of cloud at the top of the mixed layer (Fig. 5.10c), the physical picture changes significantly, with radiation taking on an important role. The ocean surface is shielded by the cloud layer and the turbulent heat flux in the subcloud layer becomes weak, or even negative. The turbulent energy generation is concentrated in the cloud layer and is produced by buoyancy as radiative cooling at cloud top destabilizes the layer of cloud. Turbulence is also generated by wind shear in the layer [term $\mathcal{C}$ in (2.86)]. The turbulent layer continues to increase in depth as a result of entrainment.

Finally, the cloud layer breaks up. Observations indicate that when the boundary layer is not extremely unstable, the cloud layer breaks up into stratocumulus

---

[123] This estimate was made by Randall *et al.* (1984).

[124] In this discussion Δ indicates the difference obtained by subtracting the lower-layer value from the upper-layer value.

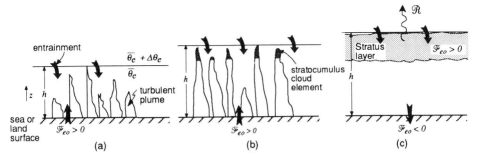

Figure 5.10 Conceptual model of the evolution of a cloud-topped mixed layer over the ocean. (a) Unsaturated turbulent plumes are occurring in a well-mixed layer of depth h. $\mathscr{F}_{eo}$ is the heat flux across the ocean surface. $\bar{\theta}_e$ is the mean equivalent potential temperature in the mixed layer. (b) Plumes reaching above the lifting condensation level have stratocumulus elements in their upper portion. Large arrows indicate entrainment across the top of the mixed layer. (c) Cloud layer becomes solid. Infrared radiative flux $\mathscr{R}$ becomes important, especially at cloud top.

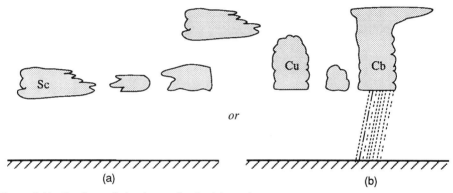

Figure 5.11 Breakup of cloud-topped mixed layer into (a) stratocumulus (Sc) elements and (b) cumulus (Cu) and cumulonimbus (Cb) elements.

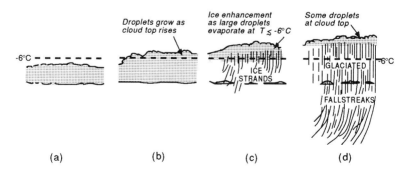

Figure 5.12 Empirical model of the formation of ice in stratocumulus clouds. The model is based on extensive probing of clouds by research aircraft. (From Hobbs and Rangno, 1985. Reprinted with permission from the American Meteorological Society.)

or small cumulus elements, as suggested by Fig. 5.11a. However, when the boundary layer is especially unstable, the elements in the breakup phase may take the form of larger cumulus or even small cumulonimbus (Fig. 5.11b).

The mechanism of the breakup process is not well understood. It has been suggested that it is related to increased entrainment across the top of the mixed layer. The entrainment might be expected to become large and run out of control when the $\Delta\theta_e$ between the boundary layer and the layer above is negative and exceeds a certain large magnitude. According to this idea, a parcel of air entrained into the mixed layer from above, upon mixing with the cloudy air, would become denser than its surroundings. Because of the negative buoyancy, the air from aloft would be rapidly mixed through a portion of the cloud, leaving patchy, dissipating stratocumulus. Observational studies show, however, that the cloud layer persists even when the $\Delta\theta_e$ would appear to be sufficiently negative for instability. The problem appears to be that the potential energy expended in pulling down a tongue of stable air from above exceeds the energy released in the mixing.[125] The cloud-layer breakup problem will be discussed further in Sec. 5.2.3.

When the tops of mixed-layer stratocumulus elements extend to heights above the $-6°C$ level, they sometimes produce high concentrations of ice particles by rapid ice enhancement (Sec. 3.2.6). When the drop size spectrum in the cloud is broad, with largest particles $\sim 20~\mu m$ in dimension, high concentrations of ice are observed to form in ~ 10 min. These particles fall out in the form of ice strands in the cloud (Fig. 5.12). It has been argued that the most likely ice enhancement mechanism near cloud top is contact nucleation [mechanism (iii) in Sec. 3.2.6].[126] Once the droplets are frozen, they grow by vapor diffusion, while smaller super-cooled droplets in their vicinity evaporate or are collected by the growing ice particles. Thus, the cloud element is rapidly glaciated. Aircraft observations show that the strands of ice particles extend down through the cloud, rapidly removing the liquid cloud particles. Finally, all that remain are a few remnants of the original cloud and streamers of falling ice particles extending below the original cloud base.

5.2.2.2 *Mathematical Modeling*

The essential dynamics of the development of the cloud atop the mixed layer (depicted schematically in Fig. 5.10a–c) can be examined quantitatively through a consideration of the thermodynamic and water-continuity equations integrated over the depth h of the boundary layer.[127] For simplicity, we will consider the cloud to be warm (no ice) and nonprecipitating, although in reality drizzle may sometimes form and contribute significantly to the water budget of the stratus

---

[125] For a discussion of this problem, see Siems *et al.* (1990).

[126] See Hobbs and Rangno (1985) for details of the argument.

[127] The treatment of the cloud-topped boundary layer presented here was introduced in a seminal paper by Lilly (1968).

layer. The total water-mixing ratio can be subdivided simply into liquid and vapor form:

$$q_T = q_v + q_L \tag{5.18}$$

where q_v is the vapor-mixing ratio and q_L is the liquid water-mixing ratio. Since the clouds are assumed to be nonprecipitating, there is no need to subdivide the liquid water content further into cloud and precipitation drops, and q_T is a conservative variable. As a thermodynamic variable, we use the equivalent potential temperature θ_e, defined by (2.15) and (2.17).

The basic assumption of the mixed layer is that within the layer horizontal averages of q_T and θ_e are independent of height (Fig. 5.13). The average value $\bar{\theta}_e$ remains constant because the heating at the lower boundary (and/or cooling at the top by radiation once the cloud layer forms) continually produces a conditionally unstable lapse rate, which is practically instantly removed by vertical mixing and release of the instability. The total water is constant with height as a simple result of the mixing and the fact that q_T is a conservative quantity. The mean values $\bar{\theta}_e$ and $\bar{q}_T$ change suddenly at the top of the boundary layer, as subsidence in the lower free atmosphere maintains both high potential temperature and low dew point just above the boundary layer. There is a net increase in $\bar{\theta}_e$ from just below to just above h.

In the development and maintenance of a cloud layer under a level where a positive $\Delta\theta_e$ occurs, conservation of mass requires the rate of change of depth of the mixed layer to be

$$\frac{dh}{dt} = w_e + w(h) \tag{5.19}$$

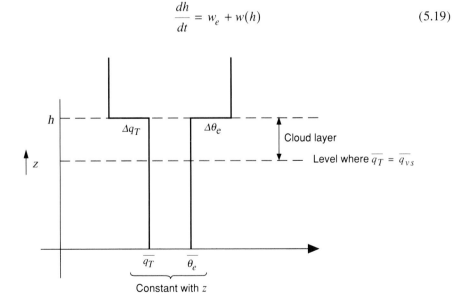

Figure 5.13   Assumed variation of q_T and θ_e with height in modeling of the cloud-topped mixed layer.

where $w(h)$ is the mean vertical velocity of the air at $z = h$ (i.e., the rate of change of height associated with net horizontal convergence or divergence in the mixed layer), while w_e is the *entrainment velocity*, or the rate at which less dense, laminar air is incorporated into the denser, turbulent layer. Typically, the cloud-topped boundary layer is divergent so that $w(h) < 0$, while the entrainment velocity is positive as the boundary layer grows by incorporation of air from above.

According to (2.16) and (5.18), calculation of averages $\bar{q}_T$ and $\bar{\theta}_e$ within the layer $z = 0$ to h yields the vertical distribution of the horizontal averages $\bar{q}_L$, $\bar{q}_v$, and $\bar{\theta}$ in the cloud, since the cloud layer is saturated and nonprecipitating. It is therefore useful to seek the predictive equations for $\bar{q}_T$ and $\bar{\theta}_e$. The equation for $\bar{q}_T$ is obtained from the Boussinesq version of the mean-variable form of the water-continuity equation (2.81), for the case in which there is only one category of water substance ($\bar{q}_T = \bar{q}_i$) and the sources and sinks $\bar{S}_i$ are all zero. The Boussinesq assumption implies that the density terms do not appear in (2.81). There are no horizontal or vertical advection terms since $\bar{q}_T$ varies neither horizontally nor vertically. The total derivative $\bar{D}/Dt$ in (2.81) simplifies to d/dt, and (2.81) reduces to

$$\frac{d\bar{q}_T}{dt} = -\frac{\partial}{\partial z}\left(\overline{w'q'_T}\right) \tag{5.20}$$

By similar reasoning, the equation for $\bar{\theta}_e$ is obtained from the Boussinesq version of the mean-variable form of the First Law of Thermodynamics (2.78). Both radiation and the latent heat of phase change (condensation or evaporation) are included in the heating-rate term $\mathcal{H}$ of (2.78). The horizontal and vertical advection terms again are zero. Thus, (2.78) may be written as

$$\frac{d\bar{\theta}_e}{dt} = -\frac{\partial}{\partial z}\left(\overline{w'\theta'_e} + \mathcal{R}\right) \tag{5.21}$$

where the overbar represents a horizontal average, the primed terms are turbulent fluctuations away from the average, and $\mathcal{R}$ is the net radiative flux, as in (5.4). We have written (5.21) in terms of θ_e, so that $\mathcal{R}$ is the only heat source/sink that appears explicitly.

Since, according to the mixed-layer assumption, $\bar{q}_T$ and $\bar{\theta}_e$ are constant in height, (5.20) and (5.21) imply that

$$\frac{\partial}{\partial z}\left(\overline{w'q'_T}\right) = \text{constant}, \quad 0 \leq z \leq h \tag{5.22}$$

and

$$\frac{\partial}{\partial z}\left(\overline{w'\theta'_e} + \mathcal{R}\right) = \text{constant}, \quad 0 \leq z \leq h \tag{5.23}$$

The water and heat fluxes at the earth's surface are denoted by

$$\left(\overline{w'q'_T}\right)_o \equiv \mathcal{F}_{To} \tag{5.24}$$

and

$$\left(\overline{w'\theta_e'}\right)_o \equiv \mathscr{F}_{eo} \tag{5.25}$$

The turbulent fluxes at the top of the mixed layer are assumed to be responsible for diluting the average values of q_T and θ_e within the mixed layer. Since the net rate of dilution is determined by the rate at which air from the layer above the mixed layer (whose values of q_T and θ_e differ from those in the mixed layer by amounts Δq_T and $\Delta \theta_e$) is incorporated into the mixed layer, we may set

$$\left(\overline{w'q_T'}\right)_h = -w_e \Delta \bar{q}_T \tag{5.26}$$

and

$$\left(\overline{w'\theta_e'}\right)_h = -w_e \Delta \bar{\theta}_e \tag{5.27}$$

where the subscript h indicates conditions at the top of the boundary layer. It has already been assumed that the air entrained at rate w_e is instantaneously mixed through the turbulent layer, thus maintaining constant values of $\bar{q}_T$ and $\bar{\theta}_e$ throughout the layer. Substituting (5.24)–(5.27) into (5.22) and (5.23), we obtain

$$\frac{\partial}{\partial z}\left(\overline{w'q_T'}\right) = \frac{-w_e \Delta \bar{q}_T - \mathscr{F}_{To}}{h} \tag{5.28}$$

and

$$\frac{\partial}{\partial z}\left(\overline{w'\theta_e'} + \mathscr{R}\right) = \frac{-w_e \Delta \bar{\theta}_e - \mathscr{F}_{eo} + \mathscr{R}_h - \mathscr{R}_o}{h} \tag{5.29}$$

which may be substituted into (5.20) and (5.21) to obtain the following predictive equations for $\bar{q}_T$ and $\bar{\theta}_e$:

$$\frac{d\bar{q}_T}{dt} = \frac{w_e \Delta \bar{q}_T + \mathscr{F}_{To}}{h} \tag{5.30}$$

and

$$\frac{d\bar{\theta}_e}{dt} = \frac{w_e \Delta \bar{\theta}_e + \mathscr{F}_{eo} - \mathscr{R}_h + \mathscr{R}_o}{h} \tag{5.31}$$

For our purposes, the fluxes of q_T and θ_e at the surface ($\mathscr{F}_{To}$ and $\mathscr{F}_{eo}$, respectively), the radiative flux difference ($\mathscr{R}_h - \mathscr{R}_o$), and the vertical velocity $w(h)$ may be considered as given.[128] In this case, (5.19), (5.30), and (5.31) form a set of three equations with four unknowns: h, $\bar{q}_T$, $\bar{\theta}_e$, and w_e. The problem reduces to determining the entrainment rate w_e in a physically reasonable way in order to close the set of equations. If this can be done, then the evolution of the mixed layer, as characterized by its height (h), thermodynamic structure ($\bar{\theta}_e$), and water content ($\bar{q}_T$), can be computed.

---

[128] The surface flux and radiation could also be parameterized in terms of other variables. But that extra complication is not necessary to illustrate the basic ideas of the mixed-layer model.

There are two approaches to closing the equations for the mixed layer, depending on whether they are being solved diagnostically or prognostically. In the former case, the entrainment rate and the time derivatives are determined from observations, and the turbulent fluxes implied by the equations are then compared to observed fluxes. In the prognostic case, the entrainment is determined from the turbulent kinetic energy equation, and the time derivatives are computed to determine the evolution of the boundary layer.

We will examine both the diagnostic and prognostic cases briefly. However, we first note that in either case, we first need to split the fluxes of q_T and θ_e into component parts. To do this, we make use of the approximate expression for equivalent potential temperature (2.19) and the definition of *cloud virtual potential temperature*,

$$\theta_{cv} \equiv \theta(1 + 0.61 q_v - q_L) = \theta_v - \theta q_L \tag{5.32}$$

With appropriate scaling of θ and q_v, (2.19) implies that the turbulent eddy flux of potential temperature can be approximated as

$$\overline{w'\theta'} \approx \overline{w'\theta_e'} - \frac{L\theta}{c_p T} \overline{w'q_v'} \tag{5.33}$$

For the subcloud layer, where there is no liquid water, (5.18), (5.32), and (5.33) imply

$$\overline{w'q_v'} = \overline{w'q_T'} \tag{5.34}$$

$$\overline{w'\theta_{cv}'} = \overline{w'\theta_v'} \approx \overline{w'\theta_e'} - \theta\left(\frac{L}{c_p T} - 0.61\right)\overline{w'q_T'} \tag{5.35}$$

In the cloud layer, the air is saturated and liquid water is present. The saturation mixing ratio q_{vs} is a function of both temperature and pressure. However, it is a much stronger function of temperature, and at a given height in the boundary layer the pressure is nearly constant. Hence, for the cloud layer, it is a good approximation to write the vapor flux as

$$\overline{w'q_v'} \approx \frac{T}{\theta}\left(\frac{\partial q_{vs}}{\partial T}\right)\overline{w'\theta'} \tag{5.36}$$

It follows from (5.32), (5.33), and (5.36) that

$$\overline{w'\theta_{cv}'} \approx \beta_T \overline{w'\theta_e'} - \theta\overline{w'q_T'} \tag{5.37}$$

where

$$\beta_T = \frac{1 + 1.61 T(\partial q_{vs}/\partial T)}{1 + (L/c_p)(\partial q_{vs}/\partial T)} \tag{5.38}$$

The factor β_T is a slowly varying function of temperature, which has a value of about 0.5 under cloud-topped mixed-layer conditions.

In the diagnostic case, closure of the equations is obtained from observations of the eddy flux of total water near the top of the mixed layer ($\overline{w'q_T'}$). The entrainment velocity w_e is obtained simply by substituting this observed value into (5.26). When this value of w_e is substituted in the right-hand sides of (5.30) and (5.31), then $d\bar{q}_T/dt$ and $d\bar{\theta}_e/dt$ are known and according to (5.20) and (5.21) imply the vertical distributions of the fluxes ($\overline{w'q_T'}$) and ($\overline{w'\theta_e'} + \mathcal{R}$). If a profile of the radiative flux $\mathcal{R}$ is assumed, then the equations imply values of $\overline{w'q_T'}$ and $\overline{w'\theta_e'}$. If these fluxes are further decomposed according to (5.33)–(5.37), we obtain the profiles of $\overline{w'\theta'}$, $\overline{w'\theta_v'}$, $\overline{w'q_L'}$, $(\overline{w'\theta_e'})$, $(\overline{w'q_T'})$, and $(\overline{w'q_v'})$. Thus, the equations for the mixed layer are used to diagnose the turbulent fluxes from measurements of $\Delta\theta_e$, Δq_T, h, and $\overline{w'q_T'}$ at h.

Examples of flux profiles computed diagnostically are shown in Fig. 5.14. These results illustrate the importance of radiation to the maintenance of the stratus layer, once it has become a continuous, persistent cloud sheet, as depicted in Fig. 5.10c. One set of results sets the radiative flux to zero. In this case, $(\overline{w'q_T'})$ and $(\overline{w'\theta_e'})$ are linear functions of height in the mixed layer, as required by (5.22) and (5.23). The fluxes $\overline{w'\theta'}$, $\overline{w'\theta_v'}$, $\overline{w'q_L'}$, and $\overline{w'q_v'}$ also are linear, though they are discontinuous at cloud base. An unsatisfactory aspect of these solutions is that $\overline{w'\theta_v'}$ is largely negative in the cloud layer, which would mean, according to (2.86),

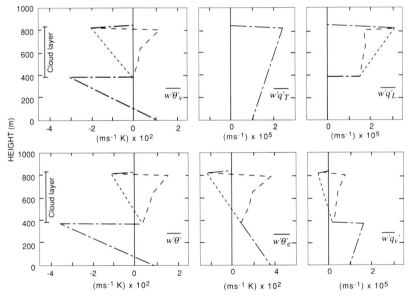

Figure 5.14 Turbulent fluxes diagnosed from aircraft measurements of $\Delta\theta_e$, Δq_T, h, and $\overline{w'q_T'}$ and the cloud-topped boundary-layer model described in text. Short-dashed line is for the case in which radiative flux is set to zero. Long-dashed line is for the case in which a radiative flux profile is assumed in the cloud layer. Dash-dot line is where the short-dashed and long-dashed lines coincide. (Adapted from Nicholls, 1984. Reprinted with permission from the Royal Meteorological Society.)

that turbulent kinetic energy is being destroyed in the cloud layer, thus contradict-ing the characterization of the cloud as a turbulent entraining layer. When a reasonable profile of upward radiative flux is assumed to apply in the cloud layer, the results are modified as indicated in Fig. 5.14. In this case, the loss of heat by radiation at cloud top is sufficiently large that the eddy flux of virtual potential temperature (i.e., the buoyancy flux) can be positive throughout most of the cloud layer. According to (2.86), the positive buoyancy flux generates turbulent kinetic energy, and the stratus is consequently maintained as an active turbulent entrain-ing cloud layer. Thus, after the cloud forms, the radiation in the cloud layer becomes a crucial factor, continually destabilizing the lapse rate to maintain the cloud as an unstable, turbulent layer.

In the prognostic application of the equations of the idealized mixed layer, one cannot defer to observations to determine the entrainment velocity w_e. The ap-proach generally used is to consider the budget of turbulent kinetic energy $\mathcal{H}$ in the mixed layer, which is expressed by (2.86). In the boundary layer heated from below and the cloud layer cooled at the top by radiation, buoyancy generation $\mathcal{B}$ is the important source term, while dissipation $\mathcal{D}$ is the only sink of turbulent kinetic energy in (2.86). Conversion from mean-flow kinetic energy $\mathcal{C}$, which also appears in (2.86) may also contribute whenever the shear in the boundary layer is large, but it is often not as significant as the buoyancy generation. In this regard, the cloud-topped mixed layer is vastly different from the fog dynamics, considered in Sec. 5.1, where buoyancy generation is zero or negative and all of the turbulence must be derived from the weak mean flow. To keep our discussion as simple as possible, we will consider the case where buoyancy generation is the only source of turbulence. Generation via $\mathcal{C}$ and through pressure–velocity correlation $\mathcal{W}$ are ignored. We also assume that $\mathcal{H}$ has no vertical or horizontal variability in the well-mixed boundary layer. The total derivative $\bar{D}/Dt$ in (2.86) then simplifies to d/dt. With these assumptions, (2.86) becomes

$$\frac{d\mathcal{H}}{dt} = \mathcal{B} - \mathcal{D} \qquad (5.39)$$

From (2.89), it is recalled that the buoyancy generation derives from the corre-lation of buoyancy and vertical velocity. The vertical integral of the buoyancy generation may be written as

$$\langle \mathcal{B} \rangle \equiv \int_0^h \mathcal{B} \; dz = \int_0^h g\left(\frac{\overline{w'\theta_v'}}{\bar{\theta}_v}\right) dz \qquad (5.40)$$

This integral has the units (velocity)^3, and w_e^3 may be thought of as the portion of the kinetic energy generation that is used to effect the entrainment. Various schemes for determining what this portion should be have been devised, and this remains a topic of research.[129]

---

[129] For a review of some of these schemes, see Nicholls and Turton (1986).

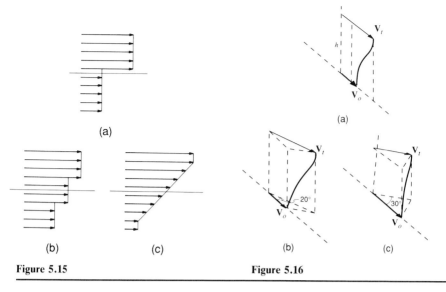

Figure 5.15

Figure 5.16

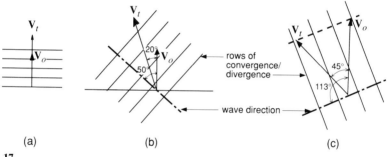

rows of
convergence/
divergence

wave direction

(a)　　　　　(b)　　　　　(c)

Figure 5.17

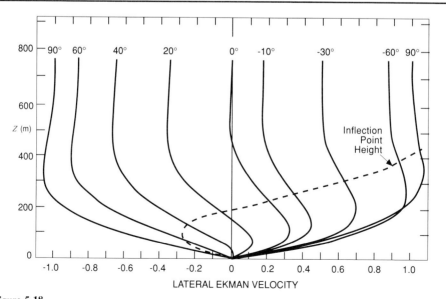

Figure 5.18

5.2.3 Mesoscale Structure of Mixed-Layer Clouds

The layer of cloud at the top of the mixed layer often consists of discrete stratocumulus elements in the form of rolls or cells. In this section, we will examine briefly the factors giving rise to this structure, which we characterize as mesoscale since the horizontal dimension of the cells and rolls ranges from 1 to 100 km. Not everything about what leads to these patterns is known. However, there are three processes that appear to play some role. These are the shear and thermal instabilities of the boundary layer and the cloud-top entrainment process. These processes will be considered in the sections below. We first consider long rolls of stratocumulus, which are called *cloud streets*.

5.2.3.1 Cloud Streets

In Sec. 2.9.2, it was shown that sheared flows are intrinsically unstable when the Richardson number Ri < 1/4 and that at the interface of two adjacent, horizontally homogeneous, two-dimensional, inviscid, incompressible flows of different density and velocity, this instability is manifest as Kelvin–Helmholtz waves (Figs. 2.6–2.8). These waves are the simplest form of a more general type of wave motion, which can arise in sheared flows whenever the vertical profile of the wind speed has an inflection point (i.e., a change of curvature).[130] Examples of two-dimensional shear profiles that have been used in more sophisticated analyses of Kelvin–Helmholtz instability are shown in Fig. 5.15. All of these profiles exhibit Kelvin–Helmholtz waves, and each has an inflection point.

A wind profile with an inflection point can be produced by either speed shear or by turning of the wind direction with height. The profiles in Fig. 5.16 are represented analytically by an arctangent profile in the vertical plane containing the tips of the velocity vectors. These inflection point profiles are associated with pure speed shear, speed plus turning, and pure turning. Stability analysis for convec-

---

[130] The discussion of inflection point instability given here is based largely on comprehensive review articles by Brown (1980, 1983).

Figure 5.15 Examples of two-dimensional shear profiles that have been used in analyses of Kelvin–Helmholtz instability. Horizontal line divides fluid into two layers. Arrows indicate fluid velocity. (From Brown, 1980. © American Geophysical Union.)

Figure 5.16 Arctangent velocity variation between constant-velocity regions $\mathbf{V}_o$ and $\mathbf{V}_t$ over depth h. (a) Pure speed shear. (b) Speed plus turning shear. (c) Pure turning shear. (From Brown, 1980. © American Geophysical Union.)

Figure 5.17 Direction and orientation of phase lines of fastest-growing waves (and cloud streets) with respect to type of velocity shear: (a) Pure speed shear. (b) Speed plus turning shear. (c) Pure turning shear. (From Brown, 1980. © American Geophysical Union.)

Figure 5.18 Theoretical Ekman-layer velocity profiles in perpendicular planes at various angles to the free-stream (geostrophic) flow. The lateral Ekman velocity shown is the component of the Ekman-layer wind perpendicular to planes oriented at various angles to the left of the geostrophic flow. The velocity scale is labeled in fractions of the geostrophic wind speed. (From Brown, 1980. © American Geophysical Union.)

tively neutral conditions shows that all these profiles are unstable and that long rolls are the preferred geometry of the most unstable solutions. The orientation of the rolls depends on whether the profiles are due to speed shear or turning (Fig. 5.17). Pure speed shear leads to rolls perpendicular to the flow (Fig. 5.17a). These are essentially Kelvin–Helmholtz billows. They have a characteristic wavelength approximately twice the layer depth. Pure turning profiles (Fig. 5.17c) yield rolls approximately parallel to the mean flow in the layer and have a characteristic wavelength two to four times the layer depth.

It is an interesting characteristic of the theoretical Ekman layer (Sec. 2.11.1) that its velocity profile has an inflection point (Fig. 5.18). The inflection point instability in the Ekman layer appears to produce disturbances which can grow to finite size and come into equilibrium with the mean flow. The secondary roll flow associated with an Ekman layer is illustrated in Fig. 5.19. The upward branches of the rolls are conducive to supporting cloud lines generally parallel to mean flow in the planetary boundary layer and may thus sometimes be responsible for organization of stratocumulus, or small cumulus, into *cloud streets*.[131]

It is doubtful that the inflection point shear instability alone accounts for cloud streets. Usually there is also some degree of thermal instability in the planetary boundary layer. When the boundary layer is heated from below, instability of the Rayleigh–Bénard type (Sec. 2.9.3) may arise. According to the instability criterion (2.180), there is no preferred geometry of the convection in a stagnant basic state. The convective overturning may be in the form of long rolls or symmetric cells. However, the situation changes if the basic state of the unstable layer of fluid is characterized by a horizontal velocity that changes with height. The convective solutions then have a preferred geometry.[132] Specifically, they are in the form of rolls parallel to the shear vector (Fig. 5.20).

We have also seen that when the thermal stratification is neutral, inflection point instability rolls transverse to the shear can appear. The general case of a thermally unstable and unstably sheared fluid layer has been considered theoretically.[133] When the basic-state wind field for a layer heated from below is prescribed to be a theoretical Ekman layer, Rayleigh–Bénard rolls parallel to the

---

[131] Long rows of clouds sitting atop the boundary layer were dubbed "cloud streets" by glider pilots. Since the top of one of these cloud rows marks the top of an updraft, and the row extends downwind, the cloud street encompasses the optimum combination of tail wind and updraft, which allows a long glide to be achieved. In 1935, Dr. E. Steinhoff reported the first glider flight in cloud streets. He left from Bayreuth in western Germany and finally landed in Brünn, Czechoslovakia, a distance of over 500 km and exceeding the world record for that time. There he met three other glider pilots, who had just accomplished the same feat. Cloud streets are described by Kuettner (1947, 1959, 1971). He noted in 1959 that although the glider pilots had become "fond" of cloud streets, "the priority of the discovery that the tedious circling technique in columnar thermals could be replaced by a comfortable tailwind flight beneath the cloud streets must be given to the seagulls ... who seem to have enjoyed this technique for millions of years." The observations of seagulls soaring was reported by Woodcock (1942).

[132] This effect was first noted when Bénard tilted his cells in the laboratory.

[133] An especially lucid account can be found in the series of articles of Asai (1964, 1970a,b, 1972) and Asai and Nakasuji (1973).

shear dominate as long as the Richardson number is low (i.e., the differential heating of the layer dominates). At higher Richardson number, inflection point instability dominates, with rolls 20° to the left of the geostrophic flow. A "parallel instability" mode, not associated with the inflection point, is also found. It is characterized by rolls 10° to the right of the geostrophic flow. However, the parallel instability occurs under conditions not likely to be significant in the atmosphere.[134]

When cold air flows off a cold land mass or region of ice over an adjacent region of warm water, well-defined cloud streets appear (Fig. 5.21; see also Figs. 1.11b and 1.30). In this situation, the shear and mean flow vectors in the boundary layer are approximately the same; therefore, it is not easy to distinguish from the orientation of the cloud streets whether they are of thermal or shear origin. However, the strong surface flux that occurs when the cold air lies over the warm water indicates that the Richardson number is quite low and that the rolls are most likely of the thermal instability type. Before the air reaches the water, inflection point instability must dominate, and it probably produces rolls in the boundary layer of size and orientation similar to the thermally induced rolls, but without the cloud tracers to make their appearance evident in the satellite picture. When the winds are moderate (>5 m s^{-2}) and the stratification is only weakly unstable, the organization of the cloud streets is more likely dominated by the Ekman-layer inflection point instability.[135]

Returning to Fig. 5.21, we note that as the air flows farther out to sea, the cloud rolls widen and deepen. Finally the cloud field turns into a more complex pattern of clouds, which have more the character of cumulus or small cumulonimbus than stratocumulus (as in Fig. 5.11b). The widening and deepening of the roll circulations are not well understood. The roll circulations may grow as a result of instability of the flow to perturbations of finite size. To the extent that cumulonimbus occurs, precipitation may also affect the structure of the cloud pattern at this stage.

Whatever the nature of the rolls, the upward branch of the roll circulation is the favored environment for the formation of the cloud street. The flow is directed along the rolls, and the stages of the life cycle of the cloud-topped mixed layer depicted in Fig. 5.10 and Fig. 5.11 can be thought of as occurring at points along the upward branch of a roll at locations successively farther out over the ocean, while cloud formation is suppressed in the downward branches of the rolls. The alternating cloud streets of stratocumulus and clear gaps comprise the cloud pattern until far out to sea, where the stratocumulus elements develop into cumulus or cumulonimbus and the pattern becomes more cellular.

5.2.3.2   Cellular Patterns

The cellular pattern seen out at sea in Fig. 5.21 is rather common for mixed-layer clouds east of cold continents in the winter as well as offshore from large ice-

---

[134] See Brown (1980) for discussion of the relative importance of the parallel and inflection point modes.

[135] LeMone (1973) has presented evidence to this effect.

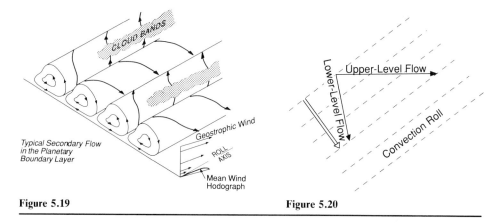

Figure 5.19   **Figure 5.20**

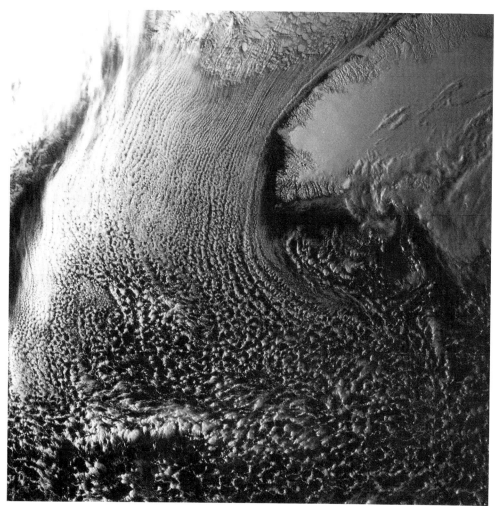

Figure 5.21

covered areas. Cellular patterns are also frequently seen in the regions of subtropical stratocumulus west of continents, especially at locations where the cloud sheets appear to be breaking up (i.e., reaching the last stage of the life cycle indicated by Fig. 5.11).

The cellular patterns seen in oceanic stratocumulus are of two types: *Open cells* are walls of cloud surrounding open areas. They are envisioned as having downward motion in the cell center. The cells at the downwind end of the cloud streets in Fig. 5.21 are of the open type. *Closed cells* are rings of open area surrounding solid cloud. They apparently have upward motion in the cell center. An example is shown in Fig. 5.22. Between the regions of open and closed cells, radial arms of cloudiness called *actinae* are sometimes found (Fig. 5.23).

A climatology of the occurrence of open and closed cells in the atmosphere is shown in Fig. 5.24. In the low-cloud fields east of continents, cellular patterns tend to occur over warm ocean currents, while west of continents they occur over cool currents. The cellular patterns in any of these regions can occur as open cells, closed cells, or coexisting open and closed cells. The open cells, however, are statistically favored over warm water, while the closed cells are preferred over colder water. These results suggest that open cells, probably often composed of rings of cumulus and small cumulonimbus, are favored when heating from below by surface sensible heat flux is dominant (Fig. 5.11b), while closed cells are more stratiform and favored when radiative cooling from the top of a more continuous cloud layer is the dominant mechanism responsible for the differential heating (Fig. 5.11a).

The open and closed cells and actinae all have an intriguing, but possibly deceptive, similarity to phenomena observed in Rayleigh–Bénard convection under laboratory conditions.[136] In particular, the atmospheric cells sometimes appear to be roughly hexagonal. Although Rayleigh's linear theory predicts no particular horizontal shape, Bénard actually observed hexagonal cells, and this tends to be the preferred geometry of the cells when there is no shear. The hexagonal case of the classical linear solution is illustrated in Fig. 5.25. The zero vertical velocity isopleth is a circle with concentrations of upward (or downward) motion in the center and compensating reverse vertical flow on the hexagonal periphery.

[136] For a discussion of the comparisons between laboratory convection and cloud structure, see the review of Agee (1982).

Figure 5.19 Schematic of secondary roll flow in an Ekman boundary layer with mean wind hodograph shown. (From Brown, 1983.)

Figure 5.20 Schematic diagram showing characteristics of thermal convection rolls in relation to the basic flow. Dashed lines indicate the axis of the preferred roll convection. Broad arrow denotes the phase velocity of the roll convection. (From Asai, 1972.)

Figure 5.21 Detailed view of cloud streets where cold air is flowing off a region of ice over the Atlantic Ocean near the southern tip of Greenland (upper right). NOAA-7 visible wavelength satellite photograph made at 1703 GMT on 19 February 1984. (Courtesy of University of Dundee, Scotland and A. Van Delden.)

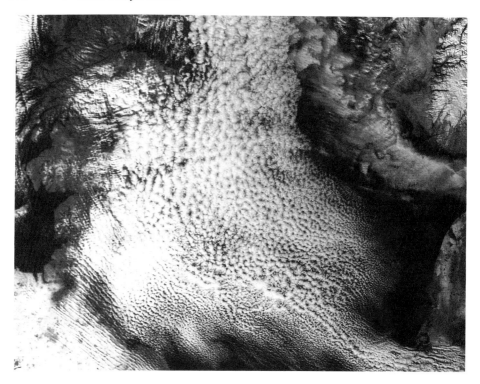

Figure 5.22   Detailed view of closed convection cells over the North Sea. Denmark is in the lower part of the picture. Norway is in the upper right and the west coast of Scotland is in the upper left to center. NOAA-9 visible wavelength satellite photograph made at 1703 GMT on 11 February 1985. (Courtesy of University of Dundee, Scotland and A. Van Delden.)

Extrema occur on the vertices, and it sometimes appears that the biggest clouds occur on the vertices of open cells.

One should be cautious in concluding that the hexagonal arrays seen in strato-cumulus are produced by an exact analog to Rayleigh–Bénard convection. As in the case of bees building a honeycomb, which is also a hexagonal array similar to that formed by the convection, nature seems to seek a geometrical configuration in which the amount of fence building between cells is minimized.[137] Although hexagonal arrays are seen both in laboratory convection and in atmospheric cloud patterns, it does not necessarily follow that the governing physics are the same in both cases, any more than the honeycomb and the convection are produced by the same process. Indeed, there are some major discrepancies between laboratory convection and atmospheric convection. One such discrepancy is that laboratory experiments produce closed cells in liquids and open cells in gases, whereas the

---

[137] For a geometrical shape of a given area, a circle has the smallest ratio of perimeter to area. The polygon that most closely approximates a circle and can also be fitted together into an array of elements with common boundaries is the hexagon.

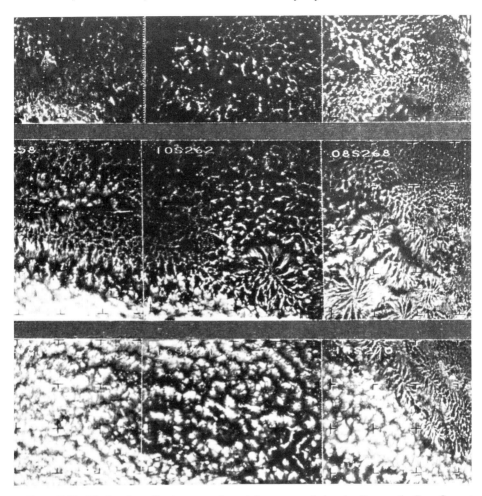

Figure 5.23   Nimbus I satellite imagery of coexisting open and closed cells over the Peru Current at 1813 GMT on 15 September 1964. Radial arms of convective cloudiness, called *actinae*, can be seen between the open- and closed-cell regions. Picture is centered at about 10°S, 95°W. Each square is ~300 km in dimension. (From Agee, 1982. Reprinted by permission from Kluwer Academic Publishers.)

atmosphere exhibits both types of structure. Since the atmosphere is a gas, the appearance of closed cells is somewhat surprising and suggests that at least the closed cells are not Rayleigh–Bénard convection. The most discomforting discrepancy is that the aspect ratio in atmospheric convection is ~10 : 1, whereas the classical theory predicts ratios ~1 : 1 (i.e., the atmospheric cells are flatter). It is mainly this discrepancy that has led to a search for other explanations for the behavior of mixed-layer stratus during its breakup stage.

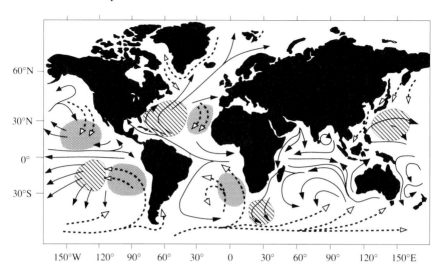

Figure 5.24 Global climatology of cellular structure of stratocumulus and small cumulus over oceans. Shaded areas are regions where closed cells predominate. Hatched areas show where open cells are more common. Solid streamlines show locations of warm ocean currents. Dashed streamlines show cold currents. Land masses are blackened. (Adapted from Agee *et al.*, 1973. Reproduced with permission from the American Meteorological Society.)

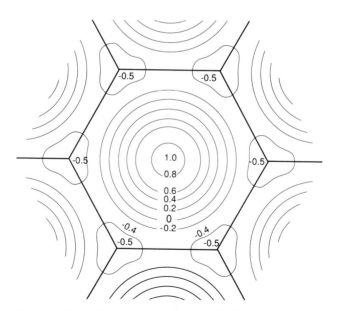

Figure 5.25 Linear solution of the relative vertical velocity field in a hexagonal convection cell. (From Pellew and Southwell, 1940.)

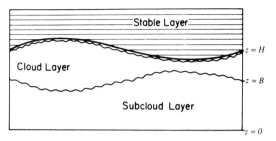

Figure 5.26 Schematic of a hypothetical mesoscale entrainment instability. A positive buoyancy fluctuation in the cloud layer rises into the stable layer and becomes adjacent to warmer air. The air entrained into the thicker cloud is relatively warmer than that where the cloud is thinner. Therefore, entrainment can reinforce the buoyancy fluctuations. The horizontal lines are lines of constant Θ_e. (From Fiedler, 1984. Reprinted with permission from the American Meteorological Society.)

The other explanations that have been attempted revolve around the process of cloud-top entrainment. We have already noted in discussing the breakup stage of mixed-layer stratus (Fig. 5.11) that early ideas attempted to relate the stability and breakup of the stratus to the $\Delta\theta_e$ across the top of the boundary layer; the idea being that a lower $\Delta\theta_e$ should be conducive to greater entrainment and breakup. However, these ideas are not confirmed by observations, which indicate a marked resistance of the stratus to mixing from above, even when the value of $\Delta\theta_e$ is low enough to suggest instability. A more recent idea is represented schematically in Fig. 5.26. It was offered to explain the large aspect ratios of cellular convection in stratocumulus. The basic idea is that an instability of the cloud-topped mixed layer comes into play whenever one part of the cloud layer is thicker than some other part and the vertical temperature gradient in the stable layer above the cloud is sufficiently great.[138] According to (5.27), the rate of dilution of the mixed layer by entrainment is proportional not only to w_e but also to $\Delta\theta_e$. If the stability of the layer above the cloud were very strong, as indicated in Fig. 5.26, the value of $\Delta\theta_e$ across the higher cloud top would be substantially greater than across the lower cloud top. It is then possible for the dilution of the cloud by entrainment to be greater over the lower cloud top than over the higher cloud top; that is, there can be positive feedback leading to the growth of the thicker elements and to the destruction of the thinner ones. Mixed-layer stratus may thus be unstable under these conditions, and theoretical calculations suggest that the resulting perturbations grow to a size whose aspect ratio is ~30:1, in reasonable agreement with observations. An argument against the applicability of this mechanism is that it requires a weak inversion at the top of the boundary layer (to get small $\Delta\theta_e$ above the thick cloud) and a strong stable layer above. These are conditions that are not normally observed.

[138] This hypothesis and the supporting theory were put forth by Fiedler (1984).

5.3 Cirriform Clouds

5.3.1 General Considerations

5.3.1.1 Descriptive Terminology

As we saw in Sec. 1.2.4, upper-tropospheric clouds that exist at temperatures of about -20 to $-85°C$ are referred to as cirrus, cirrostratus, and cirrocumulus. The various forms taken by these clouds, as seen by an observer on the ground, were shown in Figs. 1.15–1.17. In addition to these forms, the tops of deep precipitating cloud systems seen by satellite (Figs. 1.27–1.30) typically occur at these high altitudes. Strong wind shear aloft can displace these upper cloud layers, producing long streamers of high cloud extending great distances downshear from the parent phenomenon. These streamers of cloud are extensive partly because ice particles, with their low-saturation vapor pressure, are slow to sublimate and partly because they are not necessarily inactive debris of the parent phenomena, but rather can be continually regenerated by their own cloud dynamics (see Secs. 5.2.2–5.2.4). Henceforth, we will refer to all high-level clouds as *cirriform clouds*, whether they occur atop a major precipitating cloud system or as separate cloud entities.

5.3.1.2 Microphysical, Kinematic, and Radiative Characteristics

Cirriform clouds consist primarily of ice particles, although some embedded convective elements may contain supercooled water for a few minutes at a time. Aircraft observations of the composition of cirriform clouds[139] show ice contents in cirriform clouds to be in the range 0.001–0.25 g m^{-3}, with values of 0.01–0.1 g m^{-3} being typical. The sizes of the particles range from 50 to 1000 μm. The most commonly reported crystal habits are columns, bullets, bullet–rosettes, and plates. Aggregates of crystals are also seen in cirrus.[140] The vertical air motions in cirriform clouds are typically 0.1–0.2 m s^{-1}, except in floccus and uncinus elements, where the vertical velocity is $\sim$1–2 m s^{-1}.[141]

As in fog and stratus, radiation interacts strongly with the dynamics of cirriform clouds. The radiation is, however, more complicated than in fog and stratus because the particles are complex, do not scatter isotropically, and are often present in very low concentrations. A detailed treatment of the radiative transfer in cirriform cloud is beyond the scope of this text, and extensive treatments of these topics already exist.[142] The radiative transfer equations can be solved under

---

[139] Summarized in Table 1 of Liou (1986).
[140] See Heymsfield and Knight (1988).
[141] See Heymsfield (1975b, 1977) and Gultepe and Heymsfield (1988).
[142] See, for example, the book by Liou (1980) and the review articles by Stephens (1984) and Liou (1986).

various sets of assumptions. The example results shown in Fig. 5.27 assume that the crystals are columns and use empirical information on optical depth, contribution of scattering to optical depth, and the normalized scattering phase function. With these factors taken into account, the heating of air as a result of the divergence of net radiative flux through cirriform cloud of various depths has been estimated.[143] Heating by solar radiation is felt through the cloud layer, while infrared wavelengths produce a destabilizing effect by cooling the cloud top and warming the base of the cloud.

5.3.1.3   Climatology

In any global satellite picture (e.g., frontispiece), many of the photographed cloud tops are cirriform.[144] Climatologies showing the global extent and distribution of cirriform cloud have been constructed from both satellite observations (dashed curve in Fig. 5.28) and visual observations of the state of the sky (solid curve in Fig. 5.28). Both types of climatology indicate maxima of cirriform cloud in the tropics and midlatitudes and minima in the subtropics. These curves tend to mirror the global distribution of precipitation (dotted curve in Fig. 5.28), indicating that most cirriform cloud is of the type that has its origin in the upper layers of deep, precipitating cloud systems.[145]

5.3.1.4   Approach to Studying the Dynamics

A common form of cirriform cloud is a small convective entity described as consisting of a dense *head* and a long fibrous tail (or *fallstreak*) of falling snow. This form is seen in both cirrus uncinus (Fig. 1.15c) and cirrus spissatus with virga (Fig. 1.16). The uncinus or floccus elements may occur either alone or as convective elements embedded in a widespread layer of cirriform cloud. In examining the dynamics of cirriform clouds, we will first examine (in Sec. 5.3.2) the dynamics of the basic uncinus or floccus element.

We have also noted that cirriform cloud comprises the upper portions of deep precipitating cloud systems and that layers of cirriform cloud extend downwind of the deep cloud systems where the upper portion of such a deep cloud spreads into the environment. As an example of this process, we consider in Sec. 5.3.3 the dynamics of the ice-cloud layer produced by a cumulonimbus cloud (such as illustrated in Fig. 1.5).

---

[143] See Liou (1986).

[144] Actually, the area covered by cirriform cloud is even greater than that apparent in satellite images, as much cirriform cloud is too thin to be seen in satellite imagery and is barely evident to the naked eye. The existence of such thin cirriform cloud is indicated by sensitive aircraft instruments and certain ground-based remote-sensing devices.

[145] The high percent cloud cover by cirrus indicated by the visual observation in high latitudes does not contradict this conclusion, since the precipitation totals at high latitudes must be low because of the small surface areas encompassed by high-latitude circles and the low surface temperatures.

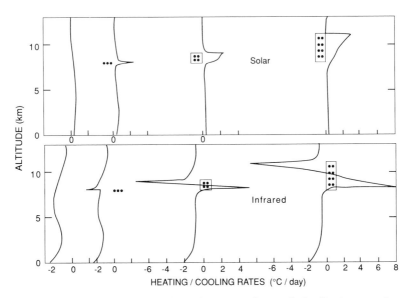

Figure 5.27   Heating of air as a result of the divergence of net radiative flux in atmospheres with cirriform cloud of thicknesses 0, 0.1, 1, and 3 km (dotted boxes). The base of the cirriform cloud is placed at 8 km. Standard atmospheric conditions were assumed. The mean ice content of the cloud layer was 0.13 g m^{-3}, and the cosine of the solar zenith angle was 0.5. Upper panel is for the solar flux. Lower panel is for the infrared. (Adapted from Liou, 1986. Reproduced with permission from the American Meteorological Society.)

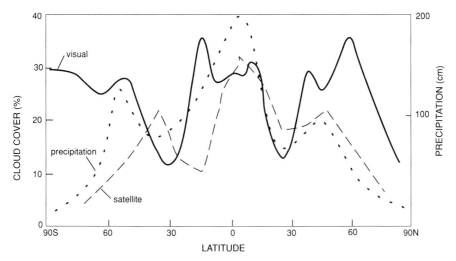

Figure 5.28   Global climatology of cloud cover by cirriform cloud. Visual observations (December, January, February conditions) are from Warren *et al.* (1988). Satellite observations (December, January, February) are from Barton (1983). Global precipitation is from Sellers (1965).

In Sec. 5.3.4 we will examine the dynamics of a thin layer of cirriform cloud in which convective elements are embedded and which forms apart from a source cloud such as a cumulonimbus.

5.3.2 Cirrus Uncinus

Cirrus uncinus (Fig. 1.15c) and cirrus spissatus with virga (Fig. 1.16) have much in common in appearance, with the spissatus being more pronounced. The similarity of appearance suggests that the two forms are dynamically similar, at least qualitatively, and that what is learned about one is probably applicable to the other. An empirical model of a cirrus uncinus element has been developed and is shown in Figs. 5.29–5.31.[146] According to this model, the head of the uncinus element

---

[146] This empirical model of cirrus uncinus was constructed by Heymsfield (1975b) on the basis of aircraft and Doppler radar observations. It is consistent with earlier work by Japanese investigators (Yagi *et al.*, 1968; Yagi, 1969; Harimaya, 1968).

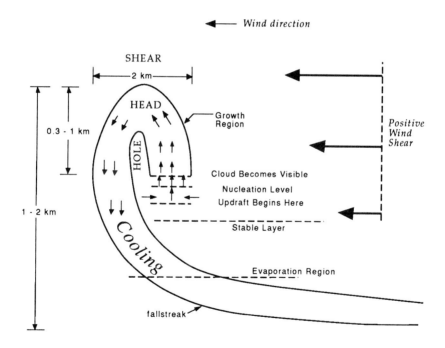

Figure 5.29 Empirical model of a cirrus uncinus element under positive wind shear conditions. (Adapted from Heymsfield, 1975b. Reproduced with permission from the American Meteorological Society.)

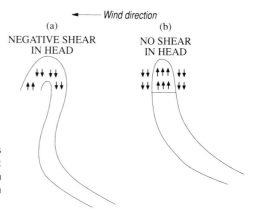

(a)
NEGATIVE SHEAR
IN HEAD

(b)
NO SHEAR
IN HEAD

Wind direction

Figure 5.30 Empirical model of a cirrus uncinus element. (a) Wind shear opposite to that in Fig. 5.29. (b) No wind shear. (From Heymsfield, 1975b. Reprinted with permission from the American Meteorological Society.)

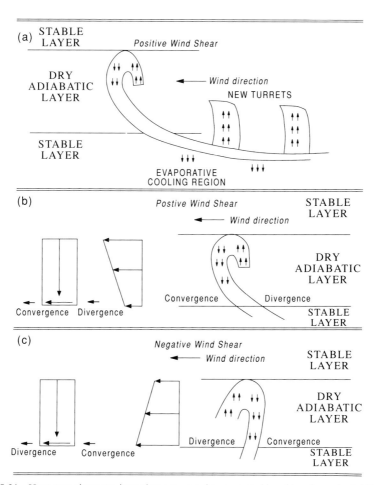

Figure 5.31 How new cirrus uncinus elements may be generated by older elements. (a) Cooling by evaporation causes turbulence in the stable layer, which in turn perturbs the unstable layer above; (b) because of the shear in the environment, horizontal momentum transported downward in the downdraft produces convergence at the bottom of the unstable layer on one side of the downdraft; (c) opposite shear to that in (b) produces shear on the other side of the downdraft. (From Heymsfield, 1975b. Reprinted with permission from the American Meteorological Society.)

occurs in a dry adiabatic layer bounded by stable layers above and below. Vertical air motions in the head are ~1 m s^{-1}. When the wind shear is in the sense shown in Fig. 5.29, the head contains a cloud-free hole. Ice particles generated and carried up to the top of the updraft are advected over the top of the hole. The fallstreak to the left of the hole contains downdraft associated with evaporation and/or drag of the falling ice particles. Evaporation of the particles continues as they fall through the stable layer below the base of the head. When the wind shear in the layer of the head is of the opposite sense, the hole and fallstreak are to the right of the updraft column (Fig. 5.30a). When there is no wind shear in the layer containing the head, the ice particles occur within the updraft, and downdraft occurs around the periphery of the updraft (Fig. 5.30b). Two ideas on how new cirrus uncinus elements could be generated by the older elements are indicated (Fig. 5.31). Either cooling by evaporation causes turbulence in the stable layer, which in turn perturbs the unstable layer above (Fig. 5.31a); or, because of the shear in the environment, horizontal momentum transported downward in the downdraft produces convergence at the bottom of the unstable layer on one side or the other of the downdraft (Fig. 5.31b and c). In either case, a new cell could be triggered in the vicinity of the older uncinus element.

5.3.3   Ice-Cloud Outflow from Cumulonimbus

As shown in Fig. 1.5, a layer of ice cloud often emanates from the upper levels of a cumulonimbus cloud. This cloud layer may take the form of cirrus spissatus, if thinner and patchier, or cirrostratus cumulonimbogenitus, if thicker and more widespread. If the base of the cloud layer is in the middle étage, it may take the form of altostratus cumulonimbogenitus. For simplicity, we will refer to these phenomena as *ice-cloud outflows* from cumulonimbus. The dynamics of these outflows can be analyzed theoretically by considering the air intruding into the environment from the cumulonimbus to be laden with ice, uniformly buoyant, isotropically turbulent, and flowing into a stably stratified environment (Fig. 5.32).[147] This outflow is envisioned to undergo a two-stage process as it moves downstream.

In the first stage, the outflow undergoes a collapse similar to that of the wake of a body moving through a stratified fluid (e.g., a submarine moving through the oceanic thermocline). During the collapse, the wake is flattened and spread laterally by the environment (Fig. 5.33). The upper part of the outflow, being denser than the environmental fluid, subsides, while the lower part of the outflow, being lighter than the ambient air, rises. At the same time that this external collapse is taking place, an internal collapse occurs, in which the turbulence in the plume is

---

[147] The treatment of this problem follows Lilly (1988), who, 20 years after his landmark paper (Lilly, 1968), which established a theoretical framework for the radiatively driven boundary-layer stratus, applied similar concepts to the turbulent layer of ice cloud emanating from the top of a cumulonimbus. This section is based on his development of the problem.

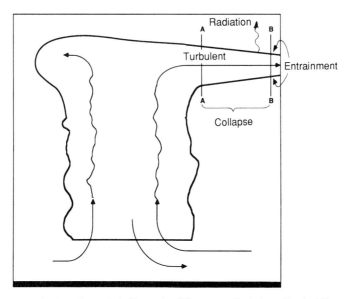

Figure 5.32   Idealized outflow of cirriform cloud from cumulonimbus. Vertical lines AA and BB indicate positions of cross sections in Fig. 5.33. (Adapted from Lilly, 1988. Reproduced with permission from the American Meteorological Society.)

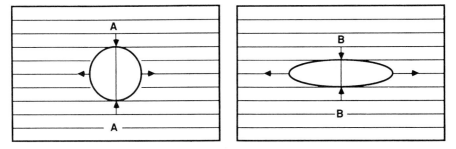

Figure 5.33   Collapse of a cumulonimbus anvil. Plume cross sections are shown for positions AA and BB in Fig. 5.32. The horizontal lines indicate potential temperature surfaces in the environment; the arrows indicate the motion field induced by the buoyancy difference between the outflow plume and the environment. (From Lilly, 1988. Reprinted with permission from the American Meteorological Society.)

slowly dissipated, its energy being transferred to waves and larger-scale two-dimensional turbulence.[148]

In the second stage of development of the outflow plume, the radiative destabilization of the outflow layer becomes important, and the ice cloud is maintained by a process similar to that of low-level mixed-layer stratus. The energy to maintain

[148] Lilly (1988) considered the external collapse quantitatively by assuming it consists of the conversion of the available potential energy to the kinetic energy of the spreading motion. In a simplified example, he found the plume radius tripled in just under 10 min and suggested that the maximum aspect ratio is achieved after about 20 min.

the turbulence is provided by radiative destabilization of a layer of cirrus—as is indicated for example by the calculations shown in Fig. 5.27. The major difference from the case of boundary-layer stratus is that the unstable layer is bounded by stable layers both above *and* below.

Radiation is the only significant heat source in the ice-cloud outflow layer since latent heat release is negligible at these altitudes. The problem thus simplifies to that of a dry mixed layer, in which case the mean potential temperature θ is constant with respect to z, and

$$\frac{d\bar{\theta}}{dt} = -\frac{\partial}{\partial z}\left(\overline{w'\theta'} + \mathcal{R}\right) = \text{constant in } z \tag{5.41}$$

which is the same as (5.21), except that the difference between θ_c and θ is now ignored. The radiative flux divergence is assumed to be a linear function

$$\frac{\partial \mathcal{R}}{\partial z} = B_o\left(z - z_m\right) - A_o \tag{5.42}$$

where A_o and B_o are positive constants and z_m is the height of the middle of the cloud layer.[149] Integration from cloud base z_b to cloud top z_t shows that the value of the constant in (5.41) is

$$\frac{d\bar{\theta}}{dt} = A_o - \frac{\left(\overline{w'\theta'}\right)_t - \left(\overline{w'\theta'}\right)_b}{2H} \tag{5.43}$$

where

$$H = \frac{z_t - z_b}{2} \tag{5.44}$$

The values of the fluxes at z_t and z_b are both considered to be determined by the rate of turbulent entrainment across the boundary of the mixed layer. Accordingly, we let the fluxes at the top and bottom of the cloud layer be

$$\left(\overline{w'\theta'}\right)_t = -w_{et}(\Delta\theta)_t \tag{5.45}$$

and

$$\left(\overline{w'\theta'}\right)_b = w_{eb}(\Delta\theta)_b \tag{5.46}$$

respectively, where $(\Delta\theta)_t$ and $(\Delta\theta)_b$ are the change in θ across the top and bottom of the layer and w_{et} and w_{eb} are entrainment velocities across the top and bottom. The changes in θ across the top and bottom of the ice-cloud layer are defined as positive numbers:

$$(\Delta\theta)_b = \bar{\theta} - \theta_{o_b}, \quad z = z_b \tag{5.47}$$

---

[149] Lilly (1988) based this assumption on the calculations of Ackerman *et al.* (1988), who considered radiative transfer theory in the context of cirrus near the equatorial tropopause.

$$(\Delta\theta)_t = -\bar{\theta} + \theta_{o_t}, \quad z = z_t \tag{5.48}$$

The sounding of potential temperature in the environment θ_o, including the specific values θ_{o_t} and θ_{o_b} at the top and bottom of the ice-cloud mixed layer, is assumed to be known.

It is readily seen that the expressions (5.45) and (5.46) are analogous to (5.27), which represents the mixing across the top of the boundary layer in the case of stratus and stratocumulus. The ice-cloud layer represented here differs from the boundary-layer cloud, however, in having entrainment across *both* the top and bottom of the cloud layer.

In the case of boundary-layer cloud, the entrainment velocity is inferred from the buoyant production of kinetic energy $\mathscr{B}$ integrated over the depth of the mixed layer. Here we follow a similar procedure, in which (5.40) is simplified to

$$\langle \mathscr{B} \rangle = \frac{g}{\bar{\theta}} \int_{z_b}^{z_t} \overline{w'\theta'} \, dz \tag{5.49}$$

The integral in this expression can be evaluated since $\partial\mathscr{R}/\partial z$ is the known linear function of height (5.42), which, together with the assumption that $(\overline{w'\theta'})_b$ is known, implies that $\overline{w'\theta'}$ is a known quadratic function of z, which is obtained by integrating (5.41) from z_b to z:

$$\overline{w'\theta'} = (\overline{w'\theta'})_b - \frac{d\bar{\theta}}{dt}(z - z_b) - \int_{z_b}^{z} \frac{\partial\mathscr{R}}{\partial z} \, dz$$

$$= (\overline{w'\theta'})_b + (z - z_b)\left[\frac{B_o}{2}(z_t - z) + A_o - \frac{d\bar{\theta}}{dt} \right] \tag{5.50}$$

When this expression is substituted into (5.49), we obtain

$$\langle \mathscr{B} \rangle = \frac{g2H}{\bar{\theta}}\left[(\overline{w'\theta'})_b + H\left(A_o - \frac{d\bar{\theta}}{dt} \right) + \frac{B_o H^2}{3} \right] \tag{5.51}$$

To complete the mathematical description of the ice-cloud layer, some assumption must be made to relate the net buoyancy generation of kinetic energy $\langle \mathscr{B} \rangle$ to the entrainment velocities at the top and bottom of the cloud layer. Just as in the case of the stratus-topped mixed layer, the entrainment velocity at the top of the mixed layer must be related to $\langle \mathscr{B} \rangle$ [as discussed in relation to Eq. (5.40)]. To establish a relationship for the ice-cloud outflow layer, the buoyancy generation of kinetic energy given by (5.49) is subdivided into positive and negative contributions

$$\langle \mathscr{B} \rangle = \langle \mathscr{G} \rangle - \langle \mathscr{N} \rangle \tag{5.52}$$

where

$$\langle \mathcal{G} \rangle = +\frac{g}{\bar{\theta}} \int_{z_b}^{z_t} \left(\overline{w'\theta'} \right)_{\mathcal{G}} dz \tag{5.53}$$

and

$$\langle \mathcal{N} \rangle = -\frac{g}{\bar{\theta}} \int_{z_b}^{z_t} \left(\overline{w'\theta'} \right)_{\mathcal{N}} dz \tag{5.54}$$

Symbol $\mathcal{G}$ indicates generation of eddy kinetic energy $[(\overline{w'\theta'}) > 0]$ while $\mathcal{N}$ indicates consumption (i.e., negative generation) of eddy kinetic energy $[(\overline{w'\theta'}) < 0]$.

This decomposition is an approach sometimes used in considering a dry (non-cloud-topped) mixed layer just above the ground. The negative fluxes are associated with entrainment and are envisaged as being most strongly felt at the top of the mixed layer, across which all the engulfed plumes must pass, while near the surface only a slight effect is felt since only the most penetrative plumes reach down to these levels (Fig. 5.34a). It is typically assumed that $(\overline{w'\theta'})_{\mathcal{G}}$ and $(\overline{w'\theta'})_{\mathcal{N}}$ both vary linearly, with $(\overline{w'\theta'})_{\mathcal{G}}$ decreasing from $\overline{w'\theta'}$ at the bottom of the layer to zero at the top and $(\overline{w'\theta'})_{\mathcal{N}}$ decreasing from 0 at the bottom to $\overline{w'\theta'}$ at the top (Fig. 5.34b). The net flux then decreases linearly from a positive value at the surface to a negative value at the top of the turbulent layer (Fig. 5.34c).

The procedure followed for the case of ice-cloud outflow is an extension of this dry mixed-layer problem. The effect of the cloud-*top* entrainment (w_{et}) is assumed to decrease linearly to zero at cloud base while the effect of the cloud-*base* entrainment (w_{eb}) is assumed to decrease linearly to zero at cloud top (Fig. 5.35). Then

$$\langle \mathcal{N} \rangle = -\frac{g}{\bar{\theta}} \int_{z_b}^{z_t} \left(\overline{w'\theta'} \right)_{\mathcal{N}} dz = -\frac{gH}{\bar{\theta}} \left[\left(\overline{w'\theta'} \right)_b + \left(\overline{w'\theta'} \right)_t \right] \tag{5.55}$$

In the case of the dry mixed planetary boundary layer, it is assumed that

$$\langle \mathcal{B} \rangle = a \langle \mathcal{G} \rangle \tag{5.56}$$

where a is a positive constant.[150] Extending this assumption to the ice-cloud outflow layer, we can combine (5.43), (5.51), (5.52), (5.55), and (5.56) to obtain

$$\left(\overline{w'\theta'} \right)_b + \left(\overline{w'\theta'} \right)_t = \frac{(a-1)2H^2 B_o}{3} \tag{5.57}$$

and

$$\langle \mathcal{B} \rangle = \frac{2 a B_o H^3 g}{3\bar{\theta}} \tag{5.58}$$

---

[150] This approach is attributed to Stage and Businger (1981a,b), who based it on empirical studies of the dry boundary layer. Lilly (1988) used a value of $a = 0.8$ for the cumulonimbus outflow calculation.

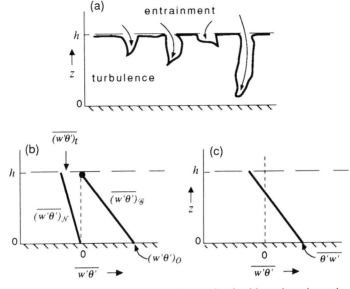

Figure 5.34   Schematic of a dry (non-cloud-topped) mixed layer just above the ground.

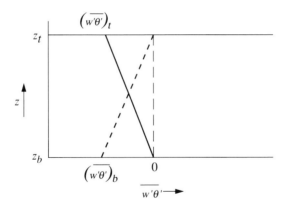

Figure 5.35   Eddy fluxes associated with cloud-top entrainment and cloud-base entrainment in ice-cloud outflow from cumulonimbus. The flux associated with cloud-top entrainment is assumed to have a maximum magnitude $(\overline{w'\theta'})_t$ at z_t and decrease to zero at z_b. The flux associated with cloud-top entrainment is assumed to have a maximum magnitude $(\overline{w'\theta'})_b$ at z_b and decrease to zero at z_t.

which show that the sum of the boundary entrainment fluxes $[(\overline{w'\theta'})_b + (\overline{w'\theta'})_t]$ and the net integrated average eddy heat flux $\langle \mathscr{B} \rangle$ are both driven by the vertical gradient of radiative heating (B_o) in the ice-cloud layer.

Since (5.57) gives only the sum of the values of the boundary fluxes, a way is needed to relate $(\overline{w'\theta'})_b$ and $(\overline{w'\theta'})_t$ in order to close the problem. For simplicity, the two fluxes are arbitrarily made proportional to each other, such that

$$\left(\overline{w'\theta'}\right)_b = \alpha_o\left[\left(\overline{w'\theta'}\right)_b + \left(\overline{w'\theta'}\right)_t\right] \tag{5.59}$$

where α_o is simply an adjustable parameter. Then (5.43) with substitution from (5.57) and (5.59) becomes

$$\frac{d\bar{\theta}}{dt} = A_o + \left(1 - 2\alpha_o\right)\left[\frac{(1-a)B_o H}{3}\right] \tag{5.60}$$

This equation predicts the mean potential temperature in the ice-cloud mixed layer. The parameter A_o represents the net radiative heating, which if positive causes the layer to warm. The jump in θ at the top of the cloud layer decreases, while that at the bottom of the layer increases. The second term is proportional to B_o, which represents the vertical gradient of the radiative heating, which drives the turbulence and entrainment. If the adjustable parameter $\alpha_o > 0.5$, then the temperature rise in the mixed layer is less than that provided by the net radiation because more potentially cool air is entrained at the bottom than warm air at the top of the cloud. The opposite situation applies if $\alpha_o < 0.5$.

Substitution of (5.57) and (5.59) into (5.45) and (5.46) gives

$$\frac{dz_b}{dt} = -\frac{2\alpha_o(1-a)B_o H^2}{3(\Delta\theta)_b} \tag{5.61}$$

and

$$\frac{dz_t}{dt} = \frac{2(1-\alpha_o)(1-a)B_o H^2}{3(\Delta\theta)_t} \tag{5.62}$$

which, together with (5.60), form a closed set of equations for calculating $\bar{\theta}$, z_b, and z_t as functions of time. For values of $A_o = 0$ (no net radiative heating) and $\alpha = 0.5$ (equal effect of entrainment at cloud top and cloud base), the change in the depth of the mixed layer is found to be slow—rather similar to boundary-layer stratus. If α is again 0.5 but $A_o > 0$ (i.e., radiation is warming the cloud layer in the mean), then $\bar{\theta}$ increases until the jump $(\Delta\theta)_t$ at cloud top becomes small. Then, according to (5.62), the rate of increase in the cloud-layer depth becomes large.[151]

5.3.4 Cirriform Cloud in a Thin Layer Apart from a Generating Source

The final type of cirriform cloud we will look at is that which occurs in a thin unstable layer aloft, apart from a generating source such as a cumulonimbus cloud. The dynamics of this type of cirriform cloud have been examined via a numerical model.[152] The model applies in the two-dimensional (x–z) spatial domain indicated in Fig. 5.36. The domain is a shallow layer, in which the two-dimensional Boussinesq vorticity equation (2.61) may be used to describe the air motions, and nothing varies in the y-direction. If turbulent mixing in the layer is

---

[151] See Lilly (1988) for further discussion of these results.
[152] This numerical model is described in a pair of papers by Starr and Cox (1985a,b). The present discussion is based on that work.

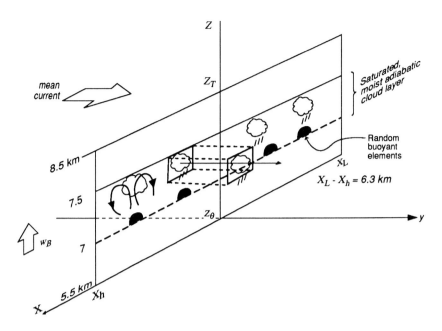

Figure 5.36 Design of a model used to study cirriform cloud that forms in a thin layer. (Adapted from Starr and Cox, 1985a. Reproduced with permission from the American Meteorological Society.)

represented by K-theory (Sec. 2.10.1), with a constant mixing coefficient K_ξ, then the mean-variable form of (2.61) may be written as

$$\bar{\xi}_t = -\bar{u}\bar{\xi}_x - \bar{w}\bar{\xi}_z - \bar{B}_x + K_\xi\nabla^2\bar{\xi} \tag{5.63}$$

where, for this two-dimensional situation, $\nabla^2 = \partial^2/\partial x^2 + \partial^2/\partial z^2$. The overbars are averages over one finite grid element. Thus, the barred quantities indicate the model-resolvable scale of motion. The mean vertical velocity is assumed to have the form

$$\bar{w} = w_B + \hat{w} \tag{5.64}$$

where w_B is a background vertical motion, which is constant throughout the layer, and $\hat{w}$ is the deviation from the background value. The mass-continuity equation for the mean flow [obtained by averaging (2.55)] is satisfied by a stream function defined such that

$$\bar{u} = \psi_z, \quad \hat{w} = -\psi_x \tag{5.65}$$

The vorticity is then given by

$$\bar{\xi} = \nabla^2\psi \tag{5.66}$$

Substituting (5.64) and (5.65) into (5.63), we obtain

$$\bar{\xi}_t = -\psi_z\bar{\xi}_x + \psi_x\bar{\xi}_z - w_B\bar{\xi}_z - \bar{B}_x + K_\xi\nabla^2\bar{\xi} \tag{5.67}$$

If (5.64) and (5.65) are used in the advection terms of the Boussinesq version of the mean-variable form of the First Law of Thermodynamics (2.78) and the eddy-flux convergence of θ is represented by K-theory with a constant mixing coefficient K_θ, then (2.78) becomes

$$\bar{\theta}_t = -\psi_z \bar{\theta}_x + \psi_x \bar{\theta}_z - w_B \bar{\theta}_z + \frac{L_s C_d}{c_p \bar{\Pi}} - \frac{\partial \mathcal{R}}{\partial z} + K_\theta \nabla^2 \bar{\theta} \qquad (5.68)$$

where $\mathcal{R}$ is the net radiative heat flux and $L_s C_d / c_p \bar{\Pi}$ is the release of latent heat as vapor is deposited on ice. The last three terms are the same as in (5.4), except that the latent heat of sublimation L_s appears in place of the latent heat of vaporization L and the deposition rate C_d appears instead of the condensation rate C. To complete the model, the Boussinesq versions of the mean-variable water-continuity equations (2.81) are written in a form analogous to (5.6) and (5.11) as

$$\bar{q}_{v_t} = -\psi_z \bar{q}_{v_x} + \psi_x \bar{q}_{v_z} - w_B \bar{q}_{v_z} - C_d + K_v \nabla^2 \bar{q}_v \qquad (5.69)$$

and

$$\bar{q}_{H_t} = -\psi_z \bar{q}_{H_x} + \psi_x \bar{q}_{H_z} - w_B \bar{q}_{H_z} + C_d + \frac{\partial}{\partial z}\left(\hat{V}_H \bar{q}_H\right) + K_H \nabla^2 \bar{q}_H \qquad (5.70)$$

where q_H is the mixing ratio of hydrometeor (i.e., ice) mass, K_v and K_H are constant mixing coefficients, and $\hat{V}_H$ is taken to be the mass-weighted mean-particle fall speed,

$$\hat{V}_H = \frac{\sum_k \int_o^\infty N_k(D) \mathrm{m}_k(D) V_k(D)\, dD}{\sum_k \int_o^\infty N_k(D)\, \mathrm{m}_k(D)\, dD} \qquad (5.71)$$

where k indicates a particular type (habit) of crystal, D represents the diameter of the crystal, $N_k(D)$ the size distribution function for crystal type k, m_k the mass of a crystal of type k and diameter D, and V_k the fall speed of crystal type k as a function of size. The expression in (5.71) is like that in (3.69), except that the falling particles are ice particles of type k instead of raindrops. Data on ice crystals in cirriform clouds can be substituted in (5.71) to obtain an empirical value of $\hat{V}_H$. Similarly the measured $N_k(D)$ may be substituted into an integral like that on the right-hand side of (4.9) to obtain a value of q_H. Repeated measurements of $N_k(D)$ can then be used to obtain an empirical correlation between $\hat{V}_H$ and q_H (Fig. 5.37). The deposition/sublimation C_d is calculated by adjusting the ambient humidity to values consistent with supersaturation with respect to ice. Radiative transfer equations are used to calculate the radiative heating term in (5.68). An absorption coefficient appropriate for cirrus is used.

   If (5.66) is substituted into (5.67) to eliminate $\bar{\xi}$, then (5.67)–(5.70) can be solved for ψ, $\bar{\theta}$, $\bar{q}_v$, and $\bar{q}_H$ as functions of time with appropriate boundary and initial conditions. The two-dimensional wind field can be diagnosed from (5.65). Most important of the boundary and initial conditions is that a 0.5-km-thick cloud layer is assumed to exist initially (Fig. 5.36). The layer is assumed to be saturated, with

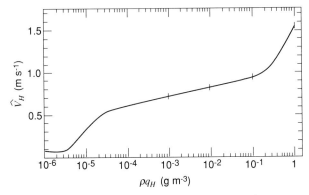

Figure 5.37   Relationship between mass-weighted particle fall speed $\hat{V}_H$ and total ice water mixing ratio q_H used in cirriform cloud model. (From Starr and Cox, 1985a. Reprinted with permission from the American Meteorological Society.)

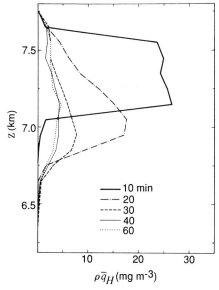

Figure 5.38   Results of model of cirriform cloud formation in a layer of air in which the background vertical motion $w_B = 2$ cm s^{-1}. Vertical profiles of horizontally averaged ice water content $\bar{q}_H$ at various times. (From Starr and Cox, 1985a. Reprinted with permission from the American Meteorological Society.)

a constant θ_e, and to contain random pulses of slightly buoyant air, which turn into convective cells as the cloud layer begins to evolve. Thus, the thin cirriform cloud layer that develops is not horizontally uniform but contains resolvable-scale convective elements—possibly like cirrus uncinus (Fig. 1.15c), cirrus floccus (Fig. 1.15b), or cells and rolls of cirrocumulus (Fig. 1.15a)—in addition to the small, subgrid-scale turbulence parameterized through the eddy mixing terms in (5.67)–(5.70).

Results obtained for a basic-state vertical motion $w_B = 2$ cm s^{-1} are shown in Figs. 5.38–5.40. Most evident in Fig. 5.38 is the strong effect of the fall speed of the crystals that comprise the cloud. With time, the height of the cloud base and the height and amount of the maximum ice concentration decrease substantially, especially during the first 20 min of the lifetime of the ice-cloud layer. Figure 5.39 shows the vertical profiles of heating in the cloud layer associated with latent heat release as well as infrared, solar, and net radiation at three different stages of the lifetime of the cirriform cloud layer. The latent heating by vapor deposition where the cloud has formed and cooling by sublimation of ice below the cloud layer are generally of the same order of magnitude as the radiative heating and cooling. A somewhat surprising result is that the net radiative heating curve indicates little if any destabilization of the cloud layer. These properties of the model of cirriform cloud formation suggest that the assumptions we made in Sec. 5.3.3, which involved viewing a cirriform cloud layer as a radiatively driven mixed layer, may not always be met. In that treatment, the fall speed of the ice particles was tacitly neglected and the profile of heating used was one appropriate for an extremely high layer of cirriform cloud.[153] From Fig. 5.40, it is evident that despite the lack of strong radiative destabilization, embedded cellular structure remains evident throughout the lifetime of the cirriform cloud layer; however, it decreases in intensity in the later stages.

5.4 Altostratus and Altocumulus

5.4.1 Altostratus and Altocumulus Produced as Remnants of Other Clouds

Like high-level cirriform cloud, middle-level altostratus and altocumulus may be produced as remnants of nimbostratus or cumulonimbus. Figure 5.41 is the result of a numerical simulation (using a model of the type to be discussed in Sec. 7.5.3) of a growing tropical cumulonimbus, which can be seen to issue protruding layers of cloud in middle (as well as lower) levels.[154] This behavior is related to the variation of the horizontal wind field in and around the cloud.

5.4.2 Altocumulus as High-Based Convective Clouds

Altostratus and altocumulus may also form in layers aloft, completely removed from any deep nimbostratus or cumulonimbus source. One important type is simply cumulus or cumulonimbus with middle-level bases (i.e., altocumulus cas-

---

[153] The fact that these assumptions may not always be met was recognized and pointed out by Lilly (1988).

[154] That these simulations represent realistic cloud structures over the equatorial Atlantic Ocean was verified by means of stereoscopic cloud photography by Warner *et al.* (1980).

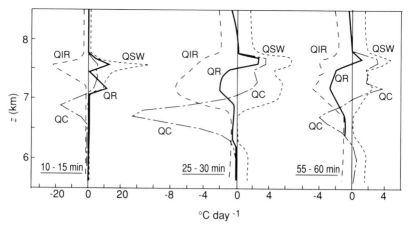

Figure 5.39

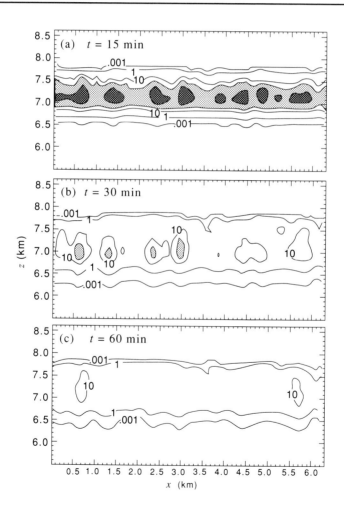

Figure 5.40

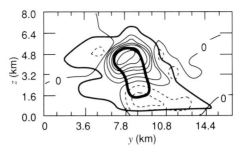

Figure 5.41 Model simulation of a cumulus congestus cloud over the tropical eastern Atlantic Ocean. Cross section runs north–south through a three-dimensional domain. Heavy contours show liquid water content in g m^{-3}. Thin isopleths are for vertical velocity at 1 m s^{-1} intervals; dashed contours are for downdrafts. (From Simpson and van Helvoirt, 1980.)

tellanus, Fig. 1.14d). The dynamics of these cloud elements are well described by the cumulus and cumulonimbus dynamics discussed in Chapters 7 and 8.

5.4.3 Altostratus and Altocumulus as Shallow-Layer Clouds Aloft

Many altostratus and altocumulus clouds are neither the remnants of other clouds nor cumulus or cumulonimbus clouds aloft. These clouds include altostratus layers such as the one seen in Fig. 1.13 and altocumulus stratiformis (Fig. 1.14a and b). These are layer clouds that closely resemble stratus and stratocumulus. However, they do not occur in the planetary boundary layer, and their dynamics are therefore not exactly analogous to those of stratus and stratocumulus (Secs. 5.1 and 5.2). They are more similar to the cirriform cloud layers described in Sec. 5.3 in that they occur in shallow layers of air aloft, separate from the boundary layer.

The dynamics of this type of altostratus and altocumulus can be investigated by means of model calculations like those used in Sec. 5.3.4 to describe a thin layer of cirriform cloud. That model has been modified to middle-level cloud conditions by changing the fall velocity relationship and radiative absorption coefficients to make them appropriate for a cloud of liquid water drops and calculating condensation and evaporation based on 100% relative humidity with respect to liquid wa-

---

Figure 5.39 Results of model of cirriform cloud formation in a layer of air in which the basic-state vertical motion $w_o = 2$ cm s^{-1}. Vertical profiles of horizontally averaged heating by phase changes of water (QC), infrared radiation (QIR), absorption of solar radiation (QSW), and net radiative processes (QR). (From Starr and Cox, 1985a. Reprinted with permission from the American Meteorological Society.)

Figure 5.40 Results of a model of cirriform cloud formation in a layer of air in which the basic-state vertical motion $w_o = 2$ cm s^{-1}. Ice–water mixing ratio field in μg g^{-1} at various times. Contours for 20 and 40 μg g^{-1} are shown by shading. (From Starr and Cox, 1985a. Reprinted with permission from the American Meteorological Society.)

ter.[155] In calculations illustrated in Fig. 5.42, the fall velocity was assumed to be 0.9 cm s^{-1}—two orders of magnitude less than that assumed in the case of cirriform cloud ($\sim$1 m s^{-1}, as in Fig. 5.37). Because of the smaller particle fall speeds, the simulated middle-level cloud retains its hydrometeors throughout the cloud lifetime, has larger mixing ratios of hydrometeor mass, and is contained in a shallower layer than the cirriform cloud (cf. Fig. 5.40 and Fig. 5.42). As in the cirriform cloud layer, the latent heat release and radiative heating are of the same order of magnitude (Fig. 5.43). However, unlike the model cirriform cloud layer, the radiative heating produces destabilization in the middle-level cloud, strongly cooling the upper part of the cloud layer and warming it below. Consequently, turbulent kinetic energy is maintained throughout the cloud lifetime (Fig. 5.44).

[155] This calculation was carried out by Starr and Cox (1985a,b) as part of the same study in which they modeled the thin layer of cirrus. Although this calculation was originally described as representing altostratus, it probably better represents altocumulus stratiformis, since altostratus is often deeper than the cloud assumed in the simulation and is more often than not glaciated while altocumulus stratiformis is shallow, contains overturning convective elements, and tends to consist of supercooled liquid water (at least during its earlier stages).

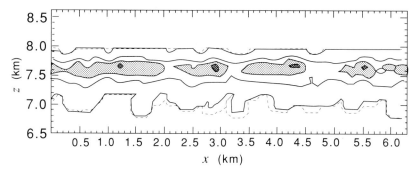

Figure 5.42

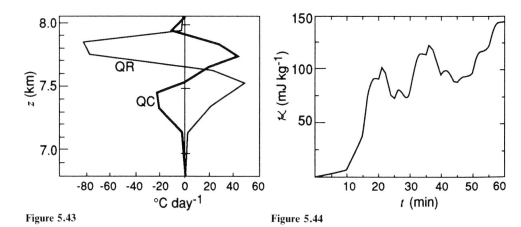

Figure 5.43

Figure 5.44

Thus, a middle-level cloud layer (altocumulus stratiformis) could be described appropriately by the theory of a cloud-filled radiatively driven mixed layer aloft, which was used in Sec. 5.3.3 to describe the layer of ice cloud generated by cumulonimbus.

The interpretation of altocumulus stratiformis as a cloud-filled radiatively driven mixed layer helps to explain its structure since the elements of altocumulus stratiformis often appear to be produced by Rayleigh–Bénard convection (as described in Sec. 2.9.3) in the form of cells (Fig. 1.14a) or rolls (Fig. 1.14b). The differential heating mechanism driving the Rayleigh–Bénard convection is evidently the infrared radiation, and the presence or absence of shear likely determines whether or not the elements take the form of rolls (Fig. 5.20). The average distance between altocumulus (and cirrocumulus) rolls has been reported to be <0.25 km in 39% of cases, <0.5 km in 78% of cases, and <0.75 km in 93% of cases.[156] The observed depths of the cloud layers moreover indicated a horizontal-to-vertical aspect ratio ~1:1, which is consistent with the rolls being of the Rayleigh–Bénard type.

In some cases, altocumulus rolls are produced by shear instability, probably of the Kelvin–Helmholtz type (Sec. 2.9.2), or they may be the result of mixed thermal and shear instability. These distinctions can be difficult to make in visual observations of clouds.

5.4.4 Ice Particle Generation by Altocumulus Elements

The individual elements (rolls or cells) of altocumulus stratiformis tend to glaciate and produce fallstreaks of precipitating ice particles in the later stages of their lifetimes (Fig. 5.45). The process appears to be similar to that which occurs in stratocumulus (Fig. 5.12). In their glaciating stage, the elements of altocumulus

---

[156] Data reported by Süring (1941) and discussed further by Borovikov *et al.* (1963). The book of Borovikov *et al.* (1963) is of interest as one of the first attempts at a comprehensive treatment of cloud dynamics.

Figure 5.42 Results of a model of middle-level cloud formation. Nighttime conditions were assumed and the basic-state vertical motion was $w_B = 2$ cm s^{-1}. Liquid–water mixing ratio field in contours of 0.001, 1, 50, 100, and 150 μg g^{-1}. (From Starr and Cox, 1985b. Reprinted with permission from the American Meteorological Society.)

Figure 5.43 Results of a model of middle-level cloud formation. Nighttime conditions were assumed and the basic-state vertical motion was $w_B = 2$ cm s^{-1}. Vertical profiles of horizontally averaged heating by phase changes of water (QC) and infrared radiation (QR). (From Starr and Cox, 1985b. Reprinted with permission from the American Meteorological Society.)

Figure 5.44 Results of a model of middle-level cloud formation. Nighttime conditions were assumed and the basic-state vertical motion was $w_B = 2$ cm s^{-1}. Domain-averaged turbulent kinetic energy as a function of time. (From Starr and Cox, 1985b. Reprinted with permission from the American Meteorological Society.)

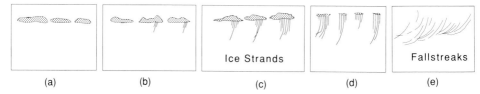

Figure 5.45 Schematic showing progressive stages in the glaciation of altocumulus clouds. The time interval between the young supercooled water clouds depicted in (a) and the ice particle fallstreaks in (e) can be several hours. (From Hobbs and Rangno, 1985. Reprinted with permission from the American Meteorological Society.)

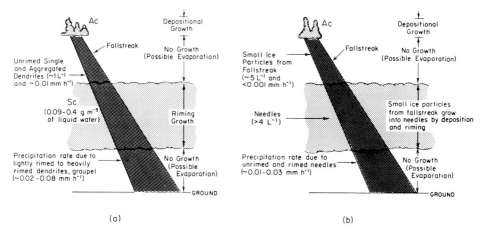

Figure 5.46 Schematic of interaction of altocumulus and stratocumulus clouds. (a) Dendritic crystals in a fallstreak from the altocumulus fall into the stratocumulus, where they grow by riming and reach the ground as light precipitation. (b) Small ice particles in a fallstreak from the altocumulus grow into needles in the stratocumulus and contribute to the precipitation at the ground. (From Locatelli *et al.*, 1983. Reproduced with permission from the American Meteorological Society.)

may have the appearance of altocumulus floccus (Fig. 1.14c), which are similar to cirrus floccus (Fig. 1.15b). Or they may develop fallstreaks and appear somewhat similar to cirrus spissatus with virga (Fig. 1.16) or, in very late stages after the original cloud element has disappeared, to cirrus uncinus (Fig. 1.15c). The more cumuliform altocumulus castellanus (Fig. 1.14d) is also observed to produce fall-streaks of precipitating ice particles in its later stages.

5.4.5 Interaction of Altocumulus and Lower Cloud Layers

When a layer of stratus or stratocumulus cloud is present below altocumulus elements, ice particles may fall into the lower cloud from the altocumulus layer. This process is illustrated in Fig. 5.46. In this case, a layer of stratocumulus was stimulated to precipitate as a result of the interaction of the two cloud layers.

Fallstreaks produced during the latter stages of the altocumulus elements (as in Fig. 5.45) penetrated through the lower stratocumulus cloud layer. The ice particles in the fallstreaks grew by riming and deposition while they passed through the supercooled stratus cloud. Thus, water condensed within the stratus layer was transferred to the ice particles produced aloft and ended up on the ground as part of the precipitation. The water in the stratus cloud would probably otherwise have never reached the ground as precipitation since the drops would have remained too small to fall out of the lower-level cloud.

Chapter 6 | Nimbostratus

"Now downwards by the world's attraction driven,
That tends to earth, which had upris'n to heaven;"[157]

The shallow-layer clouds, which we considered in Chapter 5, do not produce much precipitation because they do not have sufficient vertical extent. The few particles that grow large enough to precipitate usually fall out of the cloud layer before they exceed the size of small drizzle. In this chapter, we consider deep stratiform clouds from which significant amounts of rain or snow fall. These clouds, referred to as *nimbostratus* (see Sec. 1.2.2.2), occur extensively in both the midlatitudes and the tropics. They are associated primarily with the widespread continuous clouds of mesoscale convective systems, hurricanes, and extratropical cyclones (e.g., Figs. 1.27–1.29).[158] The structure and dynamics of these cloud systems will be considered in Chapters 9–11. These storms are large and complex, involving air motions on a range of scales and various mixtures of convective and stratiform cloud structures. As background for these later chapters, the present chapter isolates and discusses certain generic properties of the nimbostratus clouds that occur within these cloud systems. Chapters 7 and 8 will describe the properties of purely convective clouds. This preliminary examination of both the nimbostratus and convective clouds by themselves is prerequisite to our later discussions of complex cloud systems in Chapters 9–11.

Since the widespread vertical air motions essential to the production of nimbostratus cloud are largely those of the parent storm (i.e., the mesoscale convective system, hurricane, or extratropical cyclone), the dynamics of nimbostratus cannot be fully treated in this chapter. That treatment must await the later chapters. The emphasis here will be on the microphysical and convective processes that produce nimbostratus, regardless of the nature of the parent storm. What is known about these aspects of nimbostratus has been inferred largely from observations of the cloud microphysics and radar-echo structure of this cloud type and its precipitation. From radar and microphysical data, certain characteristics of the air motions that must be present within nimbostratus to produce the radar echoes and microphysical structure become evident. In particular, it can be deduced that nimbostratus does not occur alone but rather is closely associated, in each of its various contexts, with convective cloud processes.

---

[157] Goethe on precipitation falling from a cloud.

[158] It is a common misconception that nimbostratus is a midlatitude phenomenon associated primarily with extratropical cyclones.

We begin the chapter by examining the observed radar-echo structure and microphysics of the precipitation that falls from nimbostratus (Sec. 6.1). In Secs. 6.2 and 6.3, we explore the relationship of the nimbostratus to convection, which appears to be essential to the precipitation processes in nimbostratus. Finally, in Sec. 6.4, we examine briefly the radiative processes and turbulent mixing that occur within nimbostratus.

6.1 Stratiform Precipitation

6.1.1 Definition and Distinction from Convective Precipitation

Precipitation is generally considered to be of two clearly distinguishable types— *stratiform* and *convective*. Stratiform precipitation falls from nimbostratus clouds, while convective precipitation falls from cumulus and cumulonimbus clouds. A simple comparison of the salient features of these two basic types of precipitation is provided by Fig. 6.1. We will use this diagram to aid in making more precise definitions of stratiform and convective precipitation. Although our present purpose is to examine stratiform precipitation, with convective clouds and precipitation being the subject of Chapters 7–9, it is useful here to define both types of precipitation, as the definitions are better understood in contrast with each other.

Stratiform and convective precipitation can be defined in terms of their vertical velocity scales. Stratiform precipitation is defined as a precipitation process in which the vertical air motion is small compared to the fall velocity of ice crystals and snow; more specifically, the vertical velocity of the air w satisfies the condition

$$|w| < V_{\mathrm{ice}} \qquad (6.1)$$

where V_{ice} represents the scale of the terminal fall velocity of ice crystals and snow (~ 1–3 m s^{-1}). Ice particles in the upper levels of the cloud play an important role in the precipitation process, and in (6.1) the magnitude of the vertical air motion is small compared to the magnitude of the fall velocity of the ice particles.

It is possible for warm stratiform clouds (tops below the 0°C level) to produce precipitation. We have already seen in Chapter 5 that stratus and stratocumulus may produce drizzle. When low-level warm stratiform clouds are a little deeper or more vigorous, they may produce some light rain. However, by far the bulk of stratiform precipitation falls from nimbostratus that reaches well above 0°C level.

The stratiform precipitation process in deep nimbostratus containing ice particles aloft is depicted schematically in Fig. 6.1a. Precipitation particles that eventually fall to the ground as raindrops have their early history as ice particles in the upper parts of the cloud. We will defer until later a discussion of the important question of how these particles first appear in the upper reaches of the cloud. For now, we note only that they may form *in situ;* they may be introduced from a

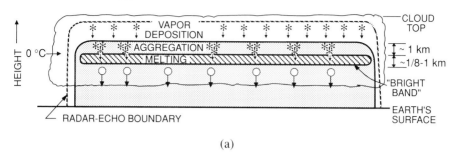

(a)

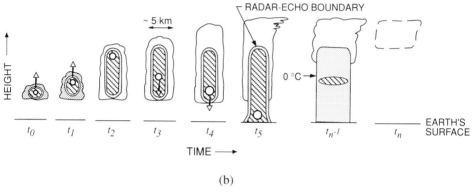

(b)

Figure 6.1 (a) Characteristics of stratiform precipitation. (b) Characteristics of convective precipitation. Shading shows higher intensities of radar echo, with hatching indicating the strongest echo. In (b) cloud is shown at a succession of times $t_o, \ldots, t_n$. Growing precipitation particle is carried upward by strong updrafts until t_2 and then falls relative to the ground, reaching the surface just after t_5. After t_5, the cloud may die or continue for a considerable time in a steady state before dissipation sets in at t_{n-1} and t_n. The dashed boundary indicates an evaporating cloud. (From Houze, 1981. © American Geophysical Union.)

source located to the side of or above the nimbostratus; or they may originate from a source embedded within the cloud itself. Once introduced into the nimbostratus cloud, the ice particles begin to grow. Upward air velocity maintains supersaturation by condensing vapor, which is deposited onto the ice particles (Sec. 3.2.3). Although w must be large enough to maintain supersaturation, it must be small enough not to violate (6.1). The ice particles in the upper levels of nimbostratus *must fall;* they cannot be suspended or carried aloft by the air motions as they grow. Thus, the general in-cloud vertical air motion in pure nimbostratus normally does not exceed a few tens of centimeters per second. These vertical velocities support growth of particles by vapor deposition.

The higher the level at which the ice particles are formed or introduced from an outside source, the longer they will be able to grow by deposition of the vapor made available as a result of mean upward air motion in the nimbostratus. The time available for growth of the ice particles falling from cloud top is $\sim$1–3 h (the time it takes a particle falling at $\sim$1–3 m s^{-1} to descend 10 km). This time is sufficient for the particles to grow by deposition. When the ice particles falling and growing by deposition descend to within about 2.5 km of the 0°C level, aggregation and riming (Sec. 3.2.4) can occur. Most importantly, the particles begin to aggregate and form large, irregularly shaped snowflakes. The particles may also grow by riming, since at these warmer levels the vertical air motions of a few tens of centimeters per second are sometimes strong enough to maintain a small number of liquid-water drops in the presence of the falling ice particles. As noted in Sec. 3.2.4, the aggregation becomes more frequent within about 1 km of the 0°C level. Aggregation, of course, does not add mass to the precipitation but rather concentrates the condensate into large particles, which, upon melting, become large, rapidly falling raindrops. As will be explained below, the layer in which the large snowflakes melt is marked on radar by a *bright band* of intense echo in a horizontal layer about 1/2 km thick located just below the 0°C level. The bright band produced by melting particles and other characteristics of the vertical structure of the radar echo in stratiform precipitation are particularly informative about the dominant precipitation mechanisms and will be discussed in more detail in Sec. 6.1.2.

The convective precipitation process is depicted in Fig. 6.1b. It differs sharply from stratiform precipitaiton. It is defined as a precipitation process in which condition (6.1) is not met. Instead it has a vertical air velocity scale of $w \sim 1$–10 m s^{-1}, which equals or exceeds the typical fall speeds of ice crystals and snow.

It is observed that the time available for the growth of precipitation particles in convective precipitation is limited; often rain reaches the ground within a half an hour of cloud formation. This time is much shorter than the 1–3 h available for growth of precipitation particles in stratiform precipitation.[159] Since the time is so short, the precipitation particles must originate and grow not far above cloud base at the time the cloud forms (time t_o in Fig. 6.1b). It is possible for the growth to begin at that time since updrafts are strong enough to carry the growing particles upward until they become heavy enough to overcome the updraft and begin to fall relative to the ground (see the particle trajectory in Fig. 6.1b). The only microphysical growth mechanism rapid enough to allow the particles to develop this quickly is accretion of liquid water. In contrast, accretion of liquid water is at most of minor importance in stratiform precipitation, where the dominant microphysical mechanisms are vapor deposition and aggregation of ice particles. Since the strong convective updrafts carrying the particles upward during the growth phase of the cloud condense large amounts of liquid water, the larger particles (whether they be liquid or ice) in the rising parcels of air can grow readily by

[159] The idea of distinguishing convective from stratiform precipitation on the basis of the magnitude of the in-cloud vertical air motion and time scale of the microphysical precipitation growth process was suggested by the pioneering cloud physicist Henry Houghton in one of his last papers (1968). An official definition of the term stratiform precipitation does not exist; it is absent from the *Glossary of Meteorology* (Huschke, 1959). Houghton's distinction, however, captures the physical essence of the two precipitation types. Further aspects of the difference were articulated by the early radar meteorologist Louis Battan in his textbooks on meteorological radar (1959, 1973). He denoted the two types of precipitation as "convective" and "continuous," the latter corresponding to our stratiform precipitation. In describing continuous precipitation, he emphasized the radar bright band, the existence of embedded nonuniform echo structure in the continuous precipitation, and how convective echoes take on a bright band structure in their late stages. The views depicted in Figs. 6.1a and b are largely a combination of those of Houghton and Battan.

accretion of cloud liquid water. Aircraft observations of ice particles in convective clouds confirm that the particles in the updrafts grow largely by riming. Since the strong updrafts in convective clouds are usually narrow (typically ~1 km or less in width), radar echoes from precipitation associated with active convection form well-defined vertical cores of maximum reflectivity, which contrast markedly with the horizontal orientation of the radar bright band seen at the melting level in stratiform precipitation (compare the reflectivity patterns in Figs. 6.1a and b).

In the dissipating stages of precipitating convective clouds (after time t_5 in Fig. 6.1b), strong upward motions cease and no longer carry precipitation particles upward or suspend them aloft. The fallout of the particles left aloft by the dying updrafts can take on a stratiform character, including a radar bright band.

6.1.2   Radar-Echo Structure

The schematic in Fig. 6.1a depicts qualitatively the radar-echo structure of stratiform precipitation, with the melting layer denoted by the radar bright band being the most prominent feature. Closer, quantitative examination of the vertical profile of radar data in stratiform precipitation allows us to divide the precipitation into distinct layers in which different microphysical processes dominate. These layers are bounded by points 0–4 in Fig. 6.2.

The zone from 0 to 1 in Fig. 6.2 is associated primarily with ice particles growing by *vapor deposition*, which is the slowest microphysical growth mode. Since the ice particles settling downward through the nimbostratus are not growing rapidly, the equivalent radar reflectivity Z_e, which is proportional to the sixth moment of the particle size distribution [recall Eqs. (4.3) and (4.8)], does not increase rapidly with decreasing altitude between 0 and 1. In the layer between 1

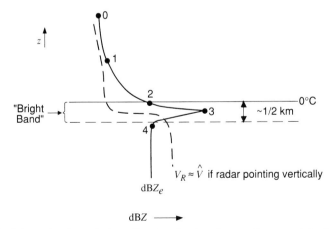

Figure 6.2 Schematic of vertical profile of radar data in stratiform precipitation. Solid curve shows reflectivity. Dashed curve shows Doppler radial velocity V_R with the antenna at vertical incidence. Under stratiform conditions V_R is related approximately to the mass-weighted terminal fall speed of the particles, $\hat{V}$. Layers in which different microphysical processes dominate are bounded by points 0–4 (see text for further discussion).

and 2, the particles continue to grow by deposition and possibly some riming. However, they also undergo aggregation with increasing frequency as they approach the melting layer. The aggregation has the effect of producing very large particles. Since Z_e depends on the sixth power of the particle dimension, the formation of these large particles sharply increases the radar reflectivity between levels 1 and 2.

The changes in the radar reflectivity profile between levels 2 (the 0°C level) and 4 are all associated with *melting*. The center of this layer (level 3) is marked by a sharp maximum of Z_e, which gives the melting layer its identity as the bright band. Several processes strongly affect the radar reflectivity profile in the melting layer. The sharp increase in the radar reflectivity downward from level 2 to 3 is thought to be the result of two effects. First, aggregation continues to occur, as in the layer from 1 to 2. This conclusion is arrived at inductively, since other effects cannot fully explain the increase in reflectivity factor from point 2 to 3. The second effect on Z_e that appears to be significant is that the magnitude of the complex index of refraction of the particles [$|K|^2$ in (4.6)] changes as they melt from 0.197 (for ice particles) to 0.93 (for liquid water drops). If in the first stage of melting, the particles take on the character of water but do not collapse to form smaller drops until the end of the melting process, then Z, which is proportional to the sixth moment of the particle size distribution [Eq. (4.3)], remains constant. However, Z_e increases by a factor of $0.93/0.197 \approx 5$ since, according to (4.3) and (4.6), $Z_e = (|K|^2/0.93)Z$. As noted above, this amount of increase in Z_e between points 2 and 3 is insufficient by itself to explain the magnitude of the peak of reflectivity in the bright band, and it is for this reason that it is thought that the aggregation acts in concert with the index of refraction change to produce the peak at 3.

The sharp dropoff of Z_e in the lower portion of the melting layer between levels 3 and 4 is produced by two effects. If the melting is completed at point 3 and all the particles collapse to form small raindrops, two things happen to Z_e. First, the particles are now all smaller, and Z_e is decreased in accordance with the sixth power of all the particle diameters. Second, the fall speed of the particles suddenly increases from ~ 1–3 m s^{-1} for snow to ~ 5–10 m s^{-1} for raindrops. If the downward flux of precipitation mass is the same at levels 3 and 4, as it would be in steady stratiform rain, then the mean concentration of rainwater (mass of water per volume of air) must decrease sharply from level 3 to level 4. Since the concentration of rainwater mass is the third moment of the drop size distribution and Z_e is proportional to the sixth moment of the distribution [recall (4.3) and (4.10)], the decrease in mass concentration between levels 3 and 4 corresponds to a decrease in Z_e through the same layer.

The layer below point 4 in Fig. 6.2 is characterized by rain. The microphysical processes that can occur in this lowest layer are quite varied and depend on the meteorological context. In some cases, the precipitation particles in this lower region simply fall to the ground at a constant rate. In other cases, the rain falls through a lower layer of cloud being continually regenerated by upward air motion, and the raindrops falling below the melting layer continue to grow by vapor diffusion and collection of cloud droplets. In still other cases, the rain falling out of the melting layer falls into a layer of dry air into which the drops partially evaporate. Thus, the flux of rain reaching the surface is less than that exiting the melting layer, and the air below the melting layer is cooled by the evaporation. The cooling of the air both by this evaporation and by the melting can have important feedbacks to cloud and storm dynamics, as we will see in later chapters.

6.1.3 Microphysical Observations

Observations of the sizes and shapes of precipitating ice particles above the 0°C level have been made aboard research aircraft flying in nimbostratus associated with mesoscale convective systems, fronts, and hurricanes. These *in situ* measurements verify the microphysical layering of the cloud that we have inferred from the radar echo profile depicted in Fig. 6.2. An example of data obtained in nimbostratus associated with tropical mesoscale convective systems is shown in Fig. 6.3. Particles were observed at flight levels ranging in temperature from −23° to +4°C. Two-dimensional images in the form of shadows were formed by the ice particles as they flowed through a laser-illuminated space under the wing of the aircraft. The crystalline structures of most of the images seen by this device are impossible to determine. However, occasionally a definable particle shape is noted. The frequency with which particles of various types could be recognized was noted and plotted as a function of flight-level temperature.

The particle types indicated in Fig. 6.3a and b (needles, columns, plates, and dendrites) are known to grow at certain ambient temperatures (Table 3.1). If the observed particles fell about 1/2–1 km from their altitude of growth before being encountered by the aircraft,[160] then their growth temperature (T_G) would be expected to be ~3–6°C lower than the flight-level temperatures indicated in Fig. 6.3. Taking this difference into account, we see that the maxima and minima of the curves in Fig. 6.3a and b for needles, columns, plates, and dendrites all occur at flight-level temperatures that are consistent with their respective T_G indicated in Table 3.1. This result indicates that, as we concluded from the radar-echo profile shown in Fig. 6.2, growth of ice particles by vapor diffusion occurs above the 0°C level. Some riming could also have been occurring, as riming would be consistent with the unidentifiable shapes of many of the particle images. No attempt was made, however, to determine whether the irregularities in particle image shape were caused by riming. In Fig. 6.3c, it can be seen that the frequency of large aggregates of crystals exhibits a broad maximum strongly overlapping with the peak of frequency of dendritic crystals seen in Fig. 6.3b. This result is consistent with aggregation becoming more frequent as the downward-settling crystals approach the 0°C level. In addition, it appears that a primary aggregation mechanism, at least in the tropical nimbostratus represented by these particular data, is the arms of the dendrites becoming entangled as the ice crystals drift downward. The nearly round particles indicated in Fig. 6.3c increase in frequency suddenly as the ice particles melt and form raindrops below the 0°C level.

The microphysical picture of nimbostratus indicated by the *in situ* data summa-

---

[160] This amount of downward displacement would require only 5–15 min at fall speeds of 1–3 m s⁻¹.

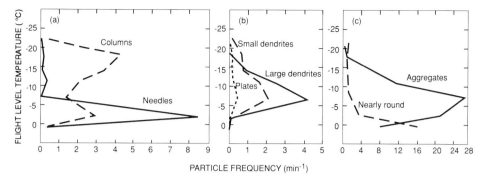

Figure 6.3 Ice particle data obtained on aircraft flights through nimbostratus in tropical mesoscale convective systems over the Bay of Bengal. Plots show relative frequency of observation of ice particles of a particular type per minute of in-cloud flight time as a function of flight-level temperature. (From Houze and Churchill, 1987. Reproduced with permission from the American Meteorological Society.)

rized in Fig. 6.3 is the same as that inferred from the vertical profile of radar reflectivity in Fig. 6.2. The region above the melting layer is characterized by particles growing by vapor diffusion and drifting downward. Some riming could occur along with the vapor diffusion, if the vertical motion becomes strong enough to condense enough water to maintain the air at saturation with respect to liquid water. Riming is most likely to occur just above the 0°C level, since the vapor content of the air increases exponentially with decreasing altitude. Although the vertical air motions might become large enough to allow some riming to occur along with the vapor diffusion, they must remain weak enough to allow the particles to continue to fall [i.e., condition (6.1) must remain satisfied to explain the observations]. As the particles settle downward, aggregation produces large ice particles within ~2–2.5 km of the top of the melting layer, with the most aggregates being found within 1 km of the 0°C level. What this structure implies about the dynamics of the nimbostratus is that there must be widespread gentle uplift of the air throughout the region of the cloud above the 0°C level. This ascent must be strong enough to supply vigorous growth by vapor diffusion (and possibly some riming), but weak enough to allow the sedimentation of the ice particles. Since these particles fall at speeds ~1 m s^{-1}, it is readily concluded that nimbostratus must have typical vertical velocities of a few to a few tens of centimeters per second. Though rather weak compared to the vertical velocities in convective clouds, these gentle upward motions, unlike highly localized convective updrafts, extend over broad areas and thus condense large amounts of water.

   The widespread ascent in nimbostratus can be produced by various mechanisms, depending on the meteorological context. In mesoscale convective systems (Chapter 9), a widespread cloud deck occurs at upper levels in association with the deep convective clouds. This cloud deck is slightly buoyant and a hydrostatic pressure pattern and associated circulation develops. The upward component of this circulation is characterized by vertical velocities ~10–50 cm s^{-1}. In hurricanes (Chapter 10), the pattern of vertical overturning in the inner core of the storm is a cross-vortex circulation required to maintain gradient-wind and hydrostatic balance. After the inflow toward the storm center rises sharply in the convective eyewall, it slopes more gently outward at upper levels. This sloping ascent produces a nimbostratus cloud layer with stratiform precipitation in a region surrounding the eyewall. In frontal clouds (Chapter 11), the widespread gentle ascent in the nimbostratus is part of the vertical overturning produced by the dynamics of a baroclinic wave and frontogenesis. We will examine the dynamics of these sources of gentle ascent in nimbostratus as we consider these specific phenomena in later chapters.

6.1.4   Role of Convection

Whether nimbostratus is observed in connection with mesoscale convective systems, fronts, or hurricanes, it is always found to be associated closely with convective clouds. Thus, the dynamics of nimbostratus are not solely a matter of gentle ascent over a wide area. Rather, there appears to be a symbiotic relationship in which both strong convective-scale vertical motions and widespread ascent play a role. How the convection is important to the nimbostratus is seen by further consideration of the microphysical processes in the stratiform precipitation.

We have seen in Secs. 6.1.2 and 6.1.3 that the precipitation particles that fall to the ground from the nimbostratus grow by vapor deposition at upper levels (between levels 0 and 1 in Fig. 6.2). We have not yet considered the origin of these particles. The radar and aircraft instrumentation, which indicate the orderly and distinct physical profiles in Figs. 6.2 and 6.3, detect particles only after they have formed and reached certain threshold sizes. Nucleation of ice particles certainly occurs readily at the upper levels of the nimbostratus cloud, since the temperature is low enough for primary nucleation of ice crystals to be quite active (Sec. 3.2.2). However, there is a difficulty in envisioning nimbostratus in which particles simply nucleate in the context of the widespread gentle ascent, begin growing by vapor deposition, and eventually start to settle downward through the cloud. The stratiform precipitation areas of nimbostratus are often limited in horizontal extent, and there is usually a relative wind blowing through the cloud. Growth of the recently nucleated crystals by vapor diffusion, moreover, is a rather slow process.[161] Thus, there is a tendency for the ice particles nucleated within the nimbostratus itself to be advected across the cloud before they can grow to a large enough size to contribute to the precipitation reaching the ground below the cloud. The efficacy of the precipitation process within the nimbostratus clearly would be enhanced by any mechanism that sped up the early growth of the particles.

Convection appears to be such a mechanism. The stronger vertical motions in even the smallest of convective updrafts can produce precipitable particles relatively quickly, since large concentrations of water substance are condensed and

---

[161] At the relatively warm temperature of $-5°C$, a small ice particle can grow by deposition to drizzle size in $\sim 1/2$ h, but after that its growth rate slows down (Wallace and Hobbs, 1977).

riming and aggregation are quickly established in the convection. If particles produced in convection are transferred into the upper levels of a nimbostratus cloud, they subsequently drift downward through the region of widespread gentle ascent growing and changing phase by the sequence of processes discussed in connection with Figs. 6.1a, 6.2, and 6.3. Thus, convection can serve as the major *source* of ice particles which *subsequently* fall out of the nimbostratus in the orderly pattern indicated by these figures.

There are two major ways in which convective and nimbostratus clouds become joined to produce the cooperative mechanism of stratiform precipitation. In one case, the convection occurs in a shallow layer embedded in the upper portion of the nimbostratus and drops ice particles into the layers of nimbostratus below. In the second case, the nimbostratus is located next to an area of deep convective clouds. Ice particles grown and carried up to upper levels by the strong convective updrafts of the deep convective clouds fall out in the neighboring stratiform region. In the next two subsections we examine these two forms of nimbostratus.

6.2 Nimbostratus with Shallow Embedded Convection Aloft

The occurrence of a layer of shallow, weak convective cells in a layer aloft is frequently noted in the nimbostratus of certain parts of frontal clouds. Whether these *generating cells* also occur in the nimbostratus cloud decks of mesoscale convective systems or hurricanes is not clear at this time, although there is some

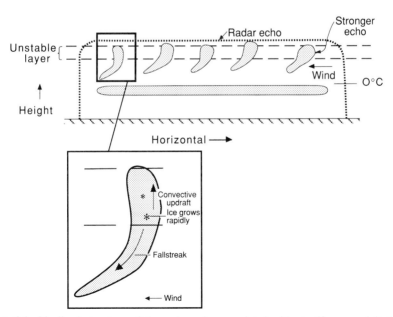

Figure 6.4 Idealized structure of the radar echo associated with stratiform precipitation with generating cells aloft.

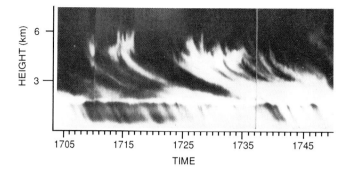

Figure 6.5

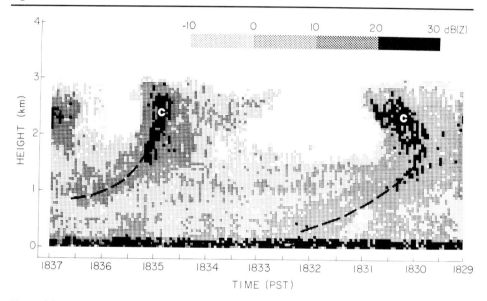

Figure 6.6

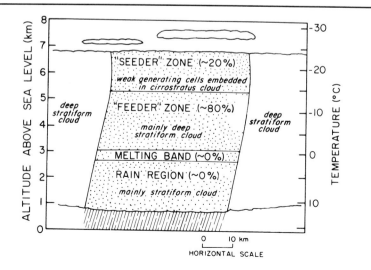

Figure 6.7

evidence to suggest that this may be the case. The present discussion is based primarily on studies of frontal clouds.

The structure of the radar echo associated with stratiform precipitation with generating cells aloft is shown schematically in Fig. 6.4. Except for the generating cells themselves, the radar echo is similar to that in Fig. 6.1a, with the basic stratiform structure of the precipitation indicated by the bright band in the melting layer. A shallow layer of potentially unstable air ($\partial\theta_e/\partial z < 0$) is located aloft, and within this layer convective cells form. As indicated in the inset of Fig. 6.4, the vigorous updrafts in these cells allow rapid growth of precipitable ice particles, which then fall out in a fallstreak below the cell. The fallstreak is typically curved as a result of the wind shear in the layer below the generating cells. The structure of the generating cell and its fallstreak are essentially the same as the cirrus uncinus described in Sec. 5.3; the main difference is that the generating cells are embedded in precipitation, while the cirrus uncinus elements exist in clear air.[162] Examples of radar data showing generating cells embedded in stratiform precipitation are shown in Figs. 6.5 and 6.6.

Two analyses of frontal stratiform precipitation regions are shown in Figs. 6.7 and 6.8. These analyses were based on Doppler radar measurements of the reflectivity and air motions, radiosonde soundings showing the thermodynamic structure, and aircraft measurements of the cloud microphysical characteristics of the nimbostratus. In both cases, the water budget of the analyzed region has been examined. In the case shown in Fig. 6.7, the melting layer was located at about the height of a cold front, below which the vertical air motion was weak. The zone of upward motion producing the condensation above this level was located entirely above the 0°C level. The region of stronger updraft was, moreover, limited in horizontal extent to the 50-km-wide zone shaded in Fig. 6.7. As we will see in Chapter 11, it is typical for frontal lifting to be concentrated into such mesoscale "rainbands" rather than to be spread evenly across the frontal cloud. In Fig. 6.7, it is indicated that 20% of the mass of precipitation that fell from the rainband was produced in the form of ice particles that fell out of the generating cells. The layer

---

[162] The generating cells in stratiform precipitation were first noted by Marshall (1953). He pointed out their similarity to cirrus uncinus by referring to the "characteristic 'mares' tail' pattern of falling snow observed in vertical section by radar..." and noting that the cells and their tails resembled the "hair-like texture of virga."

Figure 6.5   Time–height record of radar-echo intensity showing generating cells (at about 6 km) embedded in stratiform precipitation. Data were obtained with a vertically pointing radar in Quebec, Canada. (Courtesy of Roddy Rogers, Stormy Weather Group, McGill University.)

Figure 6.6   Time–height cross section of the radar reflectivity measured with an 8.6-mm-wavelength radar in stratiform precipitation on the Pacific Coast of Washington State. Generating cells are indicated by C. Dashed lines indicate fallstreaks from the cells. (From Businger and Hobbs, 1987. Reprinted with permission from the American Meteorological Society.)

Figure 6.7   Schematic cross section of a wide cold-frontal rainband in a front passing over western Washington State. (From Hobbs *et al.*, 1980. Reproduced with permission from the American Meteorological Society.)

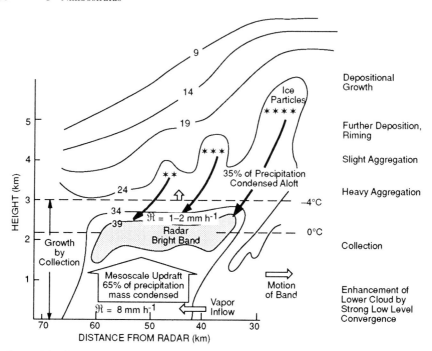

Figure 6.8   Schematic of the dynamical and microphysical processes in a stratiform frontal rainband over western Washington State. (From Houze *et al.*, 1981.)

of the generating cells is dubbed the *seeder* zone, since it seeds the layer of cloud below with ice particles. The zone below is called the *feeder* zone,[163] because the ice particles from the seeder zone increase greatly in mass as they pass through this layer of active depositional (and possibly riming) growth. The remaining 80% of the mass of precipitation was condensed in this layer, where the enhanced general uplift associated with the rainband was making vapor available to the ice crystals produced in the generating cells. Thus, the convective generating cells and the rainband dynamics, which produced the region of enhanced general ascent in the nimbostratus, combined to produce the stratiform rainfall.

The nimbostratus situation depicted in Fig. 6.8 is another example of a frontal rainband. Fallstreaks from generating cells aloft are seen above the radar bright band.[164] This case differed from Fig. 6.7 in that the mesoscale ascent in the rainband extended to low levels so that the precipitation particles continued to grow by collection of cloud water in and below the layer of the bright band. As indicated in the figure, 65% of the mass of the precipitation was added to the mass

---

[163] The basic idea of the feeder–seeder mechanism is attributed to T. Bergeron (1950). He proposed it specifically to explain the behavior of orographic clouds. The concept, however, has found wider applicability and is useful in explaining the behavior of nimbostratus in general.

[164] The fallstreaks probably all originated in generating cells at some common height (e.g., 5–6 km); however, in the cases of the first two fallstreaks on the left of Fig. 6.8, the parent cells had apparently already died or were located somewhere out of the plane of the cross section.

of the particles originating in the generating cells below the 0°C level. The vertical layering of the precipitation processes in this example is further illustrated by a numerical simulation, whose design is indicated in Fig. 6.9. Calculations were made for the feeder zone. Within this region the vertical and horizontal air motions were assumed to be those determined from Doppler radar measurements. The vertical velocity, whose vertical profile is shown in the figure, ranged from ~20 to 50 cm s^{-1}, values ideal for the stratiform precipitation process. Ice particles were assumed to be entering the feeder zone from generating cells aloft. To represent the source of ice particles from the generating cells, the mixing ratio of falling snow was held at 1 g kg^{-1} at the top boundary of the feeder zone throughout the calculation. As the snow continually fell into the region, two-dimensional (no variation in the y-direction) versions of the mean-variable thermodynamic equation (2.78) and water-continuity equations (2.81) were integrated until steady-state fields were obtained. The eddy-flux terms were omitted. The only diabatic heating effects included in $\bar{\mathcal{H}}$ were those associated with latent heat of phase change of water. Radiation was ignored. The categories of water substance assumed in the water-continuity equations were those of the cold-cloud bulk water-continuity model considered in Sec. 3.6.2: water vapor (q_v), cloud water (q_c), rain water (q_r), cloud ice (q_I), and snow (q_s). Higher-density snow (q_g and q_h) is ignored since graupel and hail particles do not exist in any significant amount in this type of nimbostratus. Results of the calculation are shown in Fig. 6.10. In the top layers of the feeder zone, vapor deposition is the only mode of growth of the snow falling from the generating cells (Fig. 6.10a). In a 1–1.5-km-thick layer lying just above the 0°C level, riming is the most important process. This result is consistent with the aircraft observations of some rimed particles in this region (Fig. 6.8). It arises because the vertical velocity (~10 cm s^{-1}) in the lower portion of the region above the 0°C level is strong enough to maintain water saturation and a small amount of liquid water in the presence of the snow falling from above. A pronounced melting layer is seen in Fig. 6.10c. Below the 0°C level (Fig. 6.10d), the raindrops that formed from the melting snow, before being generated and growing as ice particles aloft, are able to continue to grow by the collection of cloud droplets since the upward air motion and hence the feeder cloud extend into the warm air below the 0°C level.

This example illustrates the full range of precipitation growth processes that can occur in nimbostratus. A variety of other stratifications of processes could have occurred from just above the melting layer down into the lower levels of the nimbostratus if the vertical distribution of vertical air motion had been different from that in Fig. 6.9. If the region of enhanced upward motion had extended down to but not below the 0°C level, then the sequences of processes would have included those shown in Fig. 6.10a–c, but the coalescence growth in Fig. 6.10d would have been absent. If the lifting had been confined to levels above 3 km, neither the coalescence nor the riming would have occurred and the processes would have included only the depositional growth (Fig. 6.10a) followed by melting (Fig. 6.10c). If below the 0°C level downward rather than upward motion had occurred (a fairly common situation in both frontal and mesoscale convective

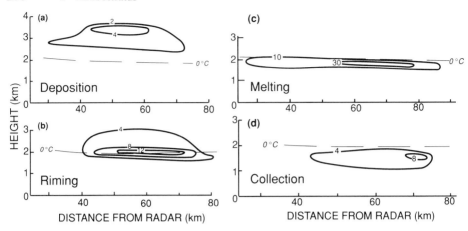

Figure 6.10 Results of numerical-model simulation of the precipitation processes in frontal stratiform precipitation. (a) Depositional growth of snow. (b) Riming of snow by collection of cloud water. (c) Melting of snow. (d) Collection of cloud water by rain. All in units of 10^{-4} g kg^{-1} s^{-1}. (From Rutledge and Hobbs, 1983. Reprinted with permission from the American Meteorological Society.)

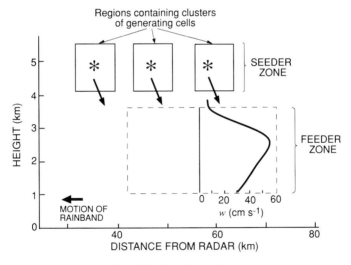

Figure 6.9 Design of numerical-model simulation of the precipitation processes in frontal stratiform precipitation. The domain of the model calculations is surrounded by the dashed lines. Asterisks represent ice particles which grow in the generating cells and fall into the region of the model domain. (From Rutledge and Hobbs, 1983. Reprinted with permission from the American Meteorological Society.)

system nimbostratus), the vapor deposition, riming, and melting seen in Fig. 6.10a–c would have been followed by a decrease in raindrop mass by evaporation rather than the growth by coalescence seen in Fig. 6.10d. Thus, the exact form of nimbostratus obtained when ice particles are provided by generating cells aloft

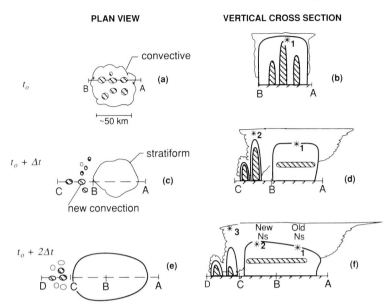

Figure 6.11   Conceptual model of the development of nimbostratus associated with deep convection. Panels (a), (c), and (e) show horizontal radar echo pattern at the earth's surface with two levels of intensity at three times, t_o, $t_o + \Delta t$, and $t_o + 2\Delta t$. Panels (b), (d), and (f) show corresponding vertical cross sections. A sketch of the visible cloud boundary has been added to the vertical cross sections. Asterisks trace the fallout of three ice particles.

depends on the vertical distribution of the widespread vertical motion in the layers below the cells. The vertical distribution of w in turn depends on the type of mesoscale dynamical setting in which the nimbostratus is occurring.

6.3   Nimbostratus Associated with Deep Convection

Precipitating ice particles can also be introduced into the upper levels of nimbostratus by deep convection. One way in which this process can occur is illustrated schematically in Fig. 6.11. All the panels in the figure are considered to be in a coordinate system moving with the clouds. Moreover, the convective and stratiform clouds and their attendant precipitation are all assumed to be moving with a constant horizontal speed relative to the ground at all levels (i.e., there is no horizontal air motion relative to the coordinate system at any altitude). This situation can obtain only if there is no vertical shear of the horizontal wind in the region of the clouds.[165] As was pointed out in Fig. 6.1b, a precipitating convective

---

[165] The assumption of absolutely no shear ignores the small horizontal wind components required by mass continuity to compensate for the vertical air motions in the cloud system. This slight inconsistency is not serious, as these small horizontal components would only slightly modify the idealized picture developed in this illustration.

cell, originally characterized as a local, intense, vertically oriented precipitation core, evolves into a stratiform structure (indicated by a radar bright band) in its later stages. It follows that if all of the convective cells initially arranged in a group at a time t_o (e.g., Fig. 6.11a and b) weaken more or less simultaneously, then by some later time, $t_o + \Delta t$, the whole region of convection turns into a stratiform precipitation area composed of the dying remains of the earlier convective cells, as shown in the right-hand portions of Fig. 6.11c and d. If during the same interval of time a new group of convective cells forms as shown in the left-hand portions of Fig. 6.11c and d, then the overall cloud structure becomes that depicted in Fig. 6.11d—a combination of cumulonimbus and nimbostratus. The right-hand side of the cloud (between A and B) is nimbostratus that has formed from the weakening of the original convection, while the left-hand side (between B and C) is composed of new cumulonimbus. If the new convection at $t_o + \Delta t$ subsequently weakens and evolves into nimbostratus and another region of convection forms farther to the left, then the cloud and precipitation structure at time $t_o + 2\Delta t$ is that shown in Fig. 6.11e and f. The region on the far right (between A and B) is composed of the old nimbostratus formed from the original convection, the middle region (between B and C) is new nimbostratus formed from the weakening of the convection at time $t_o + \Delta t$, and the region on the far left (between C and D) consists of the newest cumulonimbus.

In this way, a rather wide region of nimbostratus can form, as each successive new group of convective cells weakens and becomes part of the stratiform zone. The nimbostratus formed in this way is typically maintained for a long time by widespread gentle ascent throughout the upper levels of the nimbostratus. The widespread upward air motion that maintains the nimbostratus may be solely the effect of the small but positive buoyancy of the air in the tops of the cumulonimbus that weaken to form the stratiform cloud. This widespread ascent may also be part of the mesoscale circulation of a squall line, cold front, or hurricane, if the clouds under consideration are occurring in the context of one of these types of storms, which will be considered in more detail in Chapters 9–11.

The widespread ascent in the nimbostratus is typically not uniform. Over most of the region condition (6.1) is met, but the cloud is composed of old convective cells. Some of these cells remain moderately active. Consequently, weak, dying convective cells are typically found embedded within the generally stratiform precipitation region.

The nimbostratus formed from the weakening of old cumulonimbus is characterized, as is all nimbostratus, by ice particles drifting downward through the gently ascending air, which rises rapidly enough to provide moisture for the growth of the ice particles but slowly enough that the particles readily drift downward toward the melting layer and the ground. In this case, though, the precipitating ice particles are generated and grow to precipitable size during the cumulonimbus phase of the cloud. When the cumulonimbus dies and takes on a stratiform character, precipitating ice particles are already present. Since they thus exist at the *onset* of the nimbostratus phase, they do not need to be nucleated and grown to precipitable size in the stratiform context itself. The history of three ice particles is indicated schematically in Fig. 6.11. Particle 1 is near the top of a convective cell at time t_o (Fig. 6.11b). It has been formed at lower levels and grown to precipitable size while being carried up to high levels by the strong convective updraft in the cumulonimbus. When the convective updrafts weaken, the particle begins to settle downward through the more gentle ascent in the upper levels of the nimbostratus. By $t_o + \Delta t$, it has reached a level midway between cloud top and the melting level, as indicated in Fig. 6.11d. This point is directly below the position of the particle at t_o in the schematic, where the coordinate system moves with the clouds and precipitation, and there is no vertical wind shear. In the meantime, ice particle 2 has been generated in the new convection between B and C and has arrived at upper levels. When the convection between B and C weakens, the second particle falls in the same manner as the first one. By time $t_o + 2\Delta t$, it has fallen midway through the upper cloud within the region of newer nimbostratus. Particle 1 meanwhile has fallen further and is located just above the melting level at this time (Fig. 6.11f). Particle 3 has appeared at the top of the most recent convection (between C and D) and will fall in a pattern similar to particles 2 and 3 when this region of convection weakens.

From the positions of particles 1, 2, and 3 in Fig. 6.11f, the relation of the deep convection to the nimbostratus formed from the decay of a succession of groups of cumulonimbus is clear. Precipitable-sized ice particles generated and carried aloft in the updrafts of deep convective cells enter the nimbostratus at upper levels and slowly settle downward in the upper layers of the nimbostratus until they reach the melting level. Since we are ignoring any effects of wind shear, there is no relative air motion across the cloud system, and the trajectory of each particle is therefore vertically downward in the coordinate system moving with the cloud.

An entirely different way that the same picture as that shown in Fig. 6.11f can be obtained is seen by changing our assumptions in two ways. First, suppose that the convective region between C and D in Fig. 6.11 is in a steady state, rather than the latest manifestation of a nonsteady succession of such regions, and has been in this steady state for a while. Second, assume that there is vertical variation (i.e., shear) of the horizontal wind in the coordinate system moving with the storm. Specifically, let the shear be such that there is airflow from left to right at upper levels. In this case, the ice particles labeled 1, 2, and 3 can be regarded as successive positions of the same particle, which is generated within the updraft of the deep convection and subsequently advected horizontally rearward into the nimbostratus cloud. The sloping trajectory from 1 to 2 to 3 is the combined result of the horizontal relative air motion and the fall speed of the ice.

Whether the nimbostratus associated with deep convection evolves according to the nonsteady/no-shear mode or the steady/sheared-cloud mode varies from one meteorological context to another. In most if not all contexts, however, the nimbostratus associated with convection can be understood in terms of the depiction in Fig. 6.11f, with the nimbostratus developing by either one or the other or both of these modes. In some tropical mesoscale convective systems (Chapter 9), the nonsteady/no-shear mode predominates. In hurricane eyewall regions (Chapter 10), a steady/sheared-cloud mode predominates in conjunction with the steady eyewall convection and shear of the radial and tangential wind components. In certain types of "squall lines" (Chapter 9), a combination of the two modes is operative. The nonsteady, successive regeneration of the convective region occurs in the presence of shear. The result is that the ice particles in the nimbostratus are partly introduced by successive decay of the old convective cells and partly by shear.

There are at least two important special cases to recognize in connection with the no-shear and sheared modes of nimbostratus associated with deep convection. One of these special cases is a degenerate form of the nonsteady/no-shear mode. This case occurs when the original group of convective cells (Fig. 6.11a and b) does not regenerate, and the stratiform precipitation at time $t_o + \Delta t$ is thus not actually accompanied by new convection, as shown in Fig. 6.11b, but rather appears as a region of pure nimbostratus, unaccompanied by convection. It should be remembered, though, that the particles falling from the nimbostratus in this degenerate case were generated in the earlier convection, and the process remains exactly the same as if the convection had regenerated and continued to be present alongside the nimbostratus. The second special case worth noting occurs in the sheared mode, when the horizontal relative flow at upper levels is especially strong. In this situation, which arises sometimes in hurricanes, the stratiform and convective regions of Fig. 6.11f become separated by a wide horizontal gap. Thus, at first glance, the nimbostratus might appear to be isolated, when actually it is connected with an upstream region of deep convection.

To examine nimbostratus associated with deep convection more quantitatively, we consider the case in which the nimbostratus is associated with a squall-line mesoscale system of the type we will consider in further detail in Chapter 9. A schematic of the situation is shown in Fig. 6.12. The picture at upper levels is similar to that in Fig. 6.11f. The nimbostratus in this case is produced by a combination of the nonsteady and sheared modes. There is vertical shear of the horizontal wind such that the storm-relative wind at upper levels is directed toward the stratiform region. However, at the same time, there is discrete regeneration of the convective region and successive incorporation of old weakened convective elements into the nimbostratus. The old convective elements gradually dissipate as they advected rearward into the stratiform region. They tend to remain moderately active and give the region of mean ascent an embedded cellular structure somewhat like that in Fig. 6.4.

The ice particle trajectories in Fig. 6.12 are in the sense shown, and the primary microphysical growth processes of the ice particles are indicated. It is emphasized that the primary mechanism of growth of ice particles in the convective updrafts is riming. After the particles are in the stratiform region, the dominant growth mechanism is vapor deposition, made possible by the widespread ascent at upper levels. As the snow approaches the 0°C level, aggregation produces larger particles but does not increase the total mass of the precipitation. In this layer (0 to $-12°C$), riming may also contribute to the growth of the particles. After the particles melt, they fall through a layer of subsiding air, whose origin will be discussed in Chapter 9. The raindrops evaporate partially as they fall through this lower layer.

Our concern here is with the growth and fallout of the particles above the 0°C level. In particular, we focus on the relative importance of the convective updrafts *vis-à-vis* the widespread ascent in the nimbostratus itself in producing the stratiform precipitation that reaches the ground. To examine this question, we refer to the results of a numerical simulation similar to that described in Figs. 6.9 and 6.10. As in that simulation, the air motions were assumed to be those given by Doppler radar measurements, and the two-dimensional thermodynamic and bulk microphysical water-continuity equations were integrated until steady-state fields were obtained. The calculations in this case were made for a region corresponding to a stratiform region of the type depicted in Fig. 6.12. The convective region is imagined to exist immediately to the right of the domain of the calculations. Its influence on the stratiform region is determined by the distribution of variables assumed to exist on the right-hand boundary of the domain. This calculation thus differs from that of Figs. 6.9 and 6.10 in that the convection providing the region with precipitating ice particles consists of deep cells located to the side of the domain rather than shallow cells located above the domain. The horizontal wind and vertical air motion assumed to exist in the domain are shown in Fig. 6.13. This flow pattern was observed by Doppler radar in the stratiform region of a midlatitude squall-line mesoscale convective system. The flow is generally from right to left across the domain. Thus, cloud ice and snow produced in the convective region and assumed to be located on the right vertical wall of the domain are continually advected into the stratiform region. The upward air motion at upper levels provides an environment in which the ice particles thus entering the domain from the convective region can grow before they fall to the 0°C level (between 3 and 4 km altitude in this case). The categories of water substance assumed in the water-continuity equations were water vapor (q_v), cloud water (q_c), rainwater (q_r), cloud ice (q_I), and two categories of precipitating ice. The two categories of precipitating ice were distinguished by their assumed density and fall velocity. Less dense snow (represented by mixing ratio q_s) had typical fall speeds of ~1 m s^{-1}, while the more dense snow (represented by mixing ratio q_g) had typical fall speeds of ~3 m s^{-1}. It would be expected that the ice particles generated in deep convection, as depicted in Fig. 6.12, would have roughly this range of fall speeds, depending on the amount of riming growth they had undergone while in the convective updrafts.

The steady-state distributions of q_s, q_g, and rainwater mixing ratio q_r are shown in Fig. 6.14. From Fig. 6.14a and b, we see that the less dense snow tends to be swept out and accreted by the more dense snow. The sloping path of the fallout of the latter is evident in Fig. 6.14b. When the more dense snow reaches the melting level, the highest concentration of q_g is seen from Fig. 6.14b to be located between horizontal coordinates −70 and −90 km. This snow then melts and falls out rapidly, such that the heaviest concentration of rainwater at lower levels occurs between these same coordinates in Fig. 6.14c. From Fig. 6.15, it is confirmed that the maximum precipitation rate and radar reflectivity correspond to the fallout pattern of the more dense snow and rain. Thus, the *location* of the heaviest precipitation from the nimbostratus is determined by the pattern of fallout of snow produced in the convective region.

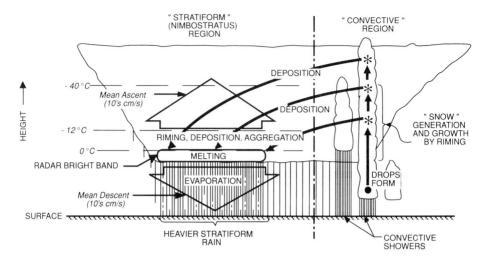

Figure 6.12 Schematic of the precipitation mechanisms in a mesoscale convective system. Solid arrows indicate particle trajectories. (From Houze, 1989. Reprinted with permission from the Royal Meteorological Society.)

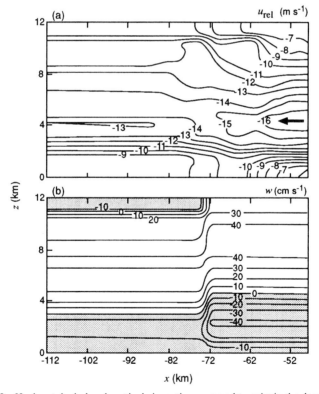

Figure 6.13 Horizontal wind and vertical air motion assumed to exist in the domain of a numerical simulation of a stratiform region. Deep convection is assumed to be located immediately to the right of the domain. Data are from a squall-line mesoscale convective system observed over Oklahoma. (From Rutledge and Houze, 1987. Reprinted with permission from the American Meteorological Society.)

Although the location of the heaviest precipitation from this type of nimbo-stratus is determined by the fallout trajectories of the snow produced in the neighboring convection, the *amount* of precipitation that falls from the stratiform cloud is determined by a combination of effects. The convection produces the initial mass of snow entering and falling through the nimbostratus. As this snow falls through the deep stratiform cloud layer, it grows by deposition of vapor (Fig. 6.16a and b) before it melts (Fig. 6.16c) and partially evaporates as it falls through the region of descent below the melting layer (Fig. 6.16d). Comparison with data shows that the model produces an amount of stratiform precipitation that is consistent with observations. Two further experiments with the numerical simulation show significant results: In the first experiment, the vertical velocity is set to zero instead of the field shown in Fig. 6.13b. In this case, the general pattern of precipitation remains qualitatively the same, but the total mass of rain reaching the surface is reduced by a factor of 4. Although the ice particles originate in convective cells, the subsequent growth of the particles in the widespread meso-scale ascent in the upper part of the nimbostratus contributes greatly to the mass of the precipitation. In the second experiment, the vertical velocity of Fig. 6.13b is retained, but the influx of ice particles across the right-hand wall of the domain is shut off. Without this influx of precipitable-sized ice particles, almost no rain reaches the surface at all. In this case, all ice particles must be nucleated and grown in the nimbostratus cloud itself. They cannot attain precipitable size before they are advected out of the model domain.

These numerical calculations clearly show the symbiotic relationship of the convective updrafts and the widespread ascent in the stratiform cloud region that is required to account for the precipitation that falls from nimbostratus associated with deep convection. The deep convection is essential to produce ice particles large enough to fall out within the mesoscale region of the nimbostratus, while the widespread ascent at upper levels in the nimbostratus is necessary for the particles to increase in mass before they fall out of the cloud.

6.4 Radiation and Turbulent Mixing in Nimbostratus

In Chapter 5, we saw that shallow layers of stratiform cloud often take the form of turbulent entraining layers. Typically the turbulence is driven by radiative destabilization of the cloud layer. Boundary-layer stratus, altostratus, and cirrostratus all provide examples of this type of cloud dynamics. The nimbostratus clouds considered in this chapter are so deep that they probably are not best described in this way. As we have seen, nimbostratus typically extends from a base at the 4-km level or lower to tops near the tropopause, which may be at an altitude of anywhere from 12 to 16 km, depending on latitude and other factors. It is hard to imagine an 8–12-km-thick cloud behaving entirely as a radiatively driven mixed-layer phenomenon.

To investigate the extent to which radiation can drive turbulence in nimbo-stratus, the model used to produce the results in Figs. 6.13–6.16 was modified to include radiation and turbulent mixing in the thermodynamic equation. The ther-

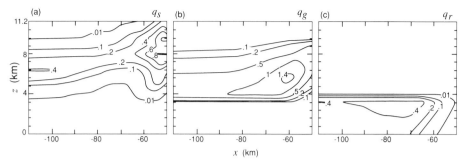

Figure 6.14

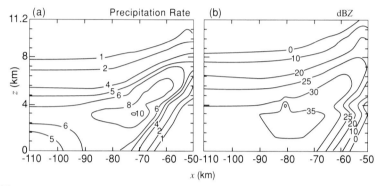

Figure 6.15

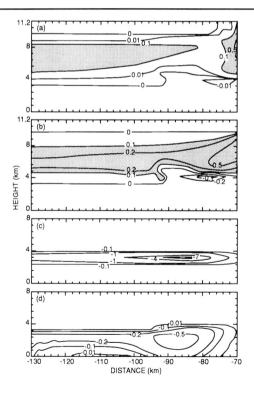

Figure 6.16

modynamic equation (2.78) in this case may be written as

$$\frac{\overline{D\theta}}{Dt} = \overline{\mathcal{H}}_L + \overline{\mathcal{H}}_I + \overline{\mathcal{H}}_S + \overline{\mathcal{H}}_T \tag{6.2}$$

where the terms on the right represent latent heating ($\overline{\mathcal{H}}_L$) associated with phase changes (condensation, evaporation, deposition, sublimation, melting, freezing), infrared heating ($\overline{\mathcal{H}}_I$), solar heating ($\overline{\mathcal{H}}_S$), and the redistribution of heating by turbulent mixing ($\overline{\mathcal{H}}_T \equiv -\rho_o^{-1}\nabla \cdot \rho_o\overline{\mathbf{v}'\theta'}$). The turbulent mixing was determined as an adjustment of the temperature lapse rate in the cloud, which was assumed to be saturated. Wherever the lapse rate becomes potentially unstable ($\partial\theta_e/\partial z < 0$) in a model time step, the temperature distribution with respect to height was immediately restored to neutrality ($\partial\theta_e/\partial z = 0$). $\overline{\mathcal{H}}_T$ is effectively the heating or cooling implied by the restoration process. The eddy flux convergence was included in the water-continuity equations with a K-theory formulation (Sec. 2.10.1) in which the value of K was chosen to be that which would produce an amount of mixing consistent with the convective adjustment $\overline{\mathcal{H}}_T$. Two-dimensional (no y-variation) versions of the thermodynamic and water-continuity equations were integrated in time until steady-state fields were obtained within the nimbostratus region of a mesoscale convective system similar to that depicted in Fig. 6.12. Air motions in the nimbostratus were given by observations and held constant at their observed values during the integration.

The horizontally averaged steady-state values of the terms in (6.2) obtained in the integration are shown in Fig. 6.17. The cloud base was at the 4-km level, which was also the 0°C level, while the cloud top was at about 13 km. Under both nighttime (Fig. 6.17a) and daytime (Fig. 6.17b) conditions, the primary diabatic-heating effect through most of the cloud layer is the latent heating associated with vapor deposition on ice. Below cloud base, melting and evaporation produce cooling, which destabilizes a shallow subcloud layer. The convective overturning that restores the lapse rate produces a shallow stratus layer, which accounts for the net warming between 2 and 3 km.

During nighttime (Fig. 6.17a), infrared radiation cools the top 2.5 km of the deep nimbostratus layer. This cooling destabilizes this upper layer, and turbulent convective overturning acts to restore the lapse rate by warming the top 1 km of the cloud above 12-km altitude and cooling a 1.5-km-deep layer immediately

---

Figure 6.14   The steady-state distributions of mixing ratios of (a) snow q_s, (b) graupel q_g, and (c) rainwater q_r in a numerical simulation of a stratiform region associated with deep convection located immediately to the right of the domain of the calculations. Units are g kg^{-1}. (From Rutledge and Houze, 1987. Reprinted with permission from the American Meteorological Society.)

Figure 6.15   As in Fig. 6.14 except for (a) rain rate (mm h^{-1}) and (b) radar reflectivity (dBZ). (From Rutledge and Houze, 1987. Reprinted with permission from the American Meteorological Society.)

Figure 6.16   As in Fig. 6.14 except for (a) depositional growth rate of less dense snow, (b) depositional growth rate of more dense snow, (c) melting rate, and (d) rate of evaporation of rain. Units are 10^{-3} g kg^{-1} s^{-1}. (Courtesy of S. A. Rutledge.)

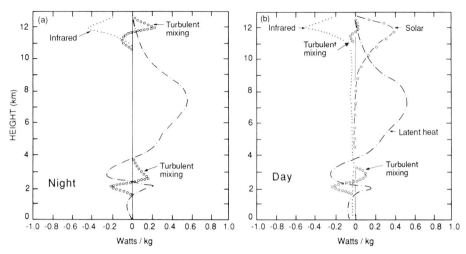

Figure 6.17   Horizontally averaged steady-state diabatic heating terms in a model of nimbostratus associated with deep convection. (a) Nightime, (b) daytime. (From Churchill and Houze, 1991. Reprinted with permission from the American Meteorological Society.)

below 12 km. Thus, the top layer of the deep nimbostratus becomes a radiatively driven mixed layer. However, it is at such a high and cold altitude that it does not produce a significant amount of snow with which to seed the lower regions of the cloud. The primary source of snow remains in the deep convection in the region adjacent to the nimbostratus (Fig. 6.12). It is possible that in nimbostratus clouds with lower tops, the radiative destabilization might have a more significant effect on precipitation.

During the day (Fig. 6.17b), with a solar zenith angle of 30°, the diabatic heating processes and profiles in the nimbostratus are similar to those during the night, except for the effect of the solar radiation, which is felt in the upper portion of the cloud. Although the short-wavelength absorption is strongest at upper levels, it extends downward much farther than the infrared cooling. The solar heating is a significant heat source throughout the upper 4–5 km of the nimbostratus layer. Despite the solar heating, the infrared cooling remains strong enough to destabilize the very top layer of cloud, and some convective overturning is present at cloud top even under these daytime conditions.

Chapter 7 | Cumulus Dynamics

"...soaring, as if some celestial call
Impell'd it to yon heaven's sublimest hall;"[166]

Clouds that occur when air becomes highly buoyant and accelerates upward in a localized region (~0.1–10 km horizontal extent) are referred to as *convective* or *cumuliform* clouds. Included in this group are both cumulus and cumulonimbus, which differ sharply from the stratiform clouds considered in Chapters 5 and 6, not only in their visual appearance (Chapter 1) but also in their dynamics and microphysics. Their vertical air motions are much stronger, and they condense and precipitate water more intensely. They all have the appearance of rapidly bubbling or "soaring" upward as they develop.

Cumuliform clouds exhibit a spectrum of forms, which include

- *Fair weather cumulus*, which are ~1 km in both horizontal and vertical scale (e.g., the cumulus humilis in Fig. 1.3a).
- *Cumulus congestus* (towering cumulus), which attain widths and depths of several kilometers as aggregates of discrete smaller buoyant bubbles within the cloud rise one after the other, reaching successively greater heights (Fig. 1.3b, Fig. 1.4a).
- *Individual cumulonimbus* (thunderstorms), which precipitate heavily, exhibit lightning and thunder, produce strong outflow winds and tornadoes, have widths of the order of tens of kilometers, and extend vertically to the tropopause, where their tops spread out and form the characteristic anvil, or thunderhead (Fig. 1.4d, Fig. 1.5, Fig. 1.6).
- *Mesoscale convective systems* (complexes of thunderstorms), which have cloud tops that extend over regions of the order of hundreds of kilometers in scale, produce large amounts of rain, contain stratiform precipitation that forms in connection with the cumulonimbus, and develop mesoscale circulation patterns in addition to the convective-scale air motions (Fig. 1.27).

This spectrum of convective cloud phenomena may be thought of as representing the relative vigor and extent of the air motions associated with the clouds. The underlying cloud dynamics are the dynamics of buoyant air; the air motions in all convective clouds originate in the form of vertical accelerations that occur when moist air becomes locally less dense than air in the surrounding larger-scale envi-

---

[166] Goethe's words to describe a cumulus cloud.

ronment. The accelerations associated with this buoyancy lead to vertical air speeds ~1–10 m s^{-1} and to several important dynamical and physical phenomena that are associated with rapid local ascent and descent of air. First of all, as parcels of air accelerate, the mass field in and around the cloud must adjust, and a distinctive pressure field is required to accomplish this adjustment. Other phenomena accompanying the buoyancy-driven vertical motions include the evolution of large liquid water content, produced by the rapid lifting, and consequent growth of hydrometeors by accretion of liquid water (recall Sec. 6.1.1 and discussion of Fig. 6.1b). In the extreme, this accretional growth leads to the formation of hailstones. Associated with the large hydrometeor concentration resulting from the strong upward air motions are feedbacks on the dynamics. The precipitation particles can initiate downward acceleration by dragging air downward. Evaporation and melting of the particles cool the air and thus also contribute to downdraft formation and maintenance. Another characteristic of the rapidly moving convective updrafts and downdrafts in convective clouds is their turbulent nature, which leads to *entrainment* of surrounding environmental air. Entrainment in turn modifies the cloud dynamics and microphysics and is thus another important feedback mechanism. Yet another important process associated with the convective drafts is the development of in-cloud *rotation*, which in extreme cases is manifested in the form of tornadoes. Rotation feeds back to the cloud dynamics by producing a dynamical pressure field, which modifies the buoyancy-produced pressure field, and by providing an additional mechanism (besides turbulence) for entrainment of environmental air.

In this chapter, we will examine these basic dynamical characteristics, which are fundamental to all convective clouds. Buoyancy and its related pressure perturbation field, turbulent entrainment, and the dynamics of rotation will be discussed in Secs. 7.1–7.4. In Sec. 7.5, we will describe how numerical models are formulated to simulate convective clouds in such a way that these important dynamical processes are taken into account, simultaneously connecting the cloud dynamics with the microphysics, so that the microphysics are calculated in a proper dynamical context and are able to interact with the dynamics. These topics are applicable to all the forms of cumulus and cumulonimbus listed above. They are introduced in this chapter at a basic level that will provide background for general study of convective clouds. In subsequent chapters some of these topics will be pursued further as we explore in more detail the dynamics of individual cumulonimbus (Chapter 8) and complexes of thunderstorms (Chapter 9).

7.1 Buoyancy

Since all convective clouds owe their existence to the fact that air becomes buoyant on a local scale (less than about 10 km), we begin by briefly recalling the nature of the buoyancy B, which appears as a contribution to vertical acceleration in the anelastic form of the equation of motion (2.47). In the absence of friction, the vertical component of the momentum equation is

$$\frac{Dw}{Dt} = -\frac{1}{\rho_o}\frac{\partial p^*}{\partial z} + B \tag{7.1}$$

where, as in Chapter 2, w is the vertical velocity, ρ_o the reference-state density, and p^* the deviation of the pressure from its reference-state value. The buoyancy B, according to (2.48), is proportional to the deviation of the density from the reference state. According to (2.50), the buoyancy may be decomposed into contributions from temperature, water vapor, pressure perturbations, and the weight of hydrometeors suspended in or falling through the air. In convective clouds all four terms in (2.50) are of the same order of magnitude. A temperature perturbation of absolute value 1 K is equivalent to a perturbation of 0.005 in water-vapor mixing ratio, 3 hPa in pressure, or 0.003 in hydrometeor mixing ratio.

7.2 Pressure Perturbation

It can be anticipated intuitively that buoyancy cannot exist without a simultaneous disruption of the pressure field. If a parcel of air of finite width and depth is less dense than the air in a surrounding horizontally uniform, otherwise undisturbed atmosphere, then at the height of the base of the parcel, the pressure is lower in the parcel than in the environment. This horizontal gradient of pressure accelerates environmental air toward the base of the buoyant parcel. This inward acceleration, moreover, is consistent with the need to replace the buoyant air at this level when it moves upward. A complete, internally consistent picture of mass, pressure, and momentum fields required to exist in association with a region of buoyant air is, of course, implied by the basic equations. We can obtain this picture by combining the horizontal and vertical equations of motion with the continuity equation. If friction and Coriolis forces are ignored, the anelastic equation of motion (2.47) can be written in Eulerian form as

$$\frac{\partial \mathbf{v}}{\partial t} = -\frac{1}{\rho_o}\nabla p^* + B\mathbf{k} - \mathbf{v} \cdot \nabla \mathbf{v} \tag{7.2}$$

If we take the three-dimensional divergence of this equation after first multiplying it by ρ_o, we obtain

$$\frac{\partial}{\partial t}\left(\nabla \cdot \rho_o \mathbf{v}\right) = -\nabla^2 p^* + \frac{\partial}{\partial z}\left(\rho_o B\right) - \nabla \cdot \left(\rho_o \mathbf{v} \cdot \nabla \mathbf{v}\right) \tag{7.3}$$

Recalling from (2.54) that the three-dimensional mass divergence in the anelastic system is zero, we see that the left-hand side of (7.3) is zero, and a diagnostic equation for the pressure perturbation is obtained:

$$\nabla^2 p^* = F_B + F_D \tag{7.4}$$

where

$$F_B \equiv \frac{\partial}{\partial z}\left(\rho_o B\right) \tag{7.5}$$

and

$$F_D \equiv -\nabla \cdot \left(\rho_o \mathbf{v} \cdot \nabla \mathbf{v}\right) \tag{7.6}$$

Thus, the Laplacian of the pressure perturbation field in an anelastic fluid must be consistent with the vertical gradient of buoyancy F_B and the three-dimensional divergence of the advection field F_D. F_B is called the *buoyancy source* and F_D the *dynamic source*. The pressure perturbation may be thought of as the sum of two partial pressures p_B^* and p_D^* such that

$$p^* = p_B^* + p_D^* \tag{7.7}$$

$$\nabla^2 p_B^* = F_B \tag{7.8}$$

$$\nabla^2 p_D^* = F_D \tag{7.9}$$

In Chapter 8 we will see that sometimes the dynamic source, and hence p_D^*, becomes dominant, especially when strong eddies form within a tornadic thunderstorm, or in a density current outflow from a thunderstorm. We will defer further discussion of the dynamic source until Chapter 8. Here we will investigate the part of p^* produced by the buoyancy source. It may be helpful in this regard to note that (7.8) is analogous to Poisson's equation in electrostatics, where $-F_B$ plays the role of a charge density, p_B^* is like the electrostatic potential, and $-\nabla p_B^*$ is equivalent to the electric field. For simple spatial arrangements of buoyancy, we can thus use known mathematical solutions of Poisson's equations to find the vector field $-\rho_o^{-1}\nabla p_B^*$. We will call this field, which is analogous to an electric field produced by a particular spatial arrangement of charge density, the *buoyancy pressure-gradient acceleration (BPGA) field*.

This solution is illustrated qualitatively in Fig. 7.1 for a uniformly buoyant parcel of finite dimensions. The plus and minus signs in the figure indicate the sign of $-F_B$ along the top and bottom of the parcel. Where $-F_B > 0$, the BPGA field (i.e., $-\rho_o^{-1}\nabla p_B^*$) diverges according to (7.8), and where $-F_B < 0$ the BPGA field converges. Everywhere except the top and bottom of the parcel, $F_B = 0$. The lines of the BPGA field are shown as streamlines, like lines of electric field for finite horizontal parallel plates of opposite charge density. Within the parcel, the lines of the BPGA field are downward. There is a divergence of the BPGA field at the top of the parcel and convergence at the bottom. Outside the parcel, lines of force are up above the parcel, downward in the regions to the sides of the parcel, and upward just below the parcel. These lines indicate the directions of forces acting to produce the compensating motions in the environment that are required to satisfy mass continuity when the buoyant parcel moves upward.

The downward lines of force shown inside the parcel in Fig. 7.1 imply that the

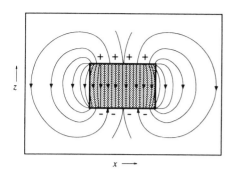 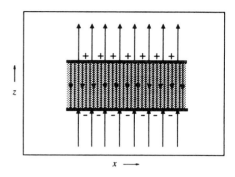

Figure 7.1   **Figure 7.2**

Figure 7.1 Vector field of buoyancy pressure-gradient force for a uniformly buoyant parcel of finite dimensions in the x–z plane. The plus and minus signs indicate the sign of the buoyancy forcing function $-\partial(\rho_o B)/\partial z$ along the top and bottom of the parcel.

Figure 7.2 Vector field of buoyancy pressure-gradient force for a uniformly buoyant parcel of infinite horizontal dimensions. The plus and minus signs indicate the sign of the buoyancy forcing function $-\partial(\rho_o B)/\partial z$ along the top and bottom of the parcel.

upward acceleration of buoyancy is counteracted to some degree by a downward BPGA. This counteraction must occur because some of the buoyancy of the parcel has to be used to move environmental air out of the way in order to preserve mass continuity while the parcel rises. The only way that downward BPGA could be absent would be to have the parcel's width shrink to zero—a nonsensical case, but nonetheless illustrative of the fact that *a given amount of buoyancy produces a larger upward acceleration the narrower the parcel.* In cumulus and cumulonimbus clouds, the distribution of B is such that BPGA is often the same order of magnitude as B. The BPGA can be especially important near the tops of growing clouds, where rising towers are actively pushing environmental air out of the way.

The maximum magnitude is achieved by the BPGA when it exactly balances B. If we let $B_o^* = 0$, then

$$B = \frac{1}{\rho_o} \frac{\partial p^*}{\partial z} = -\text{BPGA} \qquad (7.10)$$

which is the case of hydrostatic balance (Sec. 2.2.4). In the strict mathematical sense, this case occurs at the limit where the horizontal dimension of a buoyant element as in Fig. 7.1 becomes infinite in horizontal extent. This fact can be seen by multiplying (7.10) by ρ_o, taking $\partial/\partial z$ of both sides of the equation, and rearranging to obtain

$$\frac{\partial \rho_o B}{\partial z} = p_{zz}^* \qquad (7.11)$$

Then (7.4), (7.5), and (7.11) imply that

$$\nabla^2_H p^* = 0 \tag{7.12}$$

where ∇_H is the horizontal gradient operator. Thus, if the horizontal gradient of p^* is flat in at least one place, there is no horizontal variation of p^*. The counterpart of Fig. 7.1 for the hydrostatic case is shown in Fig. 7.2, which shows the lines of force for a uniformly buoyant parcel of infinite horizontal extent.

7.3   Entrainment

7.3.1   General Considerations

Another important property common to all forms of cumulus and cumulonimbus clouds is that environmental air crosses the cloud boundaries and dilutes the in-cloud air. The net buoyancy and other properties of the cloudy air are thus moderated, and the cloud is made less vigorous. In evaluating the average properties of the in-cloud air, the rate at which air is exchanged with the environment must be taken into account.

Air may be drawn laterally into cumuliform clouds to satisfy mass continuity if the mean mass flux in the cloud is increasing with respect to height. In addition, mixing across the cloud boundaries occurs as a result of the highly turbulent air motions characterizing all forms of cumulus and cumulonimbus clouds. Within convective clouds both strong shear and strong buoyancy exist. Hence, both the conversion from shear $\mathscr{C}$ and the buoyancy generation $\mathscr{B}$ in (2.86) are important sources of turbulent kinetic energy. The intensity of the turbulence in the cloud is much greater than in the surrounding environment. Some of the internal cloud motions take the form of organized overturning and rotation on the scale of the cloud itself, while some of the turbulence is of a smaller scale and more random. Mixing occurs across all the edges of the cloud, as a result of both the random and the organized motions. As a result of these motions and the general necessity to satisfy mass continuity, the cloud becomes diluted by a certain proportion of environmental air. The incorporation of environmental air into the cloud is called *entrainment*. In Chapter 5, we saw how entrainment across the tops and bottoms of mixed-layer stratiform clouds led to their dilution. In the convective clouds considered in the present chapter, the internal motions producing the entrainment are more vigorous, and mixing occurs across the sides as well as the tops and bottoms of cloud elements.

In this section, we investigate the mechanism of entrainment in convective clouds, and we shall proceed in an historical vein. First, we examine some early ideas, which present a simple, sometimes useful, but basically flawed approximation to the process of entrainment. We call this traditional view *continuous, homogeneous entrainment*, since it treats the process as continuous in time and uniform in space. We will then examine the modern view, which we will call *discontinuous, inhomogeneous entrainment*, in which entrainment is considered to be a discrete process in time and space.

7.3.2   Continuous, Homogeneous Entrainment

In the late 1940s, Henry Stommel, an oceanographer, suggested[167] that the turbulent exchange of mass between a cloud and its environment could be roughly approximated by considering a rising cloud element interacting with its environment as shown in Fig. 7.3. At time t, the rising element is considered to have mass m. Between t and $t + \Delta t$, a mass of air $(\Delta m)_\varepsilon$ is entrained from the environment and a mass of air $(\Delta m)_\delta$ is detrained to the environment. Consider some quantity $\mathscr{A}$, which has units of energy, mass, or momentum per unit mass of air. The value of $\mathscr{A}$ in the rising cloud element is represented by $\mathscr{A}_c$ and in the environment by $\mathscr{A}_e$. It is assumed that we are dealing with horizontal averages for the in-cloud and out-of-cloud regions and that

- The entrained air is brought in from the sides (i.e., *laterally*),
- The entrained air is mixed *instantaneously* and *thoroughly* across the cloud element, and
- The process occurs *continuously* as the element rises.

These assumptions are the foundation of the concept of continuous, homogeneous entrainment. With these assumptions, the conservation of $\mathscr{A}$ in the cloud parcel

[167] From aircraft data taken in and around trade-wind cumulus during a scientific expedition to the Caribbean, Stommel (1947) noted that the observed temperatures in the clouds could barely be distinguished from the temperature in the environment. This result was at odds with the substantial temperature difference expected between the undisturbed environment and a parcel of air rising under saturated adiabatic conditions, i.e., under conservation of θ_e. [See (2.18).] To explain the apparent dilution of the cloudy air, he suggested the idea of continuous lateral entrainment.

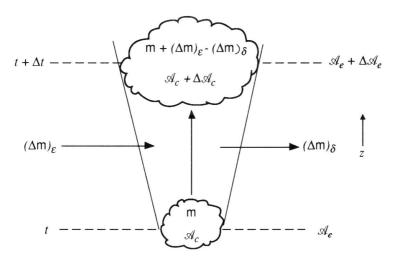

Figure 7.3   Idealization of a rising cloud element interacting with its environment.

can be written as

$$\left[m + (\Delta m)_\varepsilon - (\Delta m)_\delta \right]\left(\mathcal{A}_c + \Delta \mathcal{A}_c \right) = m\mathcal{A}_c + \mathcal{A}_e \,(\Delta m)_\varepsilon$$
$$- \mathcal{A}_c \,(\Delta m)_\delta + \left(\frac{\Delta \mathcal{A}_c}{\Delta t} \right)_S m\Delta t \qquad (7.13)$$

where $(\Delta \mathcal{A}_c / \Delta t)_S$ is the rate of change of $\mathcal{A}_c$ that would be present even if the parcel was not exchanging mass with the environment. Rearrangement of terms in (7.13) and taking the limit as $\Delta t \to 0$ leads to

$$\frac{D\mathcal{A}_c}{Dt} = \left(\frac{D\mathcal{A}_c}{Dt} \right)_S + \frac{1}{m}\left(\frac{Dm}{Dt} \right)_\varepsilon \left(\mathcal{A}_e - \mathcal{A}_c \right) \qquad (7.14)$$

where the notation D/Dt, as usual [see (2.2)], indicates a derivative following a parcel of fluid. The detrainment terms in (7.13) cancel and do not appear in (7.14) since *detrainment of mass in no way affects the mass-averaged values of variables in the cloud.* It is only entrainment that affects the in-cloud averages, since it dilutes the parcel with environmental fluid.

As we will see in Sec. 7.3.2, modern observations show that the entrainment process in cumulus clouds is *not* continuous, instantaneous, thorough, or entirely lateral; thus, an accurate representation of the process must take these facts into account. Nonetheless, the above continuous view has some value in providing a simple and tractable first approximation to cumulus dynamics.

The role of entrainment in the First Law of Thermodynamics may be seen by applying (7.14) to the *moist static energy*

$$\hbar \equiv c_p T + L q_v + g z \qquad (7.15)$$

In the absence of entrainment and diabatic processes other than release of latent heat in condensation or evaporation (ignoring ice-phase processes), the First Law is given by (2.12). If the pressure change following a parcel is to a first approximation hydrostatic, then taking $D(7.15)/Dt$, with substitution from (2.12) and (2.38), yields

$$\left(\frac{D\hbar_c}{Dt} \right)_S = 0 \qquad (7.16)$$

In this case, (7.14) becomes

$$\frac{D\hbar_c}{Dt} = \frac{1}{m}\left(\frac{Dm}{Dt} \right)_\varepsilon \left(\hbar_e - \hbar_c \right) \qquad (7.17)$$

when $\hbar$ is substituted for $\mathcal{A}$. Using the definition (7.15), we may rewrite (7.17) as

$$\frac{DT_c}{Dt} = \underset{\text{(i)}}{-\frac{g}{c_p}w_c} \; \underset{\text{(ii)}}{- \frac{L}{c_p}\frac{Dq_v}{Dt}} + \frac{1}{m}\left(\frac{Dm}{Dt} \right)_\varepsilon \underset{\text{(iii)}}{\left[(T_e - T_c) + \frac{L}{c_p}(q_{ve} - q_{vc}) \right]} \qquad (7.18)$$

where the terms on the right are recognized to be (i) the dry-adiabatic cooling, (ii) the latent heating, and (iii) the effects of entrainment.

Equations of the form (7.14) may also be written with $\mathscr{A}$ replaced by the vertical velocity or the water-continuity variables in a cumulus cloud. In the case of vertical velocity the source term in (7.14) (i.e., the change in w that would occur whether or not entrainment takes place) becomes $(Dw_c/Dt)_S$, which is given by the right-hand side of (7.1). Thus, (7.14) becomes

$$\frac{Dw_c}{Dt} = -\frac{1}{\rho_e}\frac{\partial p^*}{\partial z} + B - \frac{1}{m}\left(\frac{Dm}{Dt}\right)_\varepsilon w_c \tag{7.19}$$

which is equivalent to adding the entrainment term to (7.1). The reference-state density has been taken to be that of the environment, whose conditions can be obtained from radiosonde data. The vertical velocity in the environment does not appear in the entrainment term because it is assumed to be small compared to the vertical velocity in the cloud.

When $\mathscr{A}$ is replaced by the water substance mixing ratios, the change in the mixing ratio that would occur in the absence of entrainment $(Dq_{ic}/Dt)_S$ is given by the sources and sinks represented by S_i on the right-hand side of the water-continuity equations (2.21). The water-continuity equations then take the form

$$\frac{Dq_{vc}}{Dt} = -C + \frac{1}{m}\left(\frac{Dm}{Dt}\right)_\varepsilon (q_{ve} - q_{vc}) \tag{7.20}$$

and

$$\frac{Dq_{ic}}{Dt} = S_i + \frac{1}{m}\left(\frac{Dm}{Dt}\right)_\varepsilon (q_{ie} - q_{ic}), \quad i = 1, \ldots, k \tag{7.21}$$

where C, representing the net condensation (or evaporation) rate, is the sink (or source) term for the water vapor mixing ratio, k is the number of subdivisions of the hydrometeor content, and the source and sink terms for these mixing ratios depend upon the type of water-continuity model assumed and are therefore left in symbolic form as S_i.

Equations (7.18)–(7.21) would constitute a way to calculate the properties T_c, w_c, q_{vc}, and q_{ic} of a rising parcel in a cumulus cloud if ways of determining the pressure perturbation p^* and the entrainment rate $m^{-1}(Dm/Dt)_\varepsilon$ were available. This set of equations is the basis of the *one-dimensional Lagrangian cumulus model*. Often, this model is expressed with z rather than t as the coordinate. The vertical velocity $w = Dz/Dt$ is used as the basis for the coordinate transformation. By substituting wD/Dz for D/Dt in (7.18)–(7.21) and dividing all the equations by w (assumed to be finite and positive for a rising mass of fluid), we obtain

$$\frac{DT_c}{Dz} = -\frac{g}{c_p} - \frac{L}{c_p}\frac{Dq_{vc}}{Dz} + \Lambda\left[(T_e - T_c) + \frac{L}{c_p}(q_{ve} - q_{vc})\right] \tag{7.22}$$

$$\frac{D}{Dz}\left(\frac{1}{2}w_c^2\right) = -\frac{1}{\rho_e}\frac{\partial p^*}{\partial z} + B - \Lambda w_c^2 \tag{7.23}$$

$$\frac{Dq_{vc}}{Dz} = -\frac{C}{w_c} + \Lambda\left(q_{ve} - q_{vc}\right) \tag{7.24}$$

and

$$\frac{Dq_{ic}}{Dz} = \frac{S_i}{w_c} + \Lambda\left(q_{ie} - q_{ic}\right), \quad i = 1, \ldots, k \tag{7.25}$$

where

$$\Lambda \equiv \frac{1}{m}\left(\frac{Dm}{Dz}\right)_\varepsilon \tag{7.26}$$

It should be noted that the values of T_c, w_c, q_{vc}, and q_{ic} obtained as solutions of (7.22)–(7.25) are values of these variables at various points along the path of a cloud element that is either rising or sinking. These solutions are not instantaneous in-cloud profiles except in the special case of a steady-state cloud in which similar parcels continually follow each other upward.

Equation (7.22) can be thought of in terms of a thermodynamic diagram. The left-hand side gives the temperature change of the parcel with height. The first two terms on the right describe the temperature change with height of the parcel in the presence of condensation but in the absence of entrainment. The negative of this temperature change with height is the *moist-adiabatic lapse rate*. If entrainment is active, the lapse rate of a parcel ($-DT_c/Dz$) lies somewhere between the moist-adiabatic and environmental lapse rates, as a result of mixing environmental air of different temperature and humidity into the parcel. If the entrainment effect is strong, the parcel's temperature differs only slightly from the environmental temperature. Thus, this simple view seems to explain qualitatively the observations that Stommel was concerned about (see earlier footnote).

To close the one-dimensional Lagrangian cumulus model, some way has to be found to express the entrainment Λ and the pressure perturbation p^*. The traditional approach is to invoke some form of mass continuity, while considering the cumulus cloud to be analogous to certain laboratory phenomena. For this purpose, three types of laboratory phenomena are considered: the *jet*, the *thermal*, and the *starting plume*.

In the jet model, the updraft is considered to behave, to a first approximation, like a steady-state, mechanically driven jet (Fig. 7.4). In such a jet, environmental fluid is entrained and, since the environment is relatively laminar, there is no detrainment from the jet (i.e., the environment does not entrain air from the jet). Consider an arbitrary parcel of mass m between heights z and $z + \Delta z$ in the idealized steady-state jet depicted in Fig. 7.5. In time Δt, the original mass m is replaced by an equal mass. Therefore,

$$m = \mu_f \Delta t \tag{7.27}$$

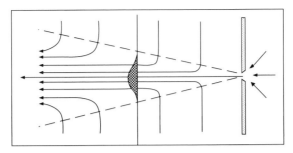

Figure 7.4 Streamlines of the flow associated with a mechanically driven fluid jet. The cross-sectional area of the jet expands downstream from its source as fluid is entrained from the environment. The shaded area midway downstream represents a velocity profile. (From Byers and Braham, 1949.)

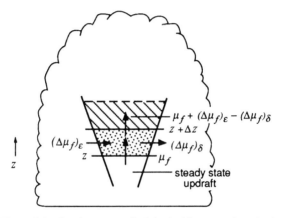

Figure 7.5 Steady-state updraft jet inside a cumulus cloud.

where μ_f is the vertical mass *flux* (in kg s^{-1}) in the jet at height z. Also in time Δt, the original parcel entrains mass

$$(\Delta m)_\varepsilon = \left(\Delta\mu_f\right)_\varepsilon \Delta t \qquad (7.28)$$

It follows from (7.27) and (7.28) that for a steady-state jet

$$\frac{1}{m}(\Delta m)_\varepsilon = \frac{1}{\mu_f}\left(\Delta\mu_f\right)_\varepsilon \qquad (7.29)$$

The entrainment rate Λ defined by (7.26) is then obtained for the steady-state jet by dividing (7.29) by Δz and taking the limit as $\Delta z \to 0$; the result is

$$\Lambda = \frac{1}{\mu_f}\left(\frac{d\mu_f}{dz}\right)_\varepsilon \qquad (7.30)$$

where we have made use of the fact that in the special case of a steady-state jet the vertical derivative following the parcel D/Dz and the vertical derivative with respect to height at an instant of time within the steady-state jet d/dz are equivalent.

The mean flow in laboratory jets is approximately steady state, incompressible, and circularly symmetric. Under these conditions the mean variable form of the Boussinesq mass continuity equation (2.75) applies. In cylindrical coordinates centered on the jet, which is assumed to be circularly symmetric about the central axis, this equation becomes

$$\frac{1}{r}\frac{\partial(r\bar{u})}{\partial r} + \frac{\partial\bar{w}}{\partial z} = 0 \tag{7.31}$$

where r is the radial coordinate and u is the radial velocity. Bars indicate time averages. Laboratory experiments with this type of jet show that

$$\bar{w} = W(z)e^{-(r/\hat{R})^2} \tag{7.32}$$

where $\hat{R}$ is a constant.[168] Other dynamical variables have similar radial profiles. For mathematical simplicity, this Gaussian profile is often replaced by a "top-hat" profile:

$$\bar{w} = \begin{cases} w_c(z), & 0 < r < b \\ w_e(z), & r > b \end{cases} \tag{7.33}$$

Substituting this profile into (7.31) and integrating over radius from 0 to b, we obtain

$$\frac{d}{dz}\left(w_c b^2\right) = -2b\bar{u}(b) \tag{7.34}$$

Thus, an increase in mass flux with height is matched by horizontal inflow.

In early experiments with laboratory jets, it was hypothesized[169] and verified experimentally that the horizontal inflow at a given altitude was proportional to the rate of upward motion at that level, that is:

$$-\bar{u}\big|_{r=\hat{R}} = \alpha_\varepsilon W \tag{7.35}$$

where α_ε stands for a constant determined from laboratory experiments. If for the top-hat approximation we associate $\hat{R}$ with b, we infer from (7.34) and (7.35) that

$$\frac{d}{dz}\left(w_c b^2\right) = 2\alpha_\varepsilon b w_c \tag{7.36}$$

Since $w_c b^2$ is proportional to the vertical mass flux in the jet (μ_f), (7.36) can be rewritten as

$$\frac{1}{\mu_f}\frac{d\mu_f}{dz} = \frac{2\alpha_\varepsilon}{b} \tag{7.37}$$

---

[168] See Turner (1973, Chapter 6).
[169] Morton et al. (1956).

Recalling (7.30) and the fact that the laboratory jets do not detrain, we see that (7.37) gives us an expression for the entrainment rate.

$$\Lambda = \frac{1}{\mu_f}\left(\frac{d\mu_f}{dz}\right)_\varepsilon = \frac{2\alpha_\varepsilon}{b} \tag{7.38}$$

The shape of the jet implied by (7.36) is given by

$$\frac{db}{dz} = -\frac{1}{2}b\frac{d\ln w_c}{dz} + \alpha_\varepsilon \tag{7.39}$$

Laboratory data show that α_ε has a value of about 0.1.

The laboratory jets to which (7.38) applies are incompressible. The analogy between the cumulus clouds and the laboratory jet is based on the similarity of the incompressible and anelastic continuity equations. The incompressible equation is identical to the Boussinesq equation (2.55). The anelastic continuity equation (2.54) differs from the others only in the inclusion of the density weighting factor ρ_o, which we set equal here to the environmental density ρ_e. For the steady-state cylindrical geometry of the jet, the anelastic continuity equation is then

$$\frac{1}{r}\frac{\partial(\rho_e r\bar{u})}{\partial r} + \frac{\partial(\rho_e \bar{w})}{\partial z} = 0 \tag{7.40}$$

which is similar to (7.31) except for the density factor. By analogy to the incompressible case, the equivalent to (7.39) is found to be

$$\frac{Db}{Dz} = -\frac{1}{2}b\frac{D}{Dz}\ln\left(\rho_e w_c\right) + \alpha_\varepsilon \tag{7.41}$$

where we have replaced the notation d/dz with D/Dz to emphasize that this equation can be solved simultaneously with (7.22)–(7.25) in the case of the steady-state jet.

To complete the jet analogy version of the one-dimensional Lagrangian cumulus equations, it is assumed that the pressure perturbation is zero, and a laboratory-derived empirical value of $\alpha_\varepsilon = 0.1$ is usually used in (7.38) and (7.41). The vertical equation of motion (7.23) then becomes

$$w_c\frac{Dw_c}{Dz} = B - \frac{0.2}{b}w_c^2 \tag{7.42}$$

It is evident from Sec. 7.2 that the assumption of zero pressure perturbation (or that its vertical gradient is zero) violates mass continuity. The assumption in this case can be partially justified because once a steady-state jet is established, a rising parcel inside the jet does not have to do much work to push the fluid ahead of the parcel out of the way since the parcel lying in its path is already in motion. Hence, the vertical pressure gradient acceleration is not large. The steady-state jet analog model of cumulus convection is then constituted by the equations (7.22), (7.24), (7.25), (7.38), (7.41), and (7.42) in the variables T_c, w_c, q_{vc}, q_{ic}, Λ, and b.

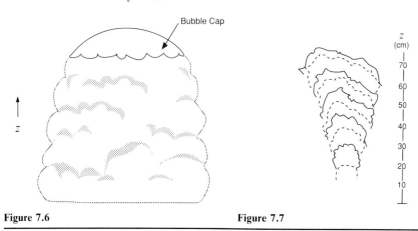

Figure 7.6   **Figure 7.7**

Figure 7.6 Bubble model of convection. (Adapted from Malkus and Scorer, 1955. Reproduced with permission from the American Meteorological Society.)

Figure 7.7 Successive outlines of laboratory thermals traced from photographs. The thermals, produced by dropping elements of salt solution into water, were negatively buoyant. The picture is inverted to indicate the analogous ascent of a positively buoyant element. (Adapted from Woodward 1959. Reproduced with permission from the Royal Meteorological Society.)

The source and sink terms C and S_i are expressed in terms of these variables to close the set of equations.

Various objections were raised to the jet analogy as a model of cumulus convection. This model in particular does not appear to account for the nonsteady aspects of many convective clouds. These objections led to the *bubble model*,[170] in which a cumulus cloud is envisioned as a series of rising bubbles of buoyant air. As each bubble rises, environmental air is pushed around the bubble and mixes into a turbulent wake behind the bubble. As the environmental air moves around the bubble, it continually erodes the surface layer of the bubble until the entire bubble disappears. A new bubble may rise through the wake of a previous bubble. Since the wake contains more water vapor than the surrounding environment, the new bubble is exposed to successively less erosion and can therefore attain a greater altitude than its predecessor. A cumulus cloud is envisioned as consisting of new bubbles rising through the wakes of old bubbles. The top of the cloud consists of the upper cap of the most highly ascending bubble in the cloud (Fig. 7.6). The bubble model has qualitative appeal in that it seems to retain the character of what many cumulus clouds look like as they grow.

A quantitative formulation of the bubble model[171] assumes that the rising bubbles are spherical and preserve their shape while diminishing in size. The vertical momentum equation (7.23) is then written as

$$w_c \frac{Dw_c}{Dz} = -D_R + B \qquad (7.43)$$

[170] Postulated by Ludlam and Scorer (1953).
[171] By Malkus and Scorer (1955).

where $-D_R$ is a parameterization of the vertical pressure-gradient acceleration term.[172] According to this model, $\Lambda = 0$. That is, the buoyant elements *do not entrain*. They only detrain, as they are eroded. The erosion rate may be related, according to a prescribed scheme, to the thermodynamic properties of the environment. The environmental air is, however, not mixed into the remaining uneroded core, and entrainment terms do not appear in any equations. Thus, buoyancy is counteracted in the vertical motion equation (7.43) entirely by the pressure-gradient acceleration, not by diluting the buoyancy by entrainment of environmental air. This characteristic of the bubble model appears to be in contradiction to the observations of diluted air in cumulus—such as those discussed by Stommel. For this reason, the bubble model was disregarded for many years. However, as we will see in Sec. 7.3.3, this judgment may have been premature.

In the mid-1950s the discussions of the jet and bubble proponents eventually led to some enlightening laboratory experiments.[173] When elements of salt solution were released into water, it was found that these (negatively) buoyant elements were not eroded, but rather expanded (Fig. 7.7). Thus, it was discovered that bubbles are not simply eroded; they also entrain. These entraining buoyant bubbles are referred to as *thermals*. The laboratory experiments showed that erosion (detrainment) occurred only in a stratified stable environment. In the atmosphere, evaporative cooling around the edges of rising cloud elements can contribute to the tendency of mixtures to be left behind, enhancing bubble model tendencies.

Laboratory thermals in a neutral environment were observed to expand along a similar cone for which the radius of a thermal was given by

$$b = \alpha_\varepsilon z \tag{7.44}$$

where z is the height of the center of the thermal and $\alpha_\varepsilon = 0.2$. Since the thermal does not detrain, its entrainment rate is

$$\frac{1}{m}\left(\frac{Dm}{Dt}\right)_\varepsilon = \frac{1}{(4/3)\pi b^3}\frac{D}{Dt}\left[(4/3)\pi b^3\right] \tag{7.45}$$

Substituting from (7.44) and dividing by w_c, we obtain

$$\Lambda = \frac{1}{m}\left(\frac{Dm}{Dz}\right)_\varepsilon = \frac{3\alpha_\varepsilon}{b} \tag{7.46}$$

where $\alpha_\varepsilon = 0.2$. Comparing (7.38) and (7.46), we note that the entrainment rates for jets and thermals are both inversely proportional to the radius of the region of

---

[172] Malkus and Scorer (1955) parameterized this acceleration in terms of "form drag," which is proportional to the square of the velocity w_c. The coefficient of proportionality is assumed to depend on the size and shape of the rising bubble. Another simpler parameterization that has been used frequently is to set $D_R = -0.33B$, the idea being that about 1/3 of the buoyancy is used to push the air ahead of the buoyant element out of the way. The fraction 1/3 appears to be rather arbitrary, but some justification is given by Turner (1962).

[173] See Scorer (1957, 1958) and Woodward (1959). Woodward's fascination with the problem was apparently spurred by her interest in gliding and observations of birds soaring in thermals.

rising fluid, but the entrainment rate for the thermal is three times larger than for the jet.

Making use of (7.46) and the empirical value of $\alpha_\varepsilon = 0.2$, we may write the vertical momentum equation (7.23) in the case of the thermal as

$$w_c \frac{Dw_c}{Dz} = -D_R + B - \frac{0.6}{b} w_c^2 \tag{7.47}$$

This relation has similarities to the momentum equations for both the jet (7.42) and the bubble (7.43). Like the jet equation, it has an entrainment term. The rate of entrainment, however, is now proportional to $0.6/b$, which is triple the rate of dilution in the jet. Moreover, the buoyancy B is much weaker since the temperature and mixing ratio equations also contain entrainment terms diluting the in-cloud thermodynamic properties—at a rate of $0.6/b$. In addition, a parameterized pressure-gradient acceleration is again included, since, like the bubble, the thermal must push the environmental air out of the way. Thus, the thermal has both high entrainment and pressure drag slowing it down and low buoyancy.

Laboratory experiments such as those illustrated in Fig. 7.7 not only reveal that the thermal entrains but also indicate the mechanism of entrainment in these phenomena. It was found that the internal circulation in laboratory thermals is similar to that of a Hill's vortex (cf. Fig. 7.8 and Fig. 7.9a). Hill vortex theory[174] has been used[175] to derive an analytic expression for the observed internal circulation in a rising cumulus-cloud element. According to this model, the upper portion of the cloud element consists of a Hill's vortex with a turbulent wake located somewhere below the center of the cloud element (Fig. 7.9b). The upward circulation in the center of the element described by the Hill's vortex is the mechanism of entrainment. This upward influx into the element is assumed to come from the wake and thus to be composed of an arbitrary mixture of environmental and undiluted cloud air.

A relationship between jets and thermals was discovered as the laboratory experiments continued into the early 1960s.[176] Specifically, it was found that as a laboratory jet becomes established, it is capped by a thermal (Fig. 7.10). This entity is called the *starting plume*. It is modeled by assuming that the cap behaves like a thermal, except that the fluid just below the cap is characterized by the solution of the steady-state jet equations. That is, fluid drawn up into the cap comes from the jet and is thus already a mixture of cloud and environmental air. The cap should therefore be diluted more slowly than an isolated thermal of the same size under similar environmental conditions. The vertical velocity in the jet, which has a Gaussian profile (7.32), almost exactly matches the analytic expression for the vertical velocity profile in the spherical vortex (Fig. 7.11). This coincidence allows one to derive a set of equations for the plume cap taking into account the influx of air from the Gaussian plume at the base of the cap.[177] The entrainment

---

[174] Elaborated in the classic *Hydrodynamics* by Sir Horace Lamb (1932).
[175] By Levine (1959).
[176] See Turner (1962).
[177] See Turner (1962) for details.

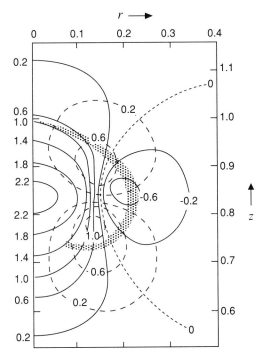

Figure 7.8 The distribution of velocity in a laboratory thermal. The outline of the buoyant fluid is shaded. The values of the vertical velocities (solid lines) and radial velocities (dashed lines) are expressed as multiples of the vertical velocity of the thermal cap. (Adapted from Woodward, 1959. Reproduced with permission from the Royal Meteorological Society.)

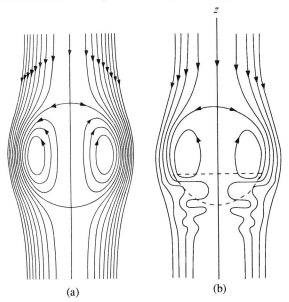

Figure 7.9 (a) Theoretical "Hill's vortex." Stream function lines both inside and outside the vortex are shown. (From Lamb, 1932. Reprinted with permission from Dover Publications, Inc.) (b) Idealization of the internal circulation in a rising cumulus-cloud element. The upper portion consists of a Hill's vortex. The lower portion, located below the center of the cloud element, is a turbulent wake. (From Levine, 1959. Reprinted with permission from the American Meteorological Society.)

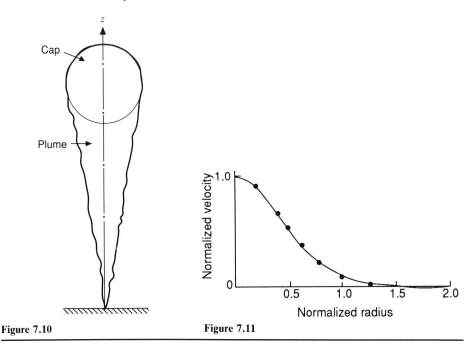

Figure 7.10

Figure 7.11

Figure 7.10   The "starting plume," which occurs as a turbulent jet begins. (Adapted from Turner, 1962. Reprinted with permission from Cambridge University Press.)

Figure 7.11   Matching of the vertical velocity in the lower jet (dots) and the upper spherical vortex (solid line) of a starting plume. (From Turner, 1962. Reprinted with permission from Cambridge University Press.)

rate predicted by these equations is inversely proportional to the cap radius. Laboratory experiments verified this relation and showed the proportionality constant was 0.2, that is:

$$\Lambda = \frac{1}{m}\left(\frac{Dm}{Dz}\right)_\varepsilon = \frac{0.2}{b} \tag{7.48}$$

for the starting plume cap. The equation for the vertical velocity of the cap can then be written as

$$w_c\frac{Dw_c}{Dz} = -D_R + B - \frac{0.2}{b}w_c^2 \tag{7.49}$$

which has the form of (7.47), the equation for the thermal, but with an entrainment rate equal to that of the steady-state jet [recall (7.38)].

During the middle to late 1960s, the one-dimensional Lagrangian model was tested extensively in the field as a way to predict the maximum height reached by a convective cloud. In these tests, the cloud element radius b was treated as an observed quantity provided by visual observation from research aircraft flying near the clouds. For simplicity, b was assumed to be a constant for a particular cloud. The thermodynamic structure of the environment was provided by radiosonde data taken in the vicinity of the clouds. It was found that the most accurate cloud-top heights were calculated for an entrainment rate of approximately $0.2/b$, which corresponds to either the jet or the starting plume analogy.[178] Cloud-top height, however, is not an especially good test of the cloud dynamics, since it is basically the stability of the environment that controls cloud height rather than any assumed model properties. Extensive experimentation has shown that one-dimensional Lagrangian models are incapable of simultaneously predicting cloud-top height and liquid-water content.[179] This realization has led to the modern notion of discontinuous, inhomogeneous entrainment, to be considered in the next section.

7.3.3 Discontinuous, Inhomogeneous Entrainment

The concept of rapid, continuous, homogeneous, lateral entrainment discussed above is useful for visualizing and predicting some of the grosser aspects of cumulus and cumulonimbus clouds. However, this view of entrainment does not hold up well under close scrutiny. As already mentioned, these models are not capable of predicting accurate hydrometeor content and cloud height simultaneously. It has also been found that droplet size spectra in cumulus cannot be calculated accurately if continuous entrainment is assumed. The calculated droplet spectrum becomes narrower as a continuously entraining parcel rises. However, observations suggest that the spectrum width in cumulus either increases or does not change very much with height.[180]

These observed differences between real cumulus clouds and clouds characterized by continuous entrainment has led to a reexamination of the basic premises of entrainment theory (i.e., that entrainment is rapid, thorough, continuous in space and time, and lateral). The discrepancies between real clouds and those described in terms of continuous entrainment theory can be traced to the fact that *none* of these assumptions are well met in real convective clouds.

Attempts to resolve this dilemma have led to the concept of discontinuous, inhomogeneous entrainment. The basic idea is that entrainment occurs not in a continuous stream but in pulses that are intermittent in both time and space. The turbulence in a cumulus updraft is now thought to be similar to a laboratory plume such as that illustrated in Fig. 7.12. Turbulent entrainment is seen to occur in strands or gulps. Data obtained on flights through cumulus are consistent with this

[178] See Weinstein and MacCready (1969) and Simpson and Wiggert (1971).

[179] This fact was first demonstrated in a seminal paper by Warner (1970), who made careful comparisons of model simulations of small nonprecipitating cumuli for which aircraft measurements of both microphysical and kinematic variables were available.

[180] This result is found in aircraft observations of cumuliform clouds obtained in a wide variety of locations, for example, Australia (Warner, 1969a and b), the High Plains of the northern United States (Rodi, 1978, 1981; Blyth and Latham, 1985), and Hawaii (Raga, 1989).

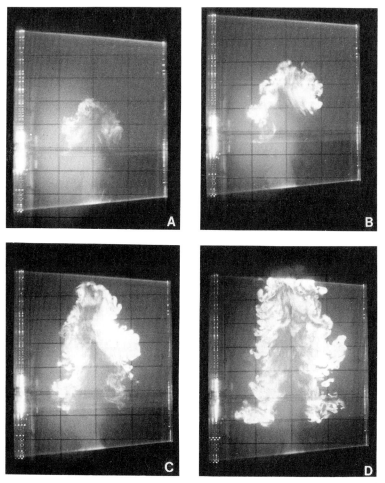

Figure 7.12   Turbulence in a laboratory plume. (Photos courtesy of Shenqyang Shy.)

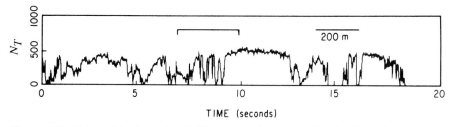

Figure 7.13   Data obtained on aircraft flights through cumulus showing intermittent regions of homogeneous properties. Here a region of highly variable droplet concentration N_T (bracketed; units are number per sample volume, which was variable but typically ≈ 0.3 cm³) was found adjacent to a region of uniform number, characteristic of a steady, nonentraining cloud. (From Austin *et al.*, 1985. Reprinted with permission from the American Meteorological Society.)

view in showing intermittent regions of homogeneous properties. An example of data in Fig. 7.13 shows a region of highly variable droplet content (bracketed) adjacent to a region of uniform maximum number, characteristic of a steady, nonentraining cloud. Thus, both laboratory and field data suggest that parcels rising in a cloud are only occasionally subjected to mixing with blobs of entrained air (i.e., the occurrence of mixing is *discontinuous* in time). The mixing events are characterized by an engulfed strand of environmental fluid. When the mixing takes place, it is not instantaneous but rather slow enough that it remains localized to the region of the engulfed air. Thus, the degree of mixing across an updraft of a cloud is expected to be spatially quite *inhomogeneous*.

This radically different view of entrainment can be illustrated mathematically first by considering the effect of the entrainment on the drop size spectrum in cumulus. The lateral mixing in the traditional view of entrainment has the property that it occurs instantly

$$\tau_m \ll \tau_e \tag{7.50}$$

where τ_m is the time scale of mixing and τ_e is the time scale of evaporation. That is, air is immediately mixed across cloud, then drops evaporate into the uniform air mixture. In inhomogeneous entrainment, just the opposite situation holds:

$$\tau_e \ll \tau_m \tag{7.51}$$

Thus, drops evaporate locally in ingested blobs, while in other areas drops are unaffected.

This view of entrainment was developed in the context of a one-dimensional Lagrangian cloud model.[181] The model consisting of Eqs. (7.22), (7.24), (7.25), (7.38), (7.41), and (7.42) was first simplified by prescribing a constant updraft motion (w_c = constant). The momentum equation (7.42) is thus eliminated. The continuous entrainment relationship (7.38) and its associated continuity equation (7.41) are disregarded in favor of the concepts to be discussed below. The equations remaining from the former model are then the thermodynamic equation (7.22) and the water-continuity equations (7.24) and (7.25). We may illustrate the important features of the discontinuous, inhomogeneous model by focusing attention on the hydrometeor equations (7.25). They are posed in terms of the explicit water-continuity scheme represented by (3.52). Only the early evolution of the cloud-droplet spectrum is considered; thus, the stochastic collection $\mathfrak{C}_i$, breakup $\mathfrak{B}_i$, and fallout $\mathfrak{F}_i$ terms in (3.52) are assumed to be negligible. It is further assumed that there is no horizontal air motion, so that with w_c = constant, the three-dimensional wind divergence is zero in (3.52). With these assumptions, together with the assumption that entrainment affects the drop size spectrum, (3.52) may be rewritten as

$$\frac{DN_i}{Dt} = \mathfrak{N}_i + \mathfrak{D}_i + \varepsilon_i, \quad i = 1, \ldots, k \tag{7.52}$$

---

[181] See the innovative paper of Baker *et al.* (1980).

where, as in (3.52), k is the number of drop size categories, $\mathfrak{N}_i$ represents the change in the number concentration N_i as a result of nucleation of drops of mass $\mathfrak{m}_i$ to $\mathfrak{m}_i + d\mathfrak{m}$, $\mathfrak{D}_i$ represents the change in N_i owing to vapor diffusion, while ε_i represents the effect of entrainment in diluting N_i. When this equation is solved simultaneously with equations for the temperature and water vapor, the drop size distribution on a parcel is determined as a result of interplay of the three terms on the right—nucleation, vapor diffusion, and entrainment.

The vapor diffusion term $\mathfrak{D}_i$ can be written as

$$\mathfrak{D}_i = \mathfrak{D}_{i_c} + \mathfrak{D}_{i_e} \tag{7.53}$$

where the subscripts c and e represent contributions associated with condensation and evaporation, respectively. Consistent with (3.54), the condensation term is given by

$$\mathfrak{D}_{i_c} = -\frac{\partial}{\partial \mathfrak{m}}\left[\dot{\mathfrak{m}}_{cond} N\right]\Bigg|_{\mathfrak{m}=\mathfrak{m}_i} \tag{7.54}$$

where $\dot{\mathfrak{m}}_{cond}$ is the rate of change of mass of a drop of mass $\mathfrak{m}$ by condensation of vapor onto the drop, as given by the right-hand side of (3.21) under conditions of supersaturation ($S > 0$). The term $\mathfrak{D}_{i_e}$ will be formulated in a special way to represent the effect of entrained air.

The entrainment term in (7.52) can be expressed as

$$\varepsilon_i = -\mathscr{L} N_i \tag{7.55}$$

where $\mathscr{L}$ (s^{-1}) is the entrainment rate. Three cases can then be considered.

In the first, there is no entrainment, so $\mathscr{L} = 0$. To follow the evolution of the initial spectrum of drops in the rising parcel of air, it is assumed that no nucleation of new drops occurs. Thus, $\mathfrak{N}_i$ is also zero. Since the parcel is always saturated in this case, $\mathfrak{D}_{i_e}$ too is zero. Equation (7.52) then reduces to

$$\frac{DN_i}{Dt} = \mathfrak{D}_{i_c} \tag{7.56}$$

Since small droplets grow faster than larger ones, this relation leads to narrowing of the drop size spectrum, which is counter to observations.

In the second case, an instantaneous, continuous lateral entrainment is assumed, like that considered in previous sections. Accordingly, we set

$$\mathscr{L} = \hat{\mathscr{L}} = \text{constant} > 0 \tag{7.57}$$

The time scale of the mixing is assumed to obey (7.50), and the value of $\hat{\mathscr{L}}$ is low enough that the entrainment does not dry out the cloud. Consequently, the cloud is everywhere at all times saturated, and $\mathfrak{D}_{i_e}$ is therefore again zero. It follows that

$$\frac{DN_i}{Dt} = \mathfrak{N}_i + \mathfrak{D}_{i_c} - \hat{\mathscr{L}} N_i, \quad i = 1, \ldots, k \tag{7.58}$$

which is of the form of the standard entrainment equation (7.14), where $\mathscr{A}_c$ in this case is N_i and the source terms are $\mathfrak{N}_i$ and $\mathfrak{D}_{i_e}$. To make this case comparable to the first case, the total concentration of drops per unit volume of air N_T is held constant:

$$N_T = \sum_{i=1}^{k} N_i = \text{constant} \tag{7.59}$$

Since condensation does not affect the total number of drops, it follows from (7.58) that the total nucleation rate must just offset the net entrainment rate. Since the nucleated particles are small, they should be placed in the smallest size category. Thus, we may write the nucleation rate for this case as

$$\mathfrak{N}_i = \begin{cases} -\sum_{j=1}^{k} \varepsilon_j, & \text{for } i = 1 \\ 0, & \text{for } i = 2, \ldots, k \end{cases} \tag{7.60}$$

Calculations of N_i using (7.54) and (7.60) in (7.58) show a drop size distribution again narrowing as the parcel rises, though not as much as in the zero-entrainment case.[182]

The third case assumes that the entrainment is both discontinuous in time and inhomogeneous in space. To represent the intermittency in time, it is assumed that the rising parcel of volume V_p is subjected to pulses of entrainment such that

$$\mathscr{L} = \frac{\mathscr{V}_\varepsilon}{\Delta t V_p} \sum_j \delta\left(t - t_j\right) \tag{7.61}$$

where $\mathscr{V}_\varepsilon$ is the entrained volume, t_j represents the discrete times at which the parcel entrains a volume of environmental air, δ is the Dirac delta function, and Δt is the time step in the numerical calculation. The entrainment term is then obtained by substituting (7.61) into (7.55) to obtain

$$\varepsilon_i = -\left[\frac{\mathscr{V}_\varepsilon}{\Delta t V_p} \sum_j^n \delta\left(t - t_j\right)\right] N_i \tag{7.62}$$

The inhomogeneity of the entrainment in space is represented by assuming that when a volume of air is entrained, all of the drops in a localized subregion of the parcel are evaporated. Thus, the total number of drops in each size category is reduced in proportion to the total reduction of liquid water mass in the parcel. Then

$$\mathfrak{D}_{i_e} = -\hat{\phi} N_i \sum_j \delta\left(t - t_j\right) \tag{7.63}$$

where $\hat{\phi}$ is the fraction

$$\hat{\phi} = \frac{\mathscr{V}_\varepsilon \left[\rho q_{vs}(T) - \rho_e q_{ve}\right]}{\Delta t V_p \rho q_L} \tag{7.64}$$

where the subscript e as usual indicates values characteristic of the environment of the cloud. Nucleation is again parameterized by requiring that (7.59) be satis-

---

[182] See Baker *et al.* (1980) for details.

fied. In this case the net drop concentration is reduced by both entrainment–dilution and the local evaporation episodes brought on by the slow, inhomogeneous mixing. If particles of the smallest size category are nucleated to offset these effects, we have

$$
\mathfrak{N}_i = \begin{cases} -\sum_{j=1}^{k}\left(\mathfrak{D}_{j_e} + \varepsilon_j\right), & \text{for } i = 1 \\ \\ 0, & \text{for } i = 2, \ldots, k \end{cases} \tag{7.65}
$$

Calculations of N_1 using (7.53), (7.54), (7.62), (7.63), and (7.65) in (7.52) lead to the finding that the width of the drop size spectrum remains roughly constant with height. This result is more consistent with observations and, when it was originally carried out,[183] it was one of the first indications that discontinuous, inhomogeneous entrainment is a more satisfactory conceptual model of entrainment than the earlier continuous, rapid, homogeneous entrainment models. More sophisticated models which predict the cloud dynamics and thermodynamics more exactly have confirmed the results of the calculations with the simple model described here.[184]

Evidence of discontinuous, inhomogeneous entrainment in cumulus is also found in certain analyses of thermodynamic data collected during aircraft penetrations of cumulus clouds. These data are exhibited in so-called Paluch diagrams[185] in which two conservative properties are plotted. For example, if $\hat{F}$ and $\hat{G}$ represent two such properties and two parcels (labeled 1 and 2) are mixed, then

$$
\hat{F} = (1 - f)\hat{F}_1 + f\hat{F}_2 \tag{7.66}
$$

and

$$
\hat{G} = (1 - \hat{f})\hat{G}_1 + f\hat{G}_2 \tag{7.67}
$$

where $\hat{f}$ is the fraction of a unit mass of the final mixture constituted by fluid originally contained in parcel 2. It follows that $\hat{F}(\hat{G})$ is a straight line with slope $(\hat{F}_2 - \hat{F}_1)/(\hat{G}_2 - \hat{G}_1)$ (see Fig. 7.14). For warm cloud elements that do not contain precipitation, it is convenient to use the total water mixing ratio q_T ($=q_v + q_c$) and the equivalent potential temperature θ_e.

Two types of data may be plotted in a Paluch diagram: observations from a vertical sounding in the environment and measurements from an aircraft penetration across the cloud at a given altitude. On some occasions, the aircraft data fall on a straight line between the cloud base and cloud top, thus indicating that the data at flight level are mixtures, in various proportions, of cloud-base air and air entrained downward from cloud top. This situation is illustrated schematically in Fig. 7.15, where data from an imaginary aircraft penetration through a cumulus cloud at a fixed altitude have been plotted. Also shown are the data from a balloon sounding in the environment of the cloud. The aircraft data are shown lying on a

[183] *Ibid.*
[184] For example, Hill and Choularton (1986).
[185] Introduced by Paluch (1979).

straight line connecting the sounding data for cloud-top and cloud-base levels. The schematic of the cloud indicates that parcels passing through flight level in this case must be either undiluted parcels from cloud top or cloud base, parcels originating at cloud top but mixing with a parcel from cloud base at some time before reaching flight level, or parcels originating at cloud base but mixing with a parcel from cloud top before reaching flight level. Data from a real case exhibiting a mixing line of this type are shown in Fig. 7.16. Such cases indicate that, at least on some occasions, vertical entrainment and mixing can be a dominant process. This picture is a clear departure from the traditional notions of lateral entrainment.

On other occasions, the aircraft data fall close to a straight line between the cloud-base point and a particular level below cloud top and near flight level. The situation resulting in such a mixing line is illustrated by the schematic in Fig. 7.17, and a real example of this type of mixing line is shown in Fig. 7.18. As shown in the schematic, parcels mix in various fractions, with entrained environmental air from near flight level, lose their buoyancy, and come to rest near this level. Of course, some parcels, by chance, come to rest at other levels. Some parcels never encounter any entrained air. It is these undiluted parcels that determine the height of the cloud top. This conceptual model thus resolves one of the major dilemmas of the traditional continuous-entrainment ideas—*it allows for the average liquid water content at a particular level in a cloud to be far from undiluted even though the cloud top is high, since the cloud-top height is determined by the maximal ascent of only the least diluted parcels*.

The model in Fig. 7.17 is essentially a modification of the old bubble model discussed in Sec. 7.3.2. Just as in the bubble model, the parcels which form the cloud towers defining the very top of the cumulus arrive there without entraining environmental air. The new model, however, differs from the old one in that those parcels that do not ascend to cloud top have their upward journey truncated by entrainment rather than by erosion.

Another type of observation obtained by aircraft is illustrated schematically in Fig. 7.19 and by a real example in Fig. 7.20. In this case, the data show a lot of scatter, indicating that the parcels at flight level are mixtures of cloud-base air with environmental air entrained from a multiplicity of altitudes. Again, cloud top would be determined by the rise of the few undiluted or nearly undiluted parcels.

Of course, the possibility exists that the parcels could undergo more than one entrainment event—as used in the above analysis of the drop size spectrum. However, it has been shown that a good reproduction of observations can be obtained by considering a cumulus cloud to consist of a collection of parcels that each undergo mixing with entrained air at one and only one level.[186] As illustrated in Fig. 7.21, equal parcels are considered to be released from cloud base to several discrete levels above. Upon reaching its designated level, each parcel is split into several subparcels, each mixing with a different fraction of environmental air. Each subparcel then rises or sinks to its level of neutral buoyancy, where it is detrained to the environment. This simple model of discontinuous, inhomoge-

---

[186] See Raymond and Blyth (1986).

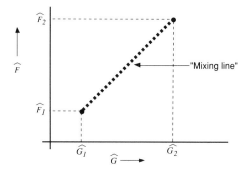

Figure 7.14

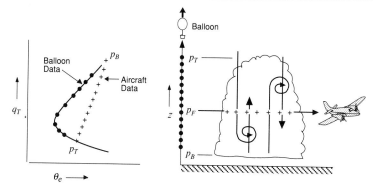

Figure 7.15

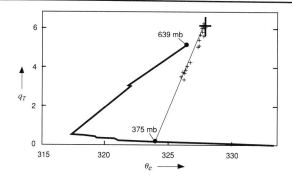

Figure 7.16

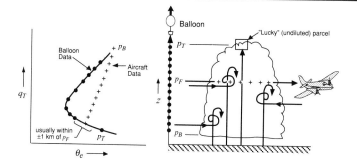

Figure 7.17

neous entrainment produced vertical profiles of detrainment similar to detrainment profiles inferred from aircraft observations of continental cumulus.

The above discussion illustrates that discontinuous, inhomogeneous entrainment accounts qualitatively for several properties of cumulus not adequately accounted for by lateral continuous entrainment with instantaneous and homogeneous mixing across the cloud. The drop size spectrum, total water content, and cloud height are all better accounted for by discontinuous, inhomogeneous entrainment. Considerable physical insight into the process of entrainment is thus gained. Much work remains, however, to make the understanding of this process quantitatively more tractable.

7.4 Vorticity

7.4.1 General Considerations

One of the important and fascinating aspects of convective clouds is that they develop internal vortical motions. We have already seen that buoyant parcels tend to overturn in the manner of a Hill's vortex (Fig. 7.9) as they ascend—like a rising smoke ring. A ring of horizontal vorticity is generated by the symmetric horizontal gradient of buoyancy centered on the buoyant parcel. We will see that the downwardly accelerating, negatively buoyant parcels of downdrafts are also bounded by a ring of vortical motions and that this ring of vorticity is a particularly dangerous part of intense downbursts (see Sec. 8.10). As downdraft air spreads out along the ground, the leading edge of the colder air continues to be characterized by the horizontal buoyancy gradient, which maintains a roll circulation just behind the gust front (Sec. 8.9). When horizontal vortex tubes associated with convective

---

Figure 7.14 Paluch diagram. $\hat{F}$ and $\hat{G}$ are conservative properties. Straight line $\hat{F}(\hat{G})$ represents mixing of parcels of air in varying proportions.

Figure 7.15 Paluch diagram of the total water mixing ratio $q_T (= q_v + q_c)$ and "equivalent potential temperature" (θ_e) for an idealized aircraft flight in a cumulus cloud characterized by mixing of air parcels from cloud base and cloud top and balloon ascent in the environment.

Figure 7.16 Paluch diagram showing aircraft measurements (small crosses) and environmental sounding data (heavy solid line) for a real cumulus cloud characterized by mixing of air parcels from cloud base and cloud top. The aircraft data were obtained at the 488-mb level. Thin solid line is theoretical mixing line for parcels of air from cloud base (heavy cross) and cloud top (375 mb). (From Jensen, 1985.)

Figure 7.17 Paluch diagram of the total water mixing ratio $q_T (= q_v + q_c)$ and "equivalent potential temperature" (θ_e) for an idealized aircraft flight in a cumulus cloud in which air parcels rising from cloud base to flight level have mixed in various fractions with air entrained from near flight level and balloon ascent in the environment. Parcels entrain environmental air from near flight level, lose their buoyancy, and come to rest near this level. Cloud top is determined by the maximum height achieved by the "lucky" parcels, which suffer no entrainment and may rise undiluted to their level of zero buoyancy.

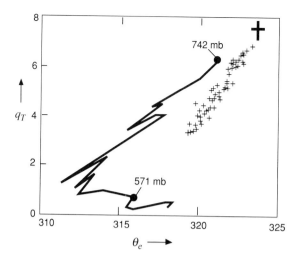

Figure 7.18

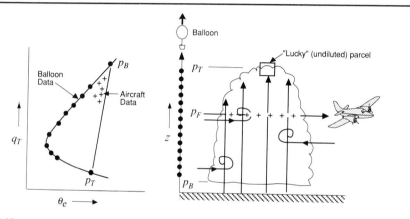

Figure 7.19

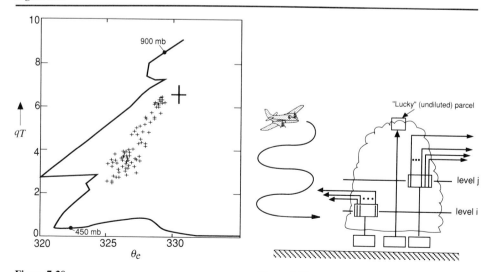

Figure 7.20

Figure 7.21

clouds become tilted out of the horizontal, some of the rotation is about a vertical axis. This vertical component of vorticity can become quite concentrated and is manifested most strongly in funnel clouds, waterspouts, and tornadoes.

While it is, of course, possible to view the fluid motions in convective clouds in terms of pressure gradients and buoyancy in the context of the equation of motion (7.2), the presence of vortical motions in clouds makes it enticing to examine the air motions via the vorticity equations (2.58) and (2.59), which eliminate the pressure-gradient terms and address the rotation directly in terms of the buoyancy source. In this section, we will examine briefly the development of vorticity in convective clouds in terms of the horizontal and vertical vorticity equations. Here we will introduce some basic concepts, which will be applied further and elaborated on in later chapters concerned with thunderstorms and mesoscale convective systems.

7.4.2 Horizontal Vorticity

The generation of horizontal vorticity ξ in a two-dimensional Boussinesq fluid is governed by (2.61), which says that *the only way that vorticity about a horizontal axis can develop within a two-dimensional parcel of air is through a horizontal gradient of buoyancy B_x* (i.e., by baroclinic generation). In the case of convective clouds, the primary situations in which ξ is generated baroclinically are illustrated in Fig. 7.22. The case of the positively buoyant rising element that overturns like a Hill's vortex (Sec. 7.3.2) is represented in Fig. 7.22a. In this case, a maximum of positive buoyancy is centered in the element, so that B_x is of equal magnitude and

---

Figure 7.18 Paluch diagram showing aircraft measurements (small crosses) and environmental sounding data (heavy solid line) for a real cumulus cloud in which air from cloud base (heavy cross) mixes in various fractions with air entrained from near flight level (about 570 mb). (From Austin *et al.*, 1985. Reprinted with permission from the American Meteorological Society.)

Figure 7.19 Paluch diagram of the total water mixing ratio $q_T (= q_v + q_c)$ and "equivalent potential temperature" (θ_e) for an idealized aircraft flight in a cumulus cloud in which the parcels of air encountered at flight level are mixtures of cloud-base air and environmental air entrained at a multiplicity of altitudes and balloon ascent in the environment. Cloud top is determined by the rise of the few "lucky" undiluted parcels.

Figure 7.20 Paluch diagram showing aircraft measurements (small crosses) and environmental sounding data (heavy solid line) for a real cumulus cloud in which the parcels of air encountered at flight level are mixtures of air from cloud base (heavy cross) and environmental air entrained at a multiplicity of altitudes. (From Taylor, 1987.)

Figure 7.21 Cumulus cloud conceptualized as consisting of a collection of parcels that each undergo mixing with entrained air at one and only one level. Equal parcels are considered to be released from cloud base to each of several discrete levels above cloud base (i, j, etc.). Upon reaching a designated level, each parcel is split into several subparcels, each mixing with a different fraction of environmental air. Cloud top is determined by the rise of the few "lucky" undiluted parcels. (Based on Raymond and Blyth, 1986.)

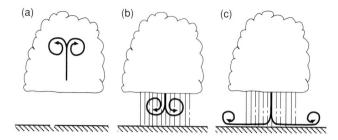

Figure 7.22

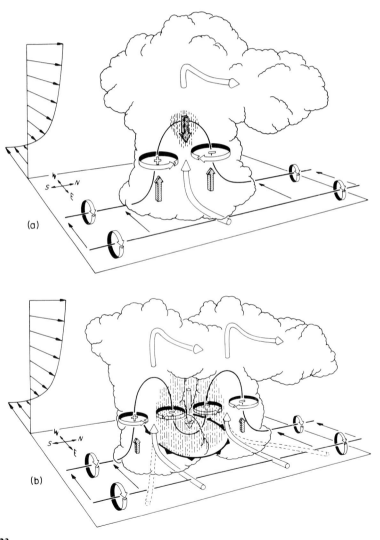

Figure 7.23

opposite sign on either side of the center line of the element. Counter-rotating vortices are thus produced on either side of the cloud. These vortices are entirely consistent with the buoyancy pressure gradient force field associated with the motions (Fig. 7.1). The case of a negatively buoyant downdraft associated with evaporative cooling and precipitation drag in the rain shower of a convective cloud is illustrated in Fig. 7.22b. It is the upside-down version of the overturning updraft. A maximum of negative buoyancy is centered in the element, so that B_x is again of equal magnitude and opposite sign on either side of the element, and counter-rotating vortices are again produced. As a downdraft of dense air spreads out along the ground, a strong buoyancy gradient and vortex is maintained at the leading edge of the outflow (Fig. 7.22c).

7.4.3 Vertical Vorticity

On the basis of numerical modeling and observations by Doppler radar of severe thunderstorms, it is now generally agreed that intense vorticity about a vertical axis (ζ) in convective clouds has its origin as horizontal vorticity. One way that the vertical vorticity in cloud arises is by converting horizontal vorticity of the *environment* to vertical vorticity in cloud. This process of conversion is easily envisaged by considering a large-scale environment in which the mean flow $\bar{u}$ is unidirectional in the x-direction and increasing with height. Then the environment has horizontal vorticity whose value is the shear $\partial\bar{u}/\partial z$. This vorticity is illustrated schematically by the north–south-aligned vortex tubes outside the cloud in Fig. 7.23a. When the updraft of the convective cloud is superimposed on the vortex tube, the tube is deformed upward such that there then exists vorticity around vertical axes in the form of counter-rotating vortices on either side of the updraft core.

This intuitive picture is formalized by ignoring the Coriolis force, which is negligible on the cumulus scale, and linearizing the vertical vorticity equation

Figure 7.22 Generation of horizontal vorticity by horizontal buoyancy gradients. (a) Positively buoyant updraft. (b) Negatively buoyant downdraft in rain. (c) Spreading of negatively buoyant downdraft along the earth's surface.

Figure 7.23 Conversion of horizontal vorticity of the environment to vertical vorticity in cloud. Mean environmental flow $\bar{u}$ (shown by thin arrows) is unidirectional in the x-direction and increasing with height. Horizontal vorticity of the environment is indicated by the north–south-aligned vortex tubes outside the cloud. Shaded arrows represent the forcings that promote new updraft and downdraft growth. Rain is shown by vertical hatching. Cylindrical arrows show the direction of cloud-relative airflow. Heavy solid lines represent vortex lines with the sense of rotation indicated by circular arrows. (a) Linear tilting of the environmental horizontal vorticity by cloud vertical air motion leads to the indicated vertical vorticity couplet. (b) Splitting of the storm as a result of nonlinear effects. Frontal symbol at the surface marks the boundary of the cold air spreading out beneath the storm. Dashed cylindrical arrows indicate shifted location of the storm inflow when updrafts become established on the storm flanks. (From Klemp, 1987. Reproduced with permission from Annual Reviews, Inc.)

(2.59) about the mean flow $\bar{\mathbf{v}} = (\bar{u},0,0)$.[187] In this case we obtain the perturbation form of the equation

$$\zeta_t' + \bar{u}\zeta_x' = \bar{u}_z w_y \tag{7.68}$$

If we consider a level where $\bar{u}$ approximates the velocity of the cloud relative to the ground, (7.68) becomes

$$\zeta_t' \approx \bar{u}_z w_y \tag{7.69}$$

in a system moving with the cloud. Thus, tilting of the vortex tubes of the mean-flow shear is the only important source of perturbation vorticity ζ'. In the case shown in Fig. 7.23a, this process leads to positive ζ' on the south side of the updraft and negative ζ' on the north side of the updraft.

Numerical models and Doppler radar data confirm that linear tilting of the environmental horizontal vorticity is an important mechanism for the formation of vertical vorticity in convective clouds that form in environments of substantial shear (large $\bar{u}_z$).[188] The ζ' couplet shown in Fig. 7.23a is indeed a characteristic of the early stage of convection that develops in shear. The counter-rotating ζ' couplet, among other things, constitutes a mechanism of entrainment. In the case shown in Fig. 7.23a, the vortices would draw environmental air into the east side of the storm at midlevels. Our previous discussion of entrainment (Sec. 7.3) considered only those entrainment processes that would occur in an environment without shear. It is evident from the present discussion that entrainment in convective clouds in strongly sheared environments will be enhanced as a result of the vortex couplet straddling the updraft.

As clouds develop beyond the stage illustrated in Fig. 7.23a, nonlinear effects become important. These effects have also been inferred from numerical-modeling results.[189] It has been found that the perturbation vorticity equation becomes

$$\zeta_t' + \bar{u}\zeta_x' \approx \bar{u}_z w_y + \zeta' w_z \tag{7.70}$$

The difference from (7.68) is that now stretching ($\zeta' w_z$) is important in addition to tilting. The regions of large $|w_z|$ at the tops and bottoms of the convective updrafts and downdrafts are especially conducive to strong vortex stretching.

Processes that occur during the later phases of cumulonimbus development will be discussed in Chapter 8. For reasons to be discussed there, we will see that a severe thunderstorm in an environment of strong unidirectional shear, like that depicted in Fig. 7.23a, splits in half, with the two halves moving in opposite directions along the y-axis, as indicated in Fig. 7.23b. The cloud motion relative to the ground at this stage of storm development may then be represented by

$$\mathbf{V}_c = u_c \mathbf{i} + v_c \mathbf{j} \tag{7.71}$$

---

[187] The mathematical arguments and deductions from numerical-model results presented in this subsection are from Rotunno (1981) and the review article of Klemp (1987).

[188] See the review of Klemp (1987).

[189] *Ibid.*

At the level of nondivergence, $\bar{u}$ is found to be approximately equal to u_c. Model results show that the only significant terms in the linearized perturbation form of the vorticity equation (2.59) at the level of nondivergence in a coordinate system moving with the cloud, with Coriolis effect neglected, are

$$\zeta'_t - v_c \zeta'_y \approx \bar{u}_z w_y \tag{7.72}$$

Under steady-state conditions,

$$\zeta' \approx -\frac{\bar{u}_z}{v_c} w \tag{7.73}$$

Thus, the maximum of ζ' coincides with the maximum w, and the southward-moving storm produced by the split has positive vorticity whenever $\bar{u}_z > 0$. In this way, a rotating updraft can develop in midlevels in severe thunderstorms. Such a rotating updraft is called a *mesocyclone* and is the main part of the storm in which intense tornadoes form. Tornadic thunderstorms will be considered in more detail in Chapter 8.

The above discussion indicates how vertical vorticity can arise in convective clouds by conversion of horizontal vorticity of the environment to in-cloud vertical vorticity. In Sec. 7.4.2 we noted how concentrated horizontal vorticity is generated by the cloud itself, especially at the edges of downdraft outflows (Fig. 7.22c). This cloud-generated horizontal vorticity can also sometimes be tilted into the vertical. Conversion of the cloud-generated horizontal vorticity can indeed be extremely important. We will see in Chapter 8 that the main tornado in a supercell thunderstorm forms *at the gust front*. Because the convergence at the gust front is so strong, the fully nonlinear form of the vertical vorticity equation (2.59) applies. The Coriolis force remains unimportant, but both tilting and stretching are important along the gust front. The horizontal vorticity generated by the strong buoyancy gradient across the gust front is advected horizontally into the center of the storm and tilted by the convective updraft and, at the same time, concentrated into an intense local vortex by the strong convergence at the base of the updraft. This strong vortex is a favorable environment in which a funnel cloud or tornado can form. These processes will be examined further in Chapter 8.

7.5   Modeling of Convective Clouds

7.5.1   General Considerations

In Sec. 7.3.2, we saw that by following a single entraining parcel of in-cloud air upward, we could estimate the properties of the rising parcel along its path by means of the closed set of equations constituting the one-dimensional Lagrangian cumulus model. In the special case of a steady-state cloud, where a series of identical parcels rise one after the other, that type of calculation also describes the properties of the cloud as a function of height. The Lagrangian model, however, is inadequate for a full description of the properties of a convective cloud in space and time. The need for a more complete description is acute. Radar, aircraft, satellite, and other special observational platforms can be employed to observe the detailed structure and evolution of certain fields in developing clouds (such as the wind and radar reflectivity). These observational technologies can be extended by methods such as the retrieval technique described in Sec. 4.4.7 to diagnose variables beyond those which can be observed directly. However, it is impossible with any foreseeable technology to observe or diagnose *all* of the thermodynamic, kinematic, and water fields simultaneously in a developing cumulus or cumulonimbus cloud with the spatial and the temporal resolution required to understand the cloud physics and dynamics. As a substitute for a complete set of observations, the science of *convective cloud modeling* is employed to calculate the relevant fields simultaneously in a physically consistent way from first principles. By seeking agreement of the model results with the few fields that are observed, one can then use the model output to extend the understanding of the clouds. In the following subsections, we summarize how the basic equations are organized to formulate the basic types of convective cloud models that are in use. The goal is to familiarize the reader with the physical basis of the models. We make no attempt to review problems related to numerical solutions, as this topic would fill a book by itself.

7.5.2 One-Dimensional Time-Dependent Model

The first type of model we will examine is the *one-dimensional time-dependent model*.[190] It is simplified in that it predicts only the horizontally averaged properties in cloud. However, these properties are computed as functions of time, thus providing an indication of the cloud's evolution. The computed variables are considered to be deviations from environmental values. If $\mathcal{A}_c$ represents the in-cloud value of any quantity $\mathcal{A}$, then

$$\mathcal{A}_c = \mathcal{A}_e + \mathcal{A}^* \tag{7.74}$$

The model is formulated in cylindrical coordinates (r,Θ,z), and the cloud region is specified *a priori* to have a radius $R(z)$. A horizontal average over the cloud is given by

$$\langle \mathcal{A} \rangle = \frac{1}{\pi R^2} \int_0^{2\pi} \int_0^R \mathcal{A} r \, dr \, d\Theta \tag{7.75}$$

while the average along the outer boundary of the cloud is

$$\overline{\mathcal{A}} = \frac{1}{2\pi} \int_0^{2\pi} \mathcal{A}(R,\Theta) \, d\Theta \tag{7.76}$$

---

[190] This type of model has appeared in the literature in several forms. The version described here was proposed by Asai and Kasahara (1967). It was further developed by Ogura and Takahashi (1971) and Scott and Hobbs (1977), who combined the dynamical equations of Asai and Kasahara with water-continuity schemes that included ice-phase microphysical processes. Ferrier and Houze (1989) added several modifications to the dynamical formulation, including new ways of representing entrainment and pressure perturbation. The discussion in this section is based mostly on their version.

The deviation from the area average of $\mathscr{A}^*$ will, for simplicity, be written as

$$\mathscr{A}' \equiv \mathscr{A}^* - \langle \mathscr{A}^* \rangle \tag{7.77}$$

Strictly, it should be written as $(\mathscr{A}^*)'$. The deviation from the boundary average is

$$\mathscr{A}'' \equiv \mathscr{A}^*(R, \Theta) - \overline{\overline{\mathscr{A}^*}} \tag{7.78}$$

The area-averaged continuity equation is obtained by applying (7.75) to the anelastic continuity equation (2.54) written in cylindrical coordinates with the density weighting factor taken to be the environmental density ρ_e. The result is

$$\frac{2}{R}\left(\overline{u^*} - \overline{w^*}\frac{\partial R}{\partial z}\right) + \frac{1}{\rho_e R^2}\frac{\partial}{\partial z}\left(\rho_e R^2 \langle w^* \rangle\right) = 0 \tag{7.79}$$

where u and w are the velocity components in the radial and vertical directions, respectively. The first term is the horizontal inflow required to balance the vertical gradient of mass flux contained in the second term. The second term in the first parentheses is the component of inflow that arises if the volume containing the cloud has nonvertical sides. For example, the radius at low levels might be specified to contract with height (Fig. 7.24b), or at high levels the cloud volume might be assumed to expand in radius with height.

The total derivative of an in-cloud variable in cylindrical coordinates can be written with the aid of (2.54) as

$$\frac{D\mathscr{A}_c}{Dt} = \frac{\partial \mathscr{A}_c}{\partial t} + \frac{1}{r}\frac{\partial}{\partial r}\left(ru^*\mathscr{A}_c\right) + \frac{1}{r}\frac{\partial}{\partial \Theta}\left(\mathscr{A}_c v^*\right) + \frac{1}{\rho_e}\frac{\partial}{\partial z}\left(\rho_e w^*\mathscr{A}_c\right) \tag{7.80}$$

where v is the velocity component in the azimuthal direction. The environment has been assumed to be motionless ($u_e = v_e = w_e = 0$). If (7.80) is multiplied by ρ_e and (7.75) is applied to it, we obtain the horizontally averaged equation

$$\frac{\partial}{\partial t}\langle \mathscr{A}^* \rangle + \langle w^* \rangle \frac{\partial}{\partial z}\langle \mathscr{A}^* \rangle = -\langle w^* \rangle \frac{\partial \mathscr{A}_e}{\partial z} + \left\langle \frac{D\mathscr{A}_c}{Dt} \right\rangle + \mathscr{E}_1 + \mathscr{E}_2 + \mathscr{E}_3 \tag{7.81}$$

where

$$\mathscr{E}_1 = \left[\frac{2}{R}\left(\overline{u^*} - \overline{w^*}\frac{\partial R}{\partial z}\right)\right]\left(\langle \mathscr{A}^* \rangle - \overline{\overline{\mathscr{A}^*}}\right) \tag{7.82}$$

$$\mathscr{E}_2 = -\frac{2}{R}\left(\overline{u''\mathscr{A}''} - \overline{w''\mathscr{A}''}\frac{\partial R}{\partial z}\right) \tag{7.83}$$

$$\mathscr{E}_3 = -\frac{1}{\rho_e R^2}\frac{\partial}{\partial z}\left(\rho_e R^2 \langle w'\mathscr{A}' \rangle\right) \tag{7.84}$$

In obtaining (7.81)–(7.84), we made use of the continuity equation (7.79) and the identity,

$$\langle w^*\mathscr{A}^* \rangle = \langle w^* \rangle \langle \mathscr{A}^* \rangle + \langle w'\mathscr{A}' \rangle \tag{7.85}$$

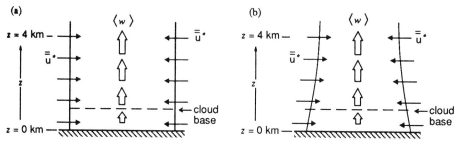

Figure 7.24 Schematic depiction of how the dynamic entrainment, represented by the horizontal flow $(\overline{u^*})$ across the cloud boundaries, is affected by the assumed vertical profile of a one-dimensional cloud model. If the vertical distribution of vertical velocity $\langle w \rangle$ is the same for both cases, then the dynamic entrainment of environmental air above cloud base is stronger in (a), where the radius is constant with height, than in (b), where the cloud radius decreases with height. (From Ferrier and Houze, 1989. Reproduced with permission from the American Meteorological Society.)

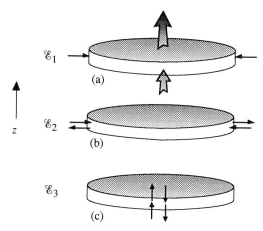

Figure 7.25 Components of entrainment in a one-dimensional time-dependent cumulus model. (a) Dynamic entrainment ($\mathscr{E}_1$). (b) Lateral eddy mixing ($\mathscr{E}_2$). (c) Convergence of vertical eddy flux ($\mathscr{E}_3$).

Equation (7.81) is identical, in the case of the circularly symmetric cloud geometry assumed here, to the one-dimensional Lagrangian entraining-parcel equation (7.14); the entrainment term, however, is now split into three components $\mathscr{E}_1$, $\mathscr{E}_2$, and $\mathscr{E}_3$, each having a distinct, physical interpretation.[191]

$\mathscr{E}_1$ is called the *dynamic entrainment*. It is the horizontal inflow or outflow of air into a layer of the cloud needed to satisfy mass continuity when there is a

[191] If $\mathscr{A}_c$ in (7.14) is written as the sum of an environmental value $\mathscr{A}_e$ and a perturbation $\mathscr{A}^*$, then (7.14) becomes

$$\frac{D\mathscr{A}^*}{Dt} = - \frac{D\mathscr{A}_e}{Dt} + \left(\frac{D\mathscr{A}_C}{Dt}\right)_s - \frac{1}{m}\left(\frac{Dm}{Dt}\right)_i \mathscr{A}^*$$

The total derivative on the left-hand side of this equation corresponds to the sum of the two terms on the left-hand side of (7.81); the first term on the right-hand side is identical to $-\langle w^* \rangle(\partial \mathscr{A}_e/\partial z)$ since $w_e = 0$ and $\partial \mathscr{A}_e/\partial t = 0$; the source term $(D\mathscr{A}_c/Dt)_s$ is identical to $\langle D\mathscr{A}_c/Dt \rangle$; and the entrainment term $-\mathrm{m}^{-1}(Dm/Dt)_i$ is equal to $\mathscr{E}_1 + \mathscr{E}_2 + \mathscr{E}_3$ in (7.81).

vertical gradient of vertical mass flux across the layer (Fig. 7.25a). *It is assumed that $\overline{\mathscr{A}^*}$ is zero when there is inflow and is equal to the in-cloud average $\langle \mathscr{A}^* \rangle$ when there is outflow.* Thus, $\mathscr{E}_1$ dilutes the cloud when entrainment is occurring, but is zero when detrainment occurs. Dynamic entrainment is a strong effect in this type of model and is not independent of the assumed model geometry. Figure 7.24 shows how the same vertical distribution of vertical velocity $\langle w^* \rangle$ has less inflow and hence less dilution of in-cloud properties when the model domain is wider at low levels. It is indeed a weakness of this type of model that the shape of the domain can determine whether or not the cloud will form and what its internal properties will be. One must be careful to specify a realistic shape for the domain if this type of model is to be useful.

$\mathscr{E}_2$ is the *lateral eddy mixing,* which is the turbulent mixing of air across the boundary without a net exchange of mass across the cell boundaries (Fig. 7.25b). There is observational evidence that the lateral mixing is smaller than vertical mixing.[192] In addition, dimensional arguments have often been invoked to indicate that

$$\mathscr{E}_2 = \alpha_L^2 \left| \langle w^* \rangle \right| \langle \mathscr{A}^* \rangle \tag{7.86}$$

where α_L^2 is a positive constant with a magnitude of about 0.1.[193] This value proves to be considerably smaller than the dynamic entrainment or vertical mixing. Thus, $\mathscr{E}_2$ appears to be the least important of the three components of entrainment and perhaps entirely negligible.

$\mathscr{E}_3$ is the *convergence of the vertical eddy flux* across the top and bottom of an infinitesimal layer of cloud (Fig. 7.25c). For this term, it is necessary to express $\langle w' \mathscr{A}' \rangle$ in meaningful physical terms. In early modeling this term was at times ignored, assumed to be implicitly accounted for by numerical smoothing or represented by simple diffusive mixing. None of these approaches turn out to be satisfactory. Somewhat better results are obtained with the following scheme, inspired by the observation that the top of a growing cumulus cloud has the form of an overturning vortex ring (Sec. 7.3.2, Figs. 7.8–7.10) and that cumulus updrafts and downdrafts tend to have triangular profiles of variables, with a peak in the center of the draft and dropping off to a minimum on the edge of the draft. These observations can be represented by letting

$$\mathscr{A}^* = \mathscr{A}_r(r) + \mathscr{A}^{\#} \tag{7.87}$$

where $\mathscr{A}_r(r)$ is the basic triangular radial distribution of $\mathscr{A}^*$, and $\mathscr{A}^{\#}$ represents a smaller-scale, turbulent fluctuation about the basic triangular profile. If the smaller-scale fluctuations are random, then we also have

$$\langle w^* \mathscr{A}^* \rangle = \langle w_r(r) \mathscr{A}_r(r) \rangle + \langle w^{\#} \mathscr{A}^{\#} \rangle \tag{7.88}$$

with the assumed triangular profiles

$$w_r = \langle w^* \rangle f(r), \quad \mathscr{A}_r = \langle \mathscr{A}^* \rangle g(r) \tag{7.89}$$

---

[192] See discussion and references in Ferrier and Houze (1989).
[193] See Asai and Kasahara (1967) or Ogura and Takahashi (1971) for a discussion of the dimensional argument.

Thus

$$\left\langle w^{*}\mathscr{A}^{*}\right\rangle = \chi_{A}\left\langle w^{*}\right\rangle\left\langle\mathscr{A}^{*}\right\rangle + \left\langle w^{\#}\mathscr{A}^{\#}\right\rangle \tag{7.90}$$

where χ_{A} is a constant. Combining this expression with (7.85), we obtain

$$\left\langle w'\mathscr{A}'\right\rangle = \left(\chi_{A} - 1\right)\left\langle w^{*}\right\rangle\left\langle\mathscr{A}^{*}\right\rangle + \left\langle w^{\#}\mathscr{A}^{\#}\right\rangle \tag{7.91}$$

Thus, the vertical eddy flux has been decomposed into a part associated with the basic triangular variation of $\mathscr{A}^{*}$ across the cloud (first term on the right) and a part arising from the smaller-scale turbulence (second term on the right). The latter is of secondary importance in the convergence of vertical eddy flux on the right-hand side of (7.84). Its effect can be estimated by a K-theory parameterization. The first term on the right of (7.91) becomes very important in (7.84) near cloud top, where it can be designed to simulate the behavior of an overturning thermal. This technique is illustrated in Fig. 7.26, where it can be seen that the triangular profile of w^{*} across the cloud is arranged and prescribed in such a way that there is a thermal cap region (enclosed by the circle) just below cloud top. From the height of the center of the thermal cap upward into the region just above cloud top z_{t}, the value of w^{*} at cloud edge is negative, while it remains positive in the center of the draft.

The vertical profile of horizontally averaged variables in cloud is computed from prognostic equations of the form (7.81) for vertical velocity, potential temperature, and a set of water-substance mixing ratios. The forcing functions $\langle D\mathscr{A}_{c}/Dt\rangle$ for these variables are obtained by applying (7.75) to the right-hand sides of (7.1), (2.9), and (2.21) over the area of the cloud. The results are, respectively,

$$\left\langle\frac{Dw_{c}}{Dt}\right\rangle = -\frac{1}{\rho_{e}}\frac{\partial}{\partial z}\langle p^{*}\rangle + \langle B\rangle \tag{7.92}$$

$$\left\langle\frac{D\theta_{c}}{Dt}\right\rangle = \langle\dot{\mathscr{H}}\rangle \tag{7.93}$$

and

$$\left\langle\frac{Dq_{ic}}{Dt}\right\rangle = \langle S_{i}\rangle, \quad i = 1,\ldots,n \tag{7.94}$$

The base-state density ρ_{o} has been taken to be the large-scale environment value ρ_{e}, $\dot{\mathscr{H}}$ is the net heating resulting from phase changes associated with the microphysical processes incorporated in the S_{i} terms, and n is the number of categories of water substance represented by the in-cloud mixing ratios q_{ic}.

The horizontally averaged pressure perturbation $\langle p^{*}\rangle$ in (7.92) can be obtained diagnostically from other equations. From Sec. 7.2, we recall that the pressure perturbation is implied by mass continuity together with the horizontal and vertical equations of motion. In the present context, the horizontal component of the equation of motion (2.47) may be written as

$$\frac{Du^{*}}{Dt} = -\frac{1}{\rho_{e}}\frac{\partial p_{c}}{\partial r} \tag{7.95}$$

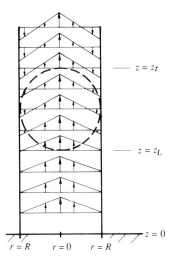

Figure 7.26 Representation of thermal-like air motions near cloud top in a one-dimensional cloud model of radius R. The thermal circulation is assumed to be a sphere located within the dashed outline. The level tangent to the top of the sphere (z_t) is cloud top. The base of the thermal is at level z_L. Triangular curves show the assumed radial distributions of vertical velocity as a function of height z. Note that the triangularity of the updraft profiles increases with height only in the thermal cap region. (From Ferrier and Houze, 1989. Reproduced with permission from the American Meteorological Society.)

To obtain this particular form, we have taken the base-state density and pressure to be ρ_e and p_e, respectively. We have further made use of the assumptions that $u_e = 0$ and that p_e does not vary in the horizontal to write $u = u_c = u^*$ and $\partial p^*/\partial r = \partial p_c/\partial r$. Integrating (7.95) from $r = R$ to an arbitrary radius r' in the cloud, we obtain

$$p_c(r') = p_c(R) - \rho_e \int_R^{r'} \frac{Du^*}{Dt} \, dr' \qquad (7.96)$$

If the environmental pressure p_e is subtracted from both sides, an average is taken over the area of the cloud, and circular symmetry ($\partial/\partial\Theta = 0$) is assumed, the above becomes

$$\langle p^* \rangle = \overline{\overline{p^*}} - \rho_e \frac{2}{R^2} \int_0^R \left[\int_R^\xi \frac{Du^*(r)}{Dt} \, dr \right] \xi \, d\xi \qquad (7.97)$$

The value of $u^*(r)$ is obtained by substituting the triangular vertical velocity profile $w_r(r)$ into the continuity equation and integrating from the center of the cloud to r. Thus, the model prediction of vertical velocity implies a prediction of u^* that must be consistent in terms of mass continuity. This prediction of u^* in turn implies the mean in-cloud pressure perturbation according to (7.97).

Results obtained with a one-dimensional time-dependent model of the type described above are shown in Fig. 7.27 for an environment characteristic of a convective region over a tropical ocean. The cloud was initiated by maintaining a

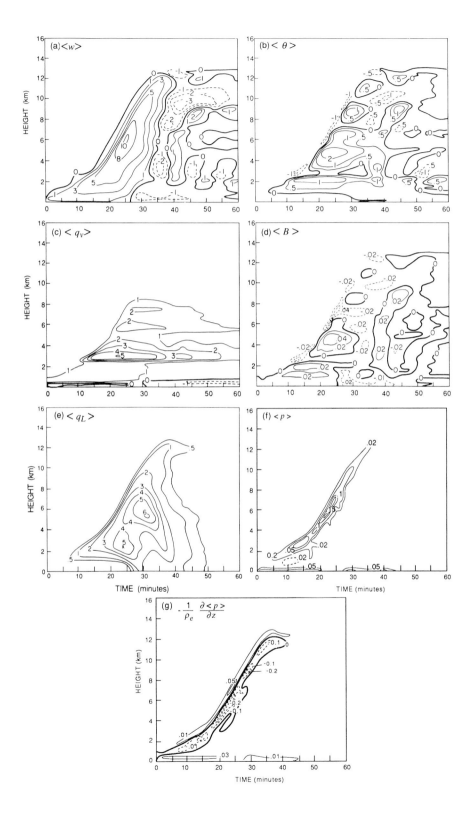

vertical velocity forcing of $w = 2$ m s^{-1} at the 0.4-km level for 20 min. The categories of water included in (7.94) were vapor, cloud liquid water, and rainwater [as defined in (3.48)]. The sources and sinks were formulated according to the warm-cloud bulk water-continuity scheme described in Sec. 3.6.1.[194]

From all the panels of Fig. 7.27, it can be seen that the model cloud grew steadily for 30–35 min, reaching a maximum altitude of 13 km. During most of this growth period, the cloud consisted of updraft at all levels, with the strongest upward motion in the middle levels of the cloud (Fig. 7.27a). In the growing updraft, the perturbations of potential temperature and water-vapor mixing ratio were generally positive (Fig. 7.27b and c), contributing to positive buoyancy (Fig. 7.27d). As the liquid water content increased (Fig. 7.27e), the buoyancy was weakened as a result of the weight of the hydrometeors [recall term q_H in (2.50)]. This negative buoyancy reversed the vertical velocity. Downdraft first formed at low levels, where the liquid water content first became large (7.27a). Eventually it was felt at higher and higher levels, following the zone of maximum liquid water content, which also was found at progressively higher levels. After 40 min, weak updraft returned after the water had mostly fallen out as precipitation. After 35–40 min, the model cloud structure is probably not very realistic.

The pressure perturbation in the model cloud is most noticeable at the earth's surface and at cloud top (Fig. 7.27f). For the first 20 min, the positive pressure perturbation at the surface was required to support the imposed updraft forcing at cloud base. The positive perturbation at low levels after 25 min was associated with the downdraft, which had to decelerate as it approached the ground. The positive pressure perturbation at cloud top is associated with the rising buoyant overturning thermal located there. The pressure-gradient acceleration associated with the thermal (Fig. 7.27g) is consistent with the idealized case illustrated in Fig. 7.1. As in the idealized case, the buoyancy gradient across the top of the cloud requires upward force above the parcel to push the environmental air out of the way of the cloud top. Model experiments verify that without this upward force above the cloud (a condition that can be produced by setting the pressure perturbation to zero), the cloud is unable to reach realistic heights. The downward pressure-gradient acceleration just below cloud top is the resistance felt by the parcel as buoyancy forces it upward. This resistance has the effect of smoothing the vertical profile of vertical velocity, since it slows the rising air down more gently in the upper reaches of the cloud than would be the case in the absence of pressure perturbation.

[194] For further details about how the calculation was set up, see Ferrier and Houze (1989).

Figure 7.27 Results of a one-dimensional time-dependent model for an environment characteristic of conditions over a tropical ocean. (a) Vertical velocity in m s^{-1}. (b) Potential temperature perturbation in °C. (c) Water-vapor mixing ratio perturbation in g kg^{-1}. (d) Buoyancy acceleration in m s^{-2}. (e) Liquid water mixing ratio in g kg^{-1}. (f) Pressure perturbation in mb. (g) Vertical acceleration owing to the vertical gradient of pressure perturbation (m s^{-2}). (From Ferrier and Houze, 1989. Reproduced with permission from the American Meteorological Society.)

7.5.3   Two- and Three-Dimensional Models

We have seen that one-dimensional Lagrangian and one-dimensional, time-dependent cumulus models illustrate some of the basic features of buoyancy, entrainment, and pressure perturbation. They also provide some interesting qualitative results and constitute rather quick and easy methods for estimating the convective response in a given thermodynamic environment.[195] On the other hand, they are highly parameterized, such that they obscure all the horizontal motions in clouds, and they reveal nothing about the effects of the wind in the environment on the convection. In view of our discussion of the development of vertical vorticity in convective clouds (Sec. 7.4), the shear of the wind in the environment of a convective cloud would seem to be quite important to the development of the cloud. Not only do vortical motions develop in response to the wind, but these vortical motions effect important feedbacks. The vorticity induces entrainment, which alters the thermal and moisture structure of the cloud and, hence, the buoyancy and vertical motion. The stronger vortices also produce dynamically induced pressure perturbations [through term F_D in (7.4)], the gradients of which strongly affect certain convective storms. To include these additional effects, it is necessary to construct two- and three-dimensional convective cloud models, which can predict accurately the horizontal as well as the vertical air motions in clouds.

In two- and three-dimensional cumulus models there are no *a priori* specified cloud boundaries, as there are in the one-dimensional models. Instead, a wind field is calculated in some larger spatial domain from initial fields of wind and thermodynamic variables. Within this model domain, clouds form whenever and wherever the predicted motions dictate.

Two- and three-dimensional cumulus models require no arbitrary assumptions about entrainment since the horizontal and vertical air motions outside and inside the cloud are resolved down to the scale of the finite-difference grid of the model. Entrainment and detrainment are automatically accounted for as air flows in and out of clouds according to the overall predicted motion field. Standard turbulence parameterizations are used to account for mixing on scales smaller than the model grid. In the terminology of the previous subsection, dynamic entrainment ($\mathscr{E}_1$) is specifically resolved, while entrainment produced by lateral eddy mixing ($\mathscr{E}_2$) and vertical eddy fluxes ($\mathscr{E}_3$) are accounted for by the parameterization of subgrid-scale turbulence.

Models are said to be two-dimensional if variations in one horizontal direction are assumed to be zero. If, in a rectangular Cartesian framework, variations in the y-direction are set to zero and motions are computed in the $x–z$ plane, the model is referred to as *slab-symmetric*. This type of geometry is sometimes useful for representing convective clouds at a front or along a squall line. If, in a cylindrical coordinate system centered on a vertical axis, variations in the azimuthal direction are set to zero and motions are computed in a radial plane, the model is called

---

[195] Note that only a temperature and humidity sounding of the environment is required as input to the one-dimensional models.

axially symmetric. This type of geometry can be useful for representing the flow in a circular vortex, such as a hurricane.

Two-dimensional models are based on the combination of the vertical equation of motion with one of the horizontal component equations of motion, for example, the x-component. To solve the x-component equation for the perturbation of velocity from a basic-state environment, the initial value of the wind in the x-direction must be provided throughout the model domain. Often this is the large-scale environment motion u_e measured by standard soundings. The initial thermo-dynamic state of the environment must be similarly provided. Evidently, since only one component of the environmental wind is an input parameter, the only situations that can be well represented by a two-dimensional model are those in which the shear of the wind is unidirectional. This limitation of two-dimensional models is severe, and consequently the three-dimensional model is the more important tool. Nonetheless, there are a few cloud systems for which a two-dimensional model can be useful.

Two-dimensional slab-symmetric motions in a vertical plane have the interest-ing characteristic that they are governed entirely by the very simple vorticity equation (2.61), in which the only forcing for the two-dimensional motions is the horizontal gradient of buoyancy B_x. The vorticity equation can be solved for ξ simultaneously with the thermodynamic and water-continuity equations, for the motions that follow from an initial perturbation of the buoyancy and/or wind field somewhere within the model domain. As shown in Sec. 5.3.4, the vorticity ξ can in turn be related through mass continuity to the velocity components u and w *via* a stream function. A two-dimensional vorticity equation in stream function form can also be derived for an axially symmetric model. This strategy of obtaining the two-dimensional velocity field avoids having to calculate the pressure perturba-tion field. However, sometimes it is desirable to know the pressure perturbation field and this technique does not provide it. Also, the stream function method is not readily extended to three-dimensional calculations. Another strategy for two-dimensional modeling is to simplify a three-dimensional model by setting horizon-tal derivatives in one direction to zero. If a three-dimensional model is available, this is the easier strategy to employ and has the advantage of providing the pressure perturbation field associated with the air motions.

Three-dimensional models are formed by combining all three components of the equation of motion (2.47) with the thermodynamic and water-continuity equa-tions. Advection terms in x, y, and z are all included in D/Dt. Initial values of both components of the horizontal wind are specified everywhere within the model domain. If atmospheric sounding data are used as input, both components of the environmental horizontal wind are used along with the measurements of the ther-modynamic state of the environment. Thus, all of the information contained in the sounding data is utilized, and the model air motions can form and evolve in response to an environment characterized by wind that varies in both direction and speed with altitude. Such a model is necessary to represent complex convec-tive phenomena such as tornadic thunderstorms, which form in highly sheared environments (Chapter 8).

One type of three-dimensional convective model is based on the anelastic equations. The first three-dimensional models were of this type.[196] In this type of cloud model, the mean variable versions of the vector equation of motion (2.83), thermodynamic equation (2.78), water-continuity equations (2.81), equation of state (2.73), and continuity equation (2.75) are solved simultaneously for a pre-scribed initial disturbance placed somewhere in the model domain. The pressure perturbation is calculated from a diagnostic equation similar to (7.4). This equa-tion is obtained by taking $\nabla \cdot \rho_o$ (2.83) and making use of the mean-variable anelas-tic continuity equation (2.75). The result is

$$\nabla^2 \overline{p^*} = \overline{F}_B + \overline{F}_D + \overline{F}_M \tag{7.98}$$

where $\overline{F}_B$ and $\overline{F}_D$ are similar to their counterparts in (7.4), except that here they are computed from the mean-variable values of buoyancy and wind, and $\overline{F}_M$ is the density-weighted divergence ($\nabla \cdot \rho_o$) of the turbulent eddy-mixing terms that ap-pear in (2.83).

Results obtained from this type of model are illustrated in Fig. 7.28 and are from one of the first articles reporting results of three-dimensional modeling. The categories of water included in (2.81) were vapor, cloud liquid water, and rainwa-ter. The warm-cloud bulk parameterization scheme described in Sec. 3.6.1 was used to relate these categories of water to one another. The model domain was 38.4 km in the x- and y-directions and 15 km in the vertical. The environment was represented by a sounding that was conditionally unstable (Sec. 2.9.1) below 5 km altitude. The relative humidity was between 80 and 100% up to 1.2 km, then decreased to 30% at 10.2 km and remained at this value up to 15 km. The environ-mental wind profile was assumed to be unidirectional, varying linearly from -6 to 6 m s^{-1} between 0.3 and 8.1 km and constant above and below these levels. An initial thermal perturbation was imposed on this environment. It consisted of a region 7.4 km in horizontal diameter and 1.2 km deep centered at an altitude of 1.5 km. The magnitude of the temperature perturbation at the center was assumed to be 0.3°C.

The structure of the model cumulonimbus cloud that evolved from this pertur-bation is shown in its mature stage in Fig. 7.28. Panel (a) is a vertical cross section along the $y = 19.2$ km line in the horizontal plane shown in panels (b)–(d). This vertical cross section shows an updraft and downdraft coexisting. The updraft slopes up and over the denser downdraft, which is located in the region of precipi-tation reaching the surface. The downdraft air can be seen flowing in from the rear at middle levels and running under the updraft. The spreading out of the down-draft air at low levels can be seen in plan view in Fig. 7.28b. At middle levels (Fig. 7.28c), the counter-rotating vortices formed by tilting of the environmental hori-zontal vorticity are seen straddling the updraft, as expected from the discussion of Fig. 7.23 in Sec. 7.4.3. This vortex pair entrains air into the downdraft on the downwind side of the storm, and middle-level environmental air splits and flows around the vortex pair. A distinctly diffluent and divergent outflow is seen near

[196] See, for example, Wilhelmson (1974), Schlesinger (1975), and Clark (1979).

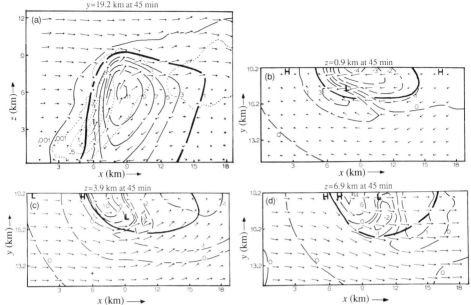

Figure 7.28 Results of an anelastic three-dimensional convective cloud model for an environment characteristic of thunderstorm conditions over the central United States. (a) Vertical cross section along the 19.2-km line in the horizontal area shown in (b)–(d). Only half the horizontal domain is shown because the results at larger y are mirror images of the fields shown. Panel (a) shows contours of the mixing ratios of cloud (dotted) and rain (solid) in g kg^{-1} superimposed on relative flow. Panels (b)–(d) show contours of vertical velocity (dashed) in m s^{-1} and relative wind vectors. The 1 g kg^{-1} contour of rainwater mixing ratio (solid) is shown, as are the positions of highs (H) and lows (L) of perturbation pressure. The magnitudes of the largest vectors in (a)–(d) are 20, 8, 7, and 10 m s^{-1}, respectively. (From Wilhelmson, 1974. Reprinted with permission from the American Meteorological Society.)

the top of the updraft of the storm in Fig. 7.28d. Although the calculations leading to Fig. 7.28 were performed some time ago and three-dimensional models have been used many times since to simulate a variety of types of cumulus and cumulonimbus, the flow patterns at low, middle, and high levels seen in this early calculation remain the basic patterns seen in many of these simulations, especially those representing precipitating clouds.

More sophisticated applications of three-dimensional cumulus modeling encounter numerical difficulties because of the form of the diagnostic equation (7.98) for the pressure perturbation. This equation, which is based on the anelastic form of mass continuity, involves higher-order spatial derivatives, which become awkward to compute if the model is applied to situations in which the model grid becomes complex. Such is the case if the flow over complex terrain is being considered, if other special boundary conditions are needed, or if a nested grid or stretched grid is desired to examine detailed features embedded in a larger-scale flow (e.g., a tornado within a thunderstorm). This difficulty led to the development of a more flexible form of the three-dimensional cumulus model, which uses the fully compressible mass continuity equation (2.20) rather than the anelastic form (2.54).

The difficulties with the model based on the anelastic form are purely numerical. The anelastic equations remain an excellent approximation of the governing dynamics of cumulus convection. The only consequence of using the continuity equation in anelastic form is that sound waves are filtered out of solutions to the model equations. These waves are of no interest in convective cloud dynamics. If uncontrolled numerically, they can obscure the solutions of interest. Therefore, the anelastic equations would seem desirable. Unfortunately, the awkward form of (7.98) severely limits their utility in numerical modeling. In the fully compressible equations [i.e., when a prognostic equation for the pressure perturbation based on (2.20) is used in place of this diagnostic relation (7.98)], all of the wind, thermodynamic, and water-continuity variables are obtained from prognostic equations of a similar form, and the numerical difficulties associated with (7.98) are removed from the model. The sound waves are retained in the equations but are controlled by a numerical technique called "splitting," in which the sound wave modes are solved for separately, using a shorter time step than used elsewhere in the model.[197] Thus, three-dimensional convective cloud models have been developed in which the fully compressible rather than the anelastic equations are integrated numerically. This type of model has been extremely successful in simulating and yielding insight into the dynamics of severe thunderstorms, especially tornadic storms and squall line thunderstorms. The dynamics of thunderstorms described in Chapter 8 have been largely learned from studies of numerical simulations of storms carried out with fully compressible three-dimensional numerical models. The same type of model has been used with great success to simulate the convection in the eyewalls of hurricanes (Chapter 10), clouds along cold fronts (Chapter 11), and mountain waves (Chapter 12).

A further degree of sophistication employed in some three-dimensional convective cloud models is in the parameterization of the subgrid-scale turbulence. The mixing coefficients for all of the variables are considered to be functions of the subgrid-scale eddy kinetic energy $\mathcal{K}$. The eddy kinetic energy equation (2.86) is introduced as an additional prognostic equation, which is solved simultaneously with the three component equations of motion, the prognostic pressure perturbation equation, the thermodynamic equation, and the water-continuity equations. In this way, the subgrid-scale turbulence is computed in a way that is internally consistent with all of the other variables that are predicted.[198] To solve the kinetic energy equation along with the other prognostic equations, expressions for the terms on the right-hand side of (2.86) must be found. One way these terms have been expressed is as follows: (i) The pressure–velocity correlation term $\mathcal{W}$ is ignored. (ii) The eddy fluxes contained in the definitions of $\mathcal{C}$ and $\mathcal{B}$ [(2.88) and

---

[197] See Klemp and Wilhelmson (1978b) for details.

[198] The type of model in which pressure perturbation and eddy kinetic energy are included as prognostic variables was introduced by Klemp and Wilhelmson (1978a) and is often referred to as the Klemp–Wilhelmson model.

(2.89)] are parameterized in terms of K-theory with mixing coefficients K_A, as in (2.185). According to this parameterization, the mixing coefficients are related to $\mathcal{K}$ according to the "inertial subrange" turbulence parameterization,[199] in which

$$K_A \propto \ell_G \mathcal{K}^{1/2} \tag{7.99}$$

where ℓ_G is the cube root of the grid volume and serves as a length scale. (iii) The dissipation $\mathcal{D}$ is also expressed according to this parameterization, as

$$\mathcal{D} \propto \ell_G^{-1} \mathcal{K}^{3/2} \tag{7.100}$$

Since the mixing terms in all the prognostic equations are parameterized as functions of mixing coefficients, which are functions of $\mathcal{K}$, and since $\mathcal{K}$ is predicted in a way that is internally consistent with the other predicted variables, the vertical and lateral eddy mixing components of entrainment are able to evolve in a way that is consistent with all the model variables.

[199] See Klemp and Wilhelmson (1978a), Deardorff (1972), and Schemm and Lipps (1976).

Chapter 8 | Thunderstorms

The previous chapter considered certain dynamical aspects that are basic to all convective clouds. Buoyancy, pressure perturbation, entrainment, and cloud vorticity were introduced and examined, and we noted how these dynamics are incorporated into cloud models. In this chapter, we focus specifically on one type of convective cloud—the cumulonimbus, or thunderstorm cloud. As noted in Chapter 1, cumulonimbus can occur either in isolation (Sec. 1.2.2.1) or as part of large mesoscale convective systems (Sec. 1.3.1). The next chapter of this book is devoted entirely to the structure and dynamics of mesoscale convective systems. Here we will concentrate on the properties of isolated thunderstorms and lines of thunderstorms. An understanding of individual storms and lines is crucial because they are important weather-producing phenomena in their own right and because they are building blocks of the larger mesoscale systems.

The isolated cumulonimbus is one of the most visually striking and photogenic of all cloud phenomena. The visual appearance of these clouds was illustrated by Figs. 1.4 and 1.5. In this chapter, however, we are not concerned so much with the exterior appearance of these clouds as with their internal structure and dynamics. We will begin by examining small, isolated cumulonimbus (Sec. 8.1). Then we will consider larger isolated thunderstorms, which are divided into two categories referred to as *multicell* and *supercell* thunderstorms (Secs. 8.2–8.6). In Secs. 8.8–8.10, we will examine three important circulation features of large thunderstorms: *tornadoes*, *gust fronts*, and *downbursts*. Finally, in Sec. 8.11, we will consider the conditions that favor the grouping of individual thunderstorms into *lines of storms*.

8.1 Small Cumulonimbus Clouds

It is possible for a warm cumulus cloud (i.e., one with no ice in it) to precipitate and hence to be classified as cumulonimbus. Such small precipitating cumuliform clouds are common in the tropics. However, most cumulonimbus clouds contain

---

[200] Goethe's reference to a thunderstorm.

ice. One of the primary distinguishing features of cumulonimbus (Sec. 1.2.2.1) is that its upper portion is usually composed of ice and is spread out in the shape of a smooth, fibrous, or striated anvil, while its lower portion exhibits the form of a mountain of bulbous towers. This basic structure is seen in both the smallest and largest of cumulonimbus clouds. An empirical model of the dynamical and micro-physical structure of a very small, isolated cumulonimbus cloud is shown in Fig. 8.1. This picture, synthesized from some 90 research flights through cumuliform clouds, shows that the cumulonimbus has a younger, developing side and an older, glaciated side. The developing side of the cloud is characterized by updraft motion. In this region are several strong, buoyant updraft cores. At the top of each updraft core is a cloud *turret* ~1–3 km in horizontal dimension. Within the turret the air overturns in the manner of a thermal (Figs. 7.8 and 7.9), where horizontal vorticity is generated to the side of a positively buoyant updraft core (Fig. 7.22a). Superimposed on each turret are spherical *tufts* ~100–200 m in diameter, where smaller-scale overturning produces entrainment of environmental air (Sec. 7.3). Across the cumulonimbus, the tops of the turrets are found at increasingly higher altitudes since the new updrafts form systematically on the developing side of the cloud. The tallest turret in Fig. 8.1 has reached above the 0°C level and attained sufficient maturity that some droplets have grown to sizes exceeding 20 μm. The conditions are thus right for ice enhancement to occur (Sec. 3.2.6). Large concentrations (≥ 1 ℓ^{-1}, often ~100 ℓ^{-1} or more) of ice particles appear within minutes of the time that the turret reaches its maximum height. The exact mechanism of enhancement [see hypotheses (i)–(iv) listed in Sec. 3.2.6] is not known. The high concentrations appear first in highly localized regions ~5–25 m wide, within the tufts. The ice particles then extend vertically through the cloud in strands of ice, which at lower levels in the cloud may be in the form of graupel. The strong updraft velocities in the cumulonimbus favor high condensation rates. The copious supercooled water condensed in the updrafts is collected by ice particles to produce the graupel. The heavier ice particles fall through the decaying updrafts and may appear as striations in the precipitation below cloud base. Smaller ice particles are carried upward and laterally outward to form the anvil portion of the cloud, which generally contains the maximum concentrations of ice particles, and where the strand structure is no longer apparent. As the ice particles in the more homogeneous anvil slowly fall out, they aggregate just above the 0°C level. The particles then melt and fall to the earth's surface in the manner of stratiform rain.

The small isolated cumulonimbus illustrated in Fig. 8.1 is sometimes called a *single-cell thunderstorm*, since it usually develops just one main precipitation shower. After the updrafts cease forming, this shower dissipates in the manner suggested by Fig. 6.16. At the end of the life cycle, all that remains is the relatively light precipitation from the anvil, which has a stratiform appearance, with a bright band at the melting level.

The radar-echo life cycle of a single-cell thunderstorm is illustrated by the example in Fig. 8.2. The similarity to the schematic life cycle in Fig. 6.1b is evident. The reflectivity increases and develops into a vertical core at 1818 GMT. It then evolves into a stratiform structure with a bright band at the melting layer.

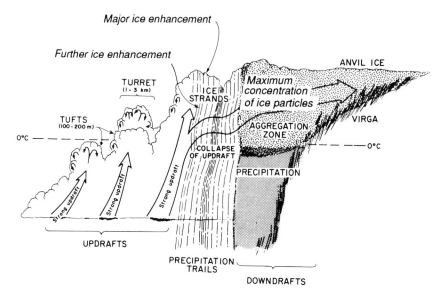

Figure 8.1   Empirical model of a small cumulonimbus cloud. Based on about 90 research aircraft penetrations of small cumulonimbus and large cumulus clouds. (From Hobbs and Rangno, 1985. Reprinted with permission from the American Meteorological Society.)

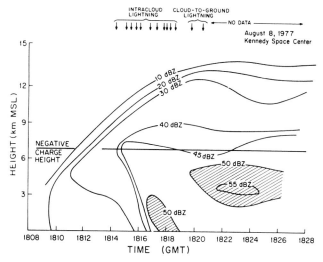

Figure 8.2   Time–height section of radar reflectivity for a thunderstorm near Cape Kennedy, Florida. Times of intracloud (IC) and cloud-to-ground (CG) lightning are indicated. (From Williams *et al.*, 1989. © American Geophysical Union.)

Also indicated in the figure is the sequence of lightning strikes produced by the cumulonimbus. The sequence is rather typical. Frequent lightning does not occur until cloud top rises above the −15 to −20°C level (about 7 km in Fig. 8.2). Intracloud (IC) lightning occurs first and at high frequency for several minutes, especially while the cloud and radar echo are still growing. Cloud-to-ground (CG) lightning activity (less frequent than IC) tends to lag the IC peak by 5–10 min and occurs after the radar echo contours become flat or, in some cases, begin to descend with time.

The lightning is a manifestation of the fact that the storm is electrified (i.e., positive and negative charges become separated within the region of cloud and precipitation, such that some regions have a net positive charge, while other regions have a net negative charge). The lightning itself is the transfer of charge from one region of a cloud to another or between the cloud and the earth.[201] The narrow channel within which the flash of lightning occurs is heated suddenly to ~30,000 K, with essentially no time to expand. The pressure in the channel is raised by an order of magnitude or two. The high-pressure channel then expands rapidly into the surrounding air and creates a shock wave (which travels faster than the speed of sound) and a sound wave. The latter is the audible signal that one hears as thunder and gives the cumulonimbus its various common names— thunderstorm, thundercloud, thunderhead, thundershower, etc.

The typical distribution of charge within a cumulonimbus is illustrated in Fig. 8.3. The main negatively charged region is sandwiched between two positively charged regions, of which the upper one is larger.[202] The main negative charge zone is notably pancake shaped, tending to be <1 km thick while extending horizontally over several kilometers or more. It is located at a level where the temperature is ~ −15°C. Negative charge is also found in a thin layer surrounding the upper part of the cumulonimbus, including the anvil.

This upper zone is thought to be produced when cosmic-ray-generated negative ions in the environment are attracted to the upper positive region of the cumulonimbus. The ions attach to small cloud particles at the edge of the cloud and form a *screening layer*. The main negative charge causes point discharge or corona from trees, vegetation, and other pointed or exposed objects on the ground below the storm, which leaves positive charge in the atmosphere above the earth's surface.

The mechanisms by which the cumulonimbus becomes electrified remain speculative and an active area of research.[203] One mechanism that appears to be important is the transfer of charge that occurs when graupel particles produced in the region of strong updraft collide with smaller ice particles. It has been demonstrated in laboratory experiments[204] that the polarity of the charge transfer in the collisions is dependent on temperature and liquid water content. Below a critical

[201] For a brief discussion of the physics of the lightning, see Wallace and Hobbs (1977, pp. 206–209).

[202] The inference of the distribution of electric charge in a thunderstorm has a colorful history. It first intrigued the American revolutionary and scientist Benjamin Franklin in the late eighteenth century. In the early twentieth century it gained the attention of the English Nobel laureate C. T. R. Wilson, who is credited with first identifying the dipole constituted by the main negative region in the middle of the cumulonimbus and the main positive region in the upper part of the cloud.

[203] For reviews of the state of knowledge of mechanisms involved in thunderstorm electrification, see Krehbiel (1986), Beard and Ochs (1986), and Williams (1988).

[204] See Sec. 3 of Williams (1988) for a review of these experiments.

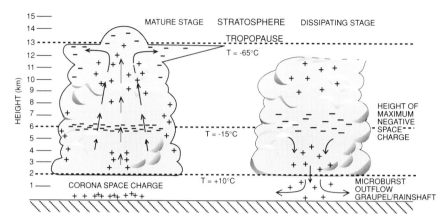

Figure 8.3 Schematic of the electrical structure of a cumulonimbus cloud. Positive and negative signs indicate the polarity of the charge at various locations. Streamlines indicate direction of airflow. (From Williams, 1988. © Scientific American, Inc. All rights reserved.)

temperature in the range -10 to $-20°C$, called the *charge-reversal temperature*, negative charge is transferred to the graupel, while in warmer air positive charge is transferred.*

The charge reversal temperature has been invoked to explain the distribution of charge depicted in Fig. 8.3, in which the main negative charge zone at $\sim-15°C$ is located between upper-level and lower-level regions of positive charge. According to the *precipitation hypothesis,* the falling graupel particles account for this struc- ture as follows. In the cold upper levels, the graupel particles take on negative charge and leave behind a cloud of small nonfalling particles, which become positively charged when negative charge is transferred to the graupel during colli- sions. The lower part of the cloud becomes dominated by the negative charge of the graupel particles descending from upper levels—hence the layer of negative charge near $-15°C$. Below this layer the graupel particles begin to take on positive charge during collisions.

An alternative explanation for the positive charge region at upper levels is the *convection hypothesis,* according to which the upper region of positive charge is accounted for by upward transport of positive charges released into the planetary boundary layer from the earth's surface from the subcloud layer to high levels by the updrafts. Which of these two hypotheses actually applies, under which condi- tions one is preferred over the other, or how the two may fit together in a common explanation of the charge distribution in a thunderstrom remains uncertain.

---

* The role of liquid water in the charge reversal is a topic of current research (Takahashi, 1978; Saunders, 1991).

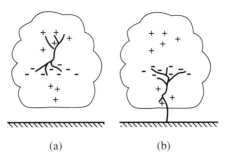

Figure 8.4 Depiction of lightning in prototype electrostatic structures: (a) intracloud; (b) cloud-to-ground. Positive and negative signs indicate the polarity of the charge at various locations. (From Williams *et al.*, 1989. © American Geophysical Union.)

The IC lightning in the earlier stages of the cumulonimbus transfers negative charge from the main negative region to the upper positive zone (Fig. 8.4a). The CG lightning, which comes later, during the mature stage of the storm, usually transfers charge from the main negative region to the ground (Fig. 8.4b). More rarely, positive charge is transferred to the ground.

There are reports of lightning from tropical warm cumulonimbus (i.e., clouds containing no ice).[205] However, this phenomenon does not appear common, and researchers presently regard warm-cloud lightning as the result of some different mechanism than that which electrifies cold clouds.

8.2 Multicell Thunderstorms

The single-cell thunderstorm described in Sec. 8.1 may actually be the most common type of thunderstorm, especially if every towering cumulus that reaches considerable height and precipitates even a small amount is considered to be a thunderstorm. However, the significance of single-cell storms in terms of precipitation or storm damage (other than lightning) is relatively small.[206] The single cell of cumulonimbus takes on more importance when it serves as a building block of a larger *multicell thunderstorm*. The internal structure of the multicell thunderstorm was revealed in the first modern field project designed for intensive storm documentation. It was called the "Thunderstorm Project" and was carried out over 40 years ago. The results of this project, which was the first to employ simulta-

---

[205] See Moore (1976).

[206] The minor role of single-cell thunderstorms as opposed to multicell storms was addressed by Simpson *et al.* (1980) in relation to precipitation production and by Chisholm and Renick (1972) in connection with storm damage.

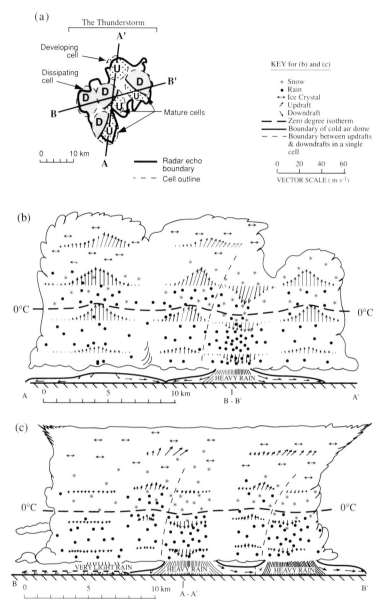

Figure 8.5   Schematic of a multicell thunderstorm in Ohio observed in the Thunderstorm Project. The storm consisted of cells in various stages of development. (a) Plan view. (b) Vertical cross section along A–A'. (c) Vertical cross section along B–B'. (From Byers, 1959. Reproduced with permission from McGraw-Hill, Inc.)

neously radar, aircraft, and other devices to observe the storms both remotely and *in situ*, were published in an important volume called *The Thunderstorm*.[207]

It was found that an individual storm ordinarily consists of a pattern of cells in various stages of development (Fig. 8.5). Cells in the early stages consist of vigorous updraft, in which hydrometeors are growing rapidly. Mature cells have both an active updraft and a downdraft, the latter coinciding with a downpour of precipitation. Dying cells contain only downdraft and precipitation that is still falling out. In *The Thunderstorm*, the term "thunderstorm" refers to the overall aggregate of cells, and its lifetime of several hours considerably exceeds that of an individual cell (~1 h). Thus, the pattern of cells within the multicell thunderstorm is continually changing.

Figure 8.6 illustrates the electrical structure of a multicell storm. An hypothetical but typical storm is shown at four successive times. At the first time (Fig. 8.6a), there are two mature cells in the storm. Each cell exhibits a distribution of charge similar to that shown in Fig. 8.3, with a region of negative charge at about the $-15°C$ level sandwiched between the upper and lower positive regions. As in Fig. 8.4a, the initial lightning is intracloud and transfers negative charge to the upper positive region. In the multicell case, however, it is possible for some of the lightning to travel from the negative region of one cell to the upper positive region of a neighboring cell. At the second time (Fig. 8.6b), cloud-to-ground strikes have begun. As in the single-cell case (Fig. 8.2), these strikes come primarily after the initial period of intracloud lightning, and they again carry negative charge from the primary negative region to the earth's surface. These discharges may also have large horizontal components, which, as shown, can extend across the main negative regions of adjacent cells. At the third time (Fig. 8.6c), the anvil has become more extensive, and intracloud discharges penetrate into it from the main part of the storm. Cloud-to-ground discharges (not shown) are also sometimes observed to emanate from the anvil. Also by this time, one of the cells has dissipated and taken on the stratiform structure characteristic of this phase of the cell's development (Fig. 6.1b and Fig. 8.2). The shading in Fig. 8.6c and d shows the location of the melting layer and radar bright band. Balloon-borne electric field measurements in stratiform precipitation of extratropical cyclones and mesoscale convective

---

[207] This report was edited by H. R. Byers and R. R. Braham (1949). The Thunderstorm Project was directed by Professor Byers, then at the University of Chicago. Before the close of World War II, it had become clear that neither military nor commercial aviation could avoid flying in and around thunderstorms. To promote the safety of such aviation "information was needed concerning the internal structure and behavior of the thunderstorm." The project was therefore organized as a joint undertaking of the U.S. Air Force, Navy, the National Advisory Committee for Aeronautics, and the Weather Bureau. It took advantage of equipment and experienced personnel that were available in great numbers at the end of the war. Twenty-two freight cars full of ground equipment (not counting numerous trucks and jeeps) and ten Northrup P-61C "Black Widow" aircraft were made available to the program. Radar equipment, first used in the war and radiosonde equipment and surface instrumentation were deployed. Upon inquiring among "highly competent instrument pilots of the Air Force for volunteers for this work. ... The response was extremely gratifying and brought to the Project experienced crews, most of whom had served as instrument flight instructors." In addition, a group from the Soaring Society of America volunteered to make instrumented sailplane flights into thunderstorms for part of the program.

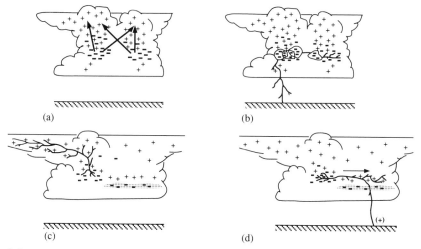

Figure 8.6

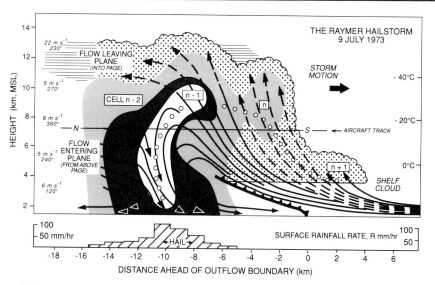

Figure 8.7

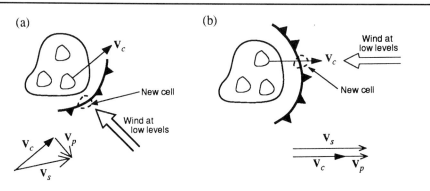

Figure 8.8

systems indicate that negative charge may accumulate in bright-band layers,[208] and this characteristic is postulated to apply in the melting layer of the dissipating cell of Fig. 8.6c and d. At the fourth time (Fig. 8.6d), horizontal intracloud lightning is occurring between the main negative area of the still-active cell and a region of positive charge at about the same level in the dissipating cell. These horizontal discharges are observed to occur repetitively at intervals of a few minutes or more. Occasionally, the dissipating cell can produce positive strokes to the ground that remove positive charge from this level.

Under certain conditions of wind shear (to be considered later in this chapter), a multicell thunderstorm takes on a form of organization illustrated in Fig. 8.7. The figure can be thought of either as an instantaneous picture of the storm, with cells in various stages of development, or as a sequence of stages in the life of one cell, which move, in a relative sense, through the storm. New cells (at $n + 1$ in Fig. 8.7) form on or just ahead of the leading edge of the storm. As cells move through the storm, they undergo their life cycles. At $n + 1$ and n, the cells are in the developing stage, with updraft air filling the cells and precipitation particles developing aloft but not yet falling to the ground. Precipitation particles are initiated near cloud base at $n + 1$ and grow by collection of cloud water. Above the 0°C level, the collectors are primarily ice particles, whose growth, after their formative stages, is dominated by the accumulation of rime ice. Continuation of this riming can build up graupel particles and hailstones, which eventually become big enough to fall relative to the ground. The schematic hail trajectory in the figure is one possibility based on an assumption that the particle, once initiated, remains

---

[208] Chauzy *et al.* (1980, 1985).

Figure 8.6 The apparent electrical structure and evolution of lightning in a multicell thunderstorm, as inferred from a variety of observations in different storms. A mature storm is illustrated in (a) and (b). A dissipating storm is depicted in (c) and (d). Branched structure of the lightning has been suggested in all cases except for the multicellular discharge shown in (a). The shaded region in (c) and (d) represents the radar bright band from melting snow. Positive and negative signs indicate the polarity of the charge at various locations. (From Krehbiel, 1986. Reprinted with permission from the National Academy Press, Washington, D.C.)

Figure 8.7 Schematic model of a multicell thunderstorm observed near Raymer, Colorado. It shows a vertical section along the storm's north-to-south (N–S) direction of travel, through a series of evolving cells. The solid lines are streamlines of flow relative to the moving system; they are broken on the left side of the figure to represent flow into and out of the plane and on the right side of the figure to represent flow remaining within a plane a few kilometers closer to the reader. The chain of open circles represents the trajectory of a hailstone during its growth from a small particle at cloud base. Lightly stippled shading represents the extent of cloud and the two darker grades of stippled shading represent radar reflectivities of 35 and 45 dBZ. The white area enclosing the hail trajectory is bounded by 50 dBZ. Environmental winds (m s^{-1}, deg) relative to the storm are shown on the left-hand side of the figure. (From Browning *et al.*, 1976. Reprinted with permission from the Royal Meteorological Society.)

Figure 8.8 Possible horizontal arrangements of cells in multicell thunderstorms. Solid contours indicate radar echoes of two different intensities. Frontal symbol denotes gust-front location. Vectors indicate the velocity of an individual cell ($\mathbf{V}_c$), storm propagation velocity resulting from new cell development ($\mathbf{V}_p$), and velocity of the storm as a whole ($\mathbf{V}_s$).

within the same cell throughout its lifetime. Other possible hail growth scenarios exist within multicell thunderstorms. For example, it has been suggested that optimal hail production in a multicell storm occurs by the initiation of graupel particles and hailstones in smaller cells and their subsequent advection into the updraft of the most intense cell of the storm.[209]

There are several horizontal configurations to which the organized multicellular vertical cross section in Fig. 8.7 can apply. The case in Fig. 8.8a, for example, shows cells forming systematically to the right of the cell-motion vector. This produces a right-moving thunderstorm. Similarly, the situation in Fig. 8.8b represents a forward-moving storm.

8.3 Supercell Thunderstorms

So far we have discussed only the single-cell and multicell thunderstorms, the latter being characterized by a fluctuating pattern of relatively short-lived cells. Another basic type of cumulonimbus structure is the *supercell thunderstorm*, which is far rarer and much more violent. Supercell thunderstorms are notorious for producing damaging hail and tornadoes. These storms could be the subject of much study because of their severe weather characteristics alone; however, further motivation is provided by the vorticity of the air in these storms. This vorticity, which gives rise to the tornadoes, is a fascinating manifestation of geophysical fluid dynamics, with many scientifically alluring facets. The name supercell[210] refers to the fact that although this type of storm is about the same size as a multicell thunderstorm, its cloud structure, air motions, and precipitation processes are dominated by a single storm-scale circulation consisting of one giant updraft–downdraft pair.

The exterior visual appearance of a supercell storm is sketched in Fig. 8.9 in the form that is usually taught to ground-based tornado spotters. The tornado vortex is visible as a funnel-shaped cloud pendent from a rotating wall cloud extending downward from the cloud base. Water condenses to form a cloud marking the funnel because of the lowering of the pressure in the intense vortex. To a first approximation, the vortex is cyclostrophic; hence, according to (2.46), the pressure decreases strongly inward toward the center of the vortex, in proportion to the square of the vortex wind speed. The tornado usually occurs near the peak of a wedge of low-level warm air, entering the region of the storm typically from the east or southeast. This warm air rises over the gust front to form the updraft of the storm-scale circulation. Dense downdraft air deposited by the storm at the surface spreads out behind the gust front. Precipitation reaching the ground behind the gust front forms a curved backdrop for the tornado. Weaker tornadoes can occur along the southwest (or rear-flank) gust front.

An idealized horizontal projection of the cloud-top topography of a supercell thunderstorm, as it would appear in a satellite picture, and the low-level precipita-

---

[209] See Heymsfield *et al.* (1980).
[210] Coined by Browning (1964).

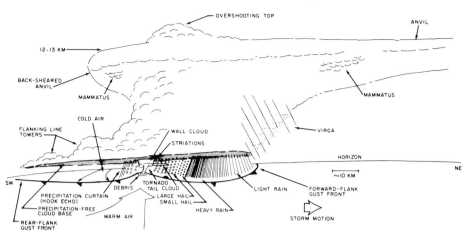

Figure 8.9 Schematic visual appearance of a supercell thunderstorm. (Based on U.S. National Severe Storms Laboratory publications and an unpublished manuscript of Howard B. Bluestein.)

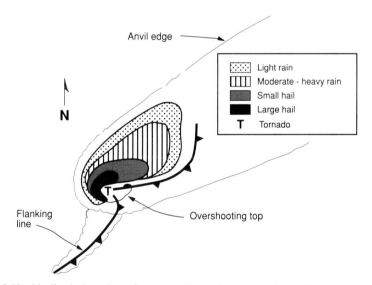

Figure 8.10 Idealized plan view of a supercell thunderstorm as it would appear in a satellite picture and in the low-level precipitation pattern that would be detected by a horizontally scanning radar. Cloud features seen by satellite include the flanking line, the edge of the anvil cloud, and the overshooting cloud top. Positions of the gust front (given by frontal symbols) and tornado are also shown. (Based on U.S. National Severe Storms Laboratory publications.)

tion pattern that would be detected by a horizontally scanning radar are shown superimposed in Fig. 8.10. The near coincidence of the tornado, the peak of the wedge of warm air, the overshooting cloud top, and the indentation in the horizontal precipitation area are evident. The horizontal distribution of precipitation at the ground is sorted according to particle size and thus produces a distinctive radar reflectivity pattern, since light rain, heavy rain, and small and large hail

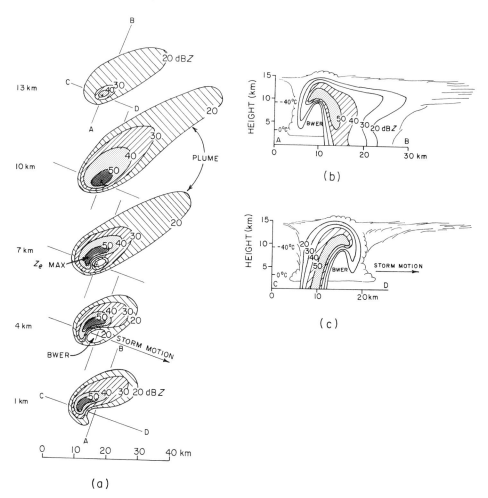

Figure 8.11 Schematic illustrating the variation of radar reflectivity patterns with height in supercell thunderstorms observed in Alberta, Canada. Horizontal sections of reflectivity (dBZ) at various altitudes are shown in (a). Vertical sections are shown in (b) and (c). Cloud boundaries are sketched. BWER refers to the bounded weak echo region. (From Chisholm and Renick, 1972.)

produce increasingly greater echo intensities. The large hail produces an extremely intense echo surrounding the notch in the precipitation pattern where the tornado is located. This radar reflectivity pattern is generally referred to as a *hook echo*.

The radar reflectivity patterns vary significantly with height in the storm (Fig. 8.11). The notch in the low-level horizontal echo pattern (1 km in Fig. 8.11a) is associated with a *bounded weak-echo region* (BWER) or *echo-free vault* that extends upward toward the overshooting top of the storm (Fig. 8.11a, 4 and 7 km; Fig. 8.11b and c).

The extremely strong updraft in the supercell thunderstorm ($\sim$10–40 m s^{-1})

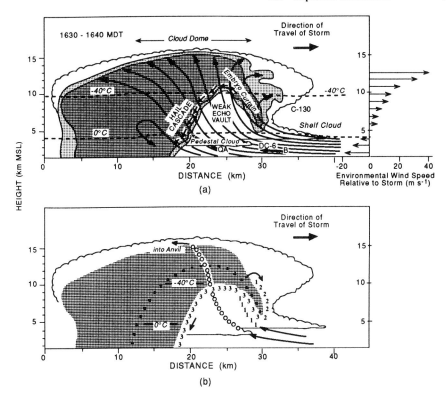

Figure 8.12 (a) Vertical cross section of cloud and radar echo structure of a supercell thunderstorm in northeastern Colorado. The section is oriented along the direction of travel of the storm, through the center of the main updraft. Two levels of radar reflectivity are represented by different densities of hatched shading. The locations of four instrumented aircraft are indicated by C-130, QA, DC-6, and B. Bold arrows denote wind vectors in the plane of the diagram as measured by two of the aircraft (scale is only half that of winds plotted on right side of diagram). Short thin arrows skirting the boundary of the vault represent a hailstone trajectory. The thin lines are streamlines of airflow relative to the storm. To the right is a profile of the wind component along the storm's direction of travel. (b) Vertical section coinciding with (a). Cloud and radar echo are the same as before. Trajectories 1, 2, and 3 represent three stages in the growth of large hailstones. The transition from stage 2 to stage 3 corresponds to the reentry of a hailstone embryo into the main updraft prior to a final up–down trajectory during which the hailstone may grow large, especially if it grows close to the boundary of the vault as in the case of the indicated trajectory 3. Other, less-favored hailstones will grow a little farther from the edge of the vault and will follow the dotted trajectory. Cloud particles growing within the updraft core are carried rapidly up and out into the anvil along trajectory 0 before they can attain precipitation size. (From Browning and Foote, 1976. Reprinted with permission from the Royal Meteorological Society.)

encourages the growth of very large hailstones. The growth process has been hypothesized to occur more or less as shown in Fig. 8.12. This figure illustrates the single, massive updraft, in which various hail trajectories can ensue, depending on the size and location of hail embryos when they first appear. The trajectories in the figure represent some plausible possibilities, given the observed radar reflectivity and air motion. Horizontal components of the trajectories, not shown

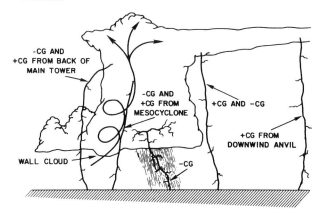

Figure 8.13 Sketch of observed locations and polarities of cloud-to-ground (CG) lightning flashes in supercell thunderstorms. The spiral (denoted mesocyclone) indicates the region of intense updraft and rotation. Only negative CGs have been observed in the precipitation core. The positive CGs seem to constitute only a very small percentage of the total flashes to ground. (From Rust *et al.*, 1981. Reprinted with permission from the American Meteorological Society.)

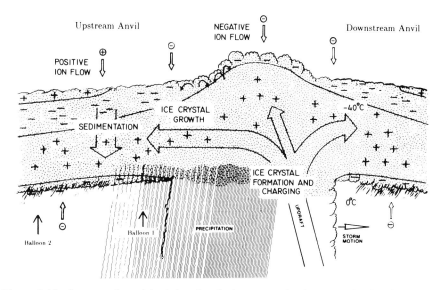

Figure 8.14 Conceptual model of the electrical structure in the upper levels of a supercell thunderstorm. Positive and negative signs indicate the polarity of the charge at various locations. Short curved streamlines at the top of the main cloud tower indicate turbulent mixing. Based on data obtained from two balloon sondes. (From Byrne *et al.*, 1989. © American Geophysical Union.)

here, can also be important to the growth of hail. The fallout of precipitation, as in all cumulonimbus, affects the downdraft for the supercell through precipitation drag and evaporation.

The supercell thunderstorm is also more active electrically than the single-cell or multicell thunderstorms. The overall lightning flash rate in supercells is ~10–40 min^{-1} overall and 5–12 min^{-1} for CG lightning. For more ordinary thunderstorms the overall rate is ~2–10 min^{-1}, while the CG rate is ~1–5 min^{-1}.[211] The observed spatial pattern of the cloud-to-ground lightning in the supercell is indicated in Fig. 8.13. In single-cell and multicell thunderstorms, positive CG strikes are associated with the dissipating cells (e.g., Fig. 8.6d). In the supercell, positive CG strikes occur in the mature as well as the dissipating stages of the storm. The positive strikes, however, account for only a small portion of the total number of strikes, and the lightning in the precipitation core is all negatively charged. Relatively little is known about the distribution of charge within the supercell thunderstorm. Measurements obtained in the anvil of one storm suggest the distribution of charge at upper levels shown in Fig. 8.14. Ice crystals transported up in the strong updraft are positively charged and are advected into both the forward and trailing anvil. This structure is not unlike the single-cell and multicell storm structures illustrated in Figs. 8.3 and 8.6. Atop the anvil is a complex double-layered screening zone.

8.4   Environmental Conditions Favoring Different Types of Thunderstorms

Whether a given thunderstorm turns out to be single-cellular, multicellular, or supercellular depends on both the wind shear and static stability of the environment. Three-dimensional numerical cloud models described in the last chapter (Sec. 7.5.3) have been particularly helpful in identifying the relationships between environmental conditions and the forms that thunderstorms take. The behavior of the model thunderstorms is related quantitatively to the stability and shear as follows.[212]

The buoyant stability of the environment can be represented by the *convective available potential energy (CAPE)*, which is given by

$$\text{CAPE} \equiv g \int_{\text{LFC}}^{z_T} \frac{\theta(z) - \bar{\theta}(z)}{\bar{\theta}(z)} \, dz \qquad (8.1)$$

where θ is the potential temperature of a parcel of air lifted from $z = 0$ to $z = z_T$ while not mixing with its environment. The parcel rises dry adiabatically [conserving its θ, according to (2.11)] until it becomes saturated and then rises moist adiabatically [conserving its θ_e, according to (2.18)] thereafter. $\bar{\theta}$ is the potential temperature of the environment (base state), the LFC (level of free convection) is the height at which the parcel becomes warmer than the environment, and the cloud top z_T is assumed to be the level where $\theta = \bar{\theta}$.[213]

---

[211] These rates were reported by Rust *et al.* (1981).

[212] This discussion is based on the work of Weisman and Klemp (1982).

[213] Synoptic meteorologists will recognize CAPE as the net *positive area* on a pseudoadiabatic chart. The positive area is the area between the parcel and environment temperature on a chart with temperature as the abscissa and height (or log pressure) as the ordinate and the scales adjusted so that the positive area is proportional to energy.

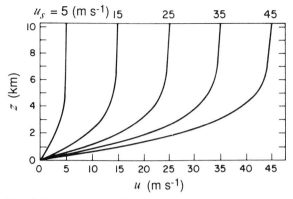

Figure 8.15   Profiles of wind speed u used in three-dimensional model simulations of multicell and supercell thunderstorms. Profiles becomes asymptotic to u_s. (From Weisman and Klemp, 1982. Reprinted with permission from the American Meteorological Society.)

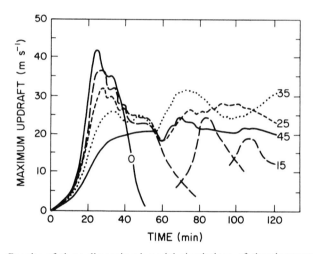

Figure 8.16   Results of three-dimensional model simulations of thunderstorms under different amounts of wind shear. The quantity plotted is the maximum vertical velocity as a function of time for different values of the wind-shear parameter u_s(m s^{-1}), which is the number plotted next to each curve. (From Weisman and Klemp, 1982. Reprinted with permission from the American Meteorological Society.)

The wind shear in an idealized environment can be represented by the asymptotic wind speed u_s, reached at the height z_s, the top of a shear layer, in an hypothetical environment in which the wind is unidirectional with magnitude,

$$u = u_s \tanh(z/z_s), \quad z_s = 3 \text{ km} \tag{8.2}$$

Examples of this wind profile are shown in Fig. 8.15.

Model results are illustrated in Fig. 8.16 for an environment whose $\bar{\theta}$ increases with height such that the environmental saturated equivalent potential temperature $\bar{\theta}_{es}$ [as defined in (2.146)] is constant.[214] Different values of $\theta(z)$ were tried in (8.1) for a parcel lifted from $z = 0$ by letting the surface humidity take on various values while keeping the surface temperature constant. The results in Fig. 8.16 are for a surface mixing ratio of 14 g kg^{-1}, a surface temperature of about 23°C, and the various wind profiles shown in Fig. 8.15. The model convection was initialized as a buoyant bubble 10 km in horizontal radius, 1.4 km in vertical radius, and 2°C in temperature excess at its center. Three different model responses were seen, depending on the strength of the environmental shear. Figure 8.16 shows the maximum vertical velocity as a function of time for different values of u_s. For zero shear, a single-cell storm occurred ($u_s = 0$). For moderate shear ($u_s = 15$ m s^{-1}), a sequence of cells occurred, indicating multicellular storm structure. For strong shear ($u_s = 25, 35,$ and 45 m s^{-1}), a single cell reached a plateau of vertical velocity and continued to be maintained through a process of redevelopment. This redevelopment, as we will see, is a characteristic of rotational supercell dynamics.

The structure of the model multicell storm obtained under moderate shear is illustrated by the superimposed low-level flow and midlevel vertical velocity in Fig. 8.17. After 40 min (Fig. 8.17a), the initial updraft (cell 1) had weakened, and cold outflow had pushed 10 km ahead of the updraft core. The updraft was thus cut off from inflow of warm air, and the maximum convergence was located at the gust front well ahead of the old updraft core. This convergence produced a new updraft (cell 2), which was ahead of cell 1 after 80 min (Fig. 8.17b). By 120 min (Fig. 8.17c), the updrafts of cells 1 and 2 had disappeared, but a third cell had formed at the gust front when the updraft of cell 2 was cut off from the inflow. Noteworthy is the consistency of this storm, with its sequential cell development, and the empirical multicell storm structure pictured in Fig. 8.7.[215]

Some characteristics of the model supercell storm obtained when the environment had higher shear are shown in Fig. 8.18, which, like Fig. 8.17, shows only half of the model domain. Since the wind shear of the environment is unidirectional, the results are symmetric about the axis $y = 0$. It is particularly important to note this symmetry in the supercell case, because two identical storms develop and move away from the y-axis. This process is referred to as *storm splitting* and is intrinsic to supercell dynamics. The storm that moves to the right of the y-axis (shown in the figure) is called the *right-moving* storm. The other member of the split (not shown) is called the *left-moving* storm. In this mode, the gust front does not outrun the updraft core. Instead, they move together in a state of near equilibrium, in which the rates of cold outflow and warm inflow are about equally matched. The movement of the updraft core away from the y-axis is a result of the rotational dynamics of the storm, which are particularly robust when the environmental shear is strong. The symmetrical behavior of storm splitting, in which left- and right-moving storms occur as mirror images of each other, is the result of unidirectional wind shear. We will see that if the direction as well as the speed of

---

[214] Such an environment is referred to as *moist adiabatic*.

[215] The behavior of the multicell storm, whereby the downdraft outflow moves so fast that it outruns the existing updraft cell and forms a new cell at the gust front, had been pointed out in an earlier modeling study by Thorpe and Miller (1978).

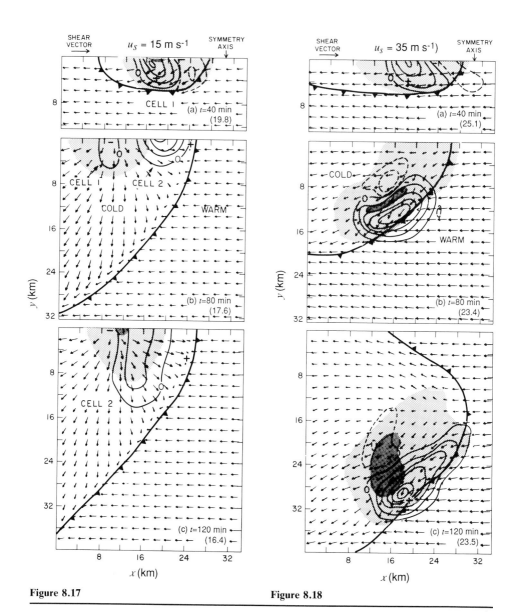

Figure 8.17   **Figure 8.18**

Figure 8.17 Results of a model simulation showing a multicell thunderstorm occurring under conditions of moderate environmental wind shear ($u_s = 15$ m s^{-1}). Fields are shown for three times during the simulation. Vectors represent storm-relative wind at an altitude of 178 m. The maximum vector magnitude (m s^{-1}) is shown in parentheses in the lower right corner of each plot. The surface rain field is indicated by stippling. The surface gust front is denoted by the frontal symbol and corresponds to the $-0.5°$C temperature perturbation contour. The midlevel (4.6 km) vertical velocity field is contoured every 5 m s^{-1} for positive values and 2 m s^{-1} for negative values. The zero contours outside the main region of storm activity have been deleted. Plus and minus signs represent the location of the low-level (178 m) vertical velocity maximum and minimum, respectively. Only the southern half of the model domain is shown. The fields in the northern half are mirror images. (From Weisman and Klemp, 1982. Reprinted with permission from the American Meteorological Society.)

Figure 8.18 Results of a model simulation showing a supercell thunderstorm occurring under conditions of strong environmental wind shear ($u_s = 35$ m s^{-1}). Format same as Fig. 8.17. (From Weisman and Klemp, 1982. Reprinted with permission from the American Meteorological Society.)

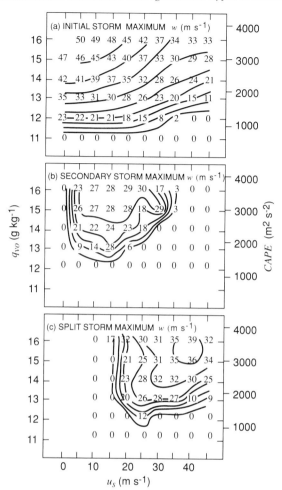

Figure 8.19 Maximum vertical velocity of model thunderstorms as a function of CAPE and wind-shear parameter u_s. Panels (a) and (b) refer to multicell cases; panel (c) refers to supercell cases. (From Weisman and Klemp, 1982. Reprinted with permission from the American Meteorological Society.)

the wind in the environment varies with height, either the left- or the right-moving storm is favored, while the other is disfavored.

Further insight into the modes of thunderstorm organization is gained by plotting the maximum vertical velocity as a function of CAPE and u_s. Figure 8.19 shows results for multicell cases. To exist, the initial cell, as would be expected, must have a threshold amount of thermodynamic instability (Fig. 8.19a). Its intensity increases with increased instability and decreases with increased shear, as a result of enhanced entrainment favored by the shear. (Recall from Sec. 7.4.3 that the counter-rotating vortices produced by the tilting of the environmental shear near the convective updraft are an important mechanism by which midlevel air is entrained into the updraft.) From Fig. 8.19b, it is evident that a second cell will not occur if there is no shear in the environment (i.e., the degenerate case of a single-cell thunderstorm occurs when there is too little shear to allow low-level inflow to match the downdraft spreading at the surface; consequently, a new cell cannot become established). When a secondary cell forms, the vertical velocity reaches a peak at low to moderate shear, where an approximate balance between inflow and outflow allows a new cell to form. When the shear becomes too strong, the new cell cannot be maintained against entrainment. The supercell cases (Fig. 8.19c) can occur only at moderate to high shear. The maxima in Fig. 8.19b and c indicate a bimodal distribution of thunderstorm type. The values of shear supporting a supercell are too high to support a secondary cell in a multicell storm. However, at the higher values of shear, internal rotation becomes stronger. Associated with the rotation are dynamic pressure perturbations. The dynamically induced pressure field, in turn, affects the movement of the storm, accounting for storm splitting and right- and left-moving storm propagation. We will now explore these and other topics associated with the rotational dynamics of the supercell thunderstorm in more detail.

8.5   Supercell Dynamics[216]

8.5.1   Storm Splitting and Propagation

The initial storm splitting and propagation of the two updraft cores of a supercell away from the along-shear axis in an environment of unidirectional shear were illustrated three-dimensionally in Fig. 7.23. The first panel shows the vortex couplet that arises when the horizontal vorticity associated with the environmental shear is tilted by the updraft. The second panel shows the storm after splitting has occurred.

Crucial to understanding the split is that both vortices contain a pressure perturbation minimum. The nature of this minimum can be seen by considering again the diagnostic equation for the pressure perturbation (7.4) and recalling that the pressure perturbation can be split into partial pressures p_B^* and p_D^*, associated with buoyancy and dynamic sources, respectively. The vertical component of the equation of motion (7.2) can then be written

$$\frac{\partial w}{\partial t} = -\frac{1}{\rho_o}\frac{\partial p_D^*}{\partial z} - \left(\frac{1}{\rho_o}\frac{\partial p_B^*}{\partial z} - B\right) - \mathbf{v}\cdot\nabla w \qquad (8.3)$$

Numerical-model calculations show that the terms in parentheses tend to balance. Substituting (7.6) into (7.9) and making use of the anelastic continuity equation (2.54) leads to

$$\nabla^2 p_D^* = -\nabla\cdot\left(\rho_o\mathbf{v}\cdot\nabla\mathbf{v}\right)$$

$$= -\rho_o\left(u_x^2 + v_y^2 + w_z^2 - \frac{d^2\ln\rho_o}{dz^2}w^2\right) - 2\rho_o\left(v_x u_y + u_z w_x + v_z w_y\right) \qquad (8.4)$$

The first term in the last parentheses is dominant in locations in the thunderstorm occupied by strong vortices. In the case of a purely rotational horizontal flow, for

---

[216] The discussion in this section closely follows the review article of Klemp (1987).

which $v_x = -u_y$, we obtain, according to the definition of the vertical vorticity given in (2.56), the relation:

$$v_x u_y = -\tfrac{1}{4}\zeta^2 \tag{8.5}$$

In a strong vortex, (8.4) then implies that

$$-\nabla^2 p_D^* \propto p_D^* \propto -\zeta^2 \tag{8.6}$$

That is, there is a dynamic pressure perturbation minimum associated with the vortex (regardless of whether the vortex is cyclonic or anticyclonic). Another way to look at this pressure perturbation development in the vortex is that, as the vortex intensifies, the pressure field within it adjusts to a state of cyclostrophic balance (2.46), which implies a pressure minimum in the center of the vortex regardless of the sense of the rotation. Thus, strong midlevel rotation acts to lower pressure at midlevels on each flank of the storm. The corresponding vertical gradient of dynamic pressure perturbation then encourages updraft growth on each flank of the storm, according to the first term on the right of (8.3). Because of this effect, updrafts on the flanks are maintained and not choked off by downdraft spreading.

Thus, the supercell type of thunderstorm organization is promoted by strong environmental shear, which leads to strong midlevel vortices on the storm flanks. The negative dynamic pressure perturbations in these vortices lead to vertical acceleration and continual regeneration of the updrafts on the two flanks of the storm. As the initial updraft shifts to the flanks of the splitting storm, the inflow pattern shifts to that shown by the dashed arrows in Fig. 7.23b. This behavior explains the quasi-steady, long-lived characteristic of the supercell, since this vertical pressure gradient provides a mechanism for the continual regeneration of the updraft on the flank of the storm. Since the vertical pressure gradient associated with the eddies occurs only on the flanks of the storm, and not on its leading edge, this same mechanism also explains the splitting of the storm. It was suspected for a long time that the storm splitting and propagation were in some way caused by the precipitation-induced downdraft. However, numerical experiments in which the precipitation is not allowed to fall also exhibit storm splitting and continuous regeneration on the storm flanks, thus confirming that it is the vertical pressure gradient associated with the eddies on the updraft flanks that is responsible for the splitting.

8.5.2 Directional Shear

Although storm splitting is essential to the formation of supercell storms, mirror-image rotating thunderstorms are rare. In most of the supercell storm cases that occur over the central United States, the right-moving storm dominates and the left-moving member of the split is seldom observed.[217] It is easily verified by numerical modeling that this behavior is not because of Coriolis effects.[218] Rather

---

[217] Davies-Jones (1986) found, using radar data taken at the U.S. National Severe Storms Laboratory in Oklahoma, that only 3 of 143 supercell storms rotated anticyclonically.

[218] See Klemp and Wilhelmson (1978b).

it is associated with the directional shear of the wind, which up to now we have ignored.

The effect of directional shear can be illustrated by linearizing (8.4) about a mean velocity of $\bar{\mathbf{v}} = (\bar{u}, \bar{v}, 0)$, in which case we obtain

$$\nabla^2 p_D^* = -2\rho_o \mathbf{S} \cdot \nabla_H w \tag{8.7}$$

where ∇_H is the horizontal gradient operator $(\mathbf{i}\partial/\partial x + \mathbf{j}\partial/\partial y)$ and $\mathbf{S}$ is the vertical shear of the mean horizontal wind, defined as

$$\mathbf{S} \equiv \frac{\partial}{\partial z}(\bar{u}\mathbf{i} + \bar{v}\mathbf{j}) \tag{8.8}$$

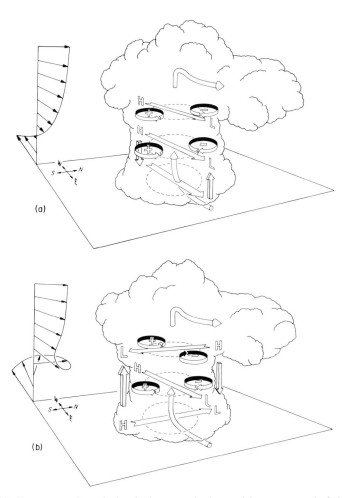

Figure 8.20 Pressure and vertical velocity perturbations arising as an updraft in a supercell thunderstorm interact with an environmental wind shear that (a) does not change direction with height and (b) turns clockwise with height. The high (H) to low (L) horizontal pressure gradients parallel to the shear vectors (flat arrows) are labeled along with the preferred location of cyclonic (+) and anticyclonic (−) vorticity. The shaded arrows depict the orientation of the resulting vertical pressure gradients. (From Klemp, 1987. Reproduced with permission from Annual Reviews, Inc.)

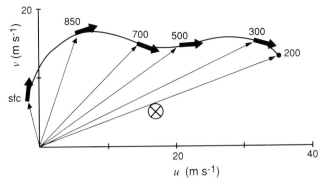

Figure 8.21 Mean wind sounding (m s^{-1}) in the vicinity of tornadic thunderstorms in the central United States. The profile is an average of 62 cases. Heavy arrows indicate the direction of the shear vector at each level labeled in millibars. The estimated mean storm motion is indicated by $\otimes$. (Adapted from Maddox, 1976. Reproduced with permission from the American Meteorological Society.)

The dot product $2\mathbf{S} \cdot \nabla_H w$ in (8.7) indicates that a maximum of pressure gradient divergence (and, hence, a minimum of pressure) occurs in the zone of strong horizontal gradient of vertical velocity on the downshear side of the updraft (Fig. 8.20a). Since the updraft (and the gradient of w bounding it) is strongest at midlevels, this pressure perturbation is strongest at midlevels, thus producing an upward pressure-gradient acceleration in the lower troposphere on the downshear side of the storm. In the unidirectional shear case, this effect induces a circulation that reinforces inflow and overturning (Fig. 8.20a). However, it has no effect on growth on the flanks of the storm, which has already been seen to be a nonlinear effect. Neither side of the storm is favored.

In the part of the United States where most tornadic thunderstorms occur, the wind shear in the storm environment is typically far from unidirectional. Normally the hodograph exhibits "clockwise" turning of the shear, in which the direction of the wind-shear vector (expressed in degrees from true north) increases with increasing height (e.g., Fig. 8.21, from the surface to ~600 mb). In this case, the high-to-low pressure gradient in the direction of the wind shear $\mathbf{S}$ favors growth of the cyclonic (right-moving) member of the split, since the shear-induced vertical gradient of pressure perturbation favors development on the south side of the storm (Fig. 8.20b).[219] This behavior is consistent with the climatological preference for right-moving supercell storms in the United States.

[219] It was once thought by some that since the right-moving member of a splitting thunderstorm has a cyclonically rotating updraft (Sec. 7.4.3), the Coriolis force might be directly responsible for favoring the further development of the right-moving storms. Numerical-model simulations with the Coriolis force included demonstrate, however, that this force has no significant effect (Klemp and Wilhelmson 1978b). Equation (8.7) further shows that the Coriolis force plays no direct role in determining whether the right- or left-moving storm is favored. However, the large-scale shear $\mathbf{S}$, which appears in (8.7) and is specified as input to the model calculations, does, in reality, depend on f. Therefore, whether right- or left-moving storms are favored does depend indirectly on f.

8.6 Transition of the Supercell to the Tornadic Phase

The early stages of development of the supercell thunderstorm have been depicted in Fig. 7.23 and Fig. 8.20. During these times, strong vortices (maxima and minima of ζ) are produced by tilting of the horizontal vorticity inherent in the environmental wind shear. These processes are depicted again in Fig. 8.22a and b. In Sec. 7.4.3 we noted that, in the splitting storm, the vortex and updraft may become collocated as a result of the combined effects of advection and tilting [recall (7.73)] and that the rotating updraft thus formed is referred to as the mesocyclone (Fig. 8.22c). The mesocyclone, moreover, extends to low levels to a location on the gust front (dotted line in Fig. 8.22d). The positioning of the center of the mesocyclone at low levels is determined by a complex combination of tilting, advection, and stretching [i.e., all the terms in (2.59), except those involving f, are important]. The vorticity development at the gust front will be discussed further below. The mesocyclone, formed between midlevels and the low-level gust front, is the environment in which the main tornado of the supercell forms. Locating the mesocyclone by Doppler radar is a useful method for short-term forecasting of tornado occurrence.

In Figs. 8.9 and 8.10, the typical location of the tornado within the supercell was seen, in plan view, to be near the leading point of the wedge of warm air entering the storm at low levels. The schematic view of a tornadic thunderstorm shown in Fig. 8.23 also shows this region as one of the most likely locations for a tornado. This empirical model of storm structure near the ground includes a second tornado location to the south as well as features labeled FFD and RFD, which are the outflows of *forward-flank* and *rear-flank downdrafts*. This schema represents the storm after its transition to the tornadic phase. This transition usually occurs relatively suddenly, on a scale $\sim$10 min, after the supercell has evolved slowly over a period of several hours. During the rapid transition, the low-level rotation increases, the rear-flank downdraft forms behind the updraft, and the cold outflow and warm inflow become intertwined at low levels.

Figure 8.22 Schematic of vorticity development in a supercell thunderstorm. Cloud boundary is sketched. Precipitation is hatched. White tube represents a vortex tube. Heavy arrows are updrafts and downdrafts. GF indicates gust front. Storms move over a horizontal surface shown in perspective. Point O is fixed to the surface and is directly under the center of the cloud in (a). Storms move away from O in time. Storm cross sections are in vertical planes outlined by dashed lines. In (b) a split occurs. In (c) and (d), the two components of the storm move away from the center line of the horizontal plane. In (c) divergence (DIV) and convergence (CONV) are indicated. In (d) the vortex at the base of the downdraft is deleted because by this time it has been greatly weakened by the strong divergence at low levels. The dotted line in (d) represents the center of the mesocyclone. (From Houze and Hobbs, 1982.)

Figure 8.23 Schematic plan view of low-level structure of a tornadic thunderstorm. The thick line surrounds the radar echo. The frontal symbol denotes the boundary between the warm inflow and cold outflow. Low-level position of the updraft is finely stippled, while the forward-flank (FFD) and rear-flank (RFD) downdrafts are coarsely stippled. Storm-relative surface flow is shown along with the likely location of tornadoes. (Adapted from Lemon and Doswell, 1979. Reproduced with permission from the American Meteorological Society.)

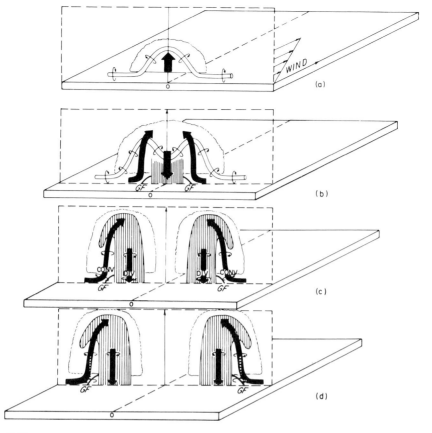

Figure 8.22

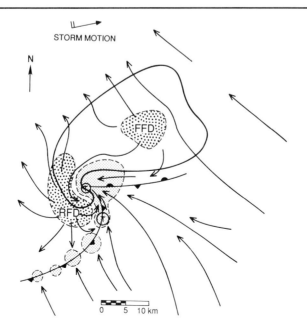

Figure 8.23

Numerical simulations show that the increased vorticity at low levels induces a minimum of pressure perturbation, according to (8.6). The low-level rotation increase and tornado formation at the peak of the warm wedge (at the lower end of the mesocyclone) are the result of a sequence of events beginning with the generation of a horizontal buoyancy gradient across the gust front separating the front-flank downdraft and the warm sector. As discussed in Sec. 7.4.2, the buoyancy gradient along the gust front is a source of horizontal vorticity along this gust front to the east of the storm. The easterly relative wind in the vicinity of the gust front advects the horizontal vorticity created along the gust front toward the center of the storm. The remainder of the sequence of events can be understood with the aid of Eq. (2.59), which expresses the time change of the vertical component of vorticity. If the Coriolis effect is ignored, (2.59) may be rewritten in Eulerian form as

$$\underbrace{\frac{\partial \zeta}{\partial t}}_{} = \underbrace{-\mathbf{v} \cdot \nabla \zeta}_{(i)} + \underbrace{\zeta w_z}_{(ii)} + \underbrace{\left(\xi w_y + \eta w_x\right)}_{(iii)} \tag{8.9}$$

where the terms on the right are the changes in the vertical component of relative vorticity ζ associated with (i) advection, (ii) intensification by convergence (stretching) of vortex tubes, and (iii) tilting of horizontal components of vorticity into the vertical. Once horizontal vorticity generated along the gust front is advected into the center of the storm, it is tilted into the vertical by the main storm updraft [term (iii)]. The vertical vorticity thus created is then concentrated by the convergence at low levels in the center of the storm to form an intense vortex [term (ii)].

Thus, the storm's own gust front is the major source of the intense circulation at low levels in the tornadic phase of the supercell. The horizontal vorticity produced at the gust front is part of a positive feedback loop intrinsic to the supercell circulation; it should not be confused with the horizontal vorticity contained in the large-scale environment shear. The large-scale shear accounts for the rotation in the upper part of the mesocyclone (upper end of dotted line in Fig. 8.22d) and is the source of rotation of the storm-wide circulation (as discussed in Sec. 7.4.3). The latter circulation first produces a gust front, which in turn produces strong horizontal vorticity in a location where it can be advected into the center of the storm circulation. There it is tilted and stretched to produce a more intense low-level circulation, which constitutes the lower end of the mesocyclone and ultimately the tornado—when the lower part of the mesocyclonic circulation is sufficiently stretched by the gust-front convergence.

Numerical simulations further show that the rear-flank downdraft is a *result* of the intensified circulation at low levels. For many years, there was a difference of opinion as to whether the rear-flank downdraft was the cause or an effect of the transition of the supercell to its tornadic phase. This dispute appears to have been resolved by modeling studies,[220] which show that the increased rotation at low levels, according to (8.6), produces a minimum of pressure at the ground. The

[220] For further discussion, see Klemp (1987).

resulting downward-directed vertical pressure-gradient acceleration induces the rear-flank downdraft. Part of the diverging outflow from the rear-flank downdraft spreads forward rapidly enough that the main updraft and its associated tornado become cut off from the warm inflow. A new storm updraft center forms to the south. This location coincides with the southern tornado location in Fig. 8.23. The cutting off of the main updraft by the outflow from the rear-flank downdraft is sometimes referred to as "occlusion." Since the processes that create the tornado also lead to the rear-flank downdraft and the cutting off of the main updraft from its source of warm air, the tornado is usually associated with a weakening super-cell updraft.

After the old updraft is cut off and the new one forms, all the processes can repeat, with new updraft, new tornado, and new rear-flank downdraft. Repetition of these processes can lead to sequences of tornadoes, as indicated in Fig. 8.24.

8.7   Nonsupercell Tornadoes and Waterspouts

So far we have considered only tornadoes that occur in the context of supercell thunderstorms. A large number of intense vortices that form in association with cumulus congestus or developing cumulonimbus are not of the supercell variety. Over water these usually take the form of waterspouts. Generally, these vortices are weak in comparison to a supercell tornado. However, especially over land, vortices associated with nonsupercell convective clouds can occasionally reach tornado strength and be quite damaging. An empirical model of the life cycle of a nonsupercell tornado over land is shown in Fig. 8.25. This type of tornado occurs when a developing convective cloud is located above a spot of intensified local vertical vorticity in the boundary layer. Often these local maxima of boundary-layer vorticity are found along convergence lines in the boundary layer, as indicated at points A, B, and C in Fig. 8.25. The same convergence line may provide the uplift for convective cloud development. However, the cloud may develop for other reasons and simply became located over the convergence line. Also, the boundary-layer vorticity maximum could be produced at other types of locations. For example, it might be topographically produced.[221] Regardless of the origins of the cloud and the boundary-layer vorticity maxima, once a developing cloud and a vorticity maximum become vertically aligned, tornado development can occur, as shown in Fig. 8.25c at location C. The mechanism of development is thought to be vortex stretching [i.e., ζw_z, term (ii) in (8.9)]. Strong convergence, and hence stretching, occurs between the strong updraft at cloud base (where $w \sim 1$–10 m s^{-1}) and the ground (where $w = 0$). The tornado formed in this manner is observed

---

[221] On 9 April 1991, a nonsupercell tornado occurred over the coast of Puget Sound, near Silver-dale, Washington. It appears that this tornado occurred when boundary-layer vorticity was produced by flow around steep jagged terrain along the coastal area (Colman, 1992). When a developing cumulus congestus or small cumulonimbus cloud moved over this region of boundary-layer vorticity, the tornado developed and was seen from the Atmospheric Sciences–Geophysics Building at the University of Washington.

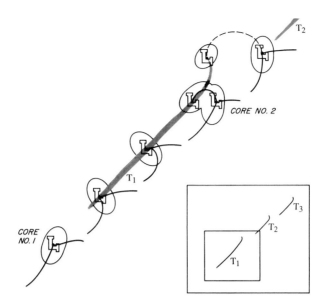

Figure 8.24 Conceptual model of the evolution of the mesovortex core (L) of a tornadic supercell thunderstorm. Thick lines show gust-front position. Shaded areas are tornado tracks. Inset shows the tracks of the tornadoes (T_1, T_2, T_3) associated with successive mesovortex redevelopment. Small square is the region expanded in the main figure. (From Burgess *et al.*, 1982. Reprinted with permission from the American Meteorological Society.)

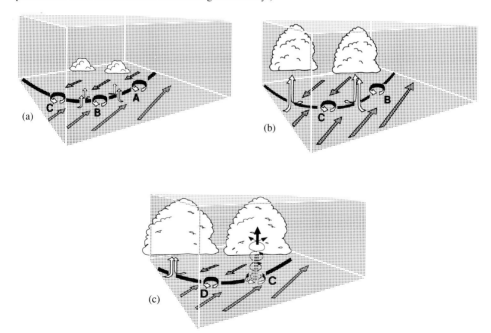

Figure 8.25 Empirical model of the life cycle of a nonsupercell tornado over land. The black line is a radar-detectable convergence zone in the boundary layer. Low-level vortices are labeled with letters. Open arrows indicate air motion. Clouds are sketched. (From Wakimoto and Wilson, 1989. Reprinted with permission from the American Meteorological Society.)

to build upward from the ground until it establishes a connection to cloud base. This behavior is also observed in dust devils, which appear to differ from the model in Fig. 8.25 only in that they form in connection with dry thermals instead of cloud updrafts.

The mechanism depicted in Fig. 8.25 requires no dynamical attribute of the cumulus cloud other than its updraft. The vortex originates in the boundary layer independent of the cloud. However, as is evident from our previous discussions of the vorticity in convective clouds (Sec. 7.4.3, Sec. 8.5, and Sec. 8.6), vortices that result from tilting of environmental horizontal vorticity into the vertical [term (iii) in (8.9)] are typically present in the vicinity of the cumulus updraft (Fig. 7.23). As a result of advection and/or stretching [terms (i) and (ii) in (8.9)], it is possible for the updraft to become collocated with one of the vorticity maxima. One example of this collocated updraft vortex is the mesocyclone of the splitting storm, described by (7.73). Just as the mesocyclone of the supercell forms a favorable environment for tornado formation, so could a rotating updraft in a smaller convective cloud. If the updraft in the cloud of Fig. 8.25 were rotating, it would provide another source of vorticity for a tornado-like vortex, in addition to independently existing boundary-layer vortices, such as those depicted along the convergence line in Fig. 8.25. Convergence, such as that along the convergence line, would concentrate the vorticity of the rotating updraft, possibly into a narrow tornado-like vortex extending below cloud base. This type of explanation has been offered for the development of waterspouts from cumulus congestus or small cumulonimbus clouds over an ocean surface (Fig. 8.26). Waterspouts tend to form in the vicinity of gust front outflows from nearby convective rain showers. The strong convergence at the gust front is envisaged to concentrate the vorticity of a rotating updraft into a narrow waterspout vortex.

The difference between Fig. 8.25 and Fig. 8.26 is that in the former the convergence line provides the vorticity and the cloud provides the convergence for the stretching [term (ii) in (8.9)], while in the latter the convergence line (gust front) provides the convergence while the cloud provides the vorticity for the stretching. Either process is feasible, and which is dominant probably has to be evaluated on a case-by-case basis.

Waterspouts appear to occur in clouds whose circulations resemble qualitatively the air motions in supercell storms. The airflow below clouds producing waterspouts disrupts the structure of the ocean surface in such a way that tracers of the air motions can be seen from a distance as shading on the water. An empirical model of a developing waterspout, its accompanying low-level flow, and the pattern of shading on the ocean is shown in Fig. 8.27 and Fig. 8.28. The proximity of the waterspout funnel to a nearby rain shower is seen in Fig. 8.27. The "shear band" indicated in both figures is a windshift line that is the analog of the gust front and flanking line of the supercell seen Figs. 8.9, 8.10, and 8.23. The waterspout vortex in Figs. 8.27 and 8.28 is also in a similar location to the supercell tornado in Figs. 8.9, 8.10, and 8.23. The scales and intensities of all of the flow features in the case of the waterspout, however, are much reduced in comparison to the supercell storm.

8.8 The Tornado

8.8.1 Observed Structure and Life Cycle

The structure of a supercell tornado observed at several stages of its life cycle is illustrated schematically in Fig. 8.29. The *organizing stage* was characterized by a visible funnel touching the ground intermittently, though the damage path was continuous. In the *mature stage*, the tornado was largest. In the *shrinking stage*, the entire funnel decreased to a thin column. The *decaying stage* was character-

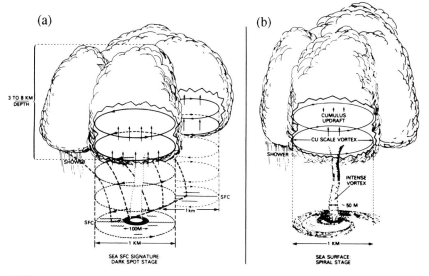

Figure 8.26

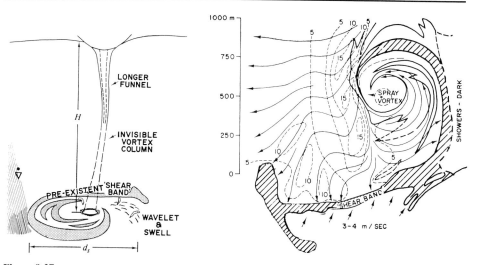

Figure 8.27   **Figure 8.28**

ized by a fragmented, contorted, but still destructive funnel. Air motions within and near the tornado (Fig. 8.30) were determined by tracing debris particles and identifiable cloud elements in motion pictures and by surveying surface damage patterns. In the mature stage, tangential velocities at a radius of 200 m and heights 60–120 m above ground were estimated to be $\sim$50–80 m s^{-1}, in general agreement with deductions from other documented cases.[222] At the wall cloud level, there was strong downward motion on the southwest side of the funnel, with upward motion on the northeast side (Fig. 8.30b). An empirical model of the life cycle of a waterspout vortex is shown in Fig. 8.31. Comparison with Fig. 8.29 shows that the waterspout life cycle resembles qualitatively that of the supercell tornado. Figure 8.31a corresponds to the organizing stage, Fig. 8.31b to the mature stage, and Fig. 8.31c to the decaying stage of the tornado.

As will be discussed further in Sec. 8.8.2 below, the tornado vortex is in cyclostrophic balance along most of its length. The condensation marking the funnel cloud is produced by the low pressure at the center of the vortex. Thus, the overall shape of the cloud is a tracer of the vortex, not fluid parcels. Recalling the component vorticity equations (2.57)–(2.59), noting that Coriolis force is negligible on the scale of the tornado, and assuming that baroclinic generation is also negligible in this situation, we may combine (2.57)–(2.59) to obtain the three-dimensional vorticity equation

$$\frac{D\boldsymbol{\omega}}{Dt} = (\boldsymbol{\omega} \cdot \nabla)\mathbf{v} \tag{8.10}$$

where the expression on the right is the vector sum of all the stretching and tilting terms on the right-hand sides of (2.57)–(2.59). The change of the shape of the tornado vortex indicated by the life cycle of its cloud form seen in Fig. 8.29 is a

---

[222] Measurements made with a small, portable Doppler radar carried to within a few kilometers of a mature tornado have confirmed estimates such as these by showing tangential velocities as large as 60 m s^{-1} (Bluestein and Unruh, 1989).

Figure 8.26 A postulated waterspout development process. (a) Early prefunnel stage. (b) Intermediate stage. The different horizontal dimensions of the parent vortices (solid circles in cloud, dashed when shown below, as on left), dark spot on ocean surface, and condensation funnel are indicated. The heavy dashed curves in (a) denote cold downdraft air which terminates at the surface as a gust front. The downdraft, the anticyclonic member of the parent vortex pair, and the subcloud extension of the parent vortices are omitted in (b). In (b), the vortex center is made visible by a condensation funnel, produced by lowering of the pressure in the central core. The dark sea-spray ring suggests tangential winds >22 m s^{-1}. (From Simpson *et al.*, 1986. Reprinted with permission from the American Meteorological Society.)

Figure 8.27 Empirical model of a developing waterspout. Cloud base height H varies from 550 to 670 m and width d_s from 100 to 920 m. Dot over triangle is the standard rain shower symbol. (From Golden, 1974a. Reprinted with permission from the American Meteorological Society.)

Figure 8.28 Horizontal pattern of wind and spray in a waterspout. Streamlines (solid) and isotachs (dashed, m s^{-1}) of boundary-layer flow are shown. (From Golden, 1974b. Reprinted with permission from the American Meteorological Society.)

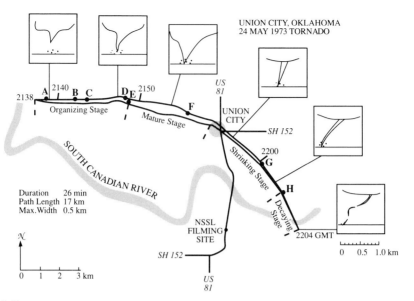

Figure 8.29

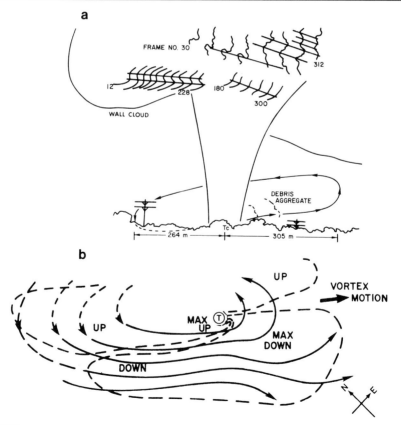

Figure 8.30

dramatic visual illustration of the effects included in (8.10). As the life cycle progresses, the initially wide, vertically oriented vortex stretches into a narrower vortex and tilts over so that the vorticity vector along portions of the vortex is highly canted, tending toward horizontal in places. The fact that the tornado remains damaging in the decaying stage is clearly the effect of stretching. Although the tornado is becoming narrower and affecting a smaller area, the rotational wind in the affected area remains strong as convergent flow at the base of the tornado continues to squeeze the vortex so that the vorticity in its interior remains strong.

The tornado funnel is sometimes observed to contain one to six smaller subvortices, ~0.5–50 m in diameter. These *suction vortices* may be stationary or orbit around the tornado center (Fig. 8.32). They contain some of the strongest winds of the tornado and leave a complex pattern of narrow trails of debris and extreme damage within the tornado's general path. As will be discussed below, the suction vortices are a result of waves superimposed on the primary tornado vortex becoming unstable and dominating the pattern. The suction vortices are observed to be particularly dangerous and damaging. Again, the effect of stretching on the intensity of vorticity is evident, as the locally intensified convergence in the small-scale perturbations concentrates the vorticity in the suction vortices.

A schematic of the flow regions in a tornado is shown in Fig. 8.33. All of region I is in cyclostrophic balance. Region Ia is called the *outer flow* and consists of air that is converging and conserving its angular momentum. Region Ib is the *core*, which immediately surrounds the axis and extends out to the radius of maximum tangential wind. The core appears to be in a state of approximate solid-body rotation (defined below). This means that, in the absence of vertical motions, the core is stable against radial displacements and supports waves. These waves are seen moving up or down tornado funnels. There is almost no entrainment into the core, and flow along the axis can be either up or downward, as we will see. Region II is the *turbulent boundary layer*, where friction upsets cyclostrophic balance. The net inward force drives strong radial inflow toward region III, called the *corner region*, where there is strong upflow into the core. Lying over the top of the vortex is region IV, the *buoyant updraft*. The tornado must end in an updraft, since a downdraft would destroy the vortex.

Figure 8.29 Structure of a tornado observed at several stages of its life cycle. Sketches show funnel cloud and associated debris. Letters A–H indicate damaged farms. (From Golden and Purcell, 1978a. Reprinted with permission from the American Meteorological Society.)

Figure 8.30 Tornado structure from photogrammetry and damage surveys of the Union City, Oklahoma, tornado of 24 May 1973. (a) Outline of tornado funnel and upper wall cloud showing features tracked on movie loops. Sequential outline of cloud tags and streamers in wall cloud and typical composite trajectory and displacement of debris aggregate are superimposed. (b) Schematic plan view of horizontal streamlines (solid lines) and low-level vertical motion patterns (dashed lines) around decaying tornado. (From Golden and Purcell, 1978b. Reprinted with permission from the American Meteorological Society.)

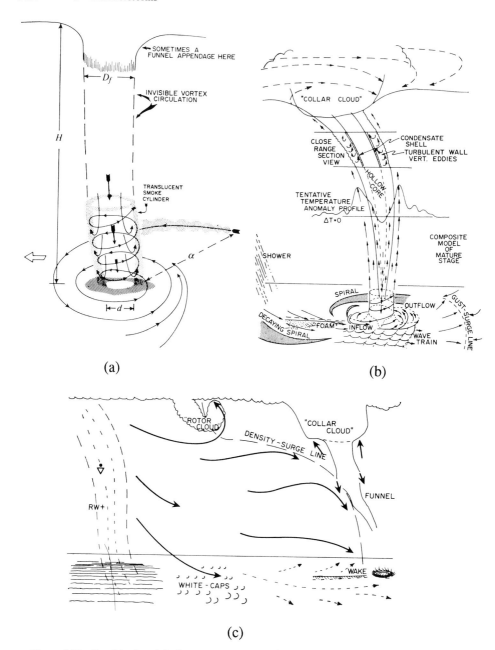

(a)

(b)

(c)

Figure 8.31 Empirical model of a waterspout. (a) Early stage. Cloud base height H varies from 550 to 670 m, maximum funnel diameter D_f from 3 to 150 m, $d = 3$–45 m, $\alpha = 15$–760 m. (b) Mature stage. Maximum funnel diameter ranges from 3 to 140 m. (c) Decay stage, during which the funnel cloud often undergoes rapid changes in shape and may become greatly contorted. Maximum funnel diameters range from 3 to 105 m. (From Golden, 1974a. Reprinted with permission from the American Meteorological Society.)

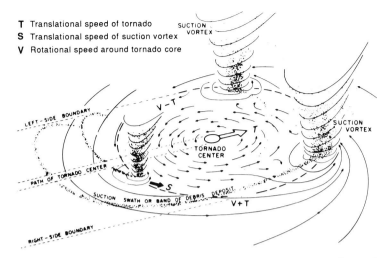

Figure 8.32   Conceptual model of the subdivision of a tornado funnel into smaller suction vortices. (From Fujita, 1981. Reprinted with permission from the American Meteorological Society.)

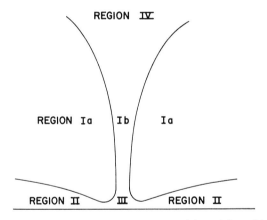

Figure 8.33   Schematic of flow regions in a tornado. (Adapted from Morton, 1970 and from Davies-Jones, 1986.)

8.8.2   Tornado Vortex Dynamics[223]

The dynamics of the core region of the tornado can be examined theoretically by considering the dry Boussinesq equations written in cylindrical polar coordinates (r, Θ, z). The z-axis is vertical and centered on the vortex, which is assumed to be axially symmetric ($\partial/\partial\Theta = 0$). Molecular friction is ignored. Under these condi-

---

[223] This discussion is based largely on a review article by Davies-Jones (1986) and papers by Davies-Jones and Kessler (1974) and Rotunno (1977, 1979, 1986).

tions, the components of the Boussinesq version of the mean-variable equation of motion (2.83) are

$$\overline{u}_t + \overline{u}\,\overline{u}_r + \overline{w}\,\overline{u}_z - \frac{\overline{v}^2}{r} = -\frac{1}{\rho_o}\frac{\partial \overline{p}^*}{\partial r} + K_m\left(\overline{u}_{rr} + \frac{1}{r}\overline{u}_r + \overline{u}_{zz} - \frac{\overline{u}}{r^2}\right) \tag{8.11}$$

$$\overline{v}_t + \overline{u}\,\overline{v}_r + \overline{w}\,\overline{v}_z + \frac{\overline{u}\,\overline{v}}{r} = K_m\left(\overline{v}_{rr} + \frac{1}{r}\overline{v}_r + \overline{v}_{zz} - \frac{\overline{v}}{r^2}\right) \tag{8.12}$$

$$\overline{w}_t + \overline{u}\,\overline{w}_r + \overline{w}\,\overline{w}_z = -\frac{1}{\rho_o}\frac{\partial \overline{p}^*}{\partial z} + \overline{B} + K_m\left(\overline{w}_{rr} + \frac{1}{r}\overline{w}_r + \overline{w}_{zz}\right) \tag{8.13}$$

The mean-variable thermodynamic equation (2.78) and continuity equation (2.75) in their Boussinesq versions (i.e., with the density factor ρ_o omitted) are

$$\overline{\theta}_t + \overline{u}\,\overline{\theta}_r + \overline{w}\,\overline{\theta}_z = K_\theta\left(\overline{\theta}_{rr} + \frac{1}{r}\overline{\theta}_r + \overline{\theta}_{zz}\right) \tag{8.14}$$

and

$$\frac{1}{r}\left(r\overline{u}\right)_r + \overline{w}_z = 0 \tag{8.15}$$

In these expressions, $u = Dr/Dt$, $v = rD\Theta/Dt$, and $w = Dz/Dt$, and the eddy-flux terms in (8.11)–(8.14) are parameterized in terms of a simplified K-theory (Sec. 2.10.1), in which the values of the mixing coefficients K_m and K_θ are constant. The expressions multiplying K_m in (8.11)–(8.13) are the axisymmetric cylindrical-coordinate components of the vector $\nabla^2\overline{\mathbf{v}}$, while the factor multiplying K_θ in (8.14) is the axisymmetric cylindrical-coordinate form of $\nabla^2\overline{\theta}$.

The key dynamical characteristic of the tornado vortex is its extremely large tangential wind component, which far exceeds its radial or vertical component (i.e., $|v| \gg |u|, |w|$). In this case, (8.11) suggests that the flow is to a first approximation in cyclostrophic balance (Sec. 2.2.6). Studies of the dynamics of the tornado vortex have therefore concentrated on simple cyclostrophic solutions to (8.11)–(8.15). We proceed below by examining the two simplest known solutions.

The very simplest solution is the *Rankine vortex*, which consists of two regions separated by radius r_c. There is assumed to be no radial or vertical motion whatsoever ($\overline{u} = \overline{w} = 0$); the flow is assumed to be frictionless ($K_m = K_\theta = 0$); and the inner region is characterized by solid-body rotation, which means the angular velocity $\overline{v}/r$ is constant. Thus, within the inner core, the tangential velocity is directly proportional to radial distance from the center of the vortex:

$$\overline{v} = ar, \quad a = \text{constant}, \; r \le r_c \tag{8.16}$$

A way of expressing the constant is obtained by first noting that, in axisymmetric cylindrical coordinates, the vertical vorticity ζ (Sec. 2.4) is given by

$$\zeta \equiv \mathbf{k} \cdot \nabla \times \mathbf{v} = \frac{1}{r}\frac{\partial(rv)}{\partial r} \tag{8.17}$$

By applying (8.17) to (8.16), we find

$$\zeta = 2a, \; r \le r_c \tag{8.18}$$

The integrated amount of vertical vorticity contained in some horizontal area A is given by

$$\iint_A \zeta \, dA = \iint_A \mathbf{k} \cdot \nabla \times \mathbf{v} \, dA = \oint v_t \, dl \equiv \Gamma \tag{8.19}$$

where the closed line integral, called the *circulation* and indicated by Γ, follows from Stokes's theorem. It is taken around the boundary of A, with v_t being the horizontal velocity component tangent to the boundary. To obtain the circulation Γ_c around the circle covered by the inner region of the Rankine vortex, we substitute (8.18) into the left-hand expression of (8.19) and integrate over the area surrounded by a circle of radius r_c. The result is

$$2a\pi r_c^2 = \Gamma_c \tag{8.20}$$

From this expression, it is evident that specification of r_c and Γ_c determine the constant in (8.16) and hence the flow across the whole inner core of the vortex.

The outer region of the Rankine vortex is characterized by *potential vortex flow*, which is a flow whose tangential velocity is inversely proportional to r. Thus, we have

$$\bar{v} = b/r, \quad b = \text{constant}, \quad r > r_c \tag{8.21}$$

This flow is *irrotational*, which means $\bar{\zeta} \equiv 0$, a fact easily verified by applying (8.17) to (8.21). Since there is no vorticity outside r_c, the circulation at any larger radius remains Γ_c; that is,

$$\Gamma = \oint_{r_c} \bar{v}(r) \, dl = \bar{v}(r) 2\pi r = \Gamma_c \tag{8.22}$$

which implies that the constant in (8.21) is

$$b = \frac{\Gamma_c}{2\pi} \tag{8.23}$$

Thus, the constants in both (8.16) and (8.21) may be expressed in terms of Γ_c.

The basic characteristics of the pressure distribution in the Rankine vortex are easily diagnosed since under the assumed velocity conditions ($\bar{u} = \bar{w} = 0$), the equation for radial momentum (8.11) reduces exactly to the cyclostrophic relation

$$\frac{\bar{v}^2}{r} = \frac{1}{\rho_o} \frac{\partial \bar{p}}{\partial r} \tag{8.24}$$

The perturbation pressure p^* has been replaced with the total pressure p in (8.24), which is allowable since the base-state pressure p_o is a function of z only. The assumed condition $\bar{w} = 0$ further implies that there is no vertical acceleration, and the total pressure is therefore governed by the hydrostatic relation (2.38), which may be written in the present case in mean-variable form as

$$\frac{\partial \bar{p}}{\partial z} = -\rho_o g \tag{8.25}$$

From (8.21) and (8.23), the maximum tangential velocity is

$$\bar{V}_{max} = \frac{\Gamma_c}{2\pi r_c} \tag{8.26}$$

The larger this velocity is, the greater the pressure deficit of the vortex. This result follows from integration of (8.24) from the center of the vortex to $r = r_c$. Making use of (8.16), (8.20), and (8.26), we obtain

$$\bar{p}(r_c) - \bar{p}_o = \int_0^{r_c} \rho_o \frac{1}{r} \left(\frac{\Gamma_c r}{2\pi r_c^2} \right)^2 dr = \frac{\rho_o \bar{V}_{max}^2}{2} \tag{8.27}$$

This relation implies that if the density of the air were 1 kg m^{-3} and the maximum wind in the tornado 50 m s^{-1}, then the pressure deficit across the vortex would be 12.5 hPa. It is also evident from (8.24), that the largest pressure gradient is located at $r = r_c$ and has the value $\rho_o \bar{v}^2/r_c$.

It has been determined from photogrammetric observations that tornadoes often appear to behave to a first approximation as a Rankine vortex. This fact has been determined by comparing the shapes of observed tornadoes to the shape of the Rankine vortex. The outline of the funnel cloud is taken to be an isobaric surface and is compared to the shape of an isobaric surface in the Rankine vortex. The slope of the isobaric surface is implied by (8.24) and (8.25):

$$\left. \frac{\partial z}{\partial r} \right|_p = - \left. \frac{\partial \bar{p}}{\partial r} \right|_z \bigg/ \left. \frac{\partial \bar{p}}{\partial z} \right|_r = \frac{\bar{v}^2}{rg} \tag{8.28}$$

Integration from a point along the funnel's edge (r,z) to the top of the funnel (∞, z_f) yields

$$z = \begin{cases} z_f - \dfrac{\bar{v}_{max}^2 r_c^2}{2gr^2}, & r > r_c \\[3mm] z_f - \dfrac{\bar{v}_{max}^2}{2g}\left(2 - \dfrac{r^2}{r_c^2} \right), & r \le r_c \end{cases} \tag{8.29}$$

If the shape of the funnel is observed, and the funnel is assumed to be a Rankine vortex, (8.29) provides a way of estimating the maximum wind in the vortex. If the tip of the funnel is observed to be at height z_o and is assumed to have radius r_c, then (8.29) implies that $\bar{v}_{max} = \sqrt{2g(z_f - z_o)}$.

The simplest vortex solution that allows a secondary circulation (i.e., motions in the r and z directions) is the *one-cell*, or Burgers–Rott vortex,[224] for which the horizontal convergence (negative divergence) is assumed to be a positive constant. In this case, the axisymmetric divergence is given by

$$\frac{1}{r}\frac{\partial}{\partial r}(r\bar{u}) = -2c = \text{constant} < 0 \tag{8.30}$$

which has the solution

$$\bar{u} = -cr \tag{8.31}$$

---

[224] The one-cell vortex was described in early papers by Burgers (1948) and Rott (1958).

The continuity equation (8.15) then becomes

$$-2c + \frac{\partial \bar{w}}{\partial z} = 0 \tag{8.32}$$

which, if $\bar{w} = 0$ at $z = 0$, implies that

$$\bar{w} = 2cz \tag{8.33}$$

With $\bar{u}$ and $\bar{w}$ given by (8.31) and (8.33), it can be verified by substitution into (8.12) that, in the absence of eddy mixing,

$$\bar{v} = d/r, \quad d = \text{constant}, \quad r > \varepsilon \tag{8.34}$$

where ε is an arbitrarily small positive number. This expression is of the same form as (8.21). As in that case, the flow is irrotational, and the circulation Γ is a constant, with all of the vorticity concentrated within $r \leq \varepsilon$. If eddy mixing is included, (8.34) is replaced by

$$\bar{v} = \frac{d}{r}\left[1 - \exp\left(-cr^2/2K_m\right)\right] \tag{8.35}$$

Thus, neither $\bar{u}$ nor $\bar{v}$ is a function of z; the horizontal circulation is independent of height. The three-dimensional circulation in the one-cell vortex is illustrated in Fig. 8.34a. The horizontal flow spirals inward, while upwelling in the center of the vortex. In the case with eddy mixing, the viscosity associated with turbulence establishes a core that rotates approximately as a solid body, as in the case of the Rankine vortex. The maximum tangential velocity occurs at a radius $r_c = 1.12\sqrt{2K_m/c}$.

Another solution, referred to as the *two-cell* or Sullivan[225] vortex, is illustrated in Fig. 8.34b. It is similar to the Burgers–Rott single vortex, except that the inner

---

[225] Discovered by Sullivan (1959).

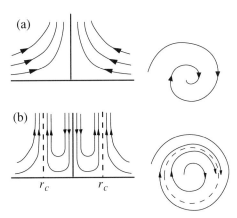

Figure 8.34 Projections of vortex streamlines onto a vertical plane through the axis of the vortex and onto a horizontal plane. (a) One-cell Burgers–Rott vortex. (b) Two-cell Sullivan vortex. (Adapted from Sullivan, 1959.)

core has a secondary circulation and the boundary between the inner and outer cells is at radius $r_c = 2.38\sqrt{2K_m/c}$. The fluid spirals inward and upward in the outer cell, downward near the axis, and outward and upward near the outer edge of the inner cell. Vorticity is concentrated (by stretching) in the annular zone of convergence straddling the boundary between the two cells. This ring of concentrated vorticity is where multiple (suction) vortices (discussed below) tend to form (Fig. 8.32).

Most of the above discussion of tornado vortex dynamics applies to regions Ia and b in Fig. 8.33. The boundary layer and corner regions (II and III) entail theoretical difficulties beyond the scope of this text. We can, however, examine qualitatively the conceptual model of tornado vortex structure in Fig. 8.35, which takes these additional regions into account. The model has been developed on the basis of the one- and two-cell theories described above, laboratory simulations, photogrammetry of real tornadoes, and numerical simulations using models like those discussed in Sec. 7.5.3.

The flow in the conceptual model is depicted as a function of a parameter called the *swirl ratio*, which is an overall measure of the ratio of tangential to vertical flow in the vortex. As the swirl ratio increases, the flow changes as depicted in

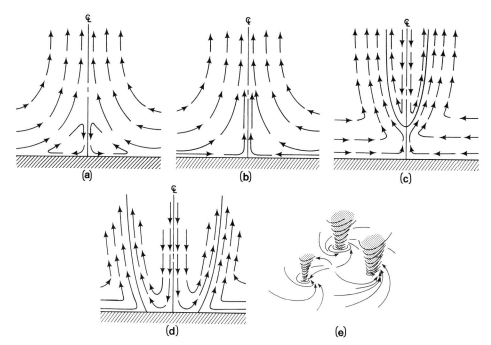

Figure 8.35 Conceptual model of tornado vortex structure. Panels (a)–(e) show structure for successively higher swirl ratio. (a) Weak swirl case: flow in boundary layer separates and passes around corner region. (b) One-cell vortex. (c) Vortex breakdown. (d) Two-cell vortex with downdraft impinging on ground. (e) Multiple vortices. The connected CL indicates the center line. (From Davies-Jones, 1986 and Lewellen, 1976.)

Fig. 8.35. For weakly rotating updrafts, it is found that the horizontal pressure gradient along the ground is adverse, and the outer flow separates from the surface and bypasses the corner region, preventing the formation of a strong vortex near the ground (Fig. 8.35a). Then, with increasing swirl ratio, a one-cell vortex becomes established (Fig. 8.35b), followed by a two-cell vortex whose downdraft does not penetrate to the ground (Fig. 8.35c), then a two-cell vortex penetrating all the way to the surface (Fig. 8.35d), and finally formation of multiple vortices (Fig. 8.35e).

The evolution of the two-cell vortex between the stages represented by Fig. 8.35c and d has been found in numerical-model simulations of laboratory experiments to be associated with the nearly cyclostrophic pressure perturbation minimum on the central axis at low levels.[226] As the swirl ratio increases this minimum becomes stronger, and a downward-directed vertical pressure perturbation gradient acceleration develops along the center axis at low levels. The vortex therefore "fills in" from above.

The intermediate configuration represented in Fig. 8.35c occurs at moderate swirl. This mode, which has the one-cell structure at lower levels and a two-cell structure higher up, has the characteristics of a general phenomenon called *vortex breakdown*. This phenomenon is seen in a variety of flow situations. It is characterized by an abrupt transition from a highly swirling narrow laminar jet to a broad turbulent flow with greatly reduced swirl and reversed (i.e., downward) vertical flow. An example of vortex breakdown in the laboratory setting is shown in Fig. 8.36. The breakdown point in Fig. 8.35c is the height at where the vortex undergoes sharp transition to larger diameter, and downward motion along the axis meets the upward motion of the intense lower-level vortex. The core is turbulent above this stagnation point.

The vortex breakdown is very complex, and several theoretical ideas are involved in its explanation.[227] One is that there is a partial analogy between a stratified flow under gravity and a rotating flow in which the role played by gravitational acceleration in the stratified flow is analogous to that of centrifugal acceleration in the rotating flow. When the horizontal flow speed in the x–z plane of a two-dimensional stratified fluid exceeds the speed of the fastest horizontally moving gravity waves, the flow is said to be "supercritical" and disturbances are unable to propagate upstream. Upstream and downstream boundary conditions are matched in steady flow by sudden reduction at some location to a subcritical flow speed and a corresponding increase in depth. This phenomenon, known as a "hydraulic jump," will be discussed in Chapter 12 (Sec. 12.2.5). Vortex breakdown is a similar transition, which can occur in rotating flows when the speed of the fluid motion along the vertical axis of the vortex exceeds the speed of the fastest of the downward-propagating "inertia waves," which form in response to the centrifugal restoring force (which plays a role analogous to the Coriolis restor-

---

[226] This physical interpretation is based on Rotunno (1977, 1986).

[227] Rotunno (1979) gives a summary of these ideas and demonstrates aspects of them with a vivid set of numerical simulations.

Figure 8.36   Example of vortex breakdown in a laboratory experiment at Purdue University. (From Rotunno, 1979. Photo courtesy of John T. Snow. Reproduced with permission from the American Meteorological Society.)

ing force in the inertial oscillations considered in Sec. 2.7.3). Vortex breakdown is characterized by an abrupt enlargement of the vortex core, like the increase of fluid depth at the hydraulic jump. However, the stagnation point and reversed velocity along the axis in the case of the vortex breakdown are different from the hydraulic jump.

It was noted above that the vortex is turbulent above the breakdown point. To see how this turbulence is achieved, an analogy to a vertically stratified flow under gravity is again relevant. According to (2.170), the stratified flow is stable and laminar (nonturbulent) as long as the Richardson number Ri $\geq$ 1/4. That criterion was obtained by considering two parcels of fluid, each of unit mass, initially separated by a distance δz in the x–z plane of a two-dimensional fluid with no variation in the y-direction and base-state shear of the horizontal wind $\partial \bar{u}/\partial z$. We now consider two parcels of unit mass initially separated by a distance δr in the r–z plane of a two-dimensional vortex (no variation in the Θ direction) and base-state shear of the z-component of the wind $\partial \bar{w}/\partial r$. In the absence of turbulence, molecular friction, and Coriolis forces, the r-component of the mean-variable equation of motion (2.83) is

$$\frac{\bar{D}u}{Dt} = -\frac{1}{\rho_o}\frac{\partial \bar{p}}{\partial r} + \frac{\bar{v}^2}{r} \tag{8.36}$$

The centrifugal term $\bar{v}^2/r$ plays the same role in this equation as does gravity g in the vertical component of (2.1). For the work done per unit volume of air in exchanging the positions of the parcels to be less than the kinetic energy released in the exchange, we must have

$$\rho_o \delta\left(\bar{v}^2/r\right)\delta r < \tfrac{1}{4}\rho_o \left(\delta W\right)^2 \tag{8.37}$$

which is the analog of (2.169) in which δU has been replaced by the difference in vertical velocity of the two parcels δW. The angular momentum is $\bar{m} = \bar{v}r$, and the vertical velocity difference may be written as $\delta W = (\partial\bar{w}/\partial r)\delta r$. With these substitutions, and as long as δr is small compared to r, the instability condition (8.37) may be rewritten as

$$\frac{r^{-3}\left(\partial\bar{m}^2/\partial r\right)}{\left(\partial\bar{w}/\partial r\right)^2} < \frac{1}{4} \tag{8.38}$$

which is the analog of (2.170). It follows that for turbulent motions to occur, the radial shear of the vertical velocity must be strong enough to overcome the inertial stability implied by a positive value of $\partial\bar{m}^2/\partial r$. A tornado core approximated by a Rankine vortex is inertially stable [i.e., (8.16) implies that $\partial\bar{m}^2/\partial r > 0$]. Thus, the observed turbulence in vortex cores is probably caused by the destabilizing influence of large shears of vertical velocity, which lower the value of the term on the left of the inequality (8.38).

After the swirl ratio increases enough that the two-cell circulation reaches the ground (Fig. 8.35d), further increase in the swirl ratio leads to a "roll up" of the vortex into multiple vortices orbiting around the axis of the original vortex (Fig. 8.35e). The sequence in Fig. 8.35 was constructed on the basis of laboratory experiments, photographic and visual observations of tornadoes, and the theory discussed above. The laboratory experiments are particularly remarkable in their similarity to real tornadoes,[228] including multiple-vortex cases, as illustrated in Fig. 8.37. The stability criterion (8.38) refers to displacements of rings of fluid in the radial direction only. The empirical results indicate that, even when the flow is stable to purely radial displacements, the vortex is unstable at very high swirl ratio to nonaxisymmetric perturbations.

---

[228] See Davies-Jones and Kessler (1974) and Davies-Jones (1986) for a discussion of the laboratory experiments.

Figure 8.37   (a) Laboratory vortex pair compared with double tornado of 11 April 1965 at Elkhart, Indiana (b). (Laboratory example from Davies-Jones, 1986; tornado photo by Paul Huffman.)

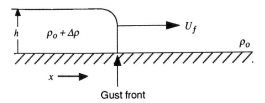

Figure 8.38   Gravity current of density $\rho_o + \Delta\rho$ and depth h moving through a fluid of density ρ_o at speed U_f in the x-direction.

8.9   Gust Fronts

The outflow of downdraft air from thunderstorms along the ground is a very important phenomenon. It can help to generate new thunderstorm cells, and it can cut old cells off from their supply of buoyant air. The boundary of the advancing outflow marks a major change in surface meteorological conditions and is thus of interest in forecasting and nowcasting. Damaging winds[229] often occur at this boundary, which is usually referred to as the *gust front*. In this section, we explore its dynamics.

The gust front is an example of the geophysical phenomenon called a *gravity current*, which may be defined as a mass of high-density fluid flowing along a horizontal bottom and displacing ambient fluid of lesser density.[230] Other examples include salt water intruding into a fresh water estuary, or muddy water displacing fresh water at the bottom of a lake. The gravity current is driven by the horizontal pressure gradient acting across the sharp lateral interface separating the two fluids. The physical situation is shown in idealized form in Fig. 8.38 for a gravity current moving at speed U_f in the x-direction. The role of the pressure gradient in determining the speed of the current can be seen from the x-component of the Boussinesq momentum equation (2.47) in the absence of friction or Coriolis forces:

$$\frac{Du}{Dt} = -\frac{1}{\rho_o}\frac{\partial p^*}{\partial x} \tag{8.39}$$

If this expression is applied to a steady-state horizontal flow in a coordinate system moving with the gust front, it may be written as

$$\frac{\partial\left(u^2/2\right)}{\partial x} = -\frac{1}{\rho_o}\frac{\partial p^*}{\partial x} \tag{8.40}$$

If the gravity current has depth h and density $\rho_o + \Delta\rho$, where $\Delta\rho > 0$, and is moving through a stagnant base-state environment, integration of (8.40) across the front edge of the gravity current gives us

$$\frac{U_f^2}{2} = g\frac{\Delta\rho}{\rho_o}h \tag{8.41}$$

Thus, the speed of the gust front is determined by the depth of the gravity current and the density difference across the interface.

A schematic cross section through the gust front of a thunderstorm is shown in Fig. 8.39. This sketch is a composite of information derived from a variety of sources, including instrumented towers, weather radar, cloud observations, and

---

[229] Sometimes called "straight-line winds."

[230] For an overview of gravity currents in earth and planetary sciences, see the book *Gravity Currents* by Simpson (1987).

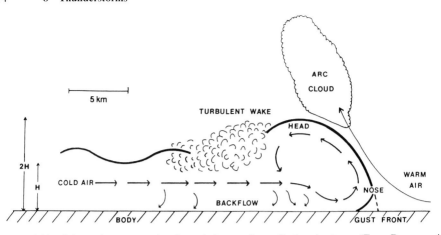

Figure 8.39 Schematic cross section through the gust front of a thunderstorm. (From Droegemeier and Wilhelmson, 1987; based on earlier studies of Charba, 1974; Goff, 1975; Wakimoto, 1982; Koch, 1984. Reprinted with permission from the American Meteorological Society.)

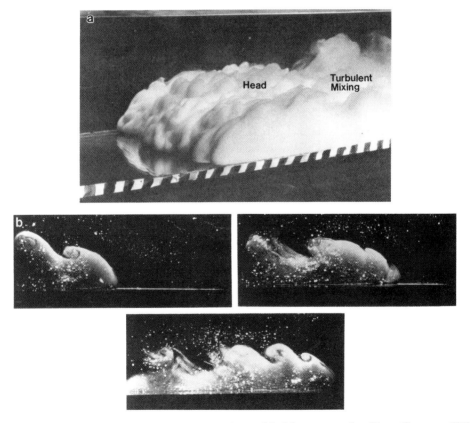

Figure 8.40 (a) and (b) Density currents observed in laboratory tanks. (From Simpson, 1969. Reprinted with permission from the Royal Meteorological Society.)

time series of surface meteorological data. The structure shown is remarkably similar to that of density currents observed in laboratory tanks (Fig. 8.40). Characteristic features include the bulbous *head* marking the leading portion of the current. Within the head is an overturning internal circulation. Following the head is a turbulent wake, which is, in turn, followed by several waves distorting the shape of the upper boundary of the dense outflow. An arc cloud sometimes forms in response to the lifting over the head.

Further details of the flow in thunderstorm outflows have been deduced from cloud models of the type discussed in Sec. 7.5.3. Results from a model simulation are shown in Figs. 8.41–8.44 .[231] A shallow layer of cold air moving along the ground enters a two-dimensional $(x–z)$ model domain from the side. The inflow enters the domain with an assigned temperature profile, depth, and vertically averaged perturbation potential temperature. As the cold layer moves across the domain, the pressure perturbation at the surface is characterized by a high at its leading edge (Fig. 8.41a). This high reflects the peak of dynamic pressure produced at the leading edge. The maximum convergence coincides with the pressure maximum (Fig. 8.41b). From the momentum equation (8.39), we can see how this pressure perturbation is derived from the kinetic energy of the low-level relative flow toward the gust front, when it is stagnated by the wall of denser fluid. If we follow a parcel of fluid moving in the x-direction at horizontal velocity $u = Dx/Dt$, then we can make the transformation $1/Dt = u/Dx$ and (8.39) can be written as

$$\frac{D}{Dx}\left(\frac{u^2}{2} + \frac{p^*}{\rho_o}\right) = 0 \qquad (8.42)$$

where Dx represents the distance between two successive locations of the fluid parcel. If this expression is integrated along the path of a parcel moving along the surface from some location far ahead of the gust front (where its relative motion is toward the gust front at speed U_f and the pressure perturbation is zero) to a point immediately ahead of the surface position of the gust front (where the relative motion of the parcel has been reduced to 0), then the pressure perturbation at the gust front is found to be

$$p^* = \rho_o \tfrac{1}{2} U_f^2 \qquad (8.43)$$

Thus, the pressure perturbation at the surface at a point just ahead of the density current results from the conversion of horizontal kinetic energy to enthalpy (p^*/ρ_o). Since the stagnation occurs *ahead* of the leading edge of the gravity current, there is no hydrostatic contribution to p^* at this point.

The second maximum of pressure to the rear of the head is hydrostatic. It simply reflects the weight of the fluid above. The minimum of pressure separating the two highs is, however, nonhydrostatic. It is associated with rotation. As can be seen from Fig. 8.42, the low-pressure center in the middle of the head coincides with a center of rotation, which can be inferred from the horizontal and vertical

---

[231] These results are from a definitive paper by Droegemeier and Wilhelmson (1987).

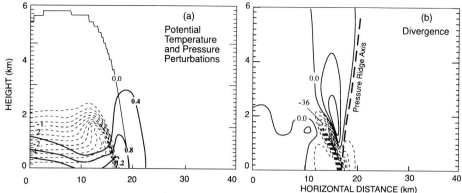

Figure 8.41 Numerical-model simulation of a thunderstorm outflow structure illustrating the nature of the surface pressure perturbation maxima associated with the outflows. (a) Potential temperature perturbation (K, thin contours, negative values dashed) and pressure perturbation (mb, heavy contours) show a maximum of pressure perturbation at the left-hand side of the domain which is hydrostatic. (b) Horizontal divergence (units of 10^{-4} s^{-1}, negative values dashed). (From Droegemeier and Wilhelmson, 1987. Reprinted with permission from the American Meteorological Society.)

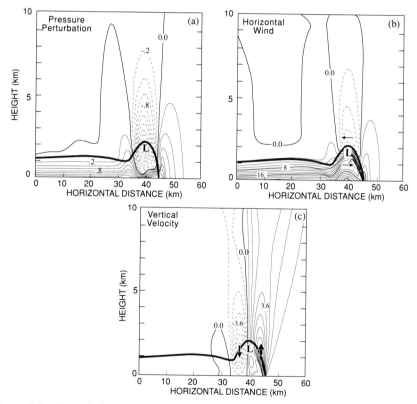

Figure 8.42 Numerical-model simulation of a thunderstorm outflow structure illustrating the pressure perturbation minimum and associated airflow in the head of the gust-front outflow. (a) Pressure perturbation (units of mb, negative values dashed). (b) Isotachs of horizontal velocity (m s^{-1}). (c) Isotachs of vertical air motion (m s^{-1}). The bold solid contour denotes the outflow boundary ($-0.1°C$ potential temperature perturbation). (From Droegemeier and Wilhelmson, 1987. Reprinted with permission from the American Meteorological Society.)

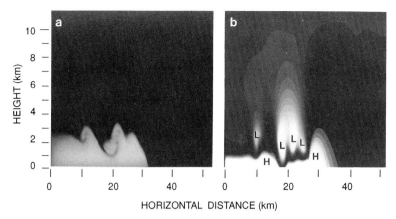

Figure 8.43 Numerical-model simulation of a thunderstorm outflow structure illustrating multiple wave structure atop the outflow: (a) Negative perturbation potential temperature (cold air) is shown by shading. The whiter the shading, the colder the air. (b) Perturbation pressure, with high (H) and low (L) anomalies indicated. (Results of Droegemeier and Wilhelmson, 1987, presented here in new graphical format by Kelvin Droegemeier.)

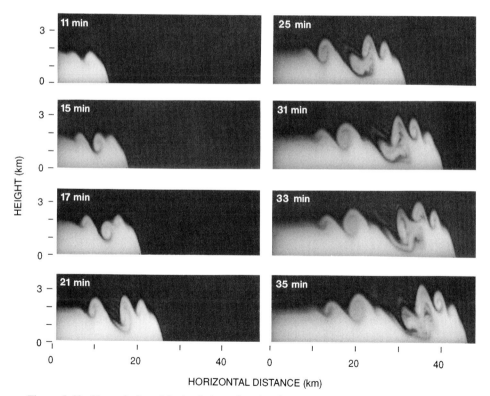

Figure 8.44 Numerical-model simulation of a thunderstorm outflow structure illustrating the sequence of wave generation on the top of the outflow layer. Negative perturbation potential temperature (cold air) is shown by shading. The whiter the shading, the colder the air. The lowest values, occurring near the surface, are about −5°C. (Results of Droegemeier and Wilhelmson, 1987, presented here in new graphical format by Kelvin Droegemeier.)

components of motion in Fig. 8.42b and c. Moreover, the pressure gradient surrounding the low in the x–z plane is in approximate cyclostrophic balance, such that

$$-\frac{1}{\rho}\frac{\partial p}{\partial n} = \frac{V_s^2}{R_s} \tag{8.44}$$

where n is a coordinate normal to the streamlines and directed radially outward, R_s is the radius of curvature of a streamline, and V_s is the wind speed at a distance R_s from the center of the circulation. Equation (8.44) is a special case of (2.46) in which the axis of the vortex is horizontal. The data in Fig. 8.42 verify the balance. If the radius is taken to be 1 km, a representative value of V_s is 6 m s^{-1}. Thus, the term on the right is 0.36 m s^{-2}. The pressure field indicates the term on the left has a value of about 0.40 m s^{-2}. The minimum of pressure is thus dynamically induced in a manner similar to the pressure minima that characterize the vortices on the flanks of the supercell thunderstorm circulation [Sec. 8.5, Eq. (8.6), Fig. 8.20] or in the center of the tornado vortex [Sec. 8.8, Eq. (8.24)]. It can be further verified from the numerical simulation that the strong horizontal vorticity of the overturning in the head is generated by the horizontal gradient of buoyancy across the gust front, as discussed in Sec. 7.4.2 and Sec. 8.6.

From the model-calculated perturbation fields of potential temperature and pressure presented in Fig. 8.43, it can be seen that each of the waves in the upper boundary of the outflow is characterized by an overturning cyclostrophic pressure perturbation similar to that in the head. The Richardson number Ri indicated by the model fields is about 0.2, which, according to (2.170), is in the range of Kelvin–Helmholtz instability. The waves evidently are Kelvin–Helmholtz "billows" of the type illustrated in Figs. 2.4–2.8. Each originates in the head and propagates rearward, gradually damping and dissipating, while a new circulation forms in the head as a result of baroclinic generation. The sequence of wave generation is illustrated by the model-simulated cold pool evolution in Fig. 8.44.

The above numerical simulations illustrate the dynamics of the mature thunderstorm outflow. However, they do not treat the life cycle of the gust front. An empirical model of the life cycle of a thunderstorm outflow has, however, been synthesized from Doppler radar and other sources of information (Fig. 8.45).[232] The overturning head is again seen as a prominent feature. Stages II and III correspond to the mature phase illustrated by the numerical-model simulations. Later, in Stage IV, the overturning head separates and can propagate far away from the parent storm. As the gust front moves away it can trigger new cumulus or cumulonimbus. Satellite studies[233] show that these lines, which have separated from the parent storm, can maintain their identities as arc-shaped lines of cumulus clouds for several hours after the storms that produced the downdrafts have dissipated and can trigger new deep convective development up to 200 km from the location of the original storm. New deep convective developments are particu-

[232] This model was developed by Wakimoto (1982).
[233] Purdom (1973, 1979), Purdom and Marcus (1982), and Sinclair and Purdom (1982).

larly favored where propagating arc lines intersect each other or where an arc line encounters pre-existing convection.

8.10 Downbursts

8.10.1 Definitions and Descriptive Models

We have seen that the downdraft is a salient feature of the thunderstorm circulation. Sometimes the downdraft becomes locally very intense over a short period of time, in which case it is referred to as a *downburst*. The following definitions are used to describe the phenomenon:[234]

Downburst—An area of strong winds produced by a downdraft over an area from <1 to 10 km in horizontal dimension.

Macroburst—A downburst that occurs over an area >4 km in dimension and is typically 5–30 min in duration.

Microburst—A downburst that covers an area <4 km in dimension and lasts 2–5 min. Differential velocity across the divergence center is >10 m s^{-1}.

Wet microburst—A microburst associated with >0.25 mm of rain or a radar echo >35 dBZ in intensity.

Dry microburst—A microburst associated with <0.25 mm of rain or a radar echo <35 dBZ in intensity.

The most scientific and public attention has been focused on the microburst, largely because of the danger it poses for aircraft operations. The existence and importance of this phenomenon were noted by Professor T. Fujita[235] in his investigations of aircraft accidents and patterns of damage to crops and property by violent winds that could not be explained by the vortical circulation patterns of tornadoes. In contrast to tornadoes, downbursts produce patterns that reflect strongly divergent and diffluent air motions at the ground. From surveying such damage patterns, Fujita was led to hypothesize the air motion pattern in a microburst shown in Fig. 8.46. As can be seen from Fig. 8.47 and Fig. 8.48, this conceptual model has proven accurate under scrutiny by Doppler radar observations and numerical modeling. Besides a shaft of strong downward velocity at its center, the microburst is characterized when it hits the ground by strong divergence at its center and an accelerating outburst of strong winds in a vigorously overturning gust-front head propagating more or less symmetrically away from the center of the microburst. When the microburst contains rain, dust, or other material from the surface, it can be seen visually (Fig. 8.49). Fujita conceptualized

---

[234] These definitions are based on the somewhat varying definitions used by Fujita (1985), Fujita and Wakimoto (1983), and Wilson *et al.* (1984).

[235] Professor T. Fujita, through his dogged and enthusiastic observations of storms and storm damage, is largely responsible for raising scientific and public consciousness about the downburst. He coined and promoted the terminology "downburst" and "microburst." Many of his investigations and adventures in the pursuit of this subject are described in his two books *The Downburst: Microburst and Macroburst* (1985) and *DFW Microburst on August 2, 1985* (1986).

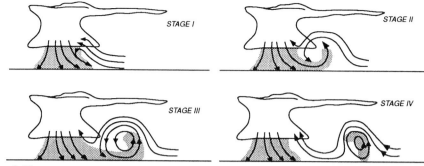

Figure 8.45

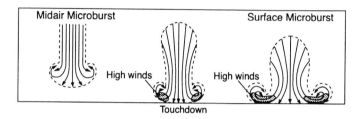

Figure 8.46

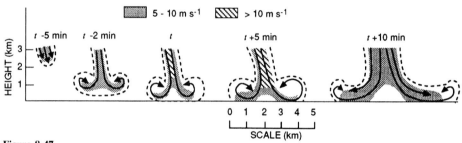

Figure 8.47

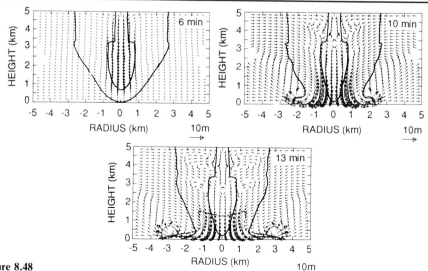

Figure 8.48

the downburst in three dimensions, as illustrated in Fig. 8.50. These sketches illustrate the highly diffluent character of the surface wind and the expanding ring vortex formed by the circular gust-front head. The purpose of Fig. 8.50b is to show that some microbursts are associated with a small-scale cyclonic circulation aloft. This cyclonic vorticity is, however, largely weakened by divergence [i.e., the stretching term ζw_z in (2.59) is negative since $w_z < 0$] before parcels reach the ground.

8.10.2 Effects of Microbursts on Aircraft

Fujita's conceptual model has been used as an explanation for many aircraft accidents. A pilot must make quick critical adjustments in flying through the wind pattern of the microburst. For example, in taking off through a microburst (Fig. 8.51), the aircraft experiences an increase of headwind as it accelerates down the runway. Then, the aircraft lifts off in the increasing headwind and begins to climb (position 1). Near position 2, it encounters the microburst downdraft, and climb performance is decreased. By position 3, the headwind is lost. Consequently, airspeed is decreased, and lift and climb performance are further reduced. Added to this is the increased downdraft at the microburst center. By position 4, all available energy is needed to maintain flight, as the tailwind continues to increase. However, there is no source on which the aircraft can draw to increase its potential energy (climb). A large aircraft is typically configured ("trimmed") such that thrust, drag, lift, and weight are all in equilibrium. Thus, no pilot input is required for the aircraft to maintain a set trajectory. Since the airspeed is below the trim airspeed (decreasing lift and drag) at position 4, the airplane system will automatically respond so as to regain the equilibrium condition by pitching the nose downward. In the illustration, the pilot intervenes and compensates for this effect. Should the pilot not fully compensate, a more radical descent could occur. The descent rate continues to increase as the airplane passes through position 5. Depending on the strength of the event, encounter altitude, aircraft performance margin, and how quickly the pilot recognizes and reacts to the hazard, the high

---

Figure 8.45 Empirical model of the life cycle of a thunderstorm outflow. (From Wakimoto, 1982. Reprinted with permission from the American Meteorological Society.)

Figure 8.46 Conceptual model of a microburst hypothesized to explain ground-damage patterns. Three stages of development are shown. A midair microburst may or may not descend to the surface. If it does, the outburst winds develop immediately after its touchdown. (From Fujita, 1985.)

Figure 8.47 Empirical model of a microburst based on Doppler radar observations. Time t refers to the arrival of divergent outflow at the surface. Shading denotes wind speed. (From Wilson et al., 1984. Reprinted with permission from the American Meteorological Society.)

Figure 8.48 Numerical-model simulation of a microburst at 6, 10, and 13 min after the initial model time. Thick lines represent the 10 and 60 dBZ radar reflectivity contours (i.e., the precipitation field). Dashed line encloses the area with temperature departures from ambient of less than -1 K. (From Proctor, 1988. Reprinted with permission from the American Meteorological Society.)

Figure 8.49 Photo of a microburst at Stapleton Airport, Denver, Colorado. The structure of the microburst is made visible by precipitation, dust, and other material from the surface. (Photo courtesy of W. Schreiber-Abshire.)

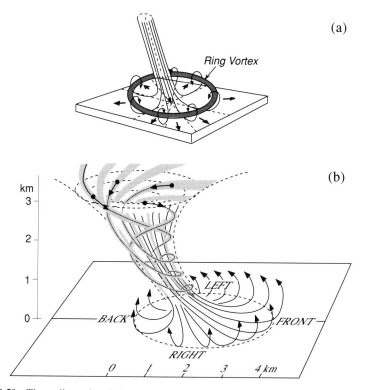

Figure 8.50 Three-dimensional visualization of a downburst. (a) Ring vortex at edge of gust front. (b) Rotation in the microburst. (From Fujita, 1985.)

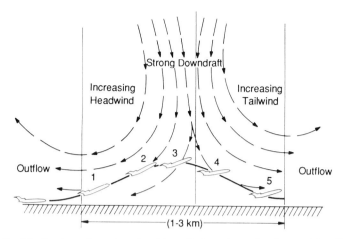

Figure 8.51 Idealization of an aircraft taking off through a microburst. After lifting off in an increasing headwind at 1, the aircraft begins to lose the headwind and enter the downdraft at 2, experiences stronger headwind loss at 3 with subsequent arrest of climb, begins descending in increasing tailwind at 4, and experiences an accelerating descent rate through 5. (From Elmore *et al.*, 1986. Reprinted with permission from the American Meteorological Society.)

descent rate may be impossible to arrest before the plane crashes into the ground.[236]

8.10.3 Dynamics of Microbursts

Investigations of the dynamics of microbursts have focused on two main issues. First, the mechanisms driving the downdraft have been sought. These mechanisms turn out to be primarily microphysical and thermodynamical. We will examine these mechanisms in Sec. 8.10.3.1. The second issue concerns the dynamics of the rotor circulation constituting the outward-propagating ring vortex, where the strong surface winds are located. We will examine the rotor dynamics in Sec. 8.10.3.2.

8.10.3.1 Mechanisms Driving Microbursts

The mechanisms of downdraft acceleration are approached through examination of the vertical component of the equation of motion (2.47), which in the absence of the frictional force may be written as

$$\frac{Dw}{Dt} = -\frac{1}{\rho_o}\frac{\partial p^*}{\partial z} + B\left(\theta_v^*, p^*, q_H\right) \tag{8.45}$$

where the dependence of the buoyancy B on the thermodynamic variables (θ_v^*, p^*, q_H) is given by (2.52). In a case study of downbursts observed by Doppler

---

[236] This illustration is taken from Elmore *et al.* (1986).

radar,[237] the terms on the right-hand side of (8.45) were estimated by performing thermodynamic retrieval analysis of the type discussed in Sec. 4.4.7.

The pressure gradient acceleration in (8.45) was found to be small compared to the buoyancy. This result could have been anticipated since the microburst is characterized by a pressure maximum at the surface, which produces an upward-directed pressure gradient force acting against the downdraft. The microburst problem thus reduces to understanding the negative contributions to B. Doppler radar data showed that the vertical acceleration [left-hand side of (8.45)] in the case study microburst was ~0.1 m s^{-2}. The precipitation mixing ratio implied by the observed radar reflectivity [Eq. (4.12)] suggested that only about 20% of the total acceleration was accounted for by precipitation drag [i.e., by the contribution of $-q_H$ to B in (2.52)]. If the contribution of p^* to B in (2.52) is small, then 80% of the negative buoyancy in the downburst must have been associated with θ_v^*. A value of $\theta_v^* = -2.5$ K is thus implied. This value is consistent with values of θ_v^* obtained by application of thermodynamic retrieval to the Doppler radar–derived flow field. If all the hydrometeors had been ice, melting could have accounted for only $\theta_v^* = -0.4$ K. Thus, most of the cooling responsible for the negative buoyancy of the downburst appears to have been brought on by evaporation of hydrometeors.

Radar evidence from other cases also suggests that microphysically induced negative buoyancy is crucial to microburst generation. However, it suggests that in some cases melting may be more important. In Fig. 8.52, dual-polarization radar measurements of the type discussed in Sec. 4.3 show a narrow region of near-zero differential reflectivity Z_{DR} [Eq. (4.25)] within the region of maximum reflectivity of the thunderstorm. This Z_{DR} hole indicates a narrow shaft of hail within the heavy shower of precipitation since, as illustrated by Fig. 4.2, the near-zero Z_{DR} is a signature of ice, while positive Z_{DR} is a signature of rain. This narrow hailshaft was coincident with the formation of a microburst, whose position was indicated by the diffluence of oppositely directed wind maxima at the surface (arrow in Fig. 8.52). Thus, we have circumstantial evidence that melting of hail was important to the dynamics of this microburst.

The observational results indicating that evaporation and melting are the key factors in producing downburst acceleration are reinforced by calculations with a one-dimensional, time-dependent nonhydrostatic model of the type described in Sec. 7.5.2.[238] The model domain is a cylinder of radius R with its top at 550 mb and its bottom at 850 mb. The bottom is open. At the top are assumed to be specified size distributions of rain and hail particles. An explicit microphysical scheme (Sec. 3.5) is used to predict the evolution of the particle distributions. The entrain-

---

[237] This outbreak of thunderstorms over Colorado containing several downbursts was documented in detail by Kessinger *et al.* (1988).

[238] This model was designed and used in two studies by Srivastava (1985, 1987). In the 1985 study, he performed calculations for downbursts containing only rain. The 1987 study extended the results to allow for the effects of hail as well as rain in a downburst. As well as being landmarks of insight into downdraft dynamics, these papers illustrate dramatically the usefulness of a simple one-dimensional model.

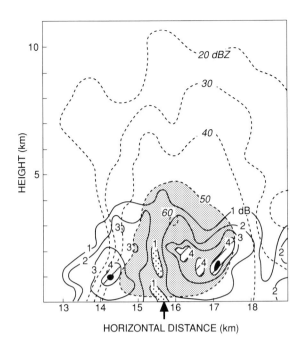

HORIZONTAL DISTANCE (km)

Figure 8.52 Vertical cross section showing dual-polarization Doppler radar measurements obtained in a thunderstorm in northern Alabama. Reflectivity data are presented in contours of dBZ. Differential reflectivity Z_{DR} are shown in dB units. Arrow indicates location of the center of a surface microburst. (From Wakimoto and Bringi, 1988. Reprinted with permission from the American Meteorological Society.)

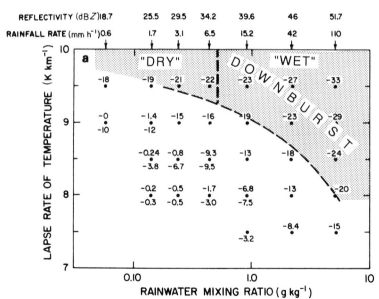

Figure 8.53 Results of a one-dimensional time-dependent nonhydrostatic cloud model of a downdraft. Plotted numbers are vertical air velocity (m s^{-1}) at a level 3.7 km below the top of the downdraft as a function of the lapse rate in the environment and total liquid water mixing ratio at the top of the downdraft. Numbers on top scale indicate the radar reflectivity and rain rate at the top of the downdraft. Curved dashed line separates downbursts (<20 m s^{-1}) from less intense downdrafts. Vertical dashed line separates dry (<35 dBZ) from wet (>35 dBZ) downbursts. Other conditions at the top of the downdraft are 550 mb, 0°C, and 100% relative humidity. The relative humidity of the environment is 70%, and there is no entrainment of environmental air. (Adapted from Srivastava, 1985. Reproduced with permission from the American Meteorological Society.)

ment rate is assumed to be inversely proportional to R [consistent with the continuous entrainment models considered in Sec 7.3.2, e.g., (7.38), (7.46), (7.48)]. The environmental lapse rate and humidity are specified. In calculations including only raindrops in the water-continuity equations, the microphysical and environmental conditions that produced downdrafts of >20 m s^{-1} in intensity were sought. Such a strong downdraft was taken to be an indication of a downburst. Results for the case of zero entrainment (achieved effectively at a radius of about 1 km) and an initial Marshall–Palmer distribution [Eq. (3.70)] of raindrops falling into the top of the domain are shown in Fig. 8.53. They indicate that downbursts are most readily obtained as the lapse rate approaches the dry adiabatic (9.8°C km^{-1}) and as the rainfall rate (or radar reflectivity) increases. Other results, not shown in the figure, are that the tendency toward downburst occurrence increases with decreasing mean raindrop size (since small drops evaporate more readily than large ones), with a *less* well-mixed boundary layer (in which relative humidity rather than mixing ratio is assumed constant in the environment) and with *increased* relative humidity in the environment. The last two results are somewhat counterintuitive, since microbursts often occur in dry and/or well-mixed boundary layers. However, the physics are clear. As the relative humidity (RH) of the environment is increased for a given temperature profile, especially near the top of the downdraft layer, buoyancy in the downdraft, measured relative to the environment, becomes more strongly negative [i.e., $q_v^* \equiv q_v - q_{ve}$ becomes more largely negative as RH increases since $q_v = q_{vs}(T)$, $q_{ve} = \text{RH} \cdot q_{vs}(T_e)$, and $T < T_e$]. The optimal conditions for downburst formation in the absence of ice are then indicated to be an environment close to dry adiabatic, with a high rainwater content near cloud base and a minimum downdraft radius of 1 km. Note that there is no indication from these calculations as to why the downburst should be small in scale.

When hail as well as rain is included in the precipitation falling into the top of the model domain, calculations show that the additional negative buoyancy provided by melting can produce downbursts when the lapse rate is more stable than dry adiabatic. At higher environmental stability, higher precipitation content in the form of ice and relatively higher concentrations of small precipitation particles are required to generate an intense downdraft. At lower environmental stability, both dry and wet downbursts are possible. As the stability increases, only progressively wetter downbursts are possible. As the environment becomes even more stable, only wet downbursts having substantial precipitation in the form of ice are possible.

Numerical modeling of downbursts with a two-dimensional model of the type discussed in Sec. 7.5.3 is consistent with the one-dimensional model results just described. In the example examined here,[239] a bulk cold-cloud scheme (Sec. 3.6.2) is used to represent microphysical processes in the water-continuity equations. The model domain is 5 km deep and 10 km wide. Calculations are initiated by prescribing a distribution of hail at the top boundary. Again, melting and evaporation are both found to be important in driving the downdraft, with evaporation

[239] From Proctor (1988).

being the stronger effect. The two-dimensional model provides additional insight into the distribution of these processes in time and space. Melting is more important in the earlier stages, with evaporation becoming dominant later and at lower levels. It is also found that the downdraft is *induced* by precipitation drag with the evaporation and melting becoming dominant subsequent to the initiation.

8.10.3.2 Rotor Circulation and Outburst Winds

The two-dimensional model also provides insight into the rotor circulation and outburst winds at the surface (Fig. 8.48). The outflow that develops is qualitatively very similar to that of thunderstorm downdraft outflows in general (Fig. 8.39). The microburst differs only in intensity, especially with the outburst of surface winds that occurs when the microburst hits the ground (Fig. 8.46 and Fig. 8.49). This behavior is seen in the two-dimensional model results in Fig. 8.54. As indicated in

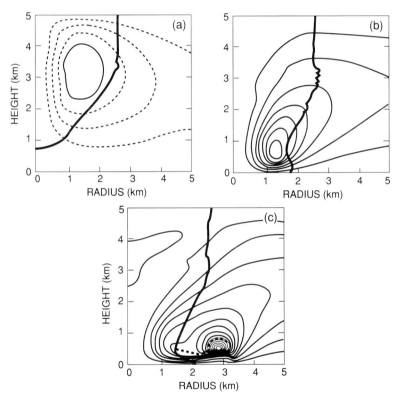

Figure 8.54 Results of a two-dimensional model showing the rotor circulation and outburst winds at the surface in a microburst. Radial–vertical cross sections for stream function for three successive times at intervals of 4 min. Thick solid line represents 10-dBZ radar reflectivity contour. The thick dashed line encloses area outside of precipitation shaft with temperature deviations from ambient of less than -1 K. The contour interval for the stream function is 8×10^5 kg s^{-1}, with intermediate contours dashed. (From Proctor, 1988. Reprinted with permission from the American Meteorological Society.)

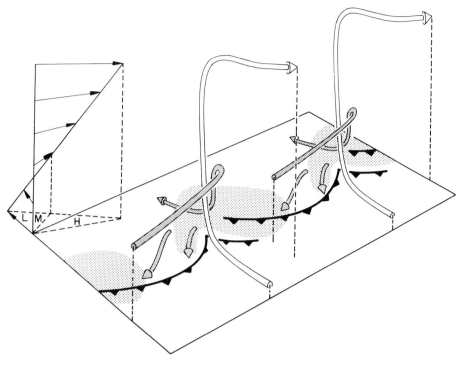

Figure 8.55

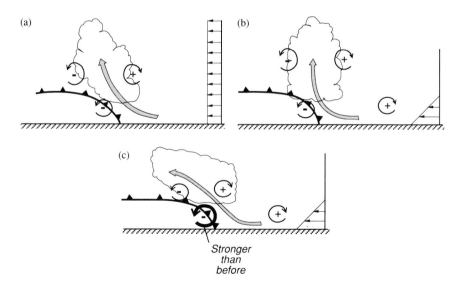

Figure 8.56

the discussion of gust fronts (Secs. 7.4.2 and 8.9), the vortex in the outflow head is the result of the horizontal gradient of buoyancy across the edge of the outflow, as expressed by the baroclinic term B_x in the horizontal vorticity equation (2.61). The baroclinic generation is largest at the surface because the negative tempera-ture anomaly in the downdraft is large only near the ground. The temperature anomaly above 1 km in height is less because the very rapid downward air motion warms the air adiabatically. Near the ground, where the downward motion is greatly slowed down, the evaporative cooling proceeds without compensating effects, thus producing a strong temperature contrast and strong baroclinic gener-ation of circulation in the head. One result is the horizontal outburst of surface wind.

8.11   Lines of Thunderstorms

Up to now we have considered thunderstorms as individual entities. Often, how-ever, thunderstorms occur in groups, and often in these groups the storms are distributed along horizontal lines—either straight or curved. Lines of thunder-storms can occur alone or as a part of a larger mesoscale convective system. In Chapter 9, we will consider lines of thunderstorms as components of larger storm systems. Here we look briefly into the environmental conditions required to main-tain a line of thunderstorms, regardless of whether or not it is part of a larger system.

It is observed that a line of thunderstorms typically has a lifetime longer than that of the individual storms making up the line. Modeling studies[240] using two- and three-dimensional models of the type discussed in Sec. 7.5.3 have addressed the question of what environmental conditions promote such a long-lived line. Results point to the importance of the role of wind shear in the environment ahead of the line in sustaining the line's organization.

[240] Rotunno *et al.* (1988), Weisman *et al.* (1988), and Fovell and Ogura (1988, 1989).

Figure 8.55   Conceptual model of a line of supercell thunderstorms in an environment of deep, strong shear. Streamlines show flow relative to the individual storms composing the line. The environmental shear vector forms a 45° angle to the orientation of the line. The storms are arranged such that their circulations do not interfere. The shear profile on the left indicates the relative winds at low (L), middle (M), and high (H) levels. The stippled regions indicate the hook echo formed by the rain areas seen on radar. (As adapted from Lilly, 1979 by Rotunno *et al.*, 1988. Reprinted with permission from the American Meteorological Society.)

Figure 8.56   Heuristic diagram indicating horizontal vorticity (+ and −) in the vicinity of a long-lived line of multicell storms. Profile of horizontal wind component normal to the line is shown on the right of each panel. Frontal symbol marks outflow boundary. (a) Case in which there is no wind shear normal to the line in the environment. (b) Early stage of line development in case of low-level shear ahead of the gust front. (c) Late stage of line development in case of low-level shear ahead of the gust front. (Panels a and b from Rotunno *et al.*, 1988. Reprinted with permission from the American Meteorological Society.)

Two types of long-lived line can occur—a line of supercell storms or a line of ordinary multicell storms. The term ordinary here refers to the fact that multicell storms are far more common than supercells and, not surprisingly, lines of multi-cell storms are the typical case, while lines of supercells are relatively rare. Nonetheless, a line of supercell storms remains of great interest, and it is easier to explain. The model calculations show that a line of supercells can exist in an environment of deep, strong shear (required to produce the highly rotational supercells), where the shear vector forms a 45° angle to the orientation of the line (Fig. 8.55). In this configuration, the storms are arranged such that their respective circulations do not interfere.[241]

The more challenging problem is to account for the long-lived line of multicell storms. Here the low-level wind shear ahead of the line appears to be the main factor. A heuristic argument[242] is based on the two-dimensional horizontal vorticity equation (2.61). Two situations are depicted in Fig. 8.56a and b. In the no-shear case (Fig. 8.56a), a buoyant parcel rising over the gust front must follow a rearward sloping path. If the buoyancy is maximum in the center of the parcel, counter-rotating vortices on either side of the center of the parcel would develop and the core of the parcel would rise vertically. However, the buoyancy gradient across the gust front constitutes an additional contribution to the buoyancy gradient at low levels. Hence the vertical rise associated with the buoyancy of the parcel is superimposed on negative vorticity generation associated with the gust front. Thus, the flow in the vicinity of the gust front is sloped, reflecting the overall dominance of negative vorticity. In the case of low-level shear ahead of the gust front (Fig. 8.56b), the initial vorticity of the inflow air (i.e., the ambient vorticity associated with the shear) is positive. It is then possible that the negative baroclinic generation by the gust front just neutralizes the initial positive vorticity, so that as the parcel rises above the boundary layer it has no predisposition toward either negative or positive vorticity, and it thus rises vertically.

There are two controversial points regarding this heuristic argument. First, it has been pointed out[243] that in the case where the line is part of a mesoscale convective system, the circulation of the mesoscale system provides additional sources of vorticity for the region of the gust front. This point will be discussed further in Chapter 9 (Sec. 9.2.2.8).

The second controversial point is concerned with the later stages of the lifetime of the convective line. The structure shown in Fig. 8.56b (i.e., vertically erect convection) is essential for a long-lived line to become established. Strong low-level shear is unquestionably important to resisting the gust front, and in the early stage of the line, parcels rise vertically as in Fig. 8.56b. Later, as the cold-pool gust front strengthens, the line structure evolves toward a structure like that in Fig. 8.56c, with parcels moving along sloping rearward trajectories, which are the result of the cold pool becoming stronger, the circulation across the gust front

[241] This form of line organization was suggested by Lilly (1979).
[242] Offered by Rotunno et al. (1988).
[243] By Lafore and Moncrieff (1989).

correspondingly strengthening, and negative vorticity beginning to outweigh positive vorticity. In some model simulations,[244] the sloping phase appears to be a period of weakening of the convection as the cold pool and gust front continue to strengthen and overwhelm the system. In other calculations,[245] the sloping phase is an equilibrium phase, which lasts indefinitely. This difference in model results may be attributable to numerical-model design. The simulations appear to be similar in other respects. The question of what leads to the death of the line remains open. Obviously, in reality, the line could move into a region where the environment was less favorable. Whether a line in a uniformly favorable environment carries the seeds of its own destruction is an unresolved question.

It should be emphasized that the sketches in Fig. 8.56 represent one individual parcel of air flowing into an updraft. The model simulations show that the thunderstorms along the line are multicellular, undergoing life cycles short in duration compared to the time scale of the line. As in individual multicell storms (Secs. 8.2 and 8.4), the spreading cold pool cuts off the mature cell and triggers a new one. So at one time, the situation might be as shown in Fig. 8.57, where a new cell is forming at the gust front. The situation depicted is one in which the negative vorticity generated across the gust front is neutralizing positive vorticity of the environment and the buoyant parcels are going straight up (i.e., the situation of Fig. 8.56b). The updraft of the older cell, which has been cut off, is located to the rear. (Note the similarity to the isolated multicell storm depicted in Fig. 8.7.) If the negative vorticity generated by the buoyancy gradient across the gust front outweighed the positive vorticity of the environment, the picture would be similar to that of Fig. 8.57, except that the updrafts of the new and old cells would be canted rearward.

The case shown in Figs. 8.56b and 8.57 has been referred to as "optimal"[246] since the parcel has no horizontal motion and hence all the convective available potential energy of the environment can be converted to kinetic energy of vertical motion. The cells in this mode are more intense (i.e., have stronger updrafts) than those which occur later, after the cold pool strengthens and the stronger horizontal vorticity associated with the gust front causes the updrafts to cant. Then some of the convective available potential energy is thought to be converted to rearward horizontal motion, and the canted stage (Fig. 8.56a and c) is referred to as "suboptimal."

An example of a model simulation of a line of thunderstorms in the optimal mode is shown in Fig. 8.58. It shows what occurs during one cycle of cell growth and the formation of a new cell. At first the precipitation cell leans slightly downshear, while the updraft is more nearly upright (Fig. 8.58a). The updraft is being fed high-θ_e boundary-layer air from ahead of the storm. At midlevels, the parcel of air labeled A, originating ahead of the mature cell, can be traced through the storm. It moves toward the cell and into the rainshaft, where it is incorporated

[244] Rotunno *et al.* (1988) and Weisman *et al.* (1988).
[245] Fovell and Ogura (1988, 1989).
[246] By Rotunno *et al.* (1988).

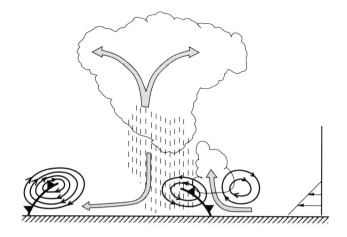

Figure 8.57

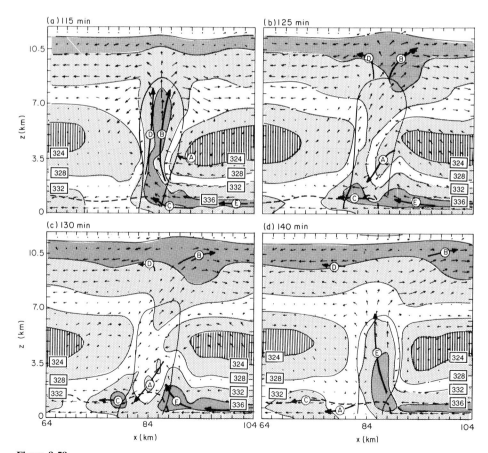

Figure 8.58

into the downdraft and descends into the developing cold pool of the cell while the old updraft is cut off from its source of high-θ_e air (Fig. 8.58b and c). A new updraft is then triggered at the gust front of the new cold pool (trajectory E in Fig. 8.58d) while parcel A joins the rearward-flowing part of the surface cold pool. This sequence, which repeats itself numerous times as the line of storms progresses, illustrates an important feature of a line of ordinary multicell thunderstorms. Part of the midlevel low-θ_e environmental air that forms the downdraft and enters the cold pool can originate *ahead* of the line, flowing from front to rear through a solid line of thunderstorms, entering at midlevels and exiting at low levels. This flow-through is possible because of the intermittent nature of the multicell development and decay process. The midlevel air is able to flow into the rainshaft for a short period of time. It then enhances the cold pool and gust front, which in turn cuts off the mature cell from its source of warm air and triggers a new updraft, which is located ahead of the older updraft, making it possible for low-level high-θ_e air to be brought upward. Thus, the line of multicell storms overturns the environment ahead of the storm by intermittently bringing midlevel low-θ_e air down in the downdrafts and low-level high-θ_e air up in the updrafts.

An important aspect of real lines of thunderstorms is that they are often a part of a mesoscale convective system, which has mesoscale circulation features intermediate in scale between the thunderstorm and the large-scale synoptic environment. One particularly important mesoscale circulation feature is the rear inflow jet, which feeds low-θ_e air into the rear of the cold pool, which is then not simply the product of the convective cells themselves. A complete description of the thunderstorm line ultimately must consider this interaction with the mesoscale circulation. We will therefore return to the convective line dynamics in the next chapter, when we examine convective lines in the context of mesoscale convective systems.

Figure 8.57 Idealization of the multicellular structure of a line of multicell storms when the low-level wind shear ahead of and perpendicular to the line is like that shown in the profile. Updraft in cell in later stage is aloft and cut off from source of low-level air by spreading of gust front. Streamlines show certain features of the airflow relative to the storm. (From Rotunno *et al.*, 1988. Reprinted with permission from the American Meteorological Society.)

Figure 8.58 Results of a two-dimensional model simulation of a line of thunderstorms. The isotherms of θ_e (intervals of 4 K) are highlighted by shading. The thick dashed line is the -1 K perturbation potential temperature contour. The solid line is the 2 g kg^{-1} rainwater contour. The vectors are scaled so that one horizontal grid interval represents 16 m s^{-1}. Selected air-parcel trajectories are followed through the time sequence; at each time, location is indicated by the identifying letter and for 5 min forward and 5 min backward the parcel follows the indicated path. Letters beginning or ending a path signify that the other half of the path was not plotted. (a) The updraft is fully developed; high-θ_e air is transported upward, and it and the rainwater field lean downshear (the parcels D and B originated from the front side earlier). (b) The downshear-leaning rainwater field evaporates into mid- and low-level air on the front side; parcel C flows through the rain, and parcel A descends from the front side. (c) The front-side air from mid- and low levels contributes to the cold pool. (d) A new cell is triggered (parcel E). (From Rotunno *et al.*, 1988. Reprinted with permission from the American Meteorological Society.)

Chapter 9 | Mesoscale Convective Systems

"The fifth night below St. Louis we had a big storm
after midnight, with a power of thunder and lightning,
and the rain poured down in a solid sheet."[247]

In Chapter 8, we examined thunderstorms occurring as isolated entities and in
lines. Thunderstorms, however, often occur in large groups and complexes, in
which individual thunderstorms and lines of thunderstorms are building blocks.
These complexes, which we refer to as *mesoscale convective systems,* are generally much larger than the individual thunderstorms and lines. In fact, they represent the largest member of the family of convective clouds (which we began
discussing in Chapter 7) and are of considerable scientific interest and practical
importance. The patterns of wind and weather associated with mesoscale convective systems are important local phenomena, which often must be forecast on
short time scales. These systems also produce a large proportion of the earth's
precipitation and thus are important from a climatological standpoint.

The dynamics of mesoscale convective systems are more complex than those
of the individual thunderstorm and lines of thunderstorms because when the individual thunderstorms and lines group together in these cloud systems, additional
phenomena appear. In particular, the mesoscale convective system often contains
a large region of stratiform precipitation. It also may contain mesoscale circulations, too large in scale to be associated with an individual thunderstorm, that are
induced by the large conglomerate of convective and stratiform clouds and precipitation.

Mesoscale convective systems occur in a variety of forms. However, they have
several features in common. For example, they all exhibit a large, contiguous area
of precipitation, which may be partly stratiform and partly convective. We can
use this observation to help define a mesoscale convective system *as a cloud
system that occurs in connection with an ensemble of thunderstorms and produces a contiguous precipitation area ~100 km or more in horizontal scale in at
least one direction.* This definition allows for the fact that some mesoscale convective systems may be long and narrow.

In Sec. 9.1, we will examine the cloud and precipitation patterns of mesoscale
convective systems. We will consider systems in both the tropics and midlati-

---

[247] From Mark Twain's *Huckleberry Finn.* The storm described by Huck could well have been a
mesoscale convective system, since these storms tend to occur at night and account for about half the
summer rainfall in the central United States.

tudes. In subsequent sections we will explore the air motions and thermodynamics associated with mesoscale convective systems. In Sec. 9.2, the focus will be on a type of mesoscale system that has been studied extensively—the *squall line with trailing stratiform precipitation*. Although this is just one type of mesoscale convective system, its relatively simple, well-defined form of organization makes it especially amenable to analysis and understanding. Moreover, many of its attributes appear to be applicable to the more complex forms of mesoscale convective system organization. In Sec. 9.3, we will return to the broader spectrum of mesoscale convective systems by examining the net divergence and vorticity associated with mesoscale convective systems of all types.

9.1   General Characteristics of the Cloud and Precipitation Patterns

9.1.1   Satellite Observations

Although we have defined a mesoscale convective system in terms of the size of its precipitation area, it is difficult, with presently available data, to obtain a statistical overview of the frequency of occurrence of these storms except from satellite cloud-top imagery, which is only a crude proxy indicator of the size of precipitation area. Larger mesoscale convective systems usually can be recognized quite easily in infrared satellite pictures by their cold, tropopause-level cloud top (e.g., Fig. 1.27). This recognition technique can be applied objectively by using a cloud-top temperature threshold to define the boundary of a cloud entity. When this technique is applied to satellite pictures from the tropics, where convective clouds occur in all sizes, results such as those in Fig. 9.1 and Fig. 9.2 are obtained. The number distribution of sizes of the cloud shields determined in this way over an equatorial region has a tendency to be lognormal, which means that the logarithm of the cloud-top size is normally distributed. A lognormal distribution has the property that the accumulated frequency is a straight line when plotted on log–probability graph paper. The number distributions in Fig. 9.1 are plotted in log–probability format for two temperature thresholds. Both curves are roughly linear (i.e., lognormal) through the middle decades of the spectrum (for cloud tops ~1000–100,000 km² in area). The curves deviate from lognormality at the largest and smallest cloud sizes. These deviations are characteristic of a truncated lognormal distribution.[248] They appear when the quantity represented has natural upper and lower limits. The large cloud shields, depending on which temperature threshold is used, do not generally exceed sizes of several hundred thousand square kilometers. Nor do these high cloud-top shields often exist at sizes less than a few hundred square kilometers.

The accumulated frequency distribution in Fig. 9.1 shows that the largest mesoscale convective systems are relatively rare. For the 198 K threshold, only 1%

---

[248] López (1977).

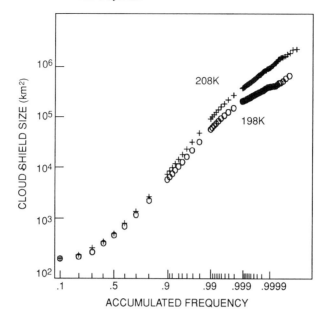

Figure 9.1

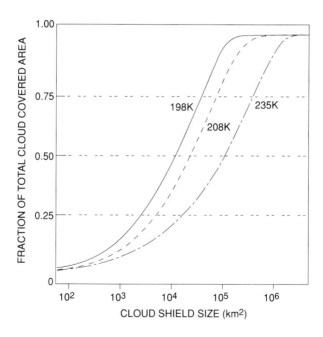

Figure 9.2

of the total number of systems have cloud shields exceeding $\sim$30,000 km^2. For the 208 K threshold, only 1% exceed $\sim$50,000 km^2. Following images of the mesoscale systems in time shows that the lifetime of the larger systems is about 10 h, typically, but can be as long as 2 or 3 days.[249]

Although, according to Fig. 9.1, the few large cloud systems are greatly outnumbered by smaller systems, the large systems are very important because they account for a very large portion of the total area covered by high, cold cloud tops. For mesoscale convective systems in the tropics, the area covered by cold cloud top (defined by some specified temperature threshold) is correlated with the amount of precipitation that falls from the cloud system. The temperature thresholds used in Fig. 9.1 and Fig. 9.2 encompass the range usually used to define the precipitating parts of tropical mesoscale convective systems.[250] Figure 9.2 shows the cumulative fraction of the cloud coverage accounted for by systems up to the indicated size. These data are for the same set of tropical Pacific cloud systems whose number distribution was shown in Fig. 9.1. By comparing the two figures, it can be seen that just 1% of the cloud systems whose areas are defined by the 208 K threshold exceed 50,000 km^2, but that this largest 1% of the cloud systems accounts for almost 40% of the total area covered by cloud tops with this temperature threshold. From the correlation between area covered by cold cloud top and rainfall, it is suggested that this largest 1% of the cloud systems also accounts for $\sim$40% of the precipitation from the population of clouds.

[249] These lifetimes can be gleaned from detailed climatologies of the satellite cloud-top imagery of near-equatorial mesoscale convective systems observed during the Global Atmospheric Research Programme (GARP) Atlantic Tropical Experiment, conducted over western Africa and the eastern Atlantic Ocean (Martin and Schreiner, 1981), and the GARP Winter Monsoon Experiment, carried out over Indonesia, Malaysia, and surrounding seas (Williams and Houze, 1987).

[250] Janowiak and Arkin (1991) used a threshold of 235 K in correlating cloud tops with precipitation over the tropical Atlantic. Williams and Houze (1987) used 213 K, which seemed to correspond well with the areas covered by precipitation observed by ground-based radar in monsoonal Malaysia and Indonesia. Airborne radar measurements in the tropical Australian monsoon (Webster and Houze, 1991; Mapes and Houze, 1993b) suggest a threshold of about 208 K.

Figure 9.1 Spectrum of sizes of the cloud shields seen in geostationary satellite infrared imagery over the region of the western tropical Pacific (bounded by 70°E and 170°W, 25°N, and 25°S) during November–February of 1986–1987, 1987–1988, and 1988–1989. The plot is in log–probability format, which means that the ordinate scale is labeled such that if the log of the quantity represented is normally distributed, the curve of the accumulated frequency distribution will be a straight line. The cloud shields for the two curves are defined by infrared temperature thresholds of 198 K and 208 K. (Provided by B. E. Mapes.)

Figure 9.2 Areal coverage by the cloud shields seen in geostationary satellite infrared imagery over the region of the western tropical Pacific (bounded by 70°E and 170°W, 25°N, and 25°S) during November–February of 1986–1987, 1987–1988, and 1988–1989. The total area covered by clouds colder than the indicated threshold temperature has been determined. The plot shows the fraction of this total area accounted for by cloud shields up to the size indicated on the ordinate of the plot. (Provided by B. E. Mapes.)

Table 9.1

Mesoscale Convective Complex (MCC)[a]

Size	A—Contiguous cold cloud shield (IR temperature ≤241 K) must have an area ≥100,000 km²
	B—Interior cold cloud region with temperature ≤221 K must have an area ≥50,000 km²
Initiate	Size definitions A and B are first satisfied
Duration	Size definitions A and B must be met for a period ≥6 h
Maximum extent	Contiguous cold cloud shield (IR temperature ≤241 K) reaches maximum size
Shape	Eccentricity (minor axis/major axis) ≥0.7 at time of maximum extent
Terminate	Size definitions A and B no longer satisfied

[a] Based upon analysis of enhanced IR satellite imagery; from Maddox (1980).

The largest member of the population of mesoscale convective systems is called the *mesoscale convective complex* (*MCC*). Parameters of the cloud-top image used to identify the MCC are listed in Table 9.1. These criteria assure that the MCC has an extremely large and cold cirriform cloud shield. This definition has been used to identify large mesoscale convective systems over the globe. The map in Fig. 9.3 shows that these very large cloud systems occur in both midlatitudes and the tropics and over both land and ocean. Some characteristics of midlatitude MCCs are listed in Table 9.2. These data indicate that the cloud shield reaches a typical size of 200,000 km² for a temperature threshold of 221 K. The MCC thus corresponds in size to the upper 1% or so of the overall size distribution of mesoscale convective systems shown in Fig. 9.1. The average lifetime is about 15 h, although on occasion an MCC can last for days.

9.1.2   Precipitation Structure

We have seen from satellite data that the cloud tops of large precipitation-producing mesoscale convective systems are ~25,000–250,000 km² in area and last tens of hours to a number of days. It is difficult, however, to derive further information

Table 9.2

Characteristics of MCCs over the Central United States

Year	Cases	Lifetime (h)	Maximum area ≤241 K (×10³ km²)	Maximum area ≤221 K (×10³ km²)
1981	23	15	310	190
1982	37	14	280	180
1983	30	16	300	160

Source: Maddox *et al.* (1982); Rodgers *et al.* (1983), (1985).

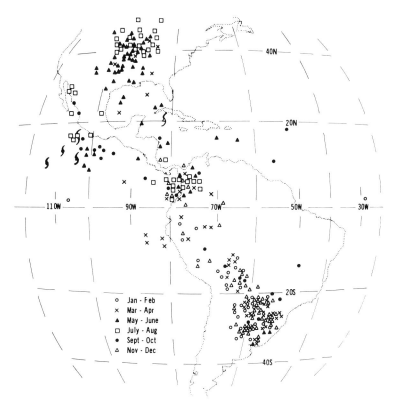

Figure 9.3 Geographic and monthly distribution of mesoscale convective complexes (MCCs) in and around the Americas. Locations are for the MCC cold-cloud shields at the time of maximum extent. Hurricane symbols indicate MCCs that developed into tropical storms. (From Velasco and Fritsch, 1987. © American Geophysical Union.)

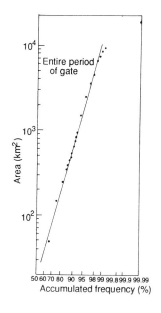

Figure 9.4 Size distribution of radar echoes over the tropical eastern Atlantic Ocean. The abscissa is a lognormal probability scale. Straight line is for a lognormal distribution of best fit to the data. (From Houze and Cheng, 1977. Reproduced with permission from the American Meteorological Society.)

on the internal structure of the mesoscale systems from satellite data. The cloud top tends to be rather uniform, shielding the structure of the storm beneath the cloud top from view. The most readily observable parameter that indicates the detailed three-dimensional structure of the cloud system below the cloud top is the precipitation, which can be scanned by radar (Chapter 4). The radar has both broad coverage and high resolution, and the spatial pattern of the radar reflectivity of the precipitation is a sensitive tracer of storm organization. Data on wind, temperature, pressure, and humidity within mesoscale convective systems are scarce and do not provide as much coverage and resolution as radar. Hence, an overview of mesoscale convective system structure is best obtained from radar data.

9.1.2.1 Tropical Cloud Systems

We will begin by looking at radar data from the tropics. The equatorial zones are good natural laboratories in which to study mesoscale convective systems since these regions are characterized by large fields of convective clouds of all sizes. Tropical clouds form in environments that are rather uniform horizontally. The air masses are continuously heated from below, and the convection is always mixing the air vertically. The environment is generally conditionally neutral or slightly conditionally unstable (Sec. 2.9.1). The horizontal uniformity over wide regions is maintained in part because the tropics are relatively unaffected by strong baroclinic waves and fronts such as those that dominate the midlatitudes (Chapter 11).[251] In midlatitudes, mesoscale convective systems can become entangled with clouds and precipitation produced by baroclinic and frontal processes, and the purely convective processes thereby become more difficult to isolate and interpret. The tropics present the better opportunity to assess ensemble properties of convective clouds and precipitation.

The patterns of rain in tropical mesoscale convective systems have been examined statistically as part of the overall spectrum of convective phenomena in the tropics. As in the case of satellite cloud images (Fig. 9.1), the sizes of rain areas seen on radar follow a lognormal distribution (Fig. 9.4). Although the large rain areas (i.e., those located in mesoscale convective systems) are few in number according to the lognormal distribution, they account for most of the precipitation in the tropics. It has been found that the largest 10% of the rain areas account for about 90% of the precipitation over equatorial oceanic regions.[252] It is clear from such statistics that for understanding the precipitation-producing phenomena in the tropics, the rain areas of the large mesoscale convective systems are of prime importance.

Further aspects of the internal structure of the mesoscale convective system are indicated by finer details of the radar echoes representing the large rain areas.

---

[251] Synoptic-scale waves do occur in the tropical easterlies. Their effect on cloudiness, however, is primarily to modulate the frequency of occurrence of mesoscale convective systems in a given region. They are generally not characterized by fronts or other strong mesoscale circulations that directly produce clouds other than cumulus and cumulonimbus.

[252] See Houze and Cheng (1977) and López (1978).

The locations of strong convective-scale air motions within the rain areas are indicated in horizontal maps of radar reflectivity by intense spots of reflectivity within the broader echo pattern. When the number of convective echo cores were counted within the rain areas of a large number of tropical mesoscale convective systems, it was found that, on average, only about 10 convective echo cores were embedded in echoes exceeding 10,000 km² in area (last column on the right in Fig. 9.5). The convective echo cores counted were typically 100 km² in area. Thus, it is concluded that *only about 10% of the rain area in a mesoscale convective system is covered by convective rain showers.* The remainder of the area is covered by stratiform rain (of the type discussed in Chapter 6).

The spatial pattern of convective cores in relation to the surrounding stratiform portion of the rain area in a mesoscale convective system varies from case to case and is a sensitive indicator of the storm organization. The example in Fig. 9.6 shows the pattern of convective areas within the otherwise stratiform rain area of a tropical mesoscale convective system. The centers of the convective showers are indicated by an X, and the region of rain thought to be directly associated with the convective cores is enclosed by a line surrounding the core centers. During the 75-min period shown, the pattern of convective cores varied. The distribution of cells was somewhat random but displayed a preference for occurring along a rough line on the southern side of the rain area. The pattern of cells can be even more random than in this example, embedded chaotically throughout the stratiform echo region. However, more often than not, they show a tendency to occur in lines or bands. Sometimes the band of convection becomes very sharply defined, as shown in the example in Fig. 9.7. In this type of case, referred to as a *tropical squall line,* the band is often arc shaped and moving rapidly (10–15 m s⁻¹). The stratiform precipitation in this case trails behind the intense convective line.

The radar echoes in the convective and stratiform portions of a rain area in a mesoscale convective system are clearly distinguished in vertical cross sections through the radar reflectivity pattern. For example, Fig. 9.8 shows a vertical cross section through the tropical squall-line rain area of Fig. 9.7. A convective cell at the leading edge of the rain area is apparent as the intense vertical core of reflectivity in the left-hand side of the picture. It stands in sharp contrast to the structure of the trailing stratiform echo (right-hand portion of Fig. 9.8), which is characterized by a horizontally oriented bright band in the melting layer just below the 0°C isotherm (Sec. 6.1.2). In this type of squall line, the upper portion of the radar echo of the convective line slopes back toward the trailing stratiform region, as seen in the region between 0 and 40 km on the horizontal scale in Fig. 9.8. In nonsquall rain areas, such as in Fig. 9.6, the radar echoes of the convective cells tend to be more upright at upper levels.

9.1.2.2   Midlatitude Convective Systems

The precipitation areas of midlatitude mesoscale convective systems are similar to those in the tropics in that the rain areas are of similar size and, at their mature stage of development, consist of a group or line of convective showers embedded in or adjoining a large region of stratiform precipitation. A common

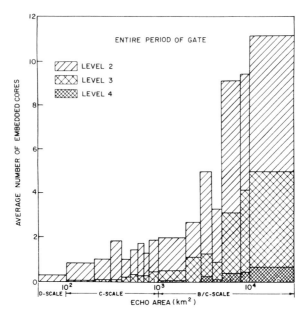

Figure 9.5 Average number of embedded cores of maximum radar reflectivity located within radar echoes in different size ranges for radar echoes located over the tropical eastern Atlantic Ocean. Shading indicates the fraction of the total number of cores in each size range which exceeded reflectivity threshold levels 2 (31 dBZ), 3 (30 dBZ), and 4 (47 dBZ). (From Houze and Cheng, 1977. Reproduced with permission from the American Meteorological Society.)

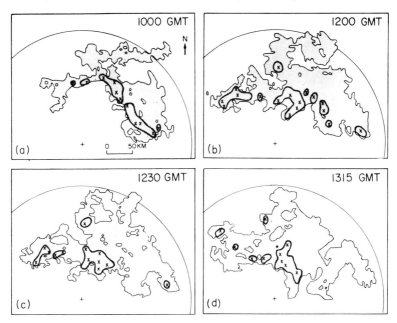

Figure 9.6 Radar echo pattern showing convective areas (surrounded by heavy solid lines) within the otherwise stratiform rain area of a tropical mesoscale convective system over the tropical eastern Atlantic Ocean. The partial circle is a range ring of the shipborne radar (located at +). Peaks of reflectivity are indicated by ×. (From Cheng and Houze, 1979. Reproduced with permission from the American Meteorological Society.)

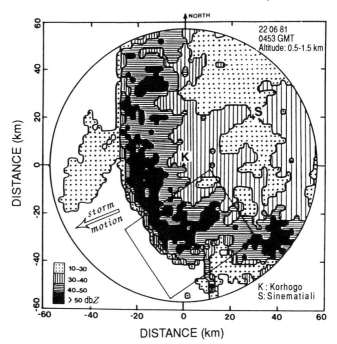

Figure 9.7 Radar echo pattern of a tropical squall line observed in Ivory Coast, western Africa, on 22 June 1981. The radar is located at Korhogo (K). A second radar was located at Sinematiali (S). The echo is derived from observations between 0.5 and 1.5 km altitude. The rectangle marks an area within which the three-dimensional wind field was derived by dual-Doppler radar synthesis of the data from the two radars. Other aspects of this storm are illustrated in Figs. 9.8, 9.20, and 9.47. (From Chong *et al.*, 1987. Reprinted with permission from the American Meteorological Society.)

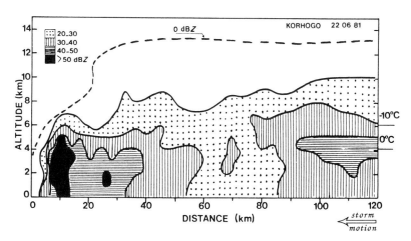

Figure 9.8 Vertical cross section of radar reflectivity in the tropical squall line represented in Figs. 9.7, 9.20, and 9.47. (From Chong *et al.*, 1987. Reprinted with permission from the American Meteorological Society.)

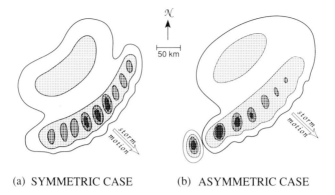

(a) SYMMETRIC CASE   (b) ASYMMETRIC CASE

Figure 9.9

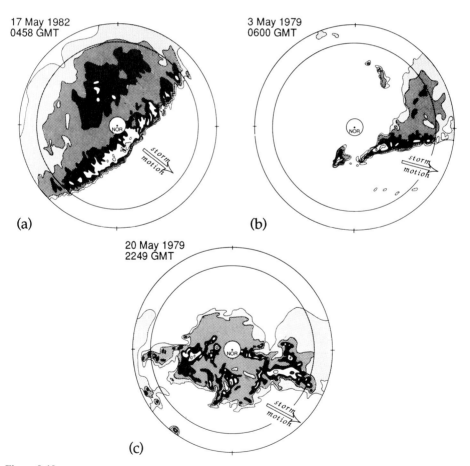

Figure 9.10

pattern is again the leading line of convective showers trailed by a region of stratiform rain. A schematic representation of the low-altitude radar echo typically associated with this pattern is shown in Fig. 9.9. The schematic is based on a study of the mesoscale systems that occurred during the springtime in Oklahoma over a 6-year period. The characteristics of the idealized radar echo pattern may be summarized as follows:

The leading *convective line* has

1. *Arc shape* (convex toward the leading edge).
2. *Generally northeast–southwest orientation.* (This is variable: some lines are nearly north–south, while others are nearly east–west. The orientations are undoubtedly determined by local climatology, and the same type of mesoscale systems in another part of the world might have different orientations than the Oklahoma systems.)
3. *Rapid movement with an eastward and/or southward component* (<10 m s^{-1} in a direction normal to the line orientation).
4. *Solid appearance* (a series of intense reflectivity cells solidly connected by an echo of more moderate intensity).
5. *Very strong reflectivity gradient at leading edge* (i.e., gradient much stronger at the leading edge than the back edge of the convective region).
6. *Serrated leading edge* (leading edge of echo is jagged, with forward-extending protrusions at an apparent wavelength ~5–10 km).
7. *Elongated cells oriented 45–90° with respect to line* (elongated cells appear to be related to the serrated leading edge).

The *trailing stratiform region* has:

8. *Large size* (>10⁴ km² in horizontal area).
9. *Notch-like concavity at rear edge* (believed to be associated with mesoscale inflow of dry air that erodes a portion of the stratiform echo).
10. *A secondary maximum of reflectivity* (separated from the convective line by a narrow channel of lower reflectivity).

Two possible manifestations of these 10 characteristics are illustrated by the schematic radar-echo patterns presented in Fig. 9.9a and b. The organization

Figure 9.9 Schematic representation of a common type of radar echo seen in midlatitude mesoscale convective systems. In both the symmetric and asymmetric cases, a leading line of convection is connected with a trailing region of stratiform precipitation. Large vectors indicate the direction of motion of the system. Levels of shading denote radar reflectivity, the most heavily shaded corresponding to convective cell cores with the most intense values. (From Houze *et al.*, 1990. Reproduced with permission from the American Meteorological Society.)

Figure 9.10 Examples of radar data from the rain areas of midlatitude mesoscale convective systems. Low-elevation reflectivity patterns from the National Severe Storms Laboratory radar located at Norman (NOR), Oklahoma, are indicated by shading levels corresponding to 20–24 dBZ (light gray), 25–34 dBZ (dark gray), 35–44 dBZ (black), 45–54 dBZ (white), 55–64 dBZ (light gray), and >65 dBZ (dark gray). Range rings are at 20, 200, and 240 km. Registration marks on outermost ring are at 90° azimuth intervals (north toward top of page). (From Houze *et al.*, 1990. Reproduced with permission from the American Meteorological Society.)

shown in Fig. 9.9a is referred to as *symmetric*, while that in Fig. 9.9b is referred to as *asymmetric*.

In the symmetric case, the convective line shows

- *no preference for the most intense cells to be found at any particular location along the leading edge of the line.* New cell growth apparently occurs all along the leading edge of the line,

while the stratiform region has its

- *centroid located directly behind the center of the convective line.*

In the asymmetric case, the convective line is

- *stronger on its southern, southwestern, or western end.* That is, younger, newly formed, and more intense cells are located toward one end of the line, while weaker, dying cells on the verge of becoming stratiform are found toward the other end of the line,

and the centroid of the stratiform region is

- *biased toward the north, northeast, or east end of the line*, rather than centered behind the line.

Actual examples of the symmetric and asymmetric type of rain area organization are shown in Fig. 9.10a and b. About two-thirds of the rain areas associated with the investigated mesoscale convective systems in Oklahoma have been found to exhibit, at least to some degree, either the symmetric or asymmetric structure at their mature stage of development. In addition to echo patterns that closely or partially resemble these prototypes, there exist mesoscale convective systems whose rain areas bear no resemblance to these prototypes. In these systems, the convective cells are distributed in a much more chaotic pattern in relation to the stratiform precipitation. An example of a rain area with this more chaotic structure is shown in Fig. 9.10c. It is somewhat reminiscent of the tropical convective system illustrated in Fig. 9.6.

9.1.3   Life Cycle of a Precipitation Area

The precipitation area of a mesoscale convective system exhibits a characteristic life cycle, which is illustrated schematically in Fig. 9.11 as it would appear on radar. In its *formative stage* (Fig. 9.11a), the mesoscale precipitation area appears on radar as a group of isolated cells, which may be randomly distributed in the horizontal or arranged in a line. The schematic follows the evolution of a line of cells. In the *intensifying stage* of the mesoscale rain area, the individual cells grow and merge (Fig. 9.11b). The feature then comprises a single contiguous rain area, in which several relatively intense cores of precipitation are interconnected by lighter precipitation. The *mature stage* of the mesoscale rain area is reached when a large stratiform area develops from older cells blending together as they begin to weaken (Fig. 9.11c). Each convective element goes through a life cycle, at the end

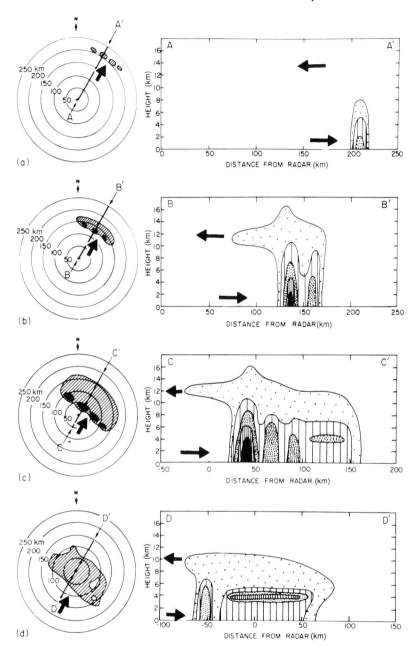

Figure 9.11   Schematic of the life cycle of the precipitation area of a mesoscale convective system as it would appear on radar in horizontal and vertical cross sections during (a) formative, (b) intensifying, (c) mature, and (d) dissipating stages. The outside contour of radar reflectivity represents the weakest detectable echo. The inner contours are for successively higher reflectivity values. Heavy arrows indicate the direction of the wind relative to the system. (From Leary and Houze, 1979a. Reproduced with permission from the American Meteorological Society.)

of which it weakens and becomes a component of a region of stratiform precipitation falling from the midlevel base of the general stratiform cloud shield of the mesoscale convective system. When several neighboring cells reach this stage, they may become indistinguishable from each other and together can form an extensive region of stratiform rain with a continuous melting layer. A stratiform region formed in this way may be as large as 200 km in horizontal dimension. This life cycle is essentially the sequence of events illustrated in Fig. 6.11, which was presented to explain how a region of nimbostratus can evolve from a group of deep convective elements.

Cloud and precipitation are detrained from the tops of active cells. If the upper-level winds carry the condensate away from the rain area, an overhang of radar echo can form, as shown in the schematic in Fig. 9.11c. If, on the other hand, the winds aloft carry the detrained condensate inward, toward the center of the rain area, the detrained condensate combines with the stratiform clouds and precipitation formed from the old cells.

As long as new cells continue to form, the precipitation region remains in a mature stage and consists of a combination of active cells, weakening cells, and stratiform precipitation. In the *dissipating stage* of the rain area (Fig. 9.11d), the formation of new convective cells diminishes, and the feature consists of a broad area of slowly weakening stratiform precipitation with weak, embedded convective cells.

The area-integrated mass of rainfall from the precipitation area of a mesoscale convective system reflects the life cycle just described. A typical example of the rain from a mesoscale convective system is shown in Fig. 9.12. During the first 6–8 h, corresponding to the formative and intensifying phases of the precipitation area, the rainfall was dominated by the convective cells. After this time, the stratiform precipitation accounted for an increasing proportion of the total rainfall until, midway through the lifetime of the rain area, the stratiform and convective regions were contributing approximately equally to the mass of rain falling from the cloud system.

Although the rain rate in the stratiform region is considerably less than in the convective cells, the great area covered by the stratiform rain implies a large total fallout of water mass over the whole region. In some cases, the stratiform rain amount actually surpasses the convective amount during the mature phase of the rain area. The stratiform component continues to be strong into the dissipating stage of the rain area. The convective and stratiform components both gradually weaken. It is typical for the stratiform rain to account for 25–50% of the total rain integrated over the lifetime of a mesoscale convective system. In the example shown in Fig. 9.12 the stratiform rain accounted for 40% of the total time- and area-integrated rainfall.

9.2 The Squall Line with Trailing Stratiform Precipitation

9.2.1 General Features

We have noted that in both the tropics and midlatitudes, one type of precipitation structure that can occur in a mesoscale convective system is a sharply defined

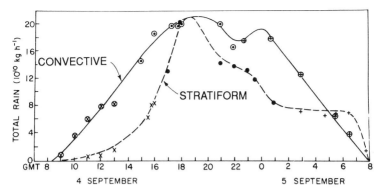

Figure 9.12   Total rain integrated over the convective (circled points) and stratiform regions of a squall-line mesoscale convective system located over the eastern tropical Atlantic Ocean. The data were obtained by three shipborne radars. The symbols ●, +, and × indicate different methods used for combining the information from the three radars. (From Houze, 1977. Reproduced with permission from the American Meteorological Society.)

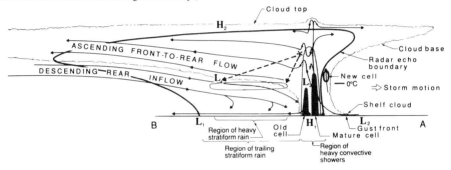

Figure 9.13   Conceptual model of the kinematic, microphysical, and radar-echo structure of a convective line with trailing stratiform precipitation viewed in a vertical cross section oriented perpendicular to the convective line (and generally parallel to its motion). Intermediate and strong radar reflectivity is indicated by medium and dark shading. The location of line AB is shown in Fig. 9.14. (From Houze *et al.*, 1989. Reproduced with permission from the American Meteorological Society.)

convective line with trailing stratiform precipitation. This structure was shown by examples in Figs. 9.7, 9.8, and 9.10a and b and in idealized form in Fig. 9.9. In this section, we examine this type of mesoscale convective system structure in more detail. Since this form is so sharply defined, it is more amenable to study and has been examined intensively in a variety of ways. Thus, at the time of this writing, more can be said about its dynamics than is possible for other forms of mesoscale convective system organization. It appears though, that many aspects of leading-line/trailing stratiform organization apply also to other forms of mesoscale organization.

A conceptual model based on many observational studies of the kinematic structure of a convective line with trailing stratiform structure is shown in Fig. 9.13. It presents a cross section oriented perpendicular to the leading line of convection. The heavy, black line indicates the boundary of the precipitation, as seen by radar. The light, scalloped line indicates the horizontal and vertical extent of the cloud, as determined from visual observation, satellite imagery, or radiosonde data. The intermediate and dark shading indicate regions of enhanced radar reflectivity. Vertically oriented cores of high reflectivity mark the showers of heavy rain in the leading convective region of the storm. The trailing region of stratiform rain shown in Fig. 9.13 is characterized by a marked radar bright band (Sec. 6.1.2).

Streamlines of the airflow in Fig. 9.13 indicate a general trend of upward motion beginning in the boundary layer near the gust front, extending up through the convective region, and sloping more gently into the trailing stratiform cloud at middle to upper levels. There is, at the same time, a general trend of downward motion in a current of rear inflow, which runs under the base of the trailing stratiform cloud and enters the stratiform region just above the 0°C level. This descending current subsides to the level of the radar bright band, passes through the melting level, and finally enters the back of the convective region at low levels, where it reinforces convergence and overturning at the leading gust front.

Superimposed on the general upflow within the convective region are intense, localized updrafts and downdrafts, associated with the intense rain showers located there. New convective cells tend to form on or just ahead of the leading edge of the region of heavy convective showers. The first radar echo appears aloft, evidently associated with a strong, convective updraft. This developing cell is followed by a mature cell, which has a deep, strong reflectivity core and is associated with heavy surface rainfall. The mature cell contains an intense narrow updraft that can penetrate above the top of the broad cirriform cloud shield. This updraft is often followed by a convective-scale downdraft at middle to upper levels. Following the mature cell is an older cell. Though in a weakening stage, the older cell is also characterized by an updraft core, which is, in turn, followed by another mid- to upper-level convective-scale downdraft. Older cells are advected rearward over a layer of dense, subsiding, storm-relative rear inflow. In the heavy rain from the mature and older cells, low-level convective-scale downdrafts spread out in the boundary layer behind the gust front and toward the rear of the system. They are of the type usually associated with precipitation drag and evaporation in isolated thunderstorms and lines of thunderstorms (Secs. 8.2, 8.3, 8.9, 8.10, and 8.11). The arrangement of cells in the convective region, in order of their stage of development, is similar to that of the multicell thunderstorm (cf. Fig. 8.7) and in the line of thunderstorms discussed in Sec. 8.11 (Figs. 8.57 and 8.58). The multicellular nature of the convective region will be examined in more detail in Sec. 9.2.2.3.

The schematic cross section in Fig. 9.13 represents the mature stage in the life of the mesoscale convective system (which undergoes the type of life cycle schematized in Figs. 6.11 and 9.11). During the formative and intensifying stages, the stratiform rain region is not present, and the cells tend to be more intense. The numerical-model simulations of thunderstorm-line development discussed in Sec. 8.11 represent these earlier stages of the storm. Those calculations suggest that the cells may regenerate somewhat periodically for the first few hours of the storm's existence. As the low-level cold pool associated with precipitation fallout strengthens, a slantwise circulation develops, upon which the convective cells are superimposed as they continue to form. As the older cells are advected rearward in the sloping flow, the structure of the mature stage of the system indicated in Fig. 9.13 evolves. This structure persists for 5–10 h and is accompanied by large amounts of stratiform rain falling from the trailing region.

The forward overhang of cloud and precipitation shown in Fig. 9.13 is not always seen. Its presence depends on the wind shear normal to the line in the environment. When there is strong relative flow toward the line at upper levels, there is no upper-level overhang. The example in Fig. 9.8 is of this latter type.

The strength of the rear inflow entering the stratiform region of the mature storm varies considerably from case to case, and it is associated with a variety of horizontal circulation patterns. Figure 9.14 shows two possible horizontal configurations with which the vertical section in Fig. 9.13 could be associated. The rear inflow could be one branch of a midlevel mesoscale vortex in the trailing stratiform rain region of an asymmetric squall line (Fig. 9.14b), or it could be associated with a windshift line that is more or less uniform in the along-line direction of a symmetric-type system (Fig. 9.14a).

Above the rear inflow, within the stratiform cloud, the layer of upward-sloping front-to-rear flow emanating from the upper portions of the convective line advects rearward ice particles falling from the convective cells (asterisk in Fig. 9.13). The trajectories of these particles (which are similar to the sequence of ice particle positions 3, 2, and 1 in Fig. 6.11f) lead to the bright band. A stratiform precipitation region is thus produced in which the particles formed in the region of deep convection, by vigorous riming growth in the convective updrafts, exit the convective drafts and grow thereafter by vapor deposition as they slowly fall through the gently ascending front-to-rear flow. These ice particles also collect smaller ice particles and aggregate with each other to form large snow aggregates as they approach the 0°C level. When they fall through this level, they melt and produce the radar bright band and the region of heavier stratiform rain directly below.

The precipitation processes in a stratiform region of the type shown in Fig. 9.13 were shown schematically in Fig. 6.12, when we discussed nimbostratus of the type that occurs in the vicinity of deep convection (Sec. 6.3). We noted that in storms of the type shown in Fig. 9.13, only the *location* of the region of heavier stratiform rain in the trailing region is explained by the fallout trajectories of the ice particles produced in the convective updrafts. The correct *amount* of stratiform rain is calculated only when the growth of the particles by vapor deposition, which occurs as they pass through the region of mesoscale ascent, is taken into account. The ascending front-to-rear flow in the stratiform region is thus both qualitatively and quantitatively important since its horizontal component spreads the particles out into the region of the bright band and its vertical component supplies the moisture for the growth by vapor deposition of the ice particles. The precipitation processes in the stratiform region are discussed further in Sec. 9.2.3.1.

The pressure field associated with the squall line with trailing stratiform precipitation is marked by several characteristic lows and highs (indicated by L and H in

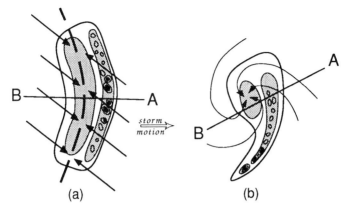

Figure 9.14 Two possible midlevel horizontal flow fields with which the vertical section between A and B in Fig. 9.13 could be associated. (a) Rear inflow is associated with a windshift line that is more or less uniform in the along-line direction of a symmetric-type system. (b) Rear inflow is one branch of a midlevel mesoscale vortex in the trailing-stratiform rain region of an asymmetric squall line. The flow fields represented by streamlines are superimposed on schematic low-level radar reflectivity fields indicated by contours and shading. Black areas are the regions of highest reflectivity. Gray areas are of intermediate intensity. (Adapted from Houze *et al.*, 1989. Reproduced with permission from the American Meteorological Society.)

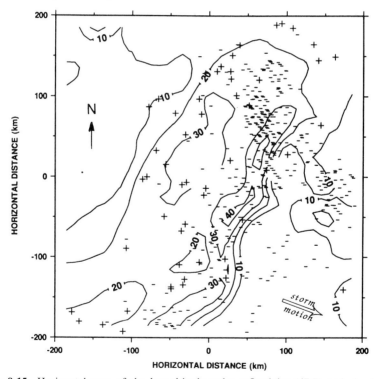

Figure 9.15 Horizontal map of the low-altitude radar reflectivity (dBZ) and cloud-to-ground lightning strike locations in a squall line with trailing stratiform region observed over Kansas at 0351 GMT 11 June 1985. Positive and negative flashes during a 30-min period centered on this time are indicated by the appropriate sign. Other aspects of this storm are illustrated in Figs. 9.18, 9.31, 9.34, 9.40, 9.41, 9.43, 9.44, 9.45, 9.48, 9.53, and 9.54. (From Rutledge and MacGorman, 1988. Reprinted with permission from the American Meteorological Society.)

Fig. 9.13). A *wake low* (L_1) occurs at the surface at the back edge of the stratiform rain, in association with warming by unsaturated descent. A high (H_1) occurs below the convective region; it is a hydrostatic gust-front high of the type discussed in Sec. 8.9. A separate, dynamically produced high is also found at the gust-front leading edge but is not shown here. Also, a surface low (L_2), associated with warming by compensating downward displacement at mid- to upper levels is often noted ahead of the convective line at the surface. In the midtroposphere, a small, apparently hydrostatic low (L_3) is located below the primary sloping buoyant convective updraft. Farther to the rear, in the vicinity of the melting layer or just above, is another low (L_4) that is larger in scale and evidently associated with the positively buoyant air in the stratiform cloud above. A high (H_2) occurs at upper levels atop the entire mesoscale cloud system in association with the predominantly buoyant air in the convective region and the ascending front-to-rear flow. Some of these pressure maxima and minima will be further elaborated on in later discussions (Secs. 9.2.2.7 and 9.2.3.2).

Like smaller thunderstorms, mesoscale convective systems are electrically active. A horizontal map of the CG lightning strikes for an actual squall line with trailing stratiform precipitation is shown in Fig. 9.15. The leading line of high radar reflectivity is characterized by a preponderance of negative strikes, while the trailing region of stratiform precipitation is characterized by positive strikes. This behavior is similar to that of the multicell thunderstorm (Fig. 8.6), where the active convective regions are characterized by negative strikes and the dissipating cell with a bright band is characterized by positive strikes.

The vertical distribution of charge in a squall line with trailing stratiform precipitation may be as shown in Fig. 9.16. The leading anvil is postulated to have a structure similar to the supercell anvil (Fig. 8.14). The part of the stratiform region immediately behind the active convective zone was found by balloon measurements in one case to have the vertical layering indicated in Fig. 9.16. A region of positive charge is found at the $-10°C$ level and has been hypothesized to be associated with the fallout of graupel from the upper levels of the convective region.[253] This temperature regime is on the warm side of the charge-reversal temperature (Sec. 8.1). Therefore, the graupel would be expected to become positively charged as a result of collisions with smaller ice particles at these levels.

The distinction between the leading convective region and the trailing stratiform region in Fig. 9.13 is not a sharp one in terms of air motions and thermodynamics. The ascending front-to-rear current of buoyant air and the descending rear-to-front current of negatively buoyant air flow continuously across the entire storm. However, the distinction in terms of the precipitation process is clear. As noted in Sec. 6.1.1, the strength of the vertical air motion as compared to the fall speeds of the precipitation particles determines whether the mechanism is convective or stratiform. The more intense, localized, and transient vertical air motions in the leading zone make the precipitation processes there tend to be of the convective type, while the predominance of gentler, more widespread, and

---

[253] See Rutledge *et al.* (1990) and Schuur *et al.* (1991).

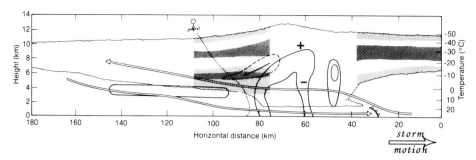

Figure 9.16 Schematic of the electrical structure of a convective line with trailing stratiform structure observed over Oklahoma. The main rear-to-front and front-to-rear circulation components are indicated by double streamlines; they are separated at the leading edge of the storm by the gust front (frontal symbol). Solid contours indicate areas of radar reflectivity at two intensity levels (both >35 dBZ). Light shading indicates regions of negative charge. The distribution of charge in the trailing anvil was deduced from measurements taken on the indicated balloon ascent. The distribution of charge in the leading anvil is assumed to be like that in other thunderstorm anvils (e.g., Fig. 8.14). The dashed contour (adapted from the model results in Fig. 6.14b) indicates a local maximum in graupel mixing ratio, with particles falling along the indicated trajectory; it is introduced to explain the layer of positive charge at the −10°C level. The distribution of charge in the convective region (large + and − signs) is assumed to be like that in a mature thunderstorm cell in a single-cell or multicell storm (Figs. 8.3 and 8.6). (From Schuur *et al.*, 1991. Reprinted with permission from the American Meteorological Society.)

longer-lasting vertical air motions to the rear comprise a zone favoring the stratiform precipitation mechanism. To facilitate the more detailed examination of the storm features indicated qualitatively in Fig. 9.13, we will divide the discussion in the remainder of this section into parts corresponding to the storm's two different precipitation regimes. In Sec. 9.2.2, we will consider details of the convective region. Then in Sec. 9.2.3, we will examine the stratiform region.

9.2.2 The Convective Region

In Figs. 9.13 and 9.14, we have seen the general structure of the squall line with trailing stratiform precipitation in a qualitative and idealized form. We now describe this type of storm more quantitatively. It is possible to proceed beyond the conceptual model because the circulations in several of these storms have been documented in special field experiments (utilizing Doppler radar, aircraft, sounding, and surface instrumentation) and because they have been successfully simulated in numerical models of the type discussed in Sec. 7.5.3. In this subsection, we will concentrate on the structure and dynamics of the convective region. We will begin by looking at the kinematic and thermodynamic structure of the convective cells. Then we will consider some aspects of the life cycles of individual cells and their interaction with the stratosphere. Finally, we will examine some of the factors relevant to the maintenance of the convective region. In Sec. 9.2.3 we will examine aspects of the stratiform region.

9.2.2.1 *Observed Airflow*

A typical example of the airflow pattern inferred by multiple Doppler radar synthesis (Sec. 4.4.6) in the convective region of a mesoscale convective system is shown in Fig. 9.17. The main current of slantwise upward flow toward the rear of the system is evident. Superimposed on this upward flow are sharp bumps corresponding to each convective cell within this cross section through the convective region. Each bump in the flow is seen to be associated with a reflectivity maximum, with each successive peak reaching a higher altitude in a stair-step fashion. The second stair step corresponds to the strongest reflectivity cell. Another example, for which isotachs of vertical velocity are shown, is in Fig. 9.18. Again the updraft peaks are found at successively higher altitudes. In this case, strong downdraft cells are also seen in the convective region. There are two types of these downdrafts. Low-level downdrafts are associated with the heavy rain showers. These evidently are the precipitation-driven downdrafts typically associated with the rainshafts and hailshafts of thunderstorms (Secs. 8.2, 8.3, 8.9, and 8.10). They feed the pool of cold air that accumulates below the convective region and whose leading edge is the gust front forming the leading edge of the mesoscale system (Fig. 9.13). In addition to the low-level downdrafts are upper-level downdrafts that tend to occur on either side of the updraft cores. They are apparently forced by the downward pressure perturbation force found on either side of a buoyant element (Fig. 7.1). It can be shown by application of the appropriate boundary condition to (7.8) that this downward force becomes intensified when the buoyant element approaches a rigid upper boundary.

9.2.2.2 *Thermodynamic Structure*

The thermodynamic structure of mesoscale convective systems is much more difficult to document than the airflow, which can be derived from Doppler radar observation. Some data on the thermodynamic structure have been obtained during aircraft missions. The most significant thermodynamic feature of the squall line revealed by aircraft data is the existence of a small, low-pressure center, about 10 km wide, located under the upward- and rearward-sloping channel of buoyant air in the convective region in Fig. 9.19. The low-pressure area was found by averaging measurements from several flights through squall lines. It is apparently centered beneath the strongest cell in the convective region. Thus, it is associated with the buoyancy component of the perturbation pressure (p_B^*), which is required to be a minimum at the base of a buoyant column as illustrated by the lines of force in Fig. 7.1. The aircraft-measured pressure perturbation minimum is roughly in hydrostatic balance with the buoyancy perturbation, also measured by aircraft in the column of air above. From the aircraft data alone, it is not apparent whether there is a dynamic contribution to this pressure perturbation minimum in addition to the buoyancy contribution. We will see below that model simulations indicate that the pressure perturbation minimum is largely hydrostatic but that in strong squall-line circulations there is also a significant contribution from the dynamic pressure perturbation (p_D^*) associated with the characteristic vortex in the gust-front head [Sec. 8.9, Eq. (8.44), Fig. 8.42].

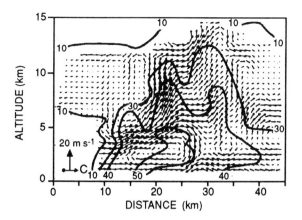

Figure 9.17 Airflow pattern inferred by multiple Doppler radar synthesis in the convective region of a tropical squall-line system observed by dual-Doppler radar in Ivory Coast, west Africa, on 23 June 1981. System is moving from right to left. Vertical arrow indicates scale of airflow vectors. Horizontal arrow C shows velocity of individual convective cells. Airflow vectors are computed relative to the cells. Contours show radar reflectivity (dBZ). This storm is also represented in Fig. 9.21. (From Roux, 1988. Reprinted with permission from the American Meteorological Society.)

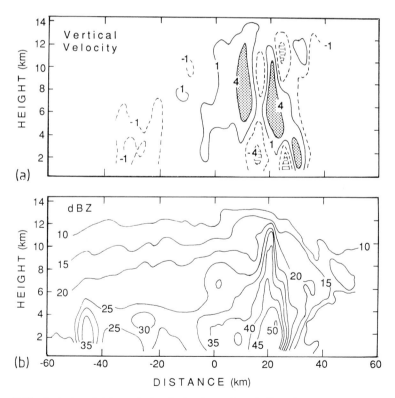

Figure 9.18 Isotachs of (a) vertical velocity (m s^{-1}) and (b) radar reflectivity (dBZ) in the convective region of a squall line with trailing stratiform precipitation. Other aspects of this storm are illustrated in Figs. 9.15, 9.31, 9.34, 9.40, 9.41, 9.43, 9.44, 9.45, 9.48, 9.53, and 9.54. (From Houze, 1989. Reprinted with permission from the Royal Meteorological Society.)

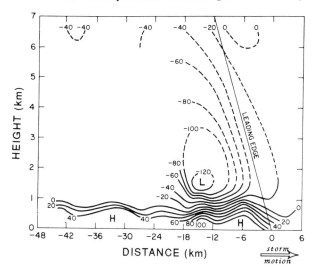

Figure 9.19 Composite pressure perturbation field (contour interval 20 Pa) based on aircraft data obtained in tropical squall lines moving faster than 7 m s^{-1}. (Adapted from LeMone *et al.*, 1984. Reproduced with permission from the American Meteorological Society.)

Since aircraft flight data are so scarce, it has been necessary to derive the thermodynamic fields in the squall line with trailing stratiform precipitation, as well as their cloud and precipitation content, by more indirect means from data that have better coverage but do not directly indicate the thermodynamic variables. A productive method has been to apply the retrieval method to Doppler radar data (Sec. 4.4.7). According to this technique, the thermodynamic and microphysical fields consistent with radar-observed reflectivity and velocity fields can be estimated from the momentum, thermodynamic, and water-continuity equations. Examples of water and thermodynamic fields derived by this methodology are shown in Fig. 9.20 for the tropical squall-line system whose radar reflectivity structure was shown in Figs. 9.7 and 9.8. The first cross section (Fig. 9.20a) shows the Doppler radar–derived wind and the observed radar reflectivity in the convective region (very similar to the case in Fig. 9.17), while Fig. 9.20b shows isotachs of the vertical velocity component (similar to Fig. 9.18, except downdrafts are not well represented in this analysis). The retrieved rainwater, cloud water, and potential temperature perturbation fields are shown in Fig. 9.20c–e, respectively. The derived rainwater field in Fig. 9.20c is seen to be similar to the reflectivity field in Fig. 9.20a, while the cloud water content in Fig. 9.20d is consistent with the vertical velocity field in Fig. 9.20b. Maxima of cloud water are located in the regions of updraft peaks. Saturation deficits (negative values) are found in the downdraft zones at low levels. The retrieved thermal field in Fig. 9.20e shows cold air with temperature perturbation ranging from 0°C to −5°C in the outflow pool. The 0°C isotherm has been superimposed on the vertical velocity pattern in Fig. 9.20b to illustrate how the updraft cells are triggered at the leading edge of the cold pool and rise over it as the air above the cold pool flows

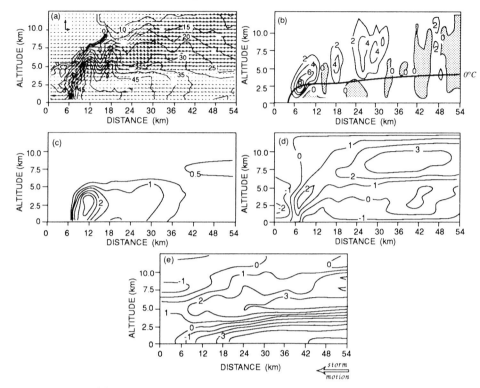

Figure 9.20   Airflow, water, and thermodynamic fields in the convective region of the west African tropical squall line with trailing stratiform precipitation shown in Figs. 9.7, 9.8, and 9.47. Airflow was determined by synthesis of dual-Doppler radar observations. The other fields were obtained from the flow pattern by the retrieval method. (a) Airflow (vectors; 15 m s^{-1} lengths indicated at upper left) and reflectivity (dBZ). (b) Vertical air velocity (m s^{-1}). (c) Precipitation water (rain below 6 km, graupel above) mixing ratio (g kg^{-1}). (d) Cloud water (positive) and saturation deficit (negative) mixing ratio (g kg^{-1}). (e) Temperature perturbation (K). (From Hauser *et al.*, 1988. Reprinted with permission from the American Meteorological Society.)

rearward. The cold air below the updrafts is seen to be sinking. Figure 9.20e shows that the air in the updraft zones has a positive temperature perturbation of up to 3°C. In cases in which the updrafts are stronger than those indicated in Fig. 9.20b, the thermal perturbation can be expected to be somewhat higher—perhaps twice the values seen in this example.

Another example of retrieved thermodynamic fields in the convective region of a tropical squall line system is shown in Fig. 9.21. The retrieved total buoyancy is in Fig. 9.21a. The total buoyancy is defined here as in (2.50), so that it contains the effect of condensed water as well as temperature, vapor, and pressure contributions. It is expressed here in temperature units by the quantity θ_a defined in (4.47). It was not decomposed further into its various components in this case. Thus, the condensate contributes negatively, counterbalancing some of the positive buoyancy associated with positive potential temperature and water vapor anomalies.

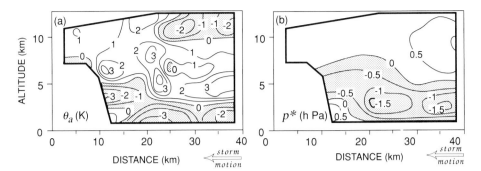

Figure 9.21 Retrieved thermodynamic fields for the tropical squall-line system shown in Fig. 9.16. (a) Buoyancy expressed in temperature units (K) by the quantity θ_a. (b) Pressure perturbation p^* (hPa). (From Roux, 1988. Reprinted with permission from the American Meteorological Society.)

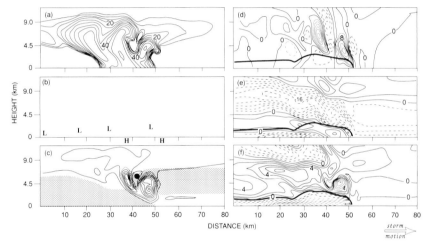

Figure 9.22 Convective portion of a two-dimensional numerically simulated squall-line system. (a) Simulated radar reflectivity (in intervals of 5 dBZ). (b) System-relative airflow. Vectors have been scaled such that one the length of the grid interval represents 14 m s^{-1}. Centers of low and high pressure are indicated by L and H, respectively. (c) Cloud water mixing ratio (contour intervals of 0.5 g kg^{-1} beginning with 0.1 g kg^{-1}. Shading indicates equivalent potential temperature $\theta_e < 327$ K. Large dot marks an air parcel that was traced. (d) Vertical air velocity in intervals of 2 m s^{-1} with negative (downward) values dashed. (e) System-relative horizontal wind perpendicular to the leading convective line in intervals of 4 m s^{-1} with negative (right-to-left) values dashed. (f) Potential temperature perturbation (K). Heavy solid contour in (b), (d), (e), and (f) outlines cold pool (region of negative potential temperature perturbation). (From Fovell and Ogura, 1988. Reprinted with permission from the American Meteorological Society.)

Hence, the total buoyancy peak, which was found to be ~4°C and collocated with the strongest updraft, reflects a temperature perturbation larger than this value since the weight of the condensate has been subtracted. A large value of radar reflectivity in the region of the maximum suggests a large negative contribution from the weight of the water. Thus, the maximum temperature perturbation could have been several degrees in excess of 4°C. Shown in Fig. 9.21b is the field of the pressure perturbation, which is seen to display a minimum directly below the region of positive buoyancy associated with the main convective cell. This re-trieved pattern is consistent with the aircraft-observed p^* pattern in Fig. 9.19.

9.2.2.3 Multicellular Aspect of the Convective Line and Cell Life Cycles

Since the data on winds, thermodynamic structure, and water content in me-soscale convective systems are rather limited, numerical cloud modeling of the type discussed in Sec. 7.5.3 again becomes an important tool in furthering the understanding of these cloud systems. Results of a two-dimensional numerical simulation of the convective portion of a squall-line system are shown in Fig. 9.22. This particular simulation did not include ice-phase microphysics. As a result, the trailing stratiform region is not well represented. Later in this chapter, we will see that when ice is allowed to occur in the model, a better representation of the stratiform region is obtained. Despite the absence of ice, the convective region is well represented in Fig. 9.22. The basic behavior of the model convective region is similar to that of the line of multicell storms discussed in Sec. 8.11. A series of convective cells, each undergoing the type of life cycle illustrated in Fig. 8.58, extends across the convective region. The cells appear as distinct cores of high radar reflectivity in Fig. 9.22a. The relative flow of air through the convective region resembles that in the observed cases shown in Figs. 9.17 and 9.20. From Fig. 9.22c, it can be seen that the slantwise upward flow over the cold pool consists of high-θ_e air originating in the boundary layer ahead of the storm. As this air first rises over the gust front, a buoyant updraft forms and develops into a cell near the gust front. According to Fig. 9.22d and f this updraft has a peak value of >8 m s^{-1} and a maximum potential temperature perturbation of about 8°C— values generally consistent with the magnitudes implied by radar-retrieval results in Figs. 9.20 and 9.21. (The case in Fig. 9.20 had thermal perturbations about half this large, but it appeared to be a weaker case than the modeled case or the case in Fig. 9.21.) Each cell is advected rearward in the slantwise relative flow. Thus, the second cell behind the leading edge of the system is older. It too is marked by a strong updraft and warm air. The third cell toward the rear is located at a still higher altitude than the first two cells and is weaker (in reflectivity, w, and thermal perturbation) since it is in a decaying stage. As illustrated in Fig. 8.58, low-θ_e air originally at midlevels ahead of the storm is pinched off between the first and second cells and is incorporated into the downdraft. The large dot in Fig. 9.22c identifies a parcel of low-θ_e air that was traced, which ultimately descended into the cold pool in the downdraft located between the first two cells. The convective-scale downdrafts located on either side of each convective updraft are a part of the convective structure associated with each cell. They correspond to the upper-level downdrafts seen between the convective updrafts in the Doppler radar–derived w field in Fig. 9.18a. The regions of downward motion seen below the second and third updraft cells in Fig. 9.22d are the lower-level, precipitation-driven downdrafts. The layer of rear-to-front flow seen below the layer of rearward-flowing updraft air in Fig. 9.22e is the descending rear inflow current depicted schematically in Fig. 9.13. It enters the convective zone from the stratiform region and descends into the cold pool below the intense convective cells.

Although the multicell sequence seen in Fig. 9.22 is taken from a two-dimensional model simulation, the results shown are very similar to the structure observed across the convective zone of a real squall line with a trailing stratiform region, especially if the cross section is taken parallel to the relative flow across a part of the line where the cells are well defined. However, there is a considerable amount of along-line variability in real squall line systems. A horizontal view from a three-dimensional simulation of a squall line with trailing stratiform precipitation is shown in Fig. 9.23. Along lines such as AB and CD in Fig. 9.23b, vertical cross sections of the results are very similar to those in Fig. 9.22. The horizontal maps in Fig. 9.23 show, however, that the only highly two-dimensional feature is the ascent forced at the gust front at low levels; in Fig. 9.23a, the updraft at the 1-km level is seen to be continuous in a narrow along-line band. The convective downdrafts to the rear of this leading updraft line are much more cellular. The continuous two-dimensional line structure of the updraft disappears quickly with height. By 2.8 km altitude (Fig. 9.23b), both the updrafts and downdrafts exhibit a highly cellular structure.

9.2.2.4   *Interaction with the Stratosphere*

We have now seen that the convective updraft cells in the convective region of the squall line intermittently form in the vicinity of the gust front, ascend, and travel rearward relative to the storm. A typical snapshot of the air motions looks like Fig. 9.17. When the updraft cells reach tropopause level, they perturb the base of the stratosphere. In doing so, they act as traveling mechanical oscillators that excite gravity waves in the nearly isothermal lower stratosphere. These waves are illustrated by the results of a model simulation[254] in Fig. 9.24. The model used is of the type discussed in Sec. 7.5.3 and is essentially similar to that used for the convective-line simulation illustrated in Fig. 8.58. In the case shown in Fig. 9.24, a base state is assumed that has no flow relative to the storm at stratospheric levels. The storm is shown in its mature stage, when the stratospheric response consists of waves whose phase lines tilt rearward and whose periods match the primary periods of the forcing by the tropospheric updrafts. Some very weak forward-tilting waves also occurred but can hardly be discerned in the figure. In the earlier stages of squall-line development, the convective updrafts generate more nearly equal amounts of forward- and rearward-propagating stratospheric gravity waves. The early period of the line is characterized by a single vertically oriented updraft that impulsively strikes the tropopause at one location, and both forward- and rearward-moving waves are excited. As discussed in Sec. 8.11, the line of convection evolves into a less vertically oriented configuration characterized by a series of weaker cells embedded in a rising, rearward-tilted airflow. The predominance of the rearward-propagating waves at later times

---

[254] Carried out by Fovell *et al.* (1992).

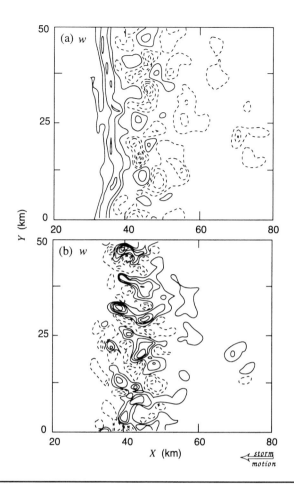

Figure 9.23

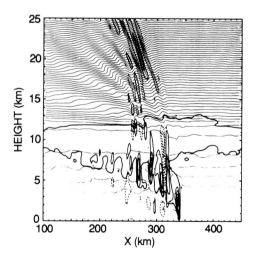

Figure 9.24

is attributed to the changeover to this latter flow configuration in which the cells impinging on the stratosphere are moving systematically rearward.

9.2.2.5   Mean Structure of the Convective Zone

When the numerical model results are averaged in time, much of the transient and spatially variable cellular structure in the convective region disappears, and the smooth patterns seen in Fig. 9.25 appear. The mean streamline pattern (Fig. 9.25b) consists of two dominant and clearly identifiable currents. The main up-draft current begins in the low to middle troposphere ahead of the storm, flows horizontally toward the storm, slopes up through the convective precipitation zone, and gradually becomes more horizontal toward the rear of the stratiform precipitation zone where it exits the storm at middle to upper levels. The main updraft current thus corresponds to the front-to-rear ascent shown in Fig. 9.13. In the upper reaches of the convective zone, a part of the main updraft current splits off and flows out of the forward side of the storm at upper levels. The second primary current seen in Fig. 9.25b is the rear inflow entering the back of the storm at the 3–4-km level and descending gradually into the cold pool in the convective region. At the surface below the rear unflow is a thin layer of front-to-rear flow.

9.2.2.6   Overturning of the Environment

The two primary mean currents seen in Fig. 9.25b, the front-to-rear ascent and the rear inflow, are two of the storm circulation features by which the storm overturns the convectively unstable prestorm environment. However, the over-turning process cannot be interpreted simply in terms of these average features. The air entering the front-to-rear ascent in the low to middle troposphere has a strong vertical gradient of θ_e (far right side of domain in Fig. 9.25c). When the low-θ_e air at the top of this inflow on the forward side enters the convective zone (at about 53 km on the horizontal scale in Fig. 9.25), it fuels convective downdrafts (as shown in Fig. 9.22c). On the other hand, the high-θ_e air at the bottom of the inflow layer runs into the leading edge of the cold pool, is forced above its level of free convection, and ascends in rapidly rising buoyant updraft cores. The convective region is thus a *crossover zone,* where the low to middle tropospheric air

---

Figure 9.23 Horizontal patterns of vertical velocity in the convective region of a three-dimensional simulation of a squall line with trailing stratiform precipitation. (a) 1-km level. Positive (solid) and negative (dashed) values are contoured respectively from 1 m s^{-1} with intervals of 2 m s^{-1} and from -0.5 m s^{-1} with intervals of 1 m s^{-1}. Zero contour not shown. (b) 2.8-km level. Positive (solid) and negative (dashed) values are contoured respectively from 2 m s^{-1} with intervals of 4 m s^{-1} and from -1 m s^{-1} with intervals of 2 m s^{-1}. Zero contour not shown. (From Redelsperger and Lafore, 1988. Reprinted with permission from the American Meteorological Society.)

Figure 9.24 Numerically simulated squall line with trailing stratiform precipitation. The calculated fields show gravity wave structure in the lower stratosphere (from about 13 km upward). Light lines are potential temperature isotherms (4 K intervals). The cloud outline is represented by a dark solid line, and heavy contours show vertical velocity (3 m s^{-1} intervals, downdrafts dashed). (From Fovell *et al.*, 1992. Reprinted with permission from the American Meteorological Society.)

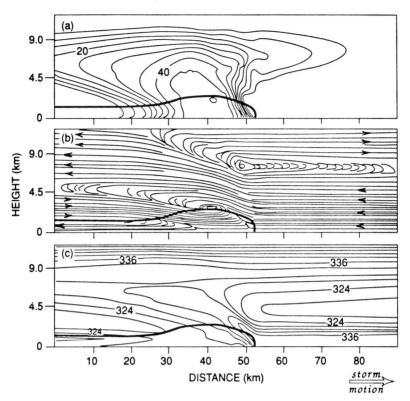

Figure 9.25 Time-averaged numerical-model simulation of a squall line with trailing stratiform precipitation. (a) Simulated radar reflectivity (in intervals of 5 dBZ). (b) Streamlines of system-relative airflow. (c) Equivalent potential temperture (intervals of 3 K). Heavy solid contour outlines cold pool (region of negative potential temperature perturbation). (From Fovell and Ogura, 1988. Reprinted with permission from the American Meteorological Society.)

entering the front side of the storm overturns in transient convective-scale drafts. This crossover is depicted in the schematic in Fig. 9.26. It is also represented in terms of model results in Fig. 9.27, where the crossover in the convective region is seen as a strong maximum-positive correlation of w and θ_e, which results from the transient convective updrafts and downdrafts having systematically positive and negative perturbations of θ_e, respectively. As a result of the crossover, θ_e is not conserved within the mean front-to-rear ascent. The stratification of θ_e seen in Fig. 9.25c changes markedly as soon as the front-to-rear flow encounters the cold pool. The maximum θ_e in the air finally flowing out the back edge of the domain at about 7 km altitude is <300 K, compared to a maximum of about 336 K in the low-level inflow.

9.2.2.7 Further Interpretation of the Pressure Perturbation Field

A general overview of the pressure perturbation field in the squall-line system was given in Sec. 9.2.1 (symbols L and H in Fig. 9.13). In Sec. 9.2.2.2, we examined the small-scale midlevel low that is found below the sloping updraft of the main convective cell in the convective region. Further aspects of the pressure perturbation field in the convective zone can be seen in the time-averaged model results. The pressure perturbation pattern corresponding to the time-averaged model circulation in Fig. 9.25 is shown in Fig. 9.28. The total perturbation p^* (Fig. 9.28a) again displays the small-scale low in the convective zone. It is the strong minimum at the top of the cold-pool head, just below the sloping front-to-rear ascent, as seen in the aircraft data composite of Fig. 9.19. In addition, there are three maxima of p^*: a small-scale peak of p^* at the surface at the leading edge of the cold-pool head, a shallow secondary maximum of p^* in the boundary layer directly below the cold-pool head, and a broad maximum aloft extending over the trailing stratiform region. In Fig. 9.28b and c, the total pressure perturbation is decomposed into its dynamic and buoyancy components p_B^* and p_D^*. From these additional cross sections, we can see the relative contribution of each component to the total pressure perturbation field in Fig. 9.28a. Comparison of Fig. 9.28a and b shows that the total field is dominated by the buoyancy component. The only exception is that the small-scale peak of p^* seen at the surface at the leading edge of the cold-pool head in Fig. 9.28a is not apparent in the buoyancy pressure-perturbation field. It is entirely a dynamically produced feature, which is strongly represented in Fig. 9.28c. This feature was described in Sec. 8.9 [Eq. (8.43)] as a dynamically produced feature that occurs where the low-level inflow current is stagnated and kinetic energy is converted to enthalpy. This interpretation is confirmed by its appearance only in the field of p_D^*. Another feature of interest in the p_D^* field is the pressure minimum that occurs in the middle of the cold-pool head. This minimum is associated with the vorticity centered at this location in the interface of the overlying rearward flow and the underlying rear-to-front flow in the cold pool. A vortex is centered at this point in the mean flow pattern shown in Fig. 9.25b. Such a vortex and an accompanying cyclostrophic pressure minimum were seen to be characteristic of a density-current outflow head in Sec. 8.9 [Eq. (8.44)]. From Fig. 9.28b and c, it is evident that this minimum of the dynamic pressure perturbation component lies just forward of the minimum in the buoyancy component, which is an effect of the deep layer of buoyant updraft air lying above this point. The superposition of the two minima leads to the broader, stronger minimum in the total field (Fig. 9.28a) than would be present from the effect of buoyancy alone. It is thus an oversimplification to think of the pressure minimum seen in data such as those in Fig. 9.19 as being either purely a hydrostatic or a purely dynamic pressure minimum.

9.2.2.8   *Influence of the Stratiform Region on the Convective Region*

The foregoing observations and model results have led to the conceptualization of the convective portion of the squall line with trailing stratiform precipitation in Fig. 9.29. The low-θ_e (dry) rear inflow air enters the convective region from the rear and descends into the cold-pool head, after passing through the trailing stratiform precipitation just below the base of the trailing stratiform cloud deck. A second source of low-θ_e air for the cold-pool head is the transient downdraft

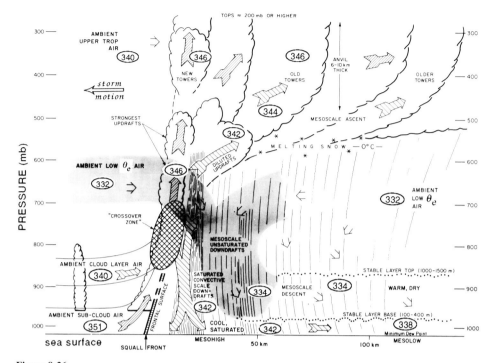

Figure 9.26

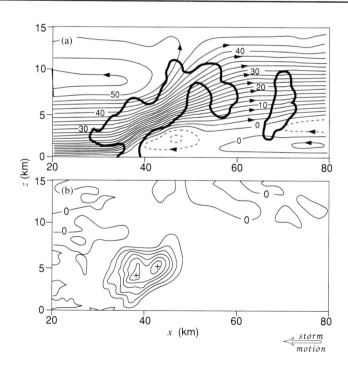

Figure 9.27

activity in the crossover zone, where dry air from ahead of the squall line also comes down into the cold-pool head (dashed arrow emanating from ahead of the storm in Fig. 9.29). In Sec. 8.11, the dynamics of the cold pool were discussed entirely in terms of this second source of cold-pool air. Model simulations, however, show that the rear inflow accounts for most of the mass of the cold-pool head, when the thunderstorm line is trailed by a stratiform precipitation region. In the case of the squall line with trailing stratiform precipitation, it is thus primarily the rear inflow that maintains the strength of the surface high in the center of the cold-pool head and thus the strength of the storm's density-current outflow. The dynamics of the cold pool should therefore be reconsidered in view of the primary source of its air being the rear inflow.

In Sec. 8.11, the maintenance of a line of multicell thunderstorms was discussed in light of the two-dimensional vorticity equation (2.61). If that equation (where the horizontal gradient of buoyancy is the only source of ξ) is integrated along a trajectory, such as the rear inflow trajectory of the conceptualized storm in Fig. 9.29, an expression for the horizontal vorticity ξ at any point along the streamline can be obtained. That vorticity is given by the upstream value of ξ, where the rear inflow enters the storm, plus the integrated effect of the baroclinic generation $(-B_x)$ experienced by the parcel of air moving along the trajectory. Since the rear inflow trajectory slopes downward through a region where there is positive buoyancy ahead and negative buoyancy to the rear, it accumulates vorticity as a result of baroclinity during the time it takes to reach the leading edge of the storm. This final vorticity contributes strongly to the vorticity budget of the cold pool and should be taken into account when considering a line of thunderstorms with a trailing stratiform region.[255]

The idealization of the flow components in the convective region, as depicted in Fig. 9.29, thus illustrates that the dynamics of the convective region of the squall line with trailing stratiform precipitation are not separable from the dynamics of the stratiform region. The rear inflow current arriving from the stratiform region is important to the cold-pool dynamics of the convective region, both as a mass

---

[255] See Lafore and Moncrieff (1989) for further discussion of this problem.

Figure 9.26 Conceptual model of a tropical oceanic squall line with trailing stratiform precipitation. All flow is relative to the squall line, which is moving from right to left. Numbers in ellipses are typical values of equivalent potential temperature (K). (Adapted from Zipser, 1977. Reproduced with permission from the American Meteorological Society.)

Figure 9.27 Updraft–downdraft crossover zone, as seen in the results of a two-dimensional numerical simulation of a squall line with trailing stratiform precipitation. (a) Stream function of two-dimensional airflow (relative units) with the outline of the region of strong positive correlation of w and θ_e perturbations superimposed (contour is for correlation of 0.6). (b) Vertical eddy flux of θ_e associated with w and θ_e perturbations in intervals of 5 K m s^{-1} with maxima indicated by a +. The crossover zone appears as the zone of strong eddy flux (implied by high correlation of w and θ_e) centered in the updraft region at $x \approx 40$ km and $z \approx 5$ km. (Adapted from Redelsperger and Lafore, 1988. Reproduced with permission from the American Meteorological Society.)

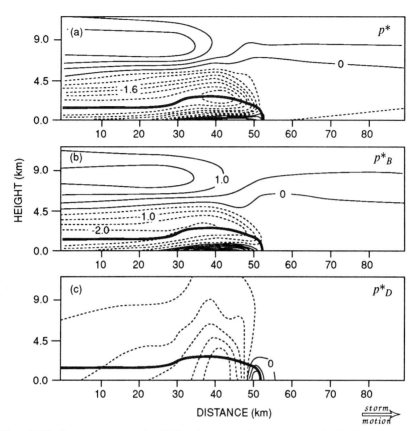

Figure 9.28 Pressure perturbation field in the convective zone as seen in the time-averaged results of a numerical-model simulation of a squall line with trailing stratiform precipitation. The patterns correspond to the fields shown in Fig. 9.25. (a) Total pressure perturbation in intervals of 0.4 mb. (b) Buoyancy component of pressure perturbation in intervals of 0.5 mb. (c) Dynamic component of pressure perturbation in intervals of 0.2 mb. Negative values dashed. (From Fovell and Ogura, 1988. Reprinted with permission from the American Meteorological Society.)

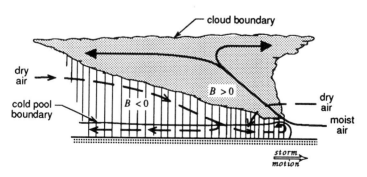

Figure 9.29 Conceptualization of the convective portion of the squall line with trailing stratiform precipitation, with emphasis on flow of water vapor into and out of the storm. B represents buoyancy. (Adapted from Fovell, 1990. Reproduced with permission from the American Meteorological Society.)

source, supplying the cold pool with negatively buoyant air to maintain the sur-face high-pressure area under the cold-pool head and thereby the strength of the surface outflow current (Fig. 9.29), and as a source of vorticity for the cold pool. In the next section, we examine the structure and dynamics of the stratiform region, of which the rear inflow is just one important aspect.

9.2.3 The Stratiform Region

We have seen in Secs. 9.2.1 and 9.2.2 that in the squall line with trailing stratiform precipitation the dynamics of the leading convective region are intimately tied to the trailing stratiform region, with the rear inflow current being the primary agent by which the stratiform region feeds back to the convection. When the circulation associated with the stratiform region appears and interacts with the convective region, the latter is no longer controlled just by the large-scale undisturbed envi-ronment but is controlled by the stratiform region dynamics as well. We will now examine the mesoscale circulation of the stratiform region, insofar as is possible from available observations and modeling results.

9.2.3.1 Upward Air Motion and Precipitation Development in the Stratiform Cloud

We begin by looking at the results of a model simulation. In Fig. 9.25, we saw the time-averaged structure of a model simulation of a squall line with a trailing stratiform region. A realistic convective region was simulated but no ice was allowed to form, and a realistic stratiform region did not appear. In Fig. 9.30, we show a time-averaged line structure for the same model but with a scheme for representing the ice-phase microphysics included. A bulk microphysical scheme of the type described in Sec. 3.6.2 was used in the water-continuity equations. The only form of precipitating ice allowed to form was snow.[256] As can be seen from Fig. 9.30a, a qualitatively realistic radar-echo pattern was obtained. The region of precipitation reaching the surface extends across a horizontal distance of about 80 km. A radar bright band and secondary maximum of surface precipitation inten-sity are found in the stratiform region. The structure of the precipitation is very similar to that in Fig. 9.13. The time-averaged kinematic and thermodynamic structure in Fig. 9.30b and c is not qualitatively different from that seen in the no-ice case in Fig. 9.25. There are some quantitative differences, however, and the case with ice included compares very well with data. For example, the radar reflectivity and vertical velocity fields derived from Doppler radar data in Fig. 9.31 are nearly identical to the model fields in Fig. 9.30a and e. The observed case in Fig. 9.31 was obtained as a composite of several volumes of Doppler radar obser-vations centered at different locations and times within the mesoscale system. All the data in a 60-km-wide strip oriented perpendicular to the convective line and extending across the system were combined, averaged, and filtered to obtain the

---

[256] Less realistic results were obtained when graupel was allowed to form because graupel falls out too quickly.

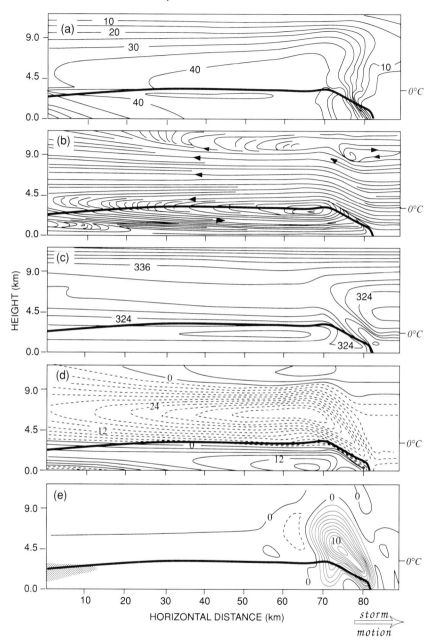

Figure 9.30 Time-averaged numerical-model simulation of a squall line with trailing stratiform region with ice-phase microphysics included: (a) Simulated radar reflectivity (in intervals of 5 dBZ). (b) Streamlines of system-relative airflow. (c) Equivalent potential temperature (intervals of 3 K). (d) System-relative horizontal wind (intervals of 3 m s^{-1}, solid left-to-right, dashed right-to-left); (e) Vertical air velocity (m s^{-1}); relative humidity <60% indicated by shading. Heavy solid contour outlines cold pool (region of negative potential temperature perturbation). (From Fovell and Ogura, 1988. Reprinted with permission from the American Meteorological Society.)

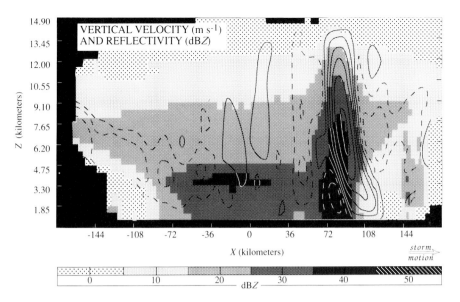

Figure 9.31 Composite radar reflectivity and vertical velocity fields constructed from Doppler radar observations obtained at different locations and times within a squall line with trailing stratiform precipitation. Other aspects of this storm are illustrated in Figs. 9.15, 9.18, 9.34, 9.40, 9.41, 9.43, 9.44, 9.45, 9.48, 9.53, and 9.54. All the data in a 60-km-wide strip oriented perpendicular to the convective line and extending across the system were combined, averaged, and filtered to obtain the mean cross section. X is the coordinate axis perpendicular to the line. The storm was moving from left to right. Radar reflectivity (dBZ) is shown by shading. Vertical wind component is shown by contours for −0.9, −0.45, −0.15, 0.15, 0.45, 0.9, 1.5, 2.4, and 3.6 m s^{-1} with negative values dashed. (From Biggerstaff and Houze, 1993. Reproduced with permission from the American Meteorological Society.)

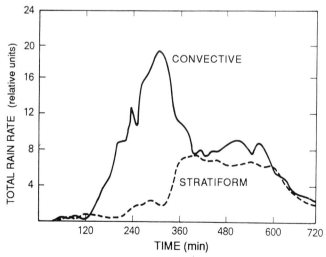

Figure 9.32 Area-integrated rain rate in a two-dimensional numerical-model simulation of a squall line with trailing stratiform region with ice-phase microphysics included. Rain amounts are summed separately for rain falling at grid points designated as convective and stratiform for each time step throughout the lifetime of the storm. (From Tao and Simpson, 1989. Reprinted with permission from the American Meteorological Society.)

mean cross section. The similarity with the model results in Fig. 9.30a and e is striking. However, the model version of the storm structure is scaled down. It is not as broad or tall as the real case. The reason for the difference in scale remains to be understood. In most other ways though, the model results are verified.

In both the time-mean model cross section in Fig. 9.30e and the example composite observed case (Fig. 9.31), the vertical air motion is seen to be positive nearly everywhere at mid- to upper levels in the stratiform region. The only exception is in the zone of minimum radar reflectivity between the convective region and the secondary maximum of stratiform precipitation below the bright band, where there is a column of downward motion. This zone is where the upper-level downdrafts tend to occur behind the strongest updraft cells (as seen in Fig. 9.18). The net effect of these strong upper-level downdrafts on the average vertical velocity field is to produce the column of mean downward motion. The general ascent found at mid- to upper levels behind the zone of mean downward motion is typical of squall lines with trailing stratiform precipitation. Generally, the mean vertical motion is zero at a height of 0–2 km above 0°C, upward below this level (corresponding to the front-to-rear ascent in Fig. 9.13), and downward below (corresponding to the subsiding rear inflow in Fig. 9.13). To understand the precipitation process in the stratiform region, it is important to note the *magnitude* of the vertical motions in the stratiform region. They are generally (though not uniformly) <0.5 m s^{-1}, whereas the fall speeds of precipitating ice particles (ice crystals, aggregates, graupel) are ~0.3–3 m s^{-1} (Sec. 3.2.7, Figs. 3.12, 3.13, and 3.14). Thus, condition (6.1) is satisfied, and the precipitation is of a stratiform character, as described in Sec. 6.1.

Results from another model simulation are shown in Figs. 9.32 and 9.33. This simulation is very similar to that illustrated in Fig. 9.30. A bulk parameterization of the ice-phase microphysics was used in the calculation and a region of stratiform precipitation formed behind the convective line. For the first 2 h the precipitation was primarily from the leading convective line, while during the last 4 h the amounts of convective and stratiform rain were about equal (Fig. 9.32). Integrated over the lifetime of the storm, about 37% of the precipitation from the storm was stratiform. This behavior compares well with the typical observed evolution and relative amounts of convective and stratiform precipitation illustrated by the example in Fig. 9.12. The observed case, however, extends over a considerably longer time period than the model case.[257] The structure of the model storm, during the time when the total convective and stratiform rain rates were approximately equal, is shown in Fig. 9.33. Surface precipitation extends over a 100-km-wide region, and the stratiform region is well represented with a radar bright band at the melting level.

Although the two model simulations we have referred to in Figs. 9.30, 9.32, and 9.33 produced storms of somewhat smaller spatial and temporal scales than their observed counterparts, they produced stratiform precipitation areas qualitatively

---

[257] The case in Fig. 9.12 was particularly long-lived. Lifetimes ~12 h are more common for squall lines with trailing-stratiform precipitation.

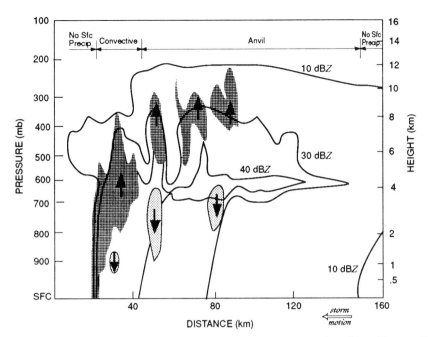

Figure 9.33 Vertical cross section of the numerically simulated squall line with trailing stratiform precipitation whose rain rate is indicated in Fig. 9.32. The structure of the storm is shown for a time (504 min after the beginning of the simulation) when the total convective and stratiform rain rates were approximately equal. The simulated radar reflectivity is shown in contours of dBZ. In the darker shaded areas, $w > 0.5$ m s^{-1}, and in the lighter shaded areas, $w < -0.5$ m s^{-1}. (From Tao and Simpson, 1989. Reprinted with permission from the American Meteorological Society.)

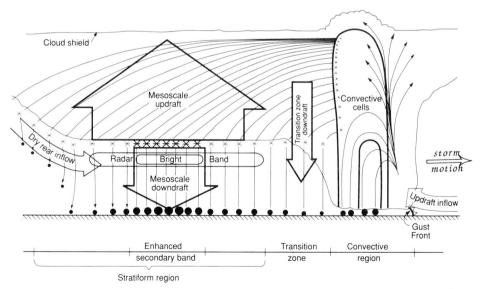

Figure 9.34 Schematic of the two-dimensional hydrometeor trajectories through the stratiform region of a squall line with trailing stratiform precipitation. Trajectories were based on fall speeds and air motions measured by Doppler radar. Other aspects of this storm are illustrated in Figs. 9.15, 9.18, 9.31, 9.40, 9.41, 9.43, 9.44, 9.45, 9.48, 9.53, and 9.54. (From Biggerstaff and Houze, 1991a. Reproduced with permission from the American Meteorological Society.)

quite similar to those observed. We have seen this similarity in both the structure of their radar reflectivity and the evolution and amount of surface precipitation, and we have noted that this similarity is achieved only when the ice-phase microphysics are included in the model simulations. Thus, it must be concluded that the production of ice in the upper levels of the convective region and its advection rearward by the front-to-rear ascent are crucial to the formation of the trailing stratiform region. As indicated by the horizontal wind pattern in Fig. 9.30d, slowly falling snow particles grown in the upper levels of the convective cells in the convective region are advected by front-to-rear flow in the manner depicted schematically in Figs. 6.11f and 9.13.

A more detailed schematic of these processes is provided in Fig. 9.34. The trailing nimbostratus cloud deck is supplied with precipitation-sized ice particles, which are spread across the trailing region. These particles fall (at speeds of ~ 1–3 m s^{-1}) through the "mesoscale updraft" zone of weak vertical air motions. They first cross the region of mean downward motion just behind the convective region. In this "transition zone," the mass of precipitation particles cannot increase, and the particles partially evaporate. The more slowly falling particles from upper levels are, however, quickly advected across the region of downward motion into the region of the mesoscale updraft, where they grow significantly by vapor deposition while continuing their slow descent to the melting layer. The upward air motions of the order of tens of centimeters per second are great enough to condense a considerable amount of vapor but small enough for all the ice particles to fall [condition (6.1) is met]. The precipitation is stratiform in character. Before reaching the melting layer, and possibly within it, the ice particles may aggregate and by virtue of the large particles produced enhance the radar reflectivity intensity in the bright band. The aggregation, however, does not increase the bulk mass of the falling snow; it only changes the particle size distribution. Once melted, the ice particles fall to the ground as rain.

As shown in Fig. 9.34, the maximum of the stratiform rain rate at the surface below the mesoscale updraft may be attributed to the enhancement of the stratiform precipitation by the mesoscale updraft above the melting layer. The minimum in rain rate and radar reflectivity at low levels between the convective region and the enhanced stratiform rain below the mesoscale updraft is associated with particles falling out of convective cells but not being able to grow effectively in the transition-zone downdraft region.

9.2.3.2   Thermodynamic Structure
of the Stratiform Region

The thermodynamic structure of the stratiform region is also indicated by model simulations. The perturbation fields of potential temperature, water-vapor mixing ratio, and pressure for the time-averaged simulation of Fig. 9.30 are shown in Fig. 9.35. A major feature is that the stratiform region is marked by a potential temperature perturbation maximum of $\sim 5°C$ all the way to the back edge of the storm (Fig. 9.35a). This feature is accompanied by a positive water vapor perturbation (Fig. 9.35b). Lying below this warm, positively buoyant layer is a cold, negatively buoyant layer, characterized by potential temperature perturbations of the order of several degrees Celsius. Sandwiched between the warm and cool layers is the pronounced, mainly hydrostatic pressure perturbation minimum (Fig. 9.35c).

The model thermodynamic fields correspond well to data analyses. Thermodynamic fields obtained by the Doppler radar retrieval technique (Sec. 4.4.7) are shown in Figs. 9.36–9.38. Figure 9.36 cuts across the leading portion of the stratiform region, just behind the convective region (which lies just to the left of the cross section). Figure 9.37 is in the middle to rear portion of the stratiform region—the bright band is evident all across this region but decreasing in intensity toward the rear of the system (i.e., toward higher x). Figure 9.38 is located near the back edge of the stratiform region. In all three figures, the vertical air motion is upward above about 4 km altitude and downward below. Inspection of panel b in each figure shows that the mesoscale updraft is ~50 cm s^{-1} at the back edge of the convective region, ~20 cm s^{-1} in the center of the stratiform region, and ~0–2 cm s^{-1} at the back edge of the system. The mesoscale downdraft at lower levels has a maximum magnitude of ~25 cm s^{-1} (Fig. 9.36b).

The c and d panels of Figs. 9.36–9.38 show the thermodynamic structure of the stratiform region. It is evident that the mesoscale updraft region aloft is positively buoyant and that a pressure perturbation minimum lies at the base of the mesoscale updraft zone aloft all across the stratiform region. This diagnosed pattern is consistent with the model results in Fig. 9.35, where the pressure perturbation minimum was seen to lie at the base of the positively buoyant layer of front-to-rear ascent atop the cold pool. The magnitude of the total buoyancy perturbation, expressed in temperature units by the quantity θ_a in the d panels of Figs. 9.36–9.38, has peak values of 2–4°C. Negative buoyancy of similar magnitude is seen in the mesoscale downdraft at lower levels. In the stratiform region, where the loading of liquid water and ice is not great, this total buoyancy perturbation approximates the potential temperature perturbation itself and may therefore be compared with the model results in Fig. 9.35a, where the peak values are 4–5°C. Since the model represents a midlatitude storm while the retrieval results are for a tropical case, and there are many uncertainties in both the model and retrieval methodologies, we may take these values as being essentially in agreement.

A conceptual model shown in Fig. 9.39 ties together the buoyancy of the trailing stratiform region, the pressure perturbation minimum at the base of the buoyant cloud, and the rear inflow current. As we have seen from Figs. 9.35–9.38, the trailing stratiform cloud layer is positively buoyant. Immediately to the rear of the convective region, the cloud layer is thicker and the buoyancy is greater. The buoyancy decreases and the cloud layer thins toward the rear of the system. Hence the magnitude of the pressure perturbation at the base of the buoyant layer is greater in the interior of the system, just behind the convective region. The resulting horizontal difference of pressure perturbation across the stratiform region accelerates air from rear to front. This mesoscale difference of pressure perturbation (Δp in Fig. 9.39) drives the rear-to-front flow across the stratiform region. As ice particles fall from the stratiform cloud layer aloft into the rear-to-front current, the air is chilled by sublimation, melting, and evaporation and subsides while it flows horizontally toward the convective region.

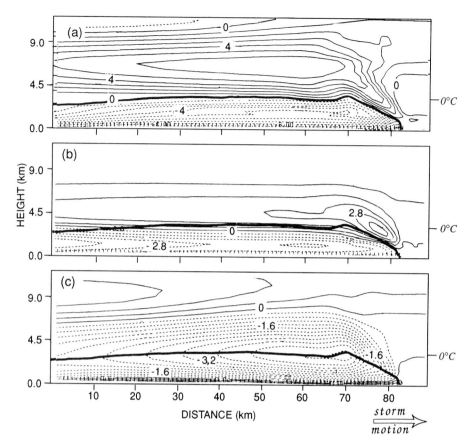

Figure 9.35

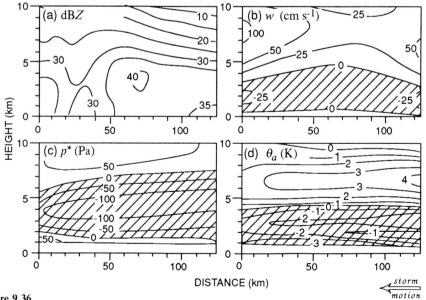

Figure 9.36

9.2.3.3 The Mesoscale Downdraft

Mesoscale convective systems tend to form in air masses characterized by low relative humidity (Sec. 3.1.1) and a minimum of θ_e in the midtroposphere. Convergence of air at midlevels, below the base of the positively buoyant upper-level stratiform cloud deck, brings this dry, low-θ_e air into the storm. The sublimation and/or evaporation of precipitation particles falling into the low-θ_e air makes the air just below cloud base negatively buoyant. In addition, melting of the ice particles falling from the upper-level cloud produces a cooling of ~ 1–$10°C$ h^{-1} in the ~ 0.5–1-km-thick melting layer (marked by the radar bright band in Fig. 9.12).[258] The response to these cooling effects is the formation of a mesoscale downdraft in the region of stratiform precipitation. Such a downdraft is found in most well-developed mesoscale convective systems, not just the squall line with trailing stratiform precipitation being considered here. In this particular type of system, the ascending front-to-rear flow spreads ice particles across the trailing portion of the storm (Fig. 9.13). A wide (~ 100 km) lower tropospheric region of negative buoyancy extends across the entire rear portion of the storm, and a *mesoscale downdraft* develops across this region.

The vertical profile of the vertical velocity w in the mesoscale downdraft (and the mesoscale updraft above) exhibits considerable similarity from one mesoscale system to another. Figure 9.40 shows the vertical profiles of w seen at three locations in the stratiform region of the storm shown in Figs. 9.15, 9.18, 9.31, 9.34, 9.41, 9.43, 9.44, 9.45, 9.48, 9.53, and 9.54. The maximum magnitudes of downward motion in the lower troposphere were 45–65 cm s^{-1}, while the peak upward motion at upper levels was 50–60 cm s^{-1}. Curves 1 and 2 were obtained near the northern edge of the stratiform region (Fig. 9.15). The top of the mesoscale downdraft was 2 km above the 0°C level in these two curves, while in curve 3, obtained farther south, nearer the center of the stratiform region, the top of the descent was at the 0°C level. These profiles are quite typical; the level separating the upward from downward motion is usually 0 to 2 km above the 0°C level, depending on

---

[258] In a study of five tropical squall lines, Leary and Houze (1979b) determined that the cooling rate by melting ranged from 1 to 7°C h^{-1}.

Figure 9.35 Time-averaged numerical-model simulation of a squall line with a trailing stratiform region with ice-phase microphysics included. (a) Potential temperature perturbation (K). (b) Water-vapor mixing ratio perturbation (intervals of 0.7 g kg^{-1}). (c) Pressure perturbation (intervals of 0.4 mb). (From Fovell and Ogura, 1988. Reprinted with permission from the American Meteorological Society.)

Figure 9.36 Thermodynamic fields obtained by the Doppler radar retrieval technique for the trailing stratiform region of the tropical squall line whose convective structure was shown in Fig. 9.20. Cross section is through the leading portion of the stratiform region, just behind the convective region, which lies just to the left of the cross section. (a) Radar reflectivity. (b) Vertical air velocity. (c) Pressure perturbation. (d) Buoyancy expressed in temperature units (K) by the quantity θ_a. (From Sun and Roux, 1988.)

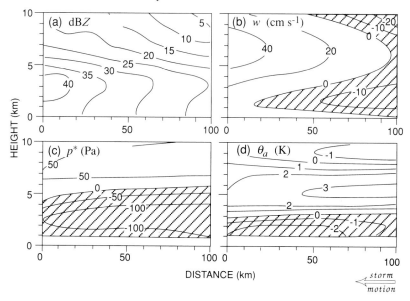

Figure 9.37

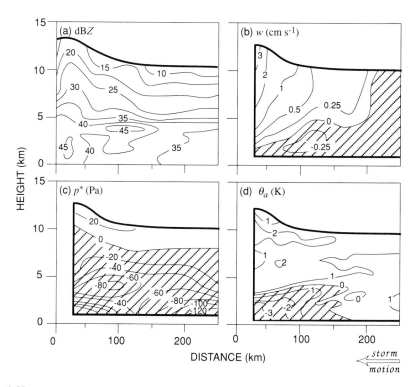

Figure 9.38

location within the stratiform region.[259] The upward velocity of ~ 10 cm s^{-1} near the ground in curve 2 appears because this profile was taken in the vicinity of the wake low (Sec. 9.2.3.5).

The association of the mesoscale downdraft with the melting layer is seen in another way in Fig. 9.41. Panel a displays areal mean vertical velocity profiles for both the convective and stratiform regions of the same mesoscale convective system for which vertical profiles for specific locations are shown in Fig. 9.40. In panel b of Fig. 9.41, the stratiform region profile is decomposed into three subregions. Curves A and B represent the core of heaviest stratiform precipitation (located in the "secondary maximum" of radar echo in the trailing stratiform region, as described schematically in Fig. 9.9, Sec. 9.1.2.2). They represent the region of stratiform precipitation with low-level radar reflectivity exceeding 25 dBZ for at least 2.5 h. This precipitation is that lying directly under the strong radar bright band seen in Fig. 9.31. Curve A is for the heaviest stratiform rain within this zone (radar echo >35 dBZ for more than 1.5 h), while curve B represents core stratiform rain of more intermediate intensity (25–35 dBZ for more than 2.5 h). Curve C is for the lighter stratiform region lying outside the secondary maximum. From these three curves it is seen that the mesoscale downdraft is concentrated within the zone of heaviest stratiform rain, namely that associated with the well-defined melting layer marked by the bright band (curves A and B). The mesoscale downdraft disappears in the average outside of the core of heavy stratiform rain (curve C). At the same time, the mesoscale updraft appears with the same intensity in all three subregions. Evidently, the horizontal scale of the mesoscale downdraft is set by the width of the region of strong melting, while the mesoscale updraft at higher levels is of a larger horizontal scale. It extends rather uniformly across the entire stratiform region, as the positively buoyant air from the upper reaches of the convective regions is spread across the stratiform region by the front-to-rear flow at upper levels.

---

[259] See Sec. 4a of Houze (1989) for further examples, including vertical profiles of w inferred from sounding analysis, Doppler radar, and profiler measurements in tropical, midlatitude, continental, and oceanic mesoscale convective systems.

Figure 9.37 Thermodynamic fields obtained by the Doppler radar retrieval technique for the trailing stratiform region of the tropical squall line whose convective structure was shown in Fig. 9.21. Cross section is through the middle to rear portion of the stratiform region. (a) Radar reflectivity. (b) Vertical air velocity. (c) Pressure perturbation. (d) Buoyancy expressed in temperature units (K) by the quantity θ_a. (From Sun and Roux, 1988.)

Figure 9.38 Thermodynamic fields obtained by the Doppler radar retrieval technique for the trailing stratiform region of a tropical squall-line system observed by dual-Doppler radar in Ivory Coast, west Africa, on 27–28 May 1981. Cross section is located near the back edge of the stratiform region. (a) Radar reflectivity. (b) Vertical air velocity. (c) Pressure perturbation. (d) Buoyancy expressed in temperature units (K) by the quantity θ_a. (From Sun and Roux, 1989. Reprinted with permission from the American Meteorological Society.)

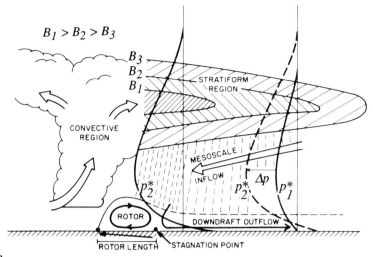

Figure 9.39

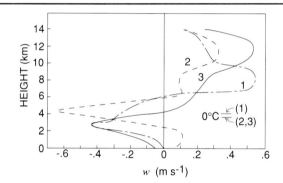

Figure 9.40

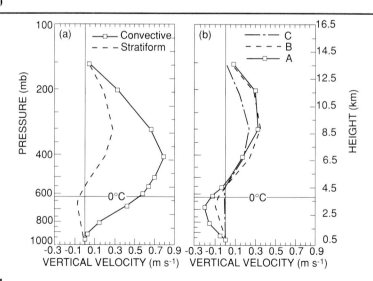

Figure 9.41

9.2.3.4 Kinematic and Thermodynamic Structure at the Top of the Stratiform Cloud

So far we have examined only the portion of the trailing stratiform cloud from which precipitation is falling. One reason for this is that radar observations are restricted to regions occupied by precipitation particles. Rawinsonde data can be used to examine the thermodynamic structure and winds in a region that extends outside the precipitation zones. However, the time and space resolution of sounding data is typically very coarse. Sometimes by making time-to-space conversions it is possible to form a composite analysis of sounding data taken in and around a storm over a period of time in which the storm was not changing its structure rapidly. Some results from such a sounding composite are shown in Fig. 9.42. The cross sections are divided into four regions: the environment ahead of the squall line, the convective precipitation region, the trailing stratiform rain area, and the poststratiform rain area. The convective and stratiform rain regions were characterized by temperature and water-vapor perturbation fields generally consistent with the results seen in Figs. 9.35–9.38. The maximum magnitude of the peak temperature perturbation in the stratiform cloud is, however, only ~1°C, which is smaller than indicated in previous discussions. However, the sounding data are of low resolution and were put into a composite coordinate framework and objectively analyzed. All of these factors contribute to smoothing out the perturbations. Thus, these fields are not really inconsistent with those shown previously. What is gained from these lower-resolution analyses is some knowledge of the regions immediately ahead of, above, and to the rear of the convective and stratiform precipitation areas, which cannot be obtained readily from radar. The main new feature seen in this analysis is in the stratiform cloud region at upper levels, toward the back of the storm in the rear half of the stratiform region and in the poststratiform zone. A negative temperature perturbation is found, which may have been the result of radiative cooling in the upper layers of the trailing stratiform cloud.

Figure 9.39 Schematic vertical cross section illustrating the relationship of the buoyancy (B) of the trailing stratiform cloud to the pressure perturbation (p^*). The difference between the pressure perturbation at the back (p_1^*) and leading portion (p_2^*) of the stratiform region is indicated by Δp. (Adapted from Lafore and Moncrieff, 1989. Reproduced with permission from the American Meteorological Society.)

Figure 9.40 Stratiform region vertical velocity profiles for a squall line with trailing stratiform precipitation. All three profiles were derived by single-Doppler radar analysis (using the VAD method described in Sec. 4.4.5). Other aspects of this storm are illustrated in Figs. 9.15, 9.18, 9.31, 9.34, 9.41, 9.43, 9.44, 9.45, 9.48, 9.53, and 9.54. (From Houze, 1989. Reprinted with permission from the Royal Meteorological Society.)

Figure 9.41 Mean vertical air motion in a squall line with trailing stratiform precipitation. In (a), the curves are for the convective (solid) and entire stratiform (dashed) regions. In (b), curves are shown for the stratiform region excluding the secondary band C, the secondary band excluding the enhanced portions B, and enhanced portions of the secondary band A. The air motions were derived by dual-Doppler radar observation. The height of the 0°C level is shown. The storm in which the data were taken is the same as that represented in Figs. 9.15, 9.18, 9.31, 9.34, 9.40, 9.43, 9.44, 9.45, 9.48, 9.53, and 9.54. (From Biggerstaff and Houze, 1991a. Reproduced with permission from the American Meteorological Society.)

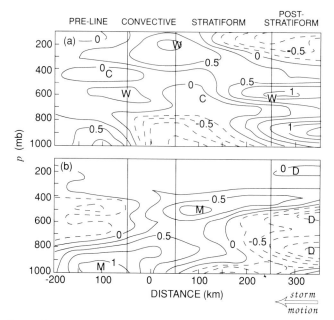

Figure 9.42 Composite analysis of radiosonde data obtained in and around a tropical squall-line system over the eastern Atlantic Ocean. Cross sections are along a line perpendicular to the leading convective region. (a) Temperature perturbation (K); maxima indicated by W, minima by C. (b) Water vapor mixing ratio perturbation (g kg^{-1}); maxima indicated by M, minima by D. Other aspects of this storm are illustrated in Figs. 9.50 and 9.59. (From Gamache and Houze, 1985. Reproduced with permission from the American Meteorological Society.)

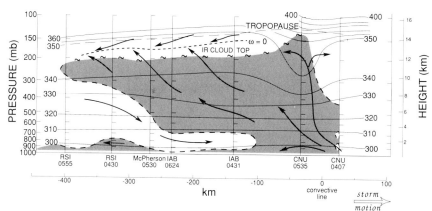

Figure 9.43 Analysis emphasizing the structure at and above the top of the squall line with trailing stratiform precipitation shown in Figs. 9.15, 9.18, 9.31, 9.40, 9.41, 9.44, 9.45, 9.48, 9.53, and 9.54. This cross section is taken along the direction of motion of the squall line during the system's later stages of development. Shown are the streamlines for system-relative flow, potential temperature isotherms (K), and relative humidity (with respect to ice at temperatures below freezing, values greater than 80% indicated by shading). Infrared (IR) satellite data were used to determine the height at cloud top. Airflow and thermal structure are deduced from soundings and assumptions about the infrared radiative cooling rate at cloud top. Dotted line indicates where vertical air motion (expressed as $\omega \equiv Dp/Dt$) is diagnosed to have been zero. (From Johnson *et al.*, 1990. Reprinted with permission from the American Meteorological Society.)

Another feature of the mesoscale trailing stratiform region for which there appears to be some observational support in rawinsonde data is a thin layer of downward air motion in the front-to-rear flow at and above cloud top (Fig. 9.43).[260] It has been hypothesized that this feature of the trailing region is associated with the tropopause settling back downward after being displaced upward by the propagating convective zone. The downward settling of the stratiform cloud top may also be a manifestation of the outflow layer from the cumulonimbus clouds in the leading convective line undergoing a "collapse" similar to that described in Sec. 5.3.3 for the upper-level outflow from an individual cumulonimbus (Figs. 5.32 and 5.33).

9.2.3.5 The Wake Low

The sounding analyses in Fig. 9.42 show that the major feature immediately to the rear of the stratiform precipitation zone is a strong positive-temperature maximum at low levels. This warming produces the wake low at the back edge of the stratiform precipitation (Fig. 9.13). A schematic view of the process producing the low is given in Fig. 9.44, which reflects the opinion that the wake low is a surface manifestation of the unsaturated descent of part of the rear inflow air. The warming appears to be maximized at the back edge of the stratiform precipitation area, where the air is subsiding especially strongly and there is insufficient evaporative cooling to offset the strong adiabatic warming.

9.2.3.6 The Rear-to-Front Current and Rear Inflow

Sounding data to the rear of the stratiform precipitation region have also extended knowledge of the rear inflow. Although the schematic in Fig. 9.13 shows descending rear inflow entering the storm at the back of the stratiform region, it does not indicate how *strong* the rear inflow is. Figure 9.45 shows an example of an especially strong rear inflow current, which is shown by Doppler radar data to be entering the back edge of the stratiform precipitation radar echo at a speed exceeding 15 m s^{-1} relative to the storm. This speed, however, is rather exceptional. When the sounding data from the environment to the rear of many squall lines with trailing stratiform precipitation are examined, it is found that only a few cases have strong inflow from the environment. Profiles of the storm-relative flow from soundings taken to the rear of such cases have been averaged and are referred to in Fig. 9.46 as strong rear inflow cases. The example in Fig. 9.45 was one of the cases included in the average. Many cases have weaker rear inflow at midlevels; many have essentially zero relative flow. These latter cases have been referred to as stagnation zone cases. All of the cases, however, have similar shear. The relative flow normal to the line tends to be front-to-rear at both upper and lower levels, with the front-to-rear flow decreasing at midlevels. The rear inflow cases are the ones in which the sign of the relative flow in the midlevel minimum is reversed, so that environmental air flows across the back edge of the precipitation area. Roughly half of the storms which have been studied are stagnation zone

---

[260] This feature has also been noted in profiler data obtained in the tropics (Balsley *et al.*, 1988).

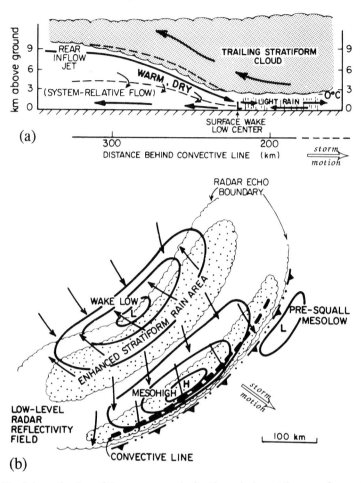

Figure 9.44   Schematic view of the process producing the wake low at the rear of a squall line with trailing stratiform precipitation. (a) Vertical cross section through wake low. (b) Plan view of surface winds and precipitation. Winds in (a) are system relative, with dashed line denoting zero relative wind. Arrows indicate streamlines, not trajectories, with those in (b) representing ground-relative wind. Note that horizontal scales are different in the two schematics. Other aspects of this storm are illustrated in Figs. 9.15, 9.18, 9.31, 9.34, 9.40, 9.41, 9.43, 9.45, 9.48, 9.53, and 9.54. (From Johnson and Hamilton, 1988. Reprinted with permission from the American Meteorological Society.)

cases. These cases are dominated by, but not restricted to, tropical squall lines, while the strong rear inflow cases have been found in midlatitudes over land.

In the stagnation zone cases, the pattern of divergence and vertical velocity of the mesoscale circulation in the trailing stratiform region does not appear to be essentially different than in the strong rear inflow cases. The ascending front-to-rear flow in the stratiform cloud itself remains in the form illustrated by Fig. 9.13, and midlevel convergence is implied in these cases by the zone of stagnation lying adjacent to the upward-sloping rearward relative flow. As in the rear inflow cases, this convergence is linked to both the mesoscale ascent in the cloud layer above and to mesoscale descent below the cloud layer. It has also been found, from aircraft and radar observations in stagnation zone cases, that rear-to-front flow develops *internally* (i.e., within the trailing rain area itself, storm frontward flow develops and penetrates forward and into the cold pool of the convective region, even though no air enters from the environment at the rear of the storm). An example of this type of storm is the tropical squall line that we have examined previously in Figs. 9.7, 9.8, and 9.20. In Fig. 9.47 we see that the profile of the relative wind component normal to the line was strongly frontward just to the rear of the convective region but decreased to a stagnation value at the back edge of the stratiform region. Thus, it is clear that the squall line system generates its own rear-to-front flow in the stratiform region, but that in some cases (especially the strong rear inflow cases) this frontward flow is enhanced by inflow from the environment.

The inflow from the environment in a strong rear inflow case has been studied in the context of a *mesoscale model*, which is a limited-area numerical weather prediction model—not a convective cloud model of the type described in Sec. 7.5. The convective cloud model starts with a single sounding (i.e., uniform and steady-state large-scale environment) and uses nonhydrostatic equations to predict the motions internal to a cloud which develops from a specified perturbation to the otherwise constant environment. The mesoscale model, in contrast, is concerned with predicting changes in the large-scale environment over an area much larger than that of the individual convective clouds. The environment may be initially nonuniform and vary with time, as predicted by the mesoscale model. The horizontal grid resolution of the model is typically 15–75 km. A variety of approaches to this type of model have been used. Often, the motions are calculated with the hydrostatically balanced equations and the convective-scale motions are parameterized, which means that the net effect of the convective-scale motions on the larger-scale environment is included in the turbulence terms in the equation of motion and the thermodynamic equation. These terms are expressed as functions of the resolvable-scale variables. A typical approach is to assume that, if the environment becomes conditionally or potentially unstable and sufficient resolvable-scale vertical motion is present to release the instability, the thermodynamic stratification is overturned by the convection, according to some prescribed scheme, to restore stability. In the case of a mesoscale convective cloud system, such as a squall line with a trailing stratiform region, this type of model resolves the air motions and cloud structure of the stratiform region, while the motions in the smaller-scale updrafts and downdrafts of the convective region must be represented by the parameterization scheme.[261]

In Fig. 9.48a, we show results of a mesoscale model simulation of a midlatitude squall line with a trailing stratiform region. This particular simulation used a nested grid, which was 75 km in resolution except in the vicinity of the convection, where the grid spacing was reduced to 25 km. The case represented is the

---

[261] See Frank (1983) for reviews of traditional methods of convective parameterization. More recent ideas are discussed by Betts (1986) and Frank and Cohen (1987). Still more recently, mesoscale models have replaced the parameterization approach with nested grids that allow explicit nonhydrostatic calculation of the convective-scale processes.

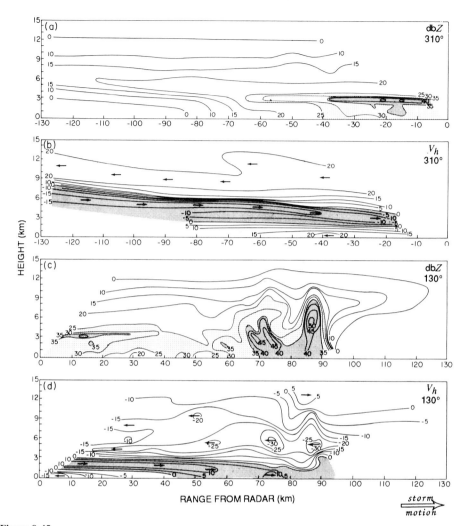

Figure 9.45

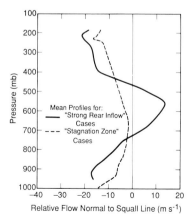

Figure 9.46

one for which the rear inflow was shown in Fig. 9.45. Other aspects of the structure of the storm have been presented in several other figures (see caption of Fig. 9.48). It was a strong rear inflow case. Comparison of Figs. 9.45 and 9.48a shows that the model represents the general structure quite well, with the rear inflow jet entering strongly from the environment to the rear of the stratiform region. From the horizontal scale of Fig. 9.48 it is evident that the model results cover a region much broader than the precipitation region shown in Fig. 9.45. Thus, it is evident that the strong rear inflow extends far out into the environment to the rear of the storm. Several model runs were made, under different sets of assumptions. These experiments reveal that, in this case, the inflow from the environment at upper levels at the back of the system was not pulled in by processes internal to the storm itself but rather was associated with the baroclinity of the large-scale flow in which it was embedded. (The squall line lay ahead of a well-defined short-wave trough in the westerlies.) With the diabatic heating set equal to zero, which is tantamount to removing all the dynamical effect of the cloud system on the resolvable-scale flow, only the upper-level (above 400 mb) part of the rear inflow (i.e., that to the rear of the rain area) developed. The mid- to lower-level part of the rear inflow (i.e., that within the rain area) was absent. This result is illustrated by Fig. 9.48b, where the result with no diabatic heating has been subtracted from the total result in Fig. 9.48a. All of the rear inflow internal to the storm remains, where the frontward flow at midlevels descended toward the convective region. The rear inflow entering from the environment at the rear of the storm is entirely absent from the difference field. Thus, it appears that the part of the rear inflow that made this case one of strong rear inflow (i.e., entering across the back edge of the storm) was not driven by processes internal to the storm but rather was determined by the large-scale environment in which the storm was embedded. The large-scale baroclinity provided deep and favorable rear-to-front flow within the upper half of the troposphere. This result is a further indication that the rear-to-front flow within the squall line with a trailing stratiform region is fundamentally the result of processes internal to the storm, while the factor which determines whether a given case has strong rear inflow, weak rear inflow, or stagnation at the back edge of the stratiform region depends on the environment in which the storm occurs.

Figure 9.45 Doppler radar data showing the strength of the rear inflow current entering the back edge of the stratiform region of the squall-line system illustrated in Figs. 9.15, 9.18, 9.31, 9.34, 9.40, 9.41, 9.43, 9.44, 9.45, 9.48, 9.53 and 9.54. This particular cross section was shown by a single radar at a particular moment. Thus, it was not subjected to smoothing as were the composites in Figs. 9.31 and 9.43, and it therefore shows more details of the flow. Panels (a) and (c) depict radar reflectivity (dBZ). Panels (b) and (d) show system-relative horizontal velocity V_h (m s^{-1}), respectively, along the 310° and 130° azimuth directions. Positive velocities represent flow away from the radar, while negative velocities denote flow toward the radar. Arrows indicating flow direction are also shown. Rear inflow is highlighted by shading in (b) and (d). (From Smull and Houze, 1987. Reproduced with permission from the American Meteorological Society.)

Figure 9.46 Profiles of the storm-relative flow from soundings taken to the rear of squall lines with trailing stratiform precipitation. (From Smull and Houze, 1987. Reproduced with permission from the American Meteorological Society.)

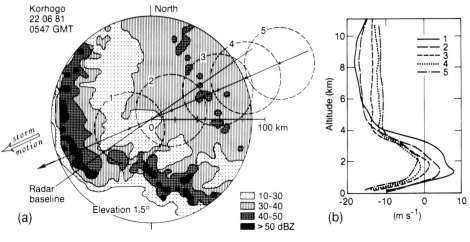

Figure 9.47

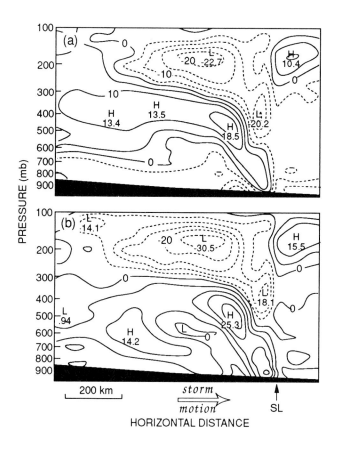

Figure 9.48

9.2.3.7   Vertical Vorticity in the Stratiform Region

The vertical vorticity ζ [defined in (2.56)] exhibits interesting behavior in the stratiform region. Often a center of cyclonic vorticity forms in association with the stratiform precipitation zone. Two observed examples are shown in Fig. 9.49 and Fig. 9.50. A simulation with a mesoscale model of the type described in Sec. 9.2.3.6 shows a similar behavior (Fig. 9.51). The cyclonic vorticity maximum is often observed to be near the back edge of the stratiform precipitation rather than in the center of the stratiform region. However, this is not true for the model example in Fig. 9.51, and the exact location of the vorticity center and other details of the vorticity pattern vary from case to case.[262]

Observations indicate that the development of the vorticity center in the trailing stratiform region is especially strong and deep in the asymmetric type of squall-line structure illustrated in Figs. 9.9b and 9.14b. The asymmetric structure of the precipitation pattern is evidently in part a consequence of the vorticity maximum. The results of one study of this type of vorticity center are indicated schematically in Fig. 9.52. Analysis of the observed wind field shows that the vertical vorticity ζ was intense at midstorm levels where strong rear inflow and convergence existed; thus, stretching appears to have been the dominant mechanism for intensifying the vorticity. A vorticity center of the observed strength could have been produced from the planetary vorticity alone in ~ 1.5 h. Pre-existing relative vorticity in the inflow current, such as might be associated with a short-wave trough in the westerlies, would accelerate the process. The short spin-up time helps explain why these mesoscale vortices are common features of the midlevel convergent layer in the trailing stratiform regions of squall lines. The strong rear inflow observed in the case shown in Fig. 9.52 was quite dry and characterized by low θ_e, and the strong midlevel vorticity in that case was associated with convergence into the mesoscale downdraft. The downdraft region was indeed dominated by cyclonic vorticity. The rear inflow air appears to have ob-

---

[262] For a concise review of mesoscale vortex development in squall lines with trailing stratiform regions, see Sec. 1 of Brandes (1990). Zhang and Fritsch (1987, 1988a,b) and Chen and Frank (1992) discuss the vortex development in model simulations of mesoscale convective systems.

Figure 9.47   Rear inflow in the tropical squall line shown in Figs. 9.7, 9.8, and 9.20. (a) Low-level radar reflectivity pattern. Superimposed circles denote locations of wind profiles shown in (b). Profiles show horizontal wind component normal to the leading convective line seen as the zone of intense echoes in (a). Positive values denote flow from rear to front of the storm. (From Chong *et al.*, 1987. Reprinted with permission from the American Meteorological Society.)

Figure 9.48   Mesoscale model simulation of the squall line with trailing stratiform region illustrated in Figs. 9.15, 9.18, 9.31, 9.34, 9.40, 9.41, 9.43, 9.44, 9.45, 9.48, 9.53, and 9.54. Vertical cross section normal to the squall line. Leading edge of the squall line at the ground is indicated by SL. Horizontal flow perpendicular and relative to the squall line is indicated in m s^{-1}. Positive values are directed from rear to front (left to right). In (b) a model result with no diabatic heating has been subtracted from the total result shown in (a). (From Zhang and Gao, 1989. Reprinted with permission from the American Meteorological Society.)

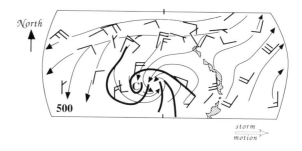

Figure 9.49

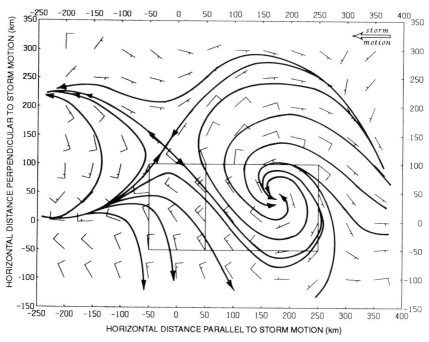

Figure 9.50

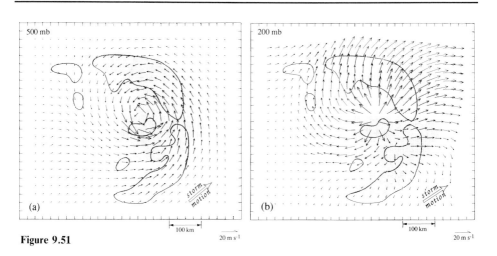

Figure 9.51

tained horizontal vorticity by baroclinic generation [B_x and B_y in (2.57) and (2.58)] upstream to the south and west of the vorticity center where warm air lay to the right facing downwind. As the inflow was tilted by the downdraft while it penetrated toward the center of the stratiform region, its horizontal vorticity was tilted into the vertical [second term on the right in (2.59)] and concentrated by convergence [first term on the right in (2.59)].

The vorticity pattern associated with the symmetric type of squall-line structure (Fig. 9.9a) has been documented in detail for one case. The horizontal pattern of vertical vorticity at midlevels was one of alternating bands: cyclonic in the convective region, anticyclonic in a zone just behind the convective region, and cyclonic at the back edge of the stratiform region (Fig. 9.53). A conceptual model illustrating how this structure developed is shown in Fig. 9.54. The vortex tubes shown were initially horizontal. The storm was located in a sheared environment characterized by along-line wind that increased (in the into-the-page sense in Fig. 9.54) with height. The banded vorticity pattern is accounted for by the tilting and stretching of the environmental vorticity. The tilting by the main up- and downdrafts of the storm bent the vortex tubes upward and downward. Selected trajectories (computed from Doppler radar–observed winds) are indicated. Where stretching favored concentration of vertical vorticity the trajectories are solid, and where concentration was disfavored they are dashed. The cyclonic vorticity in the convective region was produced by the tilting on the forward side of the main convective updraft. The anticyclonic vorticity zone immediately behind the convective line was produced by tilting on the trailing side of the main convective updraft and by tilting on the forward sides of the transition zone and mesoscale downdrafts. The cyclonic vorticity maximum at the back edge of the stratiform

Figure 9.49   Analysis of the flow shown by a composite of rawinsonde data obtained in the vicinity of a squall line with a trailing stratiform region in west Texas. Streamlines are superimposed on features of the low-level radar reflectivity field. The convective echo cores in the squall line are shaded. The curved lines trace the positions of lines of more intense echo embedded in the stratiform precipitation region. The C marks the "center of curvature" of the echo bands. (From Leary and Rappaport, 1987. Reprinted with permission from the American Meteorological Society.)

Figure 9.50   Composite analysis of the mesoscale component of the wind at the 650-mb level of a tropical squall-line system. The smaller rectangle is the region within which the leading convective line was found on radar. The larger rectangle is the region in which trailing stratiform precipitation was found. Winds from soundings and aircraft were analyzed objectively and the mesoscale component of the flow shown here was deduced by applying a mesoscale bandpass filter to the data to suppress larger- and smaller-scale wavelengths. A full wind barb represents 5 m s^{-1}. A vortex is evident toward the rear of the stratiform region. Other aspects of this storm are represented in Figs. 9.42 and 9.59. (From Gamache and Houze, 1985. Reproduced with permission from the American Meteorological Society.)

Figure 9.51   Mesoscale model result showing vortex formation in the trailing stratiform region of a midlatitude squall line. Relative wind vectors in a coordinate system moving at the speed of the vortex. Shading shows areas of stratiform (i.e., model-resolved) precipitation. Contours enclose regions of convective (parameterized) precipitation. (a) 500 mb and (b) 200 mb. (From Chen, 1990.)

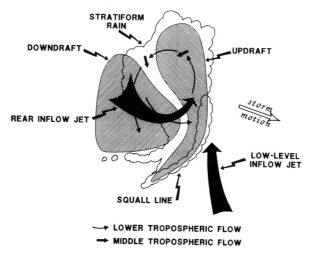

Figure 9.52 Schematic of the flow associated with an asymmetric type of squall-line system (as described in Sec. 9.2.1 and Fig. 9.9) observed in Oklahoma. Scalloped areas represent precipitation. (From Brandes, 1990. Reprinted with permission from the American Meteorological Society.)

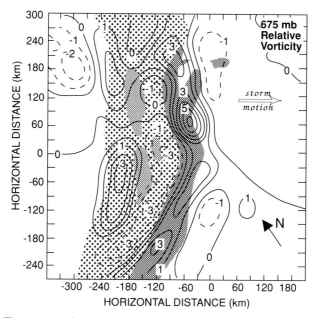

Figure 9.53 The pattern of vertical vorticity at the 675-mb level in a midlatitude squall-line system. Vorticity is contoured every 1×10^{-4} s^{-1} with negative values dashed. Background shading indicates the low-level radar echo (i.e., rainfall) pattern. Dark shading shows the convective rain area, the light shading the stratiform area, and the intermediate shading the region of heavier stratiform rain. The storm in which the data were taken is the same as that represented in Figs. 9.15, 9.18, 9.31, 9.34, 9.40, 9.41, 9.43, 9.44, 9.45, 9.48, and 9.54. (From Biggerstaff and Houze, 1991a. Reproduced with permission from the American Meteorological Society.)

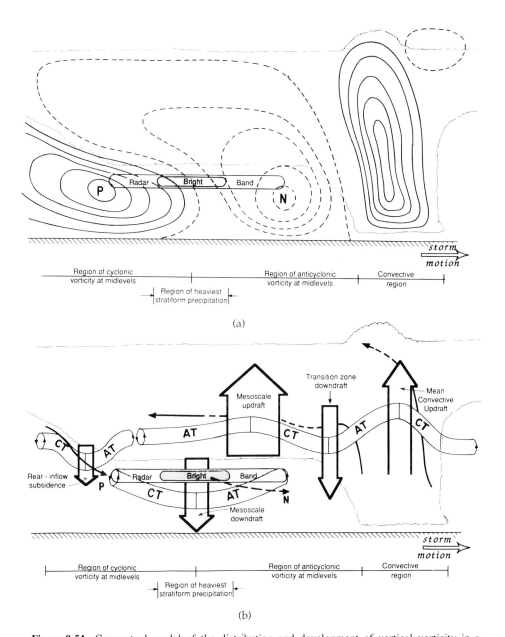

(a)

(b)

Figure 9.54 Conceptual model of the distribution and development of vertical vorticity in a midlatitude squall line with a symmetric distribution of convective and trailing stratiform precipitation. (a) Distribution of relative vertical vorticity in a cross section taken normal to the convective line with solid (dashed) lines indicating positive (negative) relative vorticity (ζ) and contours in the convective region at twice the interval used in the stratiform region. P and N indicate centers of positive and negative ζ. (b) Tilting and stretching involved in producing the observed vertical-vorticity pattern shown in (a). CT (AT) indicates where bending of initially horizontal vortex tubes upward and downward by the main up- and downdrafts of the storm produces centers of cyclonic (anticyclonic) vorticity. Air trajectories computed from Doppler radar–observed winds are indicated. Where stretching favored (disfavored) concentration of vertical vorticity, the trajectories are solid (dashed). The storm represented is the same as that in Figs. 9.15, 9.18, 9.31, 9.34, 9.40, 9.41, 9.43, 9.44, 9.45, 9.48, and 9.53. (From Biggerstaff and Houze, 1991b. Reproduced with permission from the American Meteorological Society.)

precipitation region was formed by tilting and stretching in the downward rear inflow at the back edge of the stratiform precipitation.

9.3   General Kinematic Characteristics of Mesoscale Convective Systems

In Sec. 9.1 we saw that the mesoscale convective system exhibits a variety of structures, as indicated by its pattern of precipitation. Although the pattern and organization of the precipitation vary from case to case (e.g., Fig. 9.10), the large area of rainfall of a mesoscale convective system is nearly always subdivided into clearly identifiable convective and stratiform regions. In Sec. 9.2 we examined the type of system in which the precipitation consists of a sharply defined leading line of convective precipitation trailed by a region of stratiform precipitation (e.g., Fig. 9.10a). We now consider the broader spectrum of mesoscale convective cloud systems, including ones in which the precipitation exhibits other types of meso-scale organizations (e.g., Fig. 9.10b and c). In the remainder of this chapter, we will examine the divergence and vorticity of the wind fields in and around meso-scale convective systems of various types. Despite their diversity of structure, these cloud systems exhibit rather similar gross patterns of divergence and vortic-ity. These characteristics are important because they reflect how the mesoscale convective systems are interacting with the larger-scale atmospheric circulation.

9.3.1   Divergence Associated with Mesoscale Convective Systems

The vertical distribution of divergence in a mesoscale convective system whose precipitation is organized into a convective and trailing stratiform region is illus-trated in Fig. 9.55. These profiles of divergence were determined for a storm of the type shown conceptually in Fig. 9.13. The divergence averaged over the convec-tive region is characterized by convergence in the lower troposphere and diver-gence in the upper troposphere. In the stratiform region, the profile shows conver-gence in the middle troposphere, with a maximum at about the 600–700-mb level, with divergence at upper and lower levels.[263] The profile of net divergence for the storm as a whole is the area-weighted sum of the convective and stratiform pro-files. It shows weak divergence (positive in this case, but not significantly different from zero) at low levels, where the divergence in the stratiform region tends to cancel the convergence characterizing the convective line. The level of maximum convergence in the net profile is thus elevated well above the surface. In this case, it is found between 900 and 700 mb, where both the convective and stratiform regions are convergent. The maximum divergence is at high levels, where both the convective and stratiform regions are divergent.

---

[263] These profiles are typical of mesoscale convective systems consisting of a leading line of con-vective precipitation trailed by a stratiform rain area. See Houze (1989) for a comprehensive review.

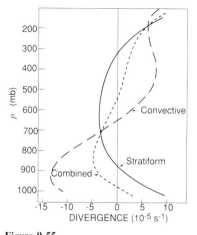

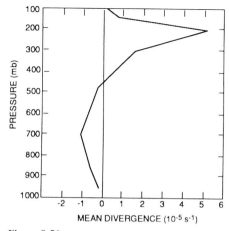

Figure 9.55

Figure 9.56

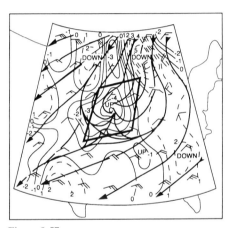

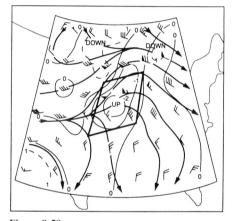

Figure 9.57

Figure 9.58

Figure 9.55 Average divergence over the convective (dashed), stratiform (solid), and combined (dotted) regions of a tropical squall-line system. Other aspects of this storm are represented in Figs. 9.42 and 9.50. (From Gamache and Houze, 1982. Reproduced with permission from the American Meteorological Society.)

Figure 9.56 Mean divergence in the vicinity of MCCs. (From Maddox, 1981.)

Figure 9.57 Composite 700-mb wind in the vicinity of MCCs. The parallelogram is the region of the MCCs, and the wind is reckoned relative to the MCCs. A full wind barb is for 10 knots or ~5 m s^{-1}. Vertical air motion (expressed as $\omega \equiv Dp/Dt$) is indicated in units of 10^{-3} mb s^{-1} (contours for positive values, indicating downward motion, are dashed). Map of the United States is shown in the background to help indicate scale. The sides of the quadrilateral are approximately 700 km in length. (From Maddox, 1981.)

Figure 9.58 Composite analysis of 200-mb wind in the vicinity of MCCs. The parallelogram is the region of the MCCs, and the wind is reckoned relative to the MCCs. A full wind barb is for 10 knots or ~5 m s^{-1}. Vertical air motion (expressed as $\omega \equiv Dp/Dt$) is indicated in units of 10^{-3} mb s^{-1} (contours for positive values, indicating downward motion, are dashed). Map of the United States is shown in the background to help indicate scale. The sides of the quadrilateral are approximately 700 km in length. (From Maddox, 1981.)

The divergence profiles in Fig. 9.55 are typical not only of the squall line with trailing stratiform structure but also of mesoscale convective systems that exhibit a wider variety of mesoscale organizations. To illustrate this fact, we now examine the largest of the mesoscale convective systems—the MCC, whose satellite-observed cloud shield satisfies the criteria listed in Table 9.1. By focusing on these large and long-lasting systems, we expect stronger and clearer kinematic signals. There is evidence, however, that the types of mesoscale organization in smaller, less intense mesoscale convective systems are qualitatively similar to those of MCCs.[264] Thus, much of what is learned from the MCCs should be applicable to smaller mesoscale convective systems.

Mean kinematic characteristics of MCCs are shown in Figs. 9.56–9.59. The mean divergence for the cloud cluster is shown by the solid curve in Fig. 9.56. It was calculated from a composite of the rawinsonde-measured winds in the vicinity of cloud clusters. The composite wind fields at 700 and 200 mb are shown in Fig. 9.57 and 9.58. The cloud cluster location is the parallelogram. The divergence in Fig. 9.56 is the average over the region of the parallelogram.

There is a strong qualitative similarity in the shape of the profile of net divergence associated with the MCC and the shape of the curve of net divergence for the squall line with trailing stratiform precipitation (cf. Figs. 9.55 and 9.56). As in the case of the squall line with trailing stratiform precipitation, the mean MCC divergence is approximately zero near the ground, where the strong convergence in the boundary layer, where air rises into the bases of the convective downdrafts at gust-front convergence lines, is offset in the mean by divergence associated with the spreading of downdraft air below cloud. Thus, the net divergence over the entire area of the MCC is near zero close to the surface. The strongest net convergence is again found to be elevated well above the surface. In the MCC profile, the maximum convergence is at 700 mb. Also similar to the squall line with trailing stratiform precipitation, the MCC exhibits a strong peak of divergence at high levels (200 mb). The strongly divergent character of the MCC winds at the 200-mb level is seen clearly in the streamline and isotach patterns of the composite relative flow shown in Fig. 9.58. When the winds are filtered to remove the synoptic-scale basic state flow, the mesoscale divergence and anticyclonic outflow are even more striking (Fig. 9.59). At this level, both the convective region and the stratiform region of a mesoscale convective system are characterized by divergence, and the two components of the MCC reinforce to produce the strong divergent signal on the mesoscale.

Further insight has been gained into the vertical profile of the divergence in mesoscale convective systems by aircraft flights into mesoscale convective systems of the type that have a more highly varied internal structure than the squall line with a trailing stratiform region. During a field experiment conducted over the ocean north of Australia, an aircraft with a Doppler radar on board flew through

---

[264] In a study of major rainstorms over Oklahoma during springtime over a 6-year period, Houze *et al.* (1990) found no difference in the mesoscale rainfall structures of MCCs and weaker mesoscale convective systems.

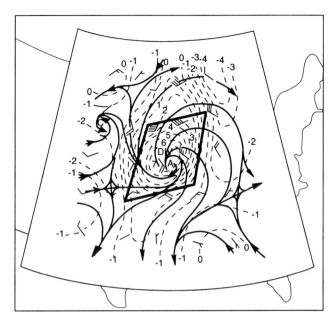

Figure 9.59   Composite analysis of 200-mb wind in the vicinity of MCCs band-pass filtered to remove the larger, synoptic-scale basic state flow and smaller, convective-scale motions. The parallelogram is the region of the MCCs, and the wind is reckoned relative to the MCCs. A full wind barb is for 10 knots or ~5 m s^{-1}. Letters A and C indicate anticyclonic and cyclonic centers, respectively. Contours indicate divergence in units of 10^{-5} s^{-1}. Map of United States is shown in the background to help indicate scale. The sides of the quadrilateral are approximately 700 km in length. (From Maddox, 1981.)

mesoscale convective systems in the manner illustrated by the example in Fig. 9.60. Generally, the mesoscale systems investigated in this study were not squall lines with trailing stratiform precipitation. However, they fit the conceptual model of rain area evolution described in Fig. 6.11, where a succession of convective rain areas occurs. As each convective area ages it turns into a region of stratiform precipitation. The entire rain area of the mesoscale convective system at a given time then consists of a nonsteady pattern of relatively new convective areas amidst a mass of stratiform precipitation formed from the decay of earlier active convection. The aircraft flew a zigzag flight track back and forth across the rain area. A Doppler radar beam pointing normal to the flight track (while scanning vertically) could then view the precipitation within each V-shaped region from two different angles.[265] The wind components along the beams were then used to calculate the divergence of the wind for a diamond-shaped region ~500 km^2 in an area centered within each V of the flight track.

[265] Current airborne Doppler radars use two beams: one pointing at an angle forward and the other pointing aft. In this way the same precipitation is viewed from two different angles while the aircraft flies a straight track.

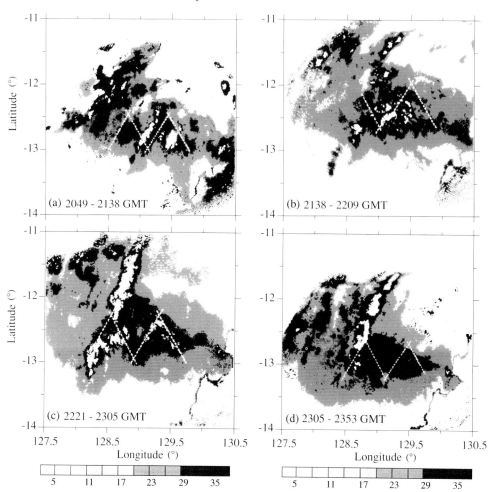

Figure 9.60 Radar reflectivity time-composite images from airborne radar observations in a tropical mesoscale convective system over the Joseph Bonaparte Gulf off the northwestern coast of Australia on 27 January 1987. The reflectivity shown at each point is the maximum observed during the period of the flight. Flight tracks are shown in white and proceed left to right in (a), right to left in (b), and so forth. Ground echo from the Australian continent appears at lower right. (From Mapes and Houze, 1993a. Reprinted with permission from the Royal Meteorological Society.)

The precipitation enclosed by each V of the flight track was classified according to whether it was *convective*, *stratiform*, or *intermediary*—meaning that it was in the process of conversion from convective to stratiform. On several occasions, the aircraft observed the same area of precipitation in all three stages of development. In the example shown in Fig. 9.60, the region at 13°S, 129.5°E was observed at four successive times. At the time shown in Fig. 9.60a, the precipitation was in a highly convective stage. By the time of Fig. 9.60d, it was completely strati-

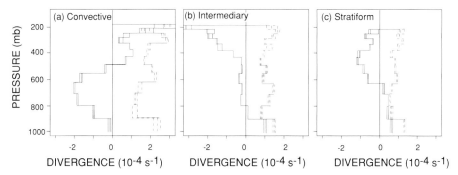

Figure 9.61 Mean (solid) and standard deviation (dotted) of divergence observed by airborne Doppler radar in mesoscale convective systems over the tropical ocean north of Australia. Profiles have been plotted against pressure so that area is proportional to mass divergence. (From Mapes and Houze, 1993a. Reprinted with permission from the Royal Meteorological Society.)

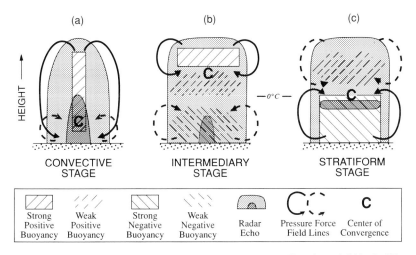

Figure 9.62 Conceptual model of the buoyancy and pressure gradient force field in the life cycle of a precipitation area within a mesoscale convective system.

form—vertical cross sections confirmed this categorization by showing a radar bright band at the melting level (Sec. 6.1.2). At the times shown in Fig. 9.60b and c, the precipitation was in the intermediary stage. By measuring the divergence in 93 V-shaped regions in nine different mesoscale convective systems, the mean divergence profiles shown in Fig. 9.61 were found. Note that sensitivity of the radar was not sufficient to observe the low intensity of the reflectivity at high levels. Other data showed that the actual cloud tops were several kilometers above the 200-mb level (~12 km), which is the highest level at which the divergence could be reliably measured by the radar. All the mean profiles show a net convergence below 12 km. On the linear pressure scale used in Fig. 9.61 the area under the divergence profiles is proportional to mass divergence. Since mass convergence and divergence must balance over a full vertical column, the net convergence seen in Fig. 9.61 implies that the higher levels must have been characterized by upward air motion and net divergence in all three cases. Considering the divergence profiles below 200 mb, we note that in the convective region (Fig. 9.61a) the maximum convergence is elevated and found at about the 700-mb level. Very little convergence is seen below 400 mb in the intermediary regions (Fig. 9.61b), while maximum convergence is found at about 500 mb in the stratiform regions (Fig. 9.61c). Although there was some underestimate of the winds at low levels as a result of echoes from the sea surface,[266] it is evident that the level of maximum net convergence into the mesoscale convective systems (convective, intermediary, and stratiform regions combined) must be elevated and found at a level somewhere in the 500–700-mb layer. This result is consistent with those contained in Fig. 9.55 and Fig. 9.56. Again, it is seen that convergence into updrafts and divergence associated with downdrafts tend to cancel in and just above the boundary layer, such that the main convergence into the mesoscale convective system is elevated. The aircraft sampling moreover shows clearly that this net divergence profile is associated not just with squall lines with trailing stratiform regions but also with mesoscale systems with more random configurations of convective and stratiform precipitation.

The very high-level (above 400 mb) maximum of convergence found in the intermediary precipitation areas (Fig. 9.61b) is of particular interest. It can be understood in terms of the typical life cycle of a subarea of rain within the mesoscale convective system. As indicated in Fig. 6.11, each subarea begins as a region of convective elements and then evolves through an intermediary stage into an area of stratiform precipitation. The subarea is thus a group of cells, each undergoing a life cycle of the type illustrated in Fig. 6.1b, where a deep cumulonimbus cell degenerates into a stratiform structure at the end of its lifetime. The divergence profile associated with a cell undergoing such a life cycle can be understood in terms of the principle illustrated in Fig. 7.1, which shows the pressure gradient force field required by mass continuity to be associated with a uniformly buoyant parcel. As discussed in Sec. 7.2, such a pressure force field is associated with any given spatial distribution of buoyancy. Figure 9.62 depicts a buoyancy pressure-gradient force field, of the type depicted in Fig. 7.1, that would be associated with a convective cell undergoing a life cycle of the type illustrated in Fig. 6.1b.

In the convective stage of development the cell is dominated by an updraft, which may be idealized as a deep vertical column of uniformly buoyant air, which has a pressure force field like that indicated in Fig. 9.62a; air is being pushed out of the path of the top of the cell and air is being forced in toward the buoyant updraft in the lower troposphere. To keep the convective-stage picture simple, the downdraft and its negative buoyancy in the lower part of the cloud have not been sketched in Fig. 9.62a; however, the pressure gradient force field associated with a low-level negatively buoyant downdraft is indicated (dashed field lines). In the intermediary stage (Fig. 9.62b), the strong buoyant updraft element has been cut off from its supply of buoyant air from below, and it has reached the stable

---

[266] See Mapes and Houze (1993a) for further discussion.

tropopause and begun to spread out. It collapses into a flattened buoyant element in the manner illustrated in Fig. 5.33. The base of the strongly buoyant element is thus found at ever higher levels as the element continues to rise and spread out. The buoyancy pressure gradient force, as shown by the field lines in Fig. 9.62b, continues to push air in under the element to maintain mass continuity. This process accounts for the strong convergence concentrated at upper levels in the intermediary stage seen in Fig. 9.61b. Below this region of maximum convergence (marked by C in Fig. 9.62b), the cloud is filled with air that has been pushed in under the rising element as its base has risen during the intermediary stage. It is shown schematically as weakly positively buoyant air in Fig. 9.62b. Still farther down, in the lower part of the cloud, downdrafts are becoming more dominant than updrafts, and the buoyancy in the lower part of the cloud is negative on average.

After the time of Fig. 9.62b, all of the region of strong local buoyancy comprising the original convective element becomes so flattened and mixed with surrounding air by entrainment that it loses its identity. Hence, no strongly buoyant element is shown at upper levels in Fig. 9.62c. However, the air pushed in under the rising element throughout the intermediary stage (Fig. 9.62b) came largely from the surrounding stratiform cloud region, which at upper levels is made up mostly of diluted but generally warm, moist air from previous convective elements. The air column above the melting layer is therefore filled with moderately buoyant air by the time of Fig. 9.62c. At lower levels, the downdraft motion has become firmly established across the cloud, from at or just above the melting layer down to near the surface, and this lower portion of the cloud is filled with negatively buoyant air, as shown by the hatching in Fig. 9.62c. Associated with this negatively buoyant downdraft region is a pressure force field like that shown in the lower part of Fig. 9.62c. This force field promotes convergence in the layer near and just above the 0°C level. As a result, the maximum convergence, found at high levels in the intermediary stage (Fig. 9.61b and Fig. 9.62b), lowers to the midtroposphere in the stratiform stage (Fig. 9.61c and Fig. 9.62c).

9.3.2 Vorticity in Regions Containing Mesoscale Convective Systems

It is not uncommon for the region containing one or more mesoscale convective systems to become characterized by cyclonic vorticity in the low to middle troposphere. Evidence of this vorticity structure is sometimes seen in satellite pictures. In particular, large mesoscale convective systems such as MCCs (Sec. 9.1.1) often exhibit a vortical cloud pattern in the middle-level cloud after the upper-level cloud shield has dissolved in the late stages of the cloud system's lifetime (Fig. 9.63). In Sec. 9.2.3.7, we saw a tendency for mesoscale cyclonic vorticity centers to form in low to middle levels *within* squall lines with trailing stratiform precipitation. Here, we refer to a cyclonic vorticity that characterizes a region somewhat larger than but containing the mesoscale convective system.

The spinup of such a vortex in low to middle levels is not surprising in view of the vertical profile of divergence observed in the vicinity of mesoscale convective

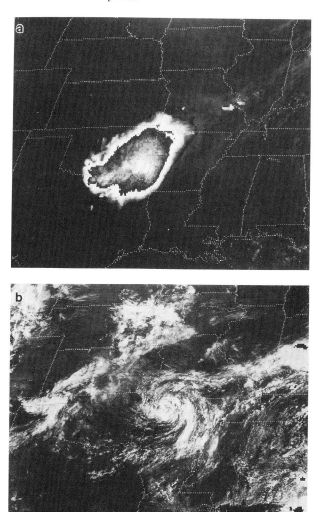

Figure 9.63   (a) Infrared satellite view of a mesoscale convective complex (MCC) centered over Oklahoma at 1131 GMT 7 July 1982. Gray shades are proportional to infrared radiative temperature at cloud top, with coldest values indicated by light shading in the interior of the cloud system. The large cold cloud shield marks the MCC. (b) Visible satellite image of the remnants of the same MCC at 1631 GMT. A cyclonic circulation is seen in the cloud pattern over northwestern Arkansas. (Photos provided by J.M. Fritsch.)

systems. The profiles of net divergence around both the squall line with trailing stratiform precipitation (Fig. 9.55) and the MCC (Fig. 9.56) exhibit a maximum of convergence in the lower troposphere, but well above the surface (between 700 and 900 mb). The three profiles of divergence for the convective, intermediary, and stratiform stages of subareas tropical oceanic of mesoscale convective systems in Fig. 9.61 combine (regardless of area weighting) to yield a net convergence in the low to middle troposphere but well above the surface. Thus, mesoscale convective systems, regardless of type, exhibit a maximum of net convergence at low levels but above the boundary layer.

The significance of this observation for the development of vorticity can be seen by first rewriting the vorticity equation (2.59) as

$$\underset{\text{(i)}}{\frac{\partial \zeta}{\partial t}} = \underset{\text{(i)}}{-\mathbf{v}_H \cdot \nabla_H (\zeta + f)} \underset{\text{(ii)}}{- w\zeta_z} + \underset{\text{(iii)}}{(\zeta + f)w_z} + \underset{\text{(iv)}}{\omega_H \cdot \nabla_H w} \tag{9.1}$$

where the symbols are as defined in Sec. 2.4. The subscript H designates horizontal components of vector quantities. Term (i) is the horizontal advection, (ii) is the vertical advection, (iii) is the stretching or intensification of existing absolute vorticity, and (iv) is the tilting of horizontal vorticity ω_H into the vertical. (9.1) is identical to (8.9), except that here the advection terms are separated into horizontal and vertical components and the earth's vorticity f is retained. In Sec. 9.2.3.7 we saw how, in the case of squall lines with trailing stratiform regions, the interplay of horizontal advection, tilting, and stretching accounts for the details of the vorticity pattern within that type of mesoscale convective system. To apply (9.1) to a somewhat larger area, containing one or more mesoscale convective systems, it is convenient to rewrite (9.1) as

$$\frac{\partial \zeta}{\partial t} = -\nabla_H \left[(\zeta + f)\mathbf{v}_H \right] - \mathbf{k} \left(\nabla_H \times w \frac{\partial \mathbf{v}_H}{\partial z} \right) \tag{9.2}$$

where the first term on the right is obtained by combining the horizontal advection and stretching terms [(i) + (iii)], with the aid of (2.55), while the second term on the right is the combination of vertical advection and tilting [(ii) + (iv)]. If (9.2) is averaged over an area A, it becomes

$$\frac{\partial \bar{\zeta}}{\partial t} = -\frac{1}{A} \oint v_n (\zeta + f) \, d\ell - \frac{1}{A} \oint w \frac{\partial v_t}{\partial z} \, d\ell \tag{9.3}$$

where the line integrals are taken along the boundary of A, $d\ell$ is an increment of distance along the boundary, and v_n and v_t are the horizontal wind components normal and tangential to the boundary, respectively. If the boundary is in the clear air surrounding the cloud system, the vertical velocity in the second term on the right will tend to be small. The first term is usually the dominant one. It is simply the net flux of vorticity across the boundary. If f is taken to be constant, it may be written as

$$-A^{-1} \oint v_n (\zeta + f) \, d\ell = -f \overline{\nabla \cdot \mathbf{v}_H} - A^{-1} \oint v_n \zeta \, d\ell \tag{9.4}$$

The first term on the right is the concentration, or stretching, of the earth's vorticity. If the mesoscale systems within the region A are the only important contributors to the average horizontal divergence in the region, then the net convergence, which we have seen prevails in the low to middle troposphere in those systems, leads to spin-up of mean cyclonic vorticity over the region at low to middle levels. For example, a convergence of $\sim 10^{-5}$ s^{-1} (like that near 700 mb in Fig. 9.55) leads to an e-folding of vorticity if it operates for a period ~ 1 day at a midlatitude location.

Chapter 10 | Clouds in Hurricanes

"... when there are no hurricanes, the weather of the hurricane months is the best of all the year"[267]

As we have proceeded through discussions of cumulus clouds, thunderstorms, and mesoscale convective systems in the previous three chapters, we have seen convective cloud dynamics involving successively larger scales of motion, ranging from the turbulent eddies associated with entrainment in even the smallest of cumulus clouds to the mesoscale updraft, downdraft, and rear inflow found in the stratiform regions of mesoscale convective systems. We now move still farther upscale and consider the clouds associated with hurricanes (Sec. 1.3.2). The clouds in these small synoptic-scale cyclones play an integral role in the storm dynamics. The clouds within a hurricane are primarily of the convective genera (cumulus and cumulonimbus) and are typically organized into large rings and bands, which have cloud and precipitation structure (including regions of nimbostratus and stratiform precipitation) similar to the mesoscale convective systems described in the last chapter. But in addition, the air motions in the clouds are strongly tied to the cyclone-scale dynamics. To see the connection between the clouds and the cyclone, we cannot depend on convective dynamics and hydrostatic reasoning alone, as we have sought to do with the mesoscale convective systems. The dynamics of the clouds in the hurricane need to be examined together with consideration of the dynamics of the cyclone itself, which is organized on a larger scale than any we have considered thus far. In this chapter, we therefore consider the clouds of hurricanes in relation to the horizontal and vertical air motions characterizing the hurricane vortex.

We begin in Sec. 10.1 with a definition of the hurricane and a review of some its generally observed features, including its regions of formation and occurrence over the globe, the general pattern of clouds and precipitation in a typical storm, and the broader aspects of storm-scale kinematics and thermodynamics. After this general background, in Sec. 10.2, we will take a more detailed look at the observed structure of the inner-core region of the storm. The dynamics of this part of the storm are particularly crucial to understanding its clouds and precipitation. In Sec. 10.3, we take a theoretical look at some basic aspects of hurricane dynamics. Then we return in Secs. 10.4 and 10.5 to the observations of clouds and precipitation to examine the consistency of the observed structures with the theoretical

---

[267] From *The Old Man and the Sea*, by Ernest Hemingway.

view. In Sec. 10.4, we focus on the observations of the inner-core region, while in Sec. 10.5, we examine the cloud and precipitation structures that occur outside the inner-core region.

10.1   General Features of Hurricanes

10.1.1   Definition and Regions of Formation

A tropical cyclone is considered to be a hurricane when the winds near the center of the vortex exceed 32 m s^{-1}. These storms form in the regions shown in Fig. 10.1. They form at low latitudes, generally between 5° and 20°, but not at the equator, where there is no Coriolis force. Some vorticity of the large-scale environment is required, and the formation regions are characterized by higher than average surface vorticity for these latitudes. The most critical characteristic of the formation zones, however, is that the sea-surface temperature must be above 26.5°C. Otherwise, there is insufficient boundary-layer energy to fuel the storm. Other characteristics of these regions are that the stratification of temperature and humidity is such that the air is moderately conditionally unstable (Sec. 2.9.1) and the vertical shear of the horizontal wind is weak. This latter property is necessary to prevent strong relative flow through developing storms. Such flow-through ventilates the storms in such a way that they cannot build up to hurricane intensity. Once hurricanes are formed, they tend to move along tracks such as those indicated in Fig. 10.2. They appear to be largely steered by midlevel winds; however, the factors that control their movement are complex. Forecasting hurricane tracks remains difficult, and empirical methods are often employed.

10.1.2   General Pattern of Clouds and Precipitation

In satellite pictures (e.g., Fig. 1.28), the clouds in a hurricane are seen to consist of low-level clouds spiraling cyclonically inward and upper-level cirriform cloud spiraling anticyclonically outward.[268] The precipitation pattern embedded within this cloud pattern is a better indication of the storm dynamics, and the rain is best observed by radar. The typical radar echo pattern in hurricanes, as seen by airborne radar during many research flights through hurricanes, is indicated schematically in Fig. 10.3. The echo pattern in a real case is shown in Fig. 10.4. The primary distinction to be made in these patterns is between the *eyewall*, which surrounds the center (or *eye*) of the storm, and the *rainbands*, which occur outside the eyewall. Further distinctions among the types of rainbands are indicated in the figure and will be discussed again in Sec. 10.5.

10.1.3   Storm-Scale Kinematics and Thermodynamics

An example of the low-level wind field in a hurricane (Gloria 1985) is shown in Fig. 10.5. The winds have been constructed from standard rawinsonde data plus spe-

---

[268] The cloud motions are easily seen when the satellite imagery is viewed in time lapse.

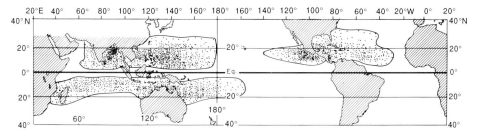

Figure 10.1 Locations of tropical cyclone formation over a 20-year period. (From Gray, 1979. Reprinted with permission from the Royal Meteorological Society.)

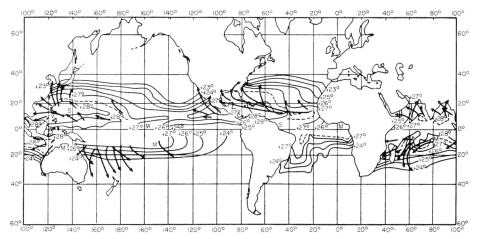

Figure 10.2 Tracks of tropical cyclones in relation to mean sea-surface temperature (°C). September temperatures are taken for the Northern Hemisphere. March temperatures are taken for the Southern Hemisphere. (From Bergeron, 1954. Reprinted with permission from the Royal Meteorological Society.)

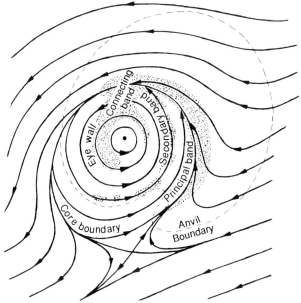

Figure 10.3 Schematic representation of the typical echo pattern seen by airborne radar in flights through hurricanes in relation to the low-level wind pattern. (From Willoughby *et al.*, 1984b. Reprinted with permission from the American Meteorological Society.)

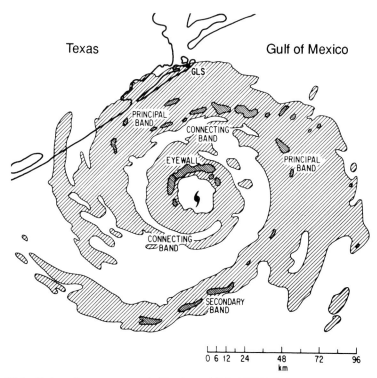

Figure 10.4 Radar echo pattern seen in Hurricane Alicia (1983) labeled according to the schematic of Fig. 10.3. Contours are for 25 and 40 dBZ. (From Marks and Houze, 1987. Reproduced with permission from the American Meteorological Society.)

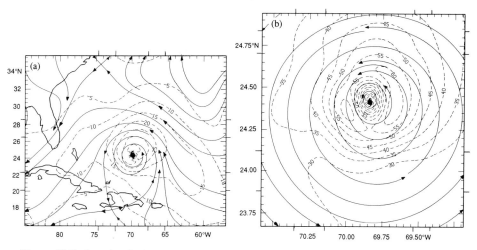

Figure 10.5 Low-level (900 mb) wind field associated with Hurricane Gloria (1985). (a) Large-scale flow analysis. Tick marks indicate boundaries of three nested rectangular domains defined for the analysis; in the inner domain, wavelengths less than about 150 km have been filtered out. In the intermediate and outer domains, wavelengths less than about 275 and 440 km have been removed. (b) High-resolution wind analysis, in which wavelengths less than about 16, 28, and 44 km have been filtered out in the three successively larger domains, whose boundaries are indicated by tick marks. Solid lines with arrows are streamlines. Dashed lines are isotachs labeled in m s^{-1}. (Courtesy of James Franklin, Hurricane Research Division, U.S. National Oceanic and Atmospheric Administration.)

cial dropwindsondes and Doppler radar data obtained aboard a research aircraft. In Fig. 10.5a, the wind data have been filtered to remove wavelengths less than about 150 km near the center of the storm and 440 km in the outer portions of the figure. This view emphasizes the large-scale flow pattern in which the storm is embedded. The hurricane appears as a cut-off low in the tropical easterlies. In Fig. 10.5b, the analysis retains wavelengths down to about 16 km in the center of the figure and down to about 44 km in the outer part of the figure. In this higher-resolution analysis, the hurricane vortex itself is highlighted. The streamlines show the boundary-layer air spiraling inward toward the center of the storm. The isotachs (Fig. 10.5b) show an annular zone of maximum wind speed, called the *radius of maximum wind*, at ~20 km (0.18° latitude) from the storm center. The air parcels spiraling inward tend to conserve their angular momentum (except for that lost by friction). The radius of maximum wind is created when the inflowing air abruptly slows down, converges, and turns suddenly upward in a ring of intense convection, which produces the eyewall. The radius of maximum wind thus closely coincides with both the eyewall precipitation maximum and the maximum updraft zone surrounding the eye of the storm. Inside the radius of maximum wind, the wind speeds drop off almost immediately to nearly zero, and the vertical air motion is downward, suppressing clouds and producing the characteristic hole marking the eye in satellite imagery (Fig. 1.28).

A full dynamical explanation for the eye of the storm is presently lacking. Discussions of the problem are given in other texts.[269] Since the eye dynamics are not directly relevant to the cloud and precipitation structure of the hurricane, we will not pursue this subject here.

Vertical cross sections of the mean radial (u) and tangential (v) components of the wind in a hurricane are shown in Figs. 10.6 and 10.7, which are composites of data collected in many storms. Most evident in the radial wind field (Fig. 10.6) is the increase in the inward-directed component at low levels as one approaches the storm center. Strong radial outflow is evident at the top of the storm, at about the 200-mb level. In the tangential wind field (Fig. 10.7), it is evident that the strongest wind at the radius of maximum wind is found in the lower troposphere, at about the 850-mb level (or about 1.5 km above the sea surface).

The pattern of horizontal winds at 200 mb, in the outflow layer near the top of the hurricane, is shown in Fig. 10.8 for Hurricane Gloria, whose low-level flow was illustrated in Fig. 10.5. The large-scale wind (Fig. 10.8a) illustrates that although the hurricane outflow is strong, it is not very symmetric. It tends to occur in one or two anticyclonic outflow jets. In the case shown, there is an outflow channel to the northeast of the storm center. It is also evident that at this high altitude the cyclonic vortex core is still evident near the storm center. The cyclonic flow changes to anticyclonic ~100 km from the storm center. The details of the cyclonic flow near the center of the storm are illustrated by the high-resolution analysis in Fig. 10.8b.

The strong low-level convergence in the eyewall and strong outflow aloft must be balanced by strong upward motion. The broad-scale pattern of vertical motion

[269] See Anthes (1982, pp. 34–36) and Palmén and Newton (1969, pp. 482–491).

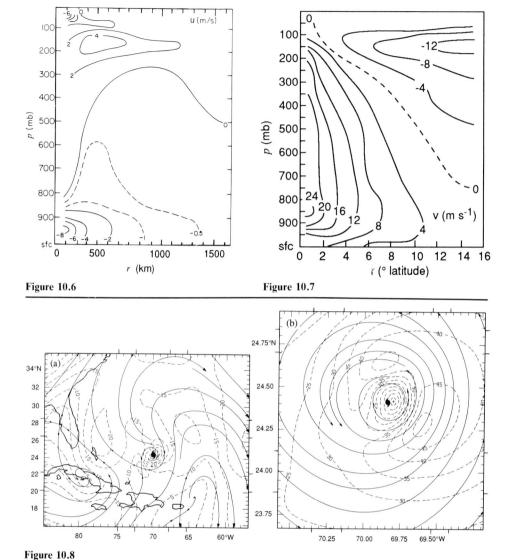

Figure 10.6

Figure 10.7

Figure 10.8

Figure 10.6 Vertical cross section of the mean radial wind (u) in western Atlantic hurricanes. Analysis is a composite of data collected in many storms. (From Gray, 1979. Reprinted with permission from the Royal Meteorological Society.)

Figure 10.7 Vertical cross section of the mean tangential component of the wind (v) in Pacific typhoons. Analysis is a composite of data collected in many storms. (From Frank, 1977. Reprinted with permission from the American Meteorological Society.)

Figure 10.8 Upper-level (200 mb) wind fields associated with Hurricane Gloria (1985). (a) Large-scale flow analysis. Tick marks indicate boundaries of three nested rectangular domains defined for the analysis; in the inner domain, wavelengths less than about 150 km have been filtered out. In the intermediate and outer domains, wavelengths less than about 275 and 440 km have been removed. (b) High-resolution wind analysis, in which wavelengths less than about 16, 28, and 44 km have been filtered out in the three domains whose boundaries are indicated by tick marks. Solid lines with arrows are streamlines. Dashed lines are isotachs labeled in m s⁻¹. (Courtesy of James Franklin, Hurricane Research Division, U.S. National Oceanic and Atmospheric Administration.)

in a hurricane is shown in Fig. 10.9. The overall cloud and precipitation amounts are determined by this vertical mass transport, which is concentrated within 400 km of the storm center. However, this mean vertical motion pattern does not have the spatial resolution to provide much insight into cloud structures, nor does it show the downward motion in the eye. To see these details, special aircraft and radar instrumentation are required. Analyses of such data will be considered in Secs. 10.2, 10.4, and 10.5.

The typical pattern of equivalent potential temperature in a hurricane, $\bar{\theta}_e$, is indicated by the example in Fig. 10.10. The overbar signifies the mean-variable field, which averages out turbulent and convective fluctuations. In the large-scale environment, far from the center of the storm, the stratification of $\bar{\theta}_e$ is typical of the tropics. Potential instability (Sec. 2.9.1) predominates in the lower troposphere, with $\bar{\theta}_e$ decreasing with height to a minimum at about the 650-mb level. Above that level, the air is potentially stable. The pattern of $\bar{\theta}_e$ changes markedly as one proceeds inward toward the center of the storm. In the low levels, the values of $\bar{\theta}_e$ increase steadily to a maximum in the eye of the storm. In the vicinity of the eyewall (~10 km from the storm center), the gradient of $\bar{\theta}_e$ is the greatest, and the isotherms of $\bar{\theta}_e$ rise nearly vertically through the lower troposphere, then flare outward as they extend into the upper troposphere. Since above the boundary layer $\bar{\theta}_e$ is conserved following a parcel [according to (2.18)], these contours reflect the flow of air upward and outward in the eyewall. In the very center of the storm, there is an especially strong decrease of $\bar{\theta}_e$ with height. The center of low $\bar{\theta}_e$ at 500 mb is evidence of the strong subsidence concentrated in the eye (not shown in Fig. 10.9), while the very large maximum at low levels is the net result of the boundary-layer turbulent mixing over the warm sea surface and inward advection of the air being affected by the mixing. The vertical circulation in relation to $\bar{\theta}_e$ is illustrated schematically in Fig. 10.11, where the low-level radial flow is depicted as converging into the center of the storm (consistent with Fig. 10.6) in the boundary layer below cloud base. As it flows inward, turbulence produces the well-mixed boundary layer of high $\bar{\theta}_e$. When this air enters cloud in the eyewall zone, it ascends undiluted to the upper troposphere along the lines of constant $\bar{\theta}_e$.

An important quantity in any consideration of hurricane dynamics is the angular momentum m about the central axis of the storm. This variable is defined by (2.35), with r in this case being the radial coordinate measured from the eye of the storm. Since above the planetary boundary layer, where friction is unimportant, m is conserved following a parcel [according to the inviscid equation (2.37)], the isotherms of $\bar{\theta}_e$ in Figs. 10.10 and 10.11 can also be regarded as lines of constant $\bar{m}$. From the discussion in Sec. 2.9.1, it is evident that the coincidence of $\bar{m}$ and $\bar{\theta}_e$ within the saturated environment of a cloud is a characteristic feature of a circulation which is in a state of conditional symmetric neutrality. This characteristic of the hurricane will be useful in our subsequent theoretical discussion of the storm dynamics.

The dynamical necessity of the outward-sloping structure of $\bar{m}$ (and $\bar{\theta}_e$) lines in the eyewall region was pointed out by early meteorological researchers. The

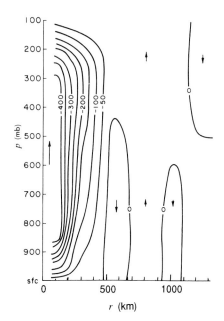

Figure 10.9 Vertical cross section of the mean vertical air motion (mb per day) in typhoons. Analysis is a composite of data collected in many storms. Radius r is measured from the eye of the storm. (From Frank, 1977. Reprinted with permission from the American Meteorological Society.)

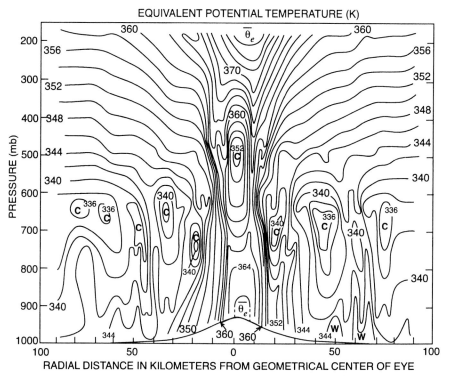

Figure 10.10   Equivalent potential temperature $\bar{\theta}_e$ in Hurricane Inez (1966). (From Hawkins and Imbembo, 1976. Reprinted with permission from the American Meteorological Society.)

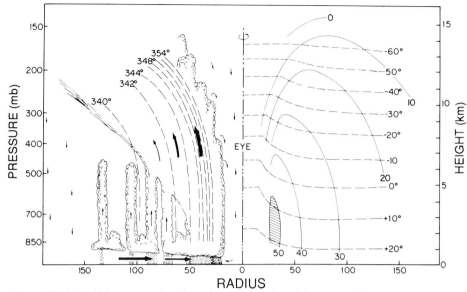

Figure 10.11   Radial cross section through an idealized, axially symmetric hurricane. On left: radial and vertical mass fluxes are indicated by arrows, equivalent potential temperature (K) by dashed lines. On right: tangential velocity in m s^{-1} is indicated by solid lines and temperature in °C by the dashed lines. (From Wallace and Hobbs, 1977, as adapted from Palmén and Newton, 1969.)

funnel shape of the storm in these fields was deduced by hydrostatic reasoning.[270] Assuming a vanishing pressure gradient at some high level, one deduces that the strong pressure gradient in a hurricane must be associated with an outward slope of the boundary of the core of warm air in the center of the storm. Another argument advanced[271] was that rising rings of air in a warm-core storm in which the radial pressure gradient decreases upward move outward for the centrifugal and Coriolis forces (corresponding to their initial angular momentum) to balance the weaker pressure gradients aloft.

Although the outward spreading of the $\bar{m}$ and $\bar{\theta}_e$ lines must be a general characteristic of balanced warm-core hurricanes, there are nonetheless observations that more vertical eyewalls occasionally exist (Fig. 10.12). The more vertical cases evidently occur when an abnormally high degree of conditional instability exists in the environment, and vertical accelerations dominate, or in storms (or portions of storms) that have not yet undergone complete adjustment to symmetric neutrality. The vertical accelerations in such cases would be dominated by buoyancy, and the momentum-conserving trajectories would therefore be more nearly vertical.[272] Occasionally, aircraft penetrations of the eyewall have shown a rather vertical convective element in a portion of an eyewall. These vertical convective towers

[270] Haurwitz (1935).
[271] Durst and Sutcliffe (1938).
[272] See Palmén and Newton (1969) for further discussion of this point.

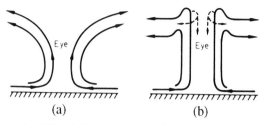

Figure 10.12 Schematic airflow in the inner parts of tropical cyclones not dominated by strong convection (a) and with intense convection in the cloud wall (b). (From Palmén and Newton, 1969.)

exhibit especially strong convective updrafts and downdrafts (~ 10 m s^{-1}) and have been called *hurricane supercells*.[273]

The degree to which an $\bar{m}$ surface in the eyewall of a balanced hurricane must slope outward can be obtained from the basic equations for a two-dimensional, axisymmetric vortex that is in hydrostatic and gradient wind balance. For such a vortex, conservation of $\bar{m}$ requires that the slope of a particular $\bar{m}$ surface be

$$\left.\frac{\partial r}{\partial p}\right|_{\bar{m}} = -\frac{\partial \bar{m}/\partial p}{\partial \bar{m}/\partial r} \tag{10.1}$$

That is, the slope of an $\bar{m}$ surface is determined by the ratio of the vertical to the horizontal shear of $\bar{m}$. In Sec. 10.3, it will be shown that with substitution from (2.35) and the thermal wind equation for a balanced vortex (2.36), (10.1) implies that the slope of the $\bar{m}$ surfaces with height must be outward, as depicted in Fig. 10.11.

10.2   The Inner-Core Region

In discussions of hurricane dynamics, attention is focused on the intense vortex associated with the eyewall and radius of maximum wind. We therefore now take a more detailed look at the observed structure of this part of the storm in preparation for the discussion of basic hurricane dynamics in Sec. 10.3.

The convention in discussing the inner-core vortex is to refer to the azimuthal (tangential) circulation as the *primary* circulation. The radial–vertical circulation transverse to the primary circulation is called the *secondary* circulation. In the inner-core region, the secondary circulation draws its energy from the sea surface and tends to be neutral with respect to conditional symmetric instability (Sec. 2.9.1). The role of vertical convection, occasionally superimposed on the secondary circulation, is not well understood but appears to be relatively minor in accounting for the basic mature storm structure, except perhaps in the rare cases of

---

[273] Two notable cases of hurricane supercells (Black *et al.*, 1986; Black and Marks, 1987) were encountered in the eyewall regions of Hurricanes Gladys (1975) and Norbert (1984). The author of this text was on the flight through the Norbert supercell and can attest to the intensity of the convection.

vertical eyewalls (Fig. 10.12b). Vertical convection may also be more important in early formative stages of the storm.

The clouds and precipitation associated with the hurricane inner core reflect aspects of both the secondary circulation, whose vertical motions produce the condensation, and the azimuthally spinning motions of the primary circulation, which strongly redistributes particles by azimuthal advection around the storm. An analysis of observations in the inner-core region of a fairly typical strong hurricane is shown in Fig. 10.13. The air motions have been reconstructed to match data obtained by airborne Doppler radar. The motions of precipitation particles were estimated by inserting hypothetical particles of various fall speeds into the observed circulation. The dashed contours indicate the primary circulation, while the wide dashed streamlines indicate the slantwise secondary circulation. Convective-scale downdrafts coincided with the heavy precipitation at low levels in the eyewall, probably as a result of the negative buoyancy associated with the weight of the hydrometeors [term $-gq_H$ in (2.50)]; however, these downdrafts were rather weak ($w \sim -1$ to -3 m s^{-1}). High-resolution radar measurements showed further that the slantwise upward current, dominating the eyewall circulation, had small-scale embedded updraft maxima ($w \sim 5$–15 m s^{-1}). These

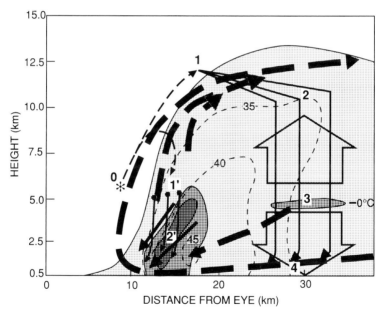

Figure 10.13 Schematic of the radius–height circulation of the inner-core region of Hurricane Alicia (1983). Shading depicts the reflectivity field, with contours at 5, 30, and 35 dBZ. The primary (tangential) circulation (v in m s^{-1}) is depicted by dashed lines and the secondary circulation by the wide dashed streamlines. The convective downdrafts are denoted by the thick solid arrows, while the mesoscale up- and downdrafts are shown by the broad arrows. Hydrometeor trajectories emanate from the * and are denoted by thin dashed and solid lines labeled with numbers corresponding to points along the horizontal projections shown in Fig. 10.14. (From Marks and Houze, 1987. Reproduced with permission from the American Meteorological Society.)

convective-scale cells of vertical motion indicate that some release of conditional instability was occurring and that the slantwise circulation was not perfectly symmetrically neutral. At radii >20 km or so the structure of the precipitation was quite stratiform, with a well-defined bright band in the radar reflectivity at the melting level. This stratiform structure adjacent to an intensely convective line (in this case the eyewall) is rather similar to the precipitation patterns in the squall lines and other mesoscale convective systems described in Chapter 9. From the particle trajectories in Fig. 10.13, it is evident that the ice particles that are not heavy enough to fall out in the eyewall convective rain shower are transported by the strong radial wind at high levels sufficiently far outward to fall out in the stratiform zone, as in the mesoscale convective systems described previously. However, in the hurricane the horizontal paths of the particles wrap around the storm as they are strongly advected azimuthally by the primary circulation, at the same time that they are carried outward by the secondary circulation (Fig. 10.14).

10.3 Basic Hurricane Dynamics

In this section, we examine some basic aspects of hurricane dynamics, with emphasis on the clouds and precipitation that form in association with the secondary circulation in the inner-core region. In a mature hurricane, the secondary circulation is the transverse circulation required to keep the primary circulation in a state of gradient-wind and hydrostatic balance. One way to see the basic character of the secondary circulation is to derive a cross-vortex stream function equation from the equations of the balanced vortex and solve for the transverse flow associated with a heat source embedded in the eyewall.[274] We will, however, take an approach more directly related to the nature of the air motions of the clouds in the inner-core region. This approach is based on the observation that the hurricane circulation in the eyewall region tends to be potentially symmetrically neutral [$\partial \bar{\theta}_{es}/\partial z = 0$ on a surface of constant $\bar{m}$, where θ_{es} is the saturation equivalent potential temperature defined by (2.146)], but not especially strongly potentially unstable ($\partial \bar{\theta}_{es}/\partial z$ only slightly negative in the vertical direction). The conditional symmetric neutrality is implied by the congruence of the $\bar{m}$ and $\bar{\theta}_{es}$ surfaces in the eyewall region (recall Fig. 10.11 and that $\bar{\theta}_{es} = \bar{\theta}_e$ in cloud). In view of this structure, we consider the dynamics of the eyewall cloud in terms of slantwise neutral motions emanating from a boundary layer whose value of $\bar{\theta}_e$ has been determined by upward flux of sensible and latent heat from the ocean surface. This approach gives insight into both the primary and secondary circulation of a mature hurricane by connecting the storm dynamics directly to the air–sea interaction that occurs in the boundary layer over the warm ocean.[275]

---

[274] For a discussion of the cross-vortex stream function equation, see Schubert and Hack (1982). A good example of the use of this approach is in the paper by Shapiro and Willoughby (1982).

[275] From this point on, the discussion in this section, detailing the air–sea interaction dynamics of the hurricane vortex, follows closely Emanuel (1986a).

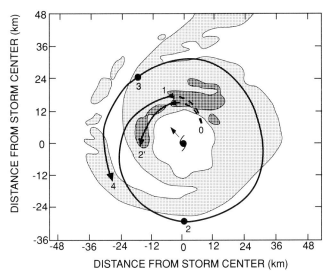

Figure 10.14 Horizontal projections of the paths of precipitation particle trajectories superimposed on the radar echo pattern of Hurricane Alicia (1983). The echo contours are for 20 and 35 dBZ. The numbers show how the trajectories correspond to the vertical cross section in Fig. 10.13. (From Marks and Houze, 1987. Reproduced with permission from the American Meteorological Society.)

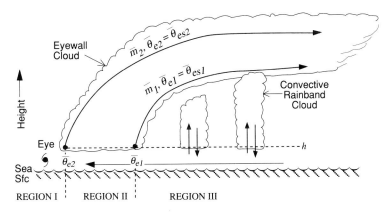

Figure 10.15 Idealization of hurricane structure for air–sea interaction hurricane model. Arrows indicate direction of airflow at selected locations. Dashed line represents the top of the boundary layer at height h. (Adapted from Emanuel, 1986a. Reproduced with permission from the American Meteorological Society.)

We will regard the vortex as two-dimensional and axisymmetric and having the structure indicated in Fig. 10.15. This conceptual model is consistent with the observed structures represented in Figs. 10.11 and 10.13. To aid discussion and analysis, we divide the boundary layer into the regions I, II, and III indicated in Fig. 10.15. All three of these regions are considered to be well mixed by turbulence. The production of the turbulent kinetic energy in the boundary layer by shear [term $\mathscr{C}$ in (2.86)] is substantial because of the strong low-level winds circulating around the center of the storm, while the generation of turbulence by buoyancy [term $\mathscr{B}$ in (2.86)] is strongly promoted by the warm ocean surface.

It is reasonable to assume that there is very strong upward eddy flux of sensible and latent heat throughout the lower part of the boundary layer. The flux at the top of the boundary layer, however, varies significantly from one region of the storm to the next. It is assumed that in region III, the region of the convective rainbands, there is a negative turbulent flux of θ_e at the top of the well-mixed layer. As will be discussed in Sec. 10.5, downdrafts from the convective rainbands are important in maintaining this negative flux of θ_e. In region II, the region of the eyewall cloud, the flux at the top of the boundary layer is assumed to be entirely positive and dominated by the mean upward flux associated with the secondary circulation of the hurricane vortex. Turbulent flux at the top of the boundary layer here is considered to be very small compared to this mean upward flux. Region I is the eye of the storm. In the present discussion, we do not consider details of the circulation in the eye.

The secondary circulation consistent with the boundary layer in Fig. 10.15 is obtained by considering the equations for a vortex in both hydrostatic and gradient-wind balance (Secs. 2.2.3–2.2.5). The horizontal equations of motion are then the gradient-wind equation (2.36) and the conservation of angular momentum (2.37). The vertical equation of motion reduces to the hydrostatic relation (2.38). The vertical gradient of angular momentum is related to the horizontal gradient of density (or specific volume) by the thermal wind equation (2.42).

The secondary circulation is indicated by mapping the isolines of angular momentum $\bar{m}$ in the $r–p$ plane. Since m is conserved, the $\bar{m}$ surfaces are also the trajectories of the mean air motion. We obtain the surfaces of angular momentum $\bar{m}(r,p)$ implied by the conservation of m by substituting in (10.1) and integrating upward, using the value of $\bar{m}$ at the top of the boundary layer as a boundary condition and assuming that the mature vortex has adjusted to a state of conditional symmetric neutrality everywhere above the boundary layer.

Since the region above the boundary layer is assumed to have adjusted to conditional symmetric neutrality, the saturation equivalent potential temperature $\bar{\theta}_{es}$ is uniform along surfaces of constant m above height h (Sec. 2.9.1). The value of $\bar{\theta}_{es}$ on the surface is assumed to be equal to the value of $\bar{\theta}_e$ where the $\bar{m}$ surface intersects the top of the boundary layer. These assumptions assure that a parcel that becomes saturated in the well-mixed boundary layer of the hurricane will experience neutral buoyancy when displaced along an angular momentum surface extending above the top of the boundary layer. The assumed relationship of $\bar{m}$ and $\bar{\theta}_{es}$ above the boundary layer to $\bar{\theta}_e$ in the boundary layer is illustrated by Fig. 10.15. To make the relationship more quantitative we make use of the following thermodynamic relationships.

The *saturated moist entropy* $\hat{S}$ is defined such that

$$T \, d\hat{S} = c_v \, dT + p \, d\alpha + L \, dq_{vs} \qquad (10.2)$$

where the saturation mixing ratio q_{vs} is used in the last term. If the air were saturated, $\hat{S}$ would be the actual entropy, and (10.2) would be a restatement of the

First Law of Thermodynamics (2.6). The quantity $\hat{h}$ is defined as

$$\hat{h} \equiv c_v T + p\alpha + Lq_{vs} \tag{10.3}$$

Its differential is seen to be

$$d\hat{h} = T\, d\hat{S} + \alpha\, dp \tag{10.4}$$

from which it follows that

$$\left.\frac{\partial \hat{h}}{\partial p}\right|_{\hat{S}} = \alpha, \qquad \left.\frac{\partial \hat{h}}{\partial \hat{S}}\right|_{p} = T \tag{10.5}$$

Since the partial derivative of the first expression with respect to $\hat{S}$ must equal the partial derivative of the second expression with respect to p, we obtain

$$\left.\frac{\partial \alpha}{\partial \hat{S}}\right|_{p} = \left.\frac{\partial T}{\partial p}\right|_{\hat{S}} \tag{10.6}$$

It follows from the definition of the saturation equivalent potential temperature θ_{es} in (2.146) that

$$c_p T\, d\ln\theta_{es} = c_p T\, d\ln\theta + L\, dq_{vs} - Lq_{vs}T^{-1}\, dT \tag{10.7}$$

The last term is negligible, and this expression may be rewritten as

$$c_p T\, d\ln\theta_{es} \approx T\, d\hat{S} \tag{10.8}$$

Thus, lines of constant $\bar{\theta}_{es}$ above the boundary layer in Fig. 10.15 may be regarded as lines of constant saturated moist entropy.

Since it has been assumed that the value of $\bar{\theta}_{es}$ on an $\bar{m}$ surface is equal to the value of $\bar{\theta}_e$ where the $\bar{m}$ surface intersects the top of the boundary layer, we may write

$$\bar{S} = c_p \ln\bar{\theta}_{es} = f(\bar{m}\ \text{only}) = c_p \ln\bar{\theta}_e(h), \quad \text{on an } \bar{m} \text{ surface at } z \geq h \tag{10.9}$$

Expressing the field of mean specific volume $\bar{\alpha}$ as a function of $\hat{S}$ and p, we obtain

$$\left.\frac{\partial\bar{\alpha}}{\partial r}\right|_{p} = \left.\frac{\partial\bar{m}}{\partial r}\right|_{p} \cdot \left.\frac{\partial\bar{\alpha}}{\partial\hat{S}}\right|_{p} \cdot \frac{d\hat{S}}{d\bar{m}} = \left.\frac{\partial\bar{m}}{\partial r}\right|_{p} \cdot \left.\frac{\partial\bar{T}}{\partial p}\right|_{\bar{m}} \cdot \frac{d\hat{S}}{d\bar{m}} \tag{10.10}$$

The expressions on the right follow from (10.6) and the condition that a surface of constant moist entropy $\bar{S}$ is also a surface of constant $\bar{m}$. With $\partial\bar{m}/\partial r$ given by (10.10) and $\partial\bar{m}/\partial p$ by the thermal wind equation (2.42), the slope of the $\bar{m}$ surface (10.1) becomes

$$\left.\frac{\partial r}{\partial p}\right|_{\bar{m}} = \left.\frac{\partial\bar{T}}{\partial p}\right|_{\bar{m}} \frac{r^3}{2\bar{m}}\frac{d\hat{S}}{d\bar{m}} \tag{10.11}$$

Integrating (10.11) along an $\bar{m}$ surface from some arbitrary radius r to $r = \infty$, we obtain

$$\frac{1}{r^2} = \frac{1}{\bar{m}} \frac{d\hat{S}}{d\bar{m}} [T_o - T_m(p)] \tag{10.12}$$

where $T_m(p)$ is the temperature on the $\bar{m}$ surface at pressure p and T_o is the outflow temperature on the m surface (i.e., the temperature at $r = \infty$). Since it is assumed that the hurricane has adjusted to a state of conditional symmetric neutrality, $T_m(p)$ is given by the temperature along the saturation moist adiabat corresponding to $\hat{S}$. In using (10.12) to construct the fields of $\bar{m}$ and $\bar{\theta}_{es}$ throughout the region above the boundary layer, we need the radial distributions of p, $\bar{m}$, and $\bar{\theta}_e$ at the top of the boundary layer ($z = h$). These distributions determine $d\hat{S}/d\bar{m}$, according to (10.9), and give us a point on the $\bar{m}$ surface from which to integrate the saturated moist-adiabatic lapse rate to obtain $T_m(p)$. It remains to make some assumption about the value of T_o; we will return to this question. First, we seek the radial distributions of p, $\bar{m}$, and $\bar{\theta}_e$ at $z = h$.

An assumption is made, on the basis of observations, that the temperature at the top of the boundary layer is a constant T_B. In this case, (10.12) applied at the top of the boundary layer is

$$(T_o - T_B) \frac{r^2}{\bar{m}} \frac{d\hat{S}}{d\bar{m}} = 1, \quad z = h \tag{10.13}$$

Since

$$\frac{d\hat{S}}{d\bar{m}} = \frac{\partial \hat{S}/\partial r}{\partial \bar{m}/\partial r} \tag{10.14}$$

(10.13) may be written as

$$(T_o - T_B) r^2 \frac{\partial \hat{S}}{\partial r} = \frac{\partial}{\partial r} \left(\frac{\bar{m}^2}{2} \right), \quad z = h \tag{10.15}$$

Substitution from the equation of state (2.3), the gradient-wind equation (2.36), and (10.9) changes (10.15) to

$$\frac{T_o - T_B}{T_B} \frac{\partial \ln \bar{\theta}_e}{\partial r} = \frac{R_d}{c_p} \frac{\partial}{\partial r} \left(\ln p + \frac{r}{2} \frac{\partial \ln p}{\partial r} \right) + \frac{f^2 r}{2 c_p T_B}, \quad z = h \tag{10.16}$$

which gives us a relation between $\bar{\theta}_e$ and p at $z = h$ if we integrate once in r. The integration is carried out from some distant radius r_a, which represents the outer boundary of the storm, where $\bar{\theta}_e = \theta_{ea}$ at $z = h$ to an arbitrary radius r within the storm. The integration is made difficult because T_o is a function of $\bar{m}$ and therefore of r. The following definition of a mean outflow temperature is used to simplify the problem:

$$\tilde{T}_o(r) \equiv \int_{\theta_{ea}}^{\bar{\theta}_e(r)} T_o(\theta_e) d \ln \theta_e \Big/ \int_{\theta_{ea}}^{\bar{\theta}_e(r)} d \ln \theta_e \tag{10.17}$$

This quantity turns out to be an insensitive function of r and is treated as a *specified constant*. Then integration of (10.16) gives

$$\frac{\tilde{T}_o - T_B}{T_B} \ln \frac{\bar{\theta}_e(r)}{\theta_{ea}} = \frac{R_d}{c_p} \ln \frac{p(r)}{p_a} + \frac{R_d}{c_p} \frac{r}{2} \frac{\partial \ln p(r)}{\partial r}$$

$$+ \frac{f^2}{4c_p T_B}\left(r^2 - r_a^2\right), \quad z = h \qquad (10.18)$$

which is one relation between $\bar{\theta}_e(r)$ and $p(r)$ at the top of the boundary layer.

At the center of the storm, where $r = 0$ and $\partial \ln p/\partial r = 0$, (10.18) becomes

$$\ln \frac{p_c}{p_a} = \frac{c_p\left(\tilde{T}_o - T_B\right)}{R_d T_B} \ln \frac{\theta_{ec}}{\theta_{ea}} + \frac{f^2 r_a^2}{4 R_d T_B}, \quad z = h \qquad (10.19)$$

where subscript c indicates the center of the storm. Since $\tilde{T}_o - T_B$ is negative, this relation implies a linear proportionality between the central pressure deficit of the storm and the equivalent potential temperature excess in the eye of the storm. The implied relationship agrees well with data in hurricanes and thus adds confidence to the applicability of (10.18).

If $\bar{\theta}_e(r)$ at the top of the boundary layer can be determined independently, then $p(r)$ at $z = h$ will be determined by (10.18), and $\bar{m}(r)$ will follow from the gradient-wind equation (2.36). Independent determination of $\bar{\theta}_e(r)$ at the top of the boundary layer is obtained by quantifying the boundary-layer model represented schematically in Fig. 10.15.

The boundary layer may be considered in terms of a quantity $\mathcal{A}$ that is conserved (like θ_e and m) in laminar inviscid flow. Under turbulent conditions, $\mathcal{A}$ is governed by a mean-variable equation of a form similar to (2.78), (2.81), and (2.83). In the Boussinesq case (which is suitable for the boundary layer), the density factor ρ_o in those equations is not present, and the equation for $\mathcal{A}$ may be written in axisymmetric cylindrical coordinates as

$$\overline{\mathcal{A}}_t + \overline{u}\,\overline{\mathcal{A}}_r + \overline{w}\,\overline{\mathcal{A}}_z = -\frac{1}{\rho}\frac{\partial \tau_A}{\partial z} \qquad (10.20)$$

where τ_A is the vertical eddy flux of $\mathcal{A}$, as defined for the examples in (2.186)–(2.189). If we now assume that the storm is in a steady state ($\overline{\mathcal{A}}_t = 0$) and the boundary layer is well mixed ($\overline{\mathcal{A}}_z = 0$), then the vertical eddy-flux convergence just balances the radial advection and (10.20) may be integrated over the depth of the boundary layer (from $z = 0$ to h) to obtain

$$\left.\frac{\partial \overline{\mathcal{A}}}{\partial r}\right|_h \psi(h) = r\left[\tau_A(h) - \tau_A(0)\right] \qquad (10.21)$$

where we have made use of the two-dimensional stream function

$$\left(\bar{\rho}\,\bar{u}\,r, \ \bar{\rho}\,\bar{w}\,r\right) \equiv \left(-\psi_z, \ \psi_r\right) \qquad (10.22)$$

which satisfies the mean-variable form of the anelastic continuity equation (2.54) in cylindrical coordinates, with $\bar{\rho}$ as the density weighting factor. The value of ψ has been set to 0 at $z = 0$.

The surface flux over the ocean, represented by term $\tau_A(0)$ in (10.21), can be calculated from the bulk aerodynamic formula[276]

$$\tau_A(0) = -\bar{\rho}\, C_A |\bar{v}| \left(\mathscr{A}_{\text{BL}} - \mathscr{A}_{\text{SFC}} \right) \tag{10.23}$$

where v is the tangential component, $\mathscr{A}_{\text{BL}}$ is the value of $\bar{\mathscr{A}}$ in the well-mixed boundary layer, $\mathscr{A}_{\text{SFC}}$ is the value of $\bar{\mathscr{A}}$ at the sea surface, and C_A is an empirical coefficient. For $\mathscr{A} = c_p \ln \theta_e$, (10.23) becomes

$$\tau_S(0) = -\bar{\rho}\, c_p C_S |\bar{v}| \left[\ln \bar{\theta}_e(h) - \ln \theta_{es}(\text{SST}) \right] \tag{10.24}$$

where $\theta_{es}(\text{SST})$ is the saturation equivalent potential temperature calculated at the sea-surface temperature (SST). For $\mathscr{A} = m$, (10.23) becomes

$$\tau_m(0) = -\bar{\rho}\, C_D |\bar{v}| \left(\bar{m} - m_{\text{SFC}} \right) = -\bar{\rho}\, C_D |\bar{v}|\, r\bar{v} \tag{10.25}$$

where the notation C_D stands for *drag coefficient*, and substitution for m from (2.35) has been made to obtain the last expression on the right. The sea-surface temperature enters the calculations as a crucial specified quantity in (10.24). It is through the surface flux of θ_e that the storm obtains its energy. In region II of the boundary layer, where fluxes at h are negligible, (10.21) for $\bar{\mathscr{A}} = \bar{\hat{S}}$ and $\bar{m}$ becomes, with substitution from (10.9), (10.24), and (10.25),

$$\left. \frac{\partial \bar{\hat{S}}}{\partial r} \right|_h \psi(h) = c_p \left. \frac{\partial \ln \bar{\theta}_e}{\partial r} \right|_h \psi(h) = r\bar{\rho}\, c_p C_S |\bar{v}| \left[\ln \bar{\theta}_e(h) - \ln \theta_{es}(\text{SST}) \right] \tag{10.26}$$

and

$$\left. \frac{\partial \bar{m}}{\partial r} \right|_h \psi(h) = \bar{\rho}\, r^2 C_D |\bar{v}|\bar{v} \tag{10.27}$$

Since

$$\left. \frac{\partial \bar{\hat{S}}}{\partial r} \right|_h \Big/ \left. \frac{\partial \bar{m}}{\partial r} \right|_h = \left. \frac{\partial \bar{\hat{S}}}{\partial \bar{m}} \right|_h = \frac{d\bar{\hat{S}}}{d\bar{m}} \tag{10.28}$$

(10.26) divided by (10.27) gives us

$$\frac{d\bar{\hat{S}}}{d\bar{m}} = \left(C_S c_p / C_D r\bar{v} \right) \ln \left[\bar{\theta}_e(h) / \theta_{es}(\text{SST}) \right] \tag{10.29}$$

Substitution of this expression and (2.35) into (10.13) leads to

$$\ln \bar{\theta}_e = \ln \theta_{es}(\text{SST}) - \frac{C_D}{C_S c_p (T_B - T_o)} \left(\bar{v}^2 + \frac{f r\bar{v}}{2} \right), \quad z = h \tag{10.30}$$

---

[276] See Roll (1965, p. 251).

When this expression is substituted into (10.18) for $\ln \bar{\theta}_e$ and $\bar{v}$ is expressed in terms of radial pressure gradient by means of the gradient-wind equation (2.36), (10.18) becomes a differential equation for $p(r)$ at $z = h$ in region II. Thus, (10.18) and (10.30) form a set of simultaneous equations for θ_e and p at the top of the boundary layer in region II.

Equation (10.30) was derived assuming that the fluxes at the top of the boundary layer $[\tau_A(h)]$ are negligible in region II. That such an assumption is inappropriate for region III can be seen from the ratio $\bar{\theta}_e/\theta_{es}$(SST) implied by (10.30). This ratio is directly proportional to the surface relative humidity.[277] It can be seen from (10.30) that $\ln \bar{\theta}_e$, and hence the surface relative humidity, is a minimum at the radius of maximum wind, implying that the relative humidity would increase with increasing radius (and decreasing $\bar{v}$) in region III. Observations indicate that the surface relative humidity is approximately constant in hurricanes, with a value of about 80%. This result evidently is not given by (10.30) because the eddy flux at the top of the boundary layer was set equal to zero. In region III, strong downdrafts are associated with convection in rainbands. These drafts evidently contribute strongly to the fluxes at the top of the boundary layer and thus keep the relative humidity lower than predicted by (10.30).

Therefore, outside the radius of maximum wind (i.e., in region III) the boundary layer is parameterized simply by setting the surface relative humidity empirically to a constant value of 80%. The radius of maximum wind is then that radius at which the relative humidity in region II, implied by (10.30), reaches 80%, and the value of $\bar{\theta}_e$ in the boundary layer outside this radius is the value corresponding to 80% relative humidity. Use of this value of $\bar{\theta}_e$ in (10.18) then gives an equation for $p(r)$ at $z = h$ in region III, which matches that in region II at the radius of maximum wind. Since (10.30) determines $p(r)$ in region II, $\bar{\theta}_e(r)$, and $\bar{m}(r)$ at $z = h$ can now be calculated throughout regions II and III. [The value of $\bar{m}(r)$ is determined from $p(r)$ and the gradient-wind equation (2.36).] Since these quantities are known at $z = h$, (10.12) can be solved for the $\bar{m}$ surfaces above the boundary layer, and according to (10.9) the m surfaces can also be labeled as $\tilde{S}$ surfaces corresponding to the value of $\bar{\theta}_e$ at $z = h$ at the radius where they intersect the boundary layer. Thus, the streamlines within the eyewall cloud can be constructed if the sea-surface temperature SST, the outflow temperature T_o, the ambient surface relative humidity RH_a (80% in the above discussion), and several less critical quantities (f, p_a, r_a, C_s/C_D, and T_B) are specified.

The results of an example of this type of calculation are shown in Figs. 10.16–10.19 for a case in which the sea-surface temperature SST = 300 K, $T_B = 295$ K, $T_o = 206$ K, $RH_a = 80\%$, $f(\phi) = f(28°)$, $p_a = 1015$ mb, $r_a = 400$ km, and $C_s = C_D$. The central pressure of the model storm is 941 mb, and the maximum tangential wind component is 58 m s^{-1}. To obtain these results, the region II boundary-layer model was assumed to extend into the eye (region I). The solutions in region I are thus not reasonable since in region I the gradient wind balance (2.36) does not hold there. This procedure was adopted only for convenience. Since our main purpose in this chapter is to gain some insight into the nature of the air motions in the

---

[277] See Emanuel (1986a) for details.

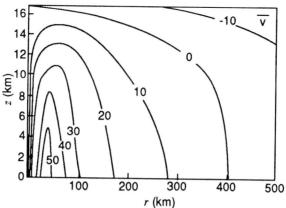

Figure 10.16 Gradient wind $\bar{v}$ (m s^{-1}) in the air–sea interaction model hurricane. (From Emanuel, 1986a. Reprinted with permission from the American Meteorological Society.)

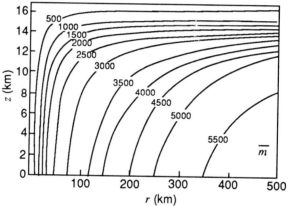

Figure 10.17 Absolute angular momentum $\bar{m}$ (10^3 m^2 s^{-1}) in the air–sea interaction model hurricane. (From Emanuel, 1986a. Reprinted with permission from the American Meteorological Society.)

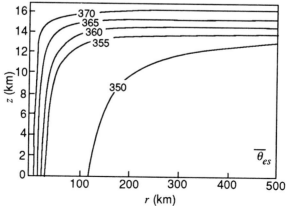

Figure 10.18 Saturation equivalent potential temperature $\bar{\theta}_{es}$ (K) in the air–sea interaction model hurricane. (From Emanuel, 1986a. Reprinted with permission from the American Meteorological Society.)

hurricane's clouds, which occur only in regions II and III, the inadequate treatment of the eye dynamics is not a serious deficiency. The primary vortex circulation is shown by Fig. 10.16. The distribution of maximum winds in the eyewall resembles that seen in Fig. 10.7, although the winds are stronger in the calculated case than in the composite section, where averaging from many cases has smeared the basic pattern. The $\bar{m}$ and $\bar{\theta}_{es}$ surfaces in Fig. 10.17 and Fig. 10.18 may be regarded as the streamlines of the flow above the boundary layer. The upward and outward flow demanded by the conservation of $\bar{m}$ is clearly evident from these cross sections. That the upward flow is highly concentrated in the eyewall region is indicated by Fig. 10.19, where the vertical velocity at the top of the boundary layer implied by the stream function in (10.26) and (10.27) is seen to have a strong maximum in an annular zone corresponding to the eyewall.

The theoretical picture just developed has been further examined by numerical model simulations, which use a nonhydrostatic cloud model of the type discussed in Sec. 7.5.3. The model is adapted to the hurricane by formulating it in cylindrical coordinates and making it two-dimensional by assuming uniformity in the azimuthal direction. Results from such a model are shown in Fig. 10.20. In this particular calculation, the undisturbed environment was assumed to be neutral to conditional instability but unstable in the conditionally symmetric sense. This assumption represents real hurricane environments fairly well, except that the latter are not actually neutral but characterized by a small degree of conditional instability. The numerical experiment was meant to minimize the role of vertical convection in the development of the storm. The idea was to see if storm development could take place without a conditionally unstable environment. The result was a rather realistic hurricane vortex. As we will see below, this model storm was similar to model storms simulated with conditionally unstable environments and resembles observations of intense hurricanes reasonably well. In Fig. 10.20 it can be seen that, after the model storm became well developed, streamlines in the eyewall region were parallel to the $\bar{m}$ and $\tilde{S}$ surfaces. Thus, it is evident that the eyewall region of the model storm developed as a conditionally symmetrically neutral circulation connected to a boundary layer in contact with the warm sea surface.

10.4 Clouds and Precipitation in the Eyewall

10.4.1 Slantwise Circulation in the Eyewall Cloud

As always in this book, our objective here is to gain insight into the air motions associated with clouds and precipitation. Since, in the hurricane, the clouds are mainly associated with the eyewall and the rainbands, we devote the last two sections of this chapter to these components of the storm. In this section we focus on the eyewall. Then in Sec. 10.5 we will examine the rainbands.

We begin our discussion of the eyewall clouds by returning to the observations of the inner-core region of the hurricane (Figs. 10.13 and 10.14) to compare them

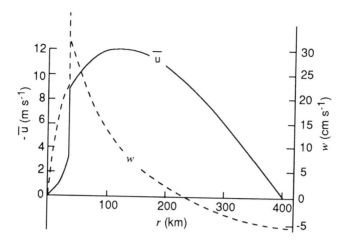

Figure 10.19

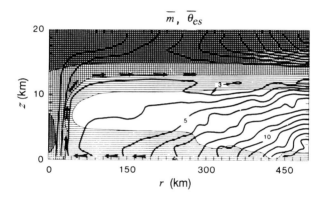

Figure 10.20

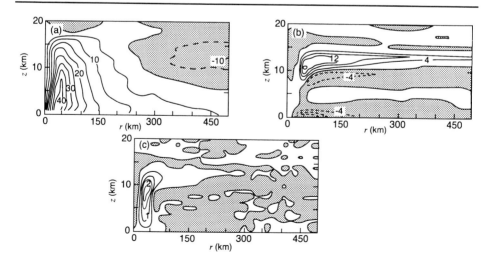

Figure 10.21

to the theoretical picture developed in Sec. 10.3 (Figs. 10.15–10.20). We have already noted that at the top of the boundary layer the strong vertical velocity in the theoretical circulation is concentrated at the location of the eyewall cloud (Fig. 10.19). Airborne Doppler radar observations, such as those illustrated in Fig. 10.13, also show the strong narrow updraft maximum in the eyewall. In addition, the radar observations indicate that the upward flux occurs in a slantwise pattern. Thus, the vertical mass transport in the eyewall cloud appears to be dominated by the momentum-conserving flow described by theory. Further consistency is seen in the upper-level radial outflow, which in both the theoretical and observed cross sections occurs in a concentrated layer aloft. The radial outflow is an important feature of the cloud and precipitation pattern since it spreads ice particles outward and thus accounts for the stratiform precipitation region lying typically just outside and adjacent to the eyewall convection. The strong tangential circulation predicted by theory also agrees well with the observed primary circulation, which in the context of the clouds and precipitation is important in distributing ice particles around the storm (Fig. 10.14).

10.4.2   Strength of the Radial–Vertical Circulation in the Eyewall Cloud

The comparisons made in Sec. 10.4.1 show *qualitatively* that the air motions observed in the eyewall region are generally similar to those computed from the theory of the symmetrically neutral circulation connected to the well-mixed boundary layer over the warm sea surface. We now explore the nature of the eyewall motions *quantitatively* by inquiring about the strength of the radial–vertical circulation responsible for determining the structure of the eyewall clouds and precipitation. We proceed by examining the results of both model simulations and data. We saw at the end of Sec. 10.3 that, when adapted to the hurricane, the nonhydrostatic cloud model of the type discussed in Sec. 7.5.3 produces a mature vortex with a conditionally symmetrically neutral eyewall circulation (Fig. 10.20). A few more results of this simulation are shown in Fig. 10.21. In addition, Fig.

Figure 10.19   Radial distribution of vertical velocity w (cm s^{-1}) at the top of the boundary layer and mean radial velocity $\bar{u}$ (m s^{-1}) within the boundary layer in the air–sea interaction model hurricane. (From Emanuel, 1986a. Reprinted with permission from the American Meteorological Society.)

Figure 10.20   Results of a two-dimensional nonhydrostatic numerical-model simulation of a hurricane. Fields were averaged over a 20-h period in which the storm was approximately steady. Shading indicates equivalent potential temperature values of 345–350 K (horizontal lines), 350–360 K (medium shading), and >360 K (heavy shading). Absolute angular momentum is shown by contours labeled in nondimensional units. Also indicated is the airflow through the updraft. (From Rotunno and Emanuel, 1987. Reprinted with permission from the American Meteorological Society.)

Figure 10.21   Results of a two-dimensional nonhydrostatic numerical-model simulation of a hurricane. Fields were averaged over a 20-h period in which the storm was approximately steady. (a) Tangential, (b) radial, and (c) vertical wind components (m s^{-1}) are shown. Negative values of fields are indicated by shading. (From Rotunno and Emanuel, 1987. Reprinted with permission from the American Meteorological Society.)

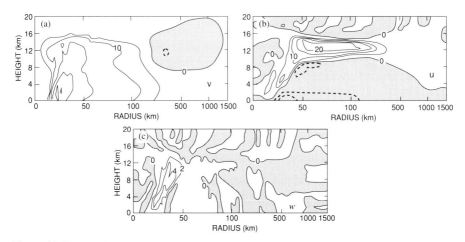

Figure 10.22 Results of a two-dimensional nonhydrostatic numerical-model simulation of a hurricane. Fields shown are (a) tangential, (b) radial, and (c) vertical wind components (m s⁻¹). Negative values of fields are indicated by shading. (From Willoughby *et al.*, 1984a. Reprinted with permission from the American Meteorological Society.)

10.22 contains results from a similar model for a simulation in which the ambient environment is not restricted to neutrality or to conditional symmetric instability, but rather is conditionally unstable. Despite this difference in large-scale environment, the model results are remarkably similar, indicating again that the eyewall dynamics are dominated by the slantwise motions. Both model simulations show maximum vertical motions in the eyewall (Figs. 10.21c and 10.22c) of 3–5 m s⁻¹, radial outflow aloft (Figs. 10.21b and 10.22b) is concentrated into an eyewall detrainment layer between 10 and 14 km in height, where u reaches values of 15–20 m s⁻¹.

These model results agree well with data from aircraft penetrations of hurricane eyewall regions. There are two principal types of data: flight track data obtained by *in situ* measurements made from the aircraft, and data from Doppler radar, airborne, which provides measurements in a large volume of space surrounding the flight path. Flight track data of the vertical velocities in the eyewall of an intense hurricane are shown in Fig. 10.23. Vertical velocities determined from the horizontal wind data by integrating the anelastic continuity equation (2.54) are shown in Fig. 10.23a, while Fig. 10.23b shows vertical velocities derived from measurements made with the inertial navigation system of the aircraft. Peak magnitudes of just over 6 m s⁻¹ are indicated by both types of measurement. Figure 10.24 shows that these vertical velocities are typical of those seen at flight level in other hurricanes. These values are slightly, though probably not significantly, larger than the values indicated by the model calculations. Vertical velocity observations by Doppler radar aboard an aircraft flying through another intense storm are shown in Fig. 10.25a. These data also show vertical velocity peaks (the largest >9 m s⁻¹) somewhat stronger than in the models or the other aircraft data. These

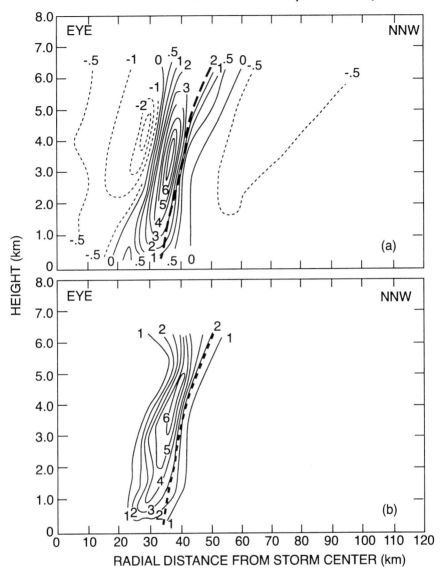

Figure 10.23 Flight-track data of the vertical velocity (m s^{-1}) in the eyewall of Hurricane Allen (1980). (a) Vertical velocity diagnosed from horizontal wind measurements by integrating the continuity equation. (b) Vertical velocity measured by the inertial navigation equipment of the aircraft. (From Jorgensen, 1984. Reprinted with permission from the American Meteorological Society.)

strongest peaks were at high levels, which can be observed remotely by radar but would not be seen at typical flight levels, which are usually below 6 km. It may be for this reason that the flight level data do not indicate more intense vertical velocities. Moreover, the model simulations referred to above assume that all of the condensation is liquid phase. When the model calculations incorporate bulk

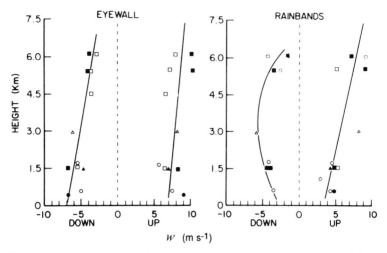

Figure 10.24   Peak updraft and downdraft speeds measured aboard research aircraft flying through hurricanes. Each style of point indicates a particular research flight. The value shown by a point is the maximum draft speed observed on that flight at that altitude. (From Jorgensen *et al.*, 1985. Reprinted with permission from the American Meteorological Society.)

ice-phase microphysics (Sec. 3.6.2), peak vertical velocities are ~ 10 m s^{-1} and occur at high levels, similar to the Doppler radar data.[278]

The Doppler radar data also confirm the radial outflow layer seen in the model results (cf. Figs. 10.21b, 10.22b, and 10.25b). The Doppler radar data do not extend as high as 14 km (the top of the radial outflow layer in the model output). Thus, Fig. 10.25b shows only the lower half of the radial outflow layer. Nonetheless, its presence and similarity to the model results are evident.

10.4.3   Vertical Convection in the Eyewall

From the model simulations and flight data, we now have an idea of the magnitudes of the in-cloud vertical motions and the upper-level radial outflow that are important in determining the eyewall cloud structure. We also have gained the impression that these motions are largely the slantwise momentum-conserving motion that has adjusted to a state of near neutrality with respect to conditional symmetric instability. We now extend our discussion of the eyewall cloud dynamics by considering whether *vertical* convection is superimposed on the slantwise convection in the real atmosphere (*vis à vis* the analytic model storm considered in Sec. 10.3). We previously pointed out that sometimes the eyewall actually exhibits a vertical structure, suggesting that the vertical convection can sometimes dominate the eyewall dynamics (Fig. 10.12). These cases are undoubtedly exceptional. Nevertheless, the ambient conditions in which hurricanes occur really are conditionally unstable, albeit not strongly so. Even when the ambient

[278] See Lord *et al.* (1984).

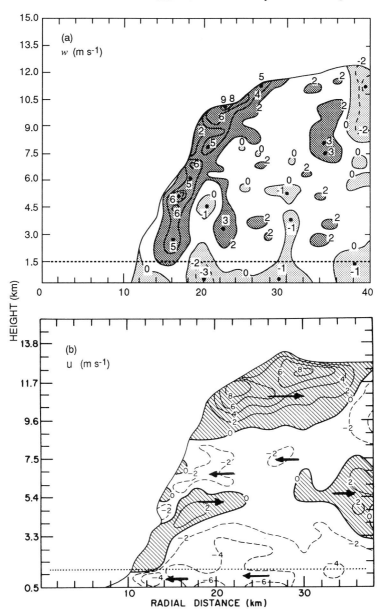

Figure 10.25 Air motions determined from airborne Doppler radar observations obtained during aircraft penetration of Hurricane Alicia (1983). Dotted line shows aircraft flight level. (a) Vertical velocity and (b) radial velocity (m s^{-1}). (From Marks and Houze, 1987. Reproduced with permission from the American Meteorological Society.)

conditions are neutral for conditional instability, the air–sea transfer, boundary-layer mixing, and hurricane circulation itself create regions of conditional instability within the storm as the hurricane develops. Thus, no matter what the case is regarding the degree of ambient conditional instability, it is reasonable to expect that some vertical convection occurs when conditioned instability is released and this convection must be superimposed on the slantwise circulation of the well-developed storm. The Doppler radar observations in Fig. 10.25a are consistent with this view in showing discrete bubbles of concentrated rising motion within the sloping upflow zone.

Model calculations also support this view. We have seen that, when the environment is assumed to be conditionally neutral, a model storm can evolve in which the streamlines in the eyewall become parallel to the $\bar{m}$ and $\bar{\theta}_{es}$ surfaces (Fig. 10.20). The process of adjustment of the eyewall cloud region to this conditionally symmetric neutral state is indicated by Fig. 10.26, which shows the fields of $\bar{m}$ and $\bar{\theta}_{es}$ at three successive times. The $\bar{m}$ and $\bar{\theta}_{es}$ lines gradually become more parallel as time passes. As the storm develops, air in the boundary layer flowing inward increases its value of $\bar{\theta}_e$, and conditional instability ($\partial \bar{\theta}_{es}/\partial z < 0$) develops near the center of the storm in the lower troposphere. Vertical convective motions take place in response to this instability and, in the process, transport high m upward. This transport produces local maxima of $\bar{m}$ and hence local areas of *inertial* instability, which is released in horizontal accelerations.[279] This combina-

---

[279] This process was pointed out by Willoughby *et al.* (1984a).

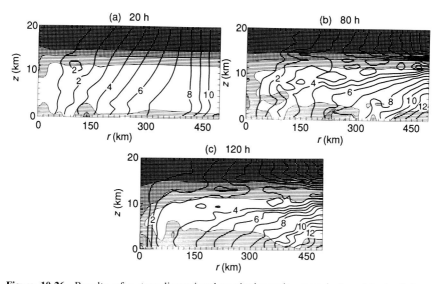

Figure 10.26 Results of a two-dimensional nonhydrostatic numerical-model simulation of a hurricane. Fields of saturation equivalent potential temperature (shaded) and absolute angular momentum (nondimensional units) are shown for three times. Closed momentum contours have a value of 3. Shading indicates equivalent potential temperature values of 345–350 K (horizontal lines), 350–360 K (medium shading), and >360 K (heavy shading). (From Rotunno and Emanuel, 1987. Reprinted with permission from the American Meteorological Society.)

tion of vertical convection in response to conditional instability and horizontal motion in response to local inertial instability evidently creates a final mixture in the eyewall region which is neutral to combined vertical and horizontal acceleration (i.e., a conditionally symmetric neutral state rather like that of the simple analytic model discussed in Sec. 10.3). In the process of achieving this final state, though, some vertical convection is superimposed on the developing slantwise circulation.

10.4.4 Downdrafts

As shown in Fig. 10.13, convective downdraft motion is found in the shower of heavy rain within the eyewall. Although the downdraft motions appear to interrupt the flow of the secondary circulation at low levels in the vicinity of the eyewall rain maximum, the peak downdraft velocities are rather weak: ~2–3 m s^{-1}. Thus, unlike thunderstorms (Chapter 8) and mesoscale convective systems (Chapter 9), the eyewall downdrafts probably do not have a strong feedback to the larger storm circulation. In the mature eyewall, the upward motion appears to be associated more with the frictional convergence of the inwardly spiraling low-level branch of the secondary circulation than with downdraft-induced convergence.

10.4.5 Eyewall Propagation

The eyewalls of highly symmetric hurricanes are observed to contract; that is, the eyewall propagates toward the center of the storm and thus shrinks in radius. As the eyewall rainband contracts, the central pressure of the storm decreases. After 1–2 days, the radius of the eyewall reaches its minimum size and dissipates. It is replaced by a new eyewall at a radius of 50–100 km and the central pressure of the storm rises. This process sometimes gives rise to multiple eyewalls. A schematic representation of a double eyewall structure is shown in Fig. 10.27, and an actual example is shown in Fig. 10.28. The old eyewall is near the storm center, while a new one is located farther out. As the life cycle of the eyewall band is repeated, the storm is characterized by a succession of shrinking eyewall rainbands and a correspondingly pulsating central pressure.

A theoretical explanation of the eyewall propagation has been attained mathematically by considering a point source of heat placed near the radius of maximum wind in a hurricane-like vortex. The secondary circulation response required to keep the vortex in gradient wind and hydrostatic balance is then described by the equations for the vortex. One of its characteristics is that temporal increases in tangential wind are greatest just inside the radius of maximum wind. This effect leads to contraction of the zone of maximum wind as the vortex intensifies.[280]

---

[280] This theory of eyewall propagation was put forth by Shapiro and Willoughby (1982). Observational support for the theory was provided by Willoughby *et al.* (1982).

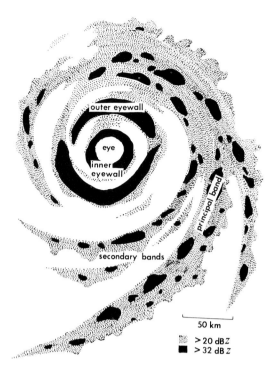

Figure 10.27 Schematic illustration of radar reflectivity in a Northern Hemisphere tropical cyclone with a double eyewall. (From Willoughby, 1988. Commonwealth of Australia copyright reproduced by permission.)

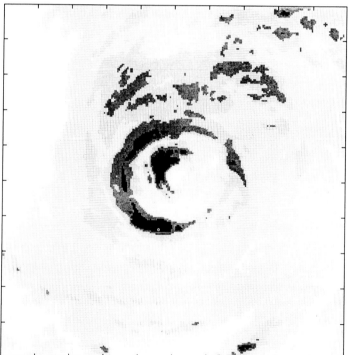

Figure 10.28 Example of double-eyewall structure. The plot is a composite of radar reflectivity from Hurricane David (1979). The darker the shading the higher the reflectivity. The size of the horizontal area of ocean enclosed by the box is 240 km × 240 km. The storm is centered at 16.34°N, 65.18°W. (From Houze and Hobbs, 1982.)

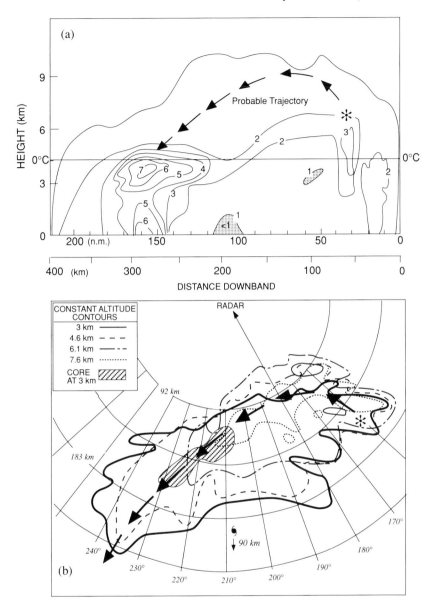

Figure 10.29 Hurricane rainband observed by a ground-based radar at North Truro, Massachusetts, as Hurricane Esther (1961) moved northward over the Atlantic Ocean just off the coast of the United States. (a) Vertical cross section showing returned power at a sequence of gain settings (numbered 1–7). The horizontal axis is along the axis of the rainband. Only the most intense echoes are seen at the highest gain setting, but the intensity has not been corrected for range from the radar. An hypothesized ice particle trajectory is indicated. (b) Horizontal projection of the hypothesized ice particle trajectory is superimposed on the plan view of the radar echo at several altitudes. For the 3-km level, the location of the intense echo core is shown in addition to the overall echo outline. (From Atlas et al., 1963.)

10.5   Rainbands

The rainbands, which are found outside the eyewall, are of more than one type. A rainband can be purely stratiform when a plume of ice particles from somewhere along the eyewall falls through the 0°C level after spiraling outward at upper levels (Fig. 10.14). Other rainbands appear to be distinct from the eyewall. These rainbands are labeled as *principal* and *secondary* rainbands in the nomenclature of Fig. 10.3 and are represented as the convective phenomena occurring in region III of the theoretical model sketched in Fig. 10.15.

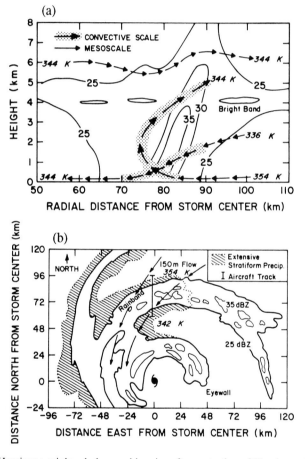

Figure 10.30   Hurricane rainband observed by aircraft penetration of Hurricane Floyd (1981). (a) Vertical cross section with horizontal axis along the aircraft track, which was across the axis of the rainband. The track is shown in (b), which is a horizontal projection of the data. In both panels, contours show radar reflectivity in dBZ, temperatures refer to values of equivalent potential temperature, and arrows indicate airflow (on either the convective scale or mesoscale, as indicated by the style of arrow). (From Barnes *et al.*, 1983. Reprinted with permission from the American Meteorological Society.)

Analyses of features that appear to have been principal rainbands are represented in Figs. 10.29 and 10.30. In the first example, a vertical cross section in the *along*-band direction is presented (Fig. 10.29a). Convective radar echo was seen on the upwind end of the rainband, while stratiform precipitation with a bright band at the melting level (Sec. 6.1.2) was seen downwind. Precipitation trajectories of the type we have seen in the eyewall (Fig. 10.13) and in mesoscale convective systems (Figs. 6.12 and 9.13) were inferred, where ice particles detrained from the upper levels of the convective towers are advected downwind while slowly falling, producing the bright band when they finally fall through the 0°C level.[281]

The analysis shown in Fig. 10.30 presents a vertical cross section in the *across*-band direction. In this cross section we again see both convective and stratiform structure in the radar echo. A sloping convective cell is surrounded by stratiform echo characterized again by a radar bright band. The cell slopes radially outward with height, in the same sense as the eyewall convection. Inward flowing boundary-layer air of high $\bar{\theta}_e$ rises out of the boundary, then reverses its radial direction and flows radially outward along a slantwise path as it continues to rise. This slantwise upward motion in relation to the sloping radar echo maximum is similar to the slantwise motion in the eyewall (Fig. 10.13). Also similar to the eyewall, downdraft motion occurs in the sloping precipitation echo core. Whereas in the case of the eyewall the downdraft appears to play a minor role in the dynamics, the downdraft in the rainband is significant. It transports low-$\bar{\theta}_e$ air from the midlevel environment into the boundary layer. This type of downdraft is evidently the primary mechanism by which negative flux of θ_e occurs at the top of the boundary layer in region III of the theoretical model hurricane (Sec. 10.3, Fig. 10.15). After entering the boundary layer as the downdraft spreads out below cloud, the low-$\bar{\theta}_e$ air flows on toward the center of the storm.

[281] Atlas *et al.* (1963), from whose paper Fig. 10.29 is taken, appears to have been the first to suggest this type of stratiform precipitation mechanism, which is now recognized to be prevalent in many types of mesoscale precipitation systems.

Chapter 11 | Precipitating Clouds in Extratropical Cyclones

"... notwithstanding the repeated fall of rain which has fallen almost constantly since ... November last."[282]

The primary cloud and precipitation producers in middle latitudes are extratropical cyclones. As noted in Sec. 1.3.3, these storms include both large-scale frontal cyclones and smaller polar lows (Figs. 1.29 and 1.30). These weather systems all contain arctic or polar air and, in this way, differ markedly from the hurricanes considered in Chapter 10, which are low-level cyclonic circulations of purely tropical air. Unlike the tropics, where the thermal contrasts are small over large horizontal distances, extratropical latitudes are characterized by zones of strong horizontal temperature contrast, referred to as *baroclinic zones*. These zones are usually associated with an upper-level jet stream, since, according to the thermal-wind relation (Sec. 2.2.5), the wind speed increases with height in a zone of strong thermal contrast. Under certain conditions, a baroclinic jet is unstable and, if perturbed in certain ways, a synoptic-scale *baroclinic wave* can develop. An important characteristic of this wave is its tendency to take the form of a closed cyclonic disturbance at low levels. The process of forming the cyclone is called *cyclogenesis*. The developing cyclone, which is the product of the cyclogenesis, is characterized, in turn, by *frontogenesis*, which is the process in which the airstreams of contrasting thermal properties within the cyclone come together in such a way that the temperature gradients and cloud-producing vertical circulations in the lower troposphere are concentrated on small scales.

The end product of baroclinic wave development is manifested in a variety of types of extratropical cyclones. The largest members of the spectrum of storms that result from this process are large frontal cyclones, such as the large cloud system shown in Fig. 1.29. In these storms, cyclogenesis produces a major low-pressure center on the surface weather map, and long fronts form and account for the extensive cloud bands extending outward from the storm center. Polar lows are smaller extratropical cyclones that occur poleward of major frontal cyclones. They are connected with the flow of cold polar or arctic air over a warm ocean

---

[282] From the journal entry of Captain Meriwether Lewis on 23 March 1806 after the expeditionary party of Lewis and Captain William Clark, dispatched to the west coast of North America by President Thomas Jefferson, had wintered on the Oregon coast. The explorers had suffered the precipitation of one extratropical cyclone after another and were evidently feeling weary as a result of the long rainy period—just as do many modern residents of the Pacific Northwest coast at the end of winter.

surface and are characterized by predominantly convective clouds at some point in their lifetimes. Comma clouds, such as the smaller cloud system to the west of the main frontal cyclonic cloud system in Fig. 1.29, are the larger of the polar lows, and they themselves may exhibit a surface low center and fronts, which are the products of cyclogenesis and frontogenesis associated with a baroclinic wave of especially short wavelength. Polar lows occurring farther from major frontal cloud systems tend to be the smallest members of the spectrum and sometimes appear to have a hurricane-like appearance (Fig. 1.30).

In this chapter, we focus on the clouds of extratropical cyclones (large frontal cyclones, comma cloud systems, and smaller polar lows). We have already considered aspects of these clouds in Chapters 1, 5, and 6. In particular, we examined the cloud streets, which form as the equatorward-moving cold-air stream of a cyclone moves out over a warmer ocean surface (Figs. 1.11b, 1.30, and 5.21). Although the individual clouds comprising the cloud streets may occasionally precipitate, they are not the primary precipitation-producing clouds of the cyclones. Rather it is the broader bands of clouds with extensive cirriform cloud tops, seen in satellite pictures such as those in Figs. 1.29 and 1.30 to be curling around and extending outward from cyclonic storm centers, that produce most of the precipitation in extratropical cyclones. In Chapter 6, we considered some basic structural aspects of these larger, precipitating clouds. In this chapter, we seek a dynamical and physical understanding of them.

Of all the cloud systems considered in this text, the precipitating clouds in extratropical cyclones have the widest range of scales of phenomena contributing to the cloud formation processes. The plan of this chapter is to proceed from the largest of these scales to the smallest. On the largest scale, the dynamics of the synoptic-scale baroclinic wave itself contribute to producing the clouds and precipitation, since it is the area of upward air motion within the wave that constitutes the broad-scale region in which clouds can form. Therefore, we will begin in Sec. 11.1 by considering the dynamics governing the vertical air motions of the wave. We will see also, in that discussion, how the wind field in the baroclinic wave favors the development of the low-level cyclone. In Sec. 11.2, we will investigate how the vertical motions associated with the wave become concentrated in the vicinity of frontal zones by examining the dynamics of frontogenesis. This discussion will serve to indicate the nature of the vertical circulation associated with the frontal zones that occur in a baroclinic wave. It is these vertical circulations that are of primary importance in focusing the cloud formation processes in the cyclone at the frontal zones. In Sec. 11.3, we will investigate how the frontal zones, produced by frontogenesis, are distributed spatially within a developing cyclone, since this spatial pattern determines, to a first approximation, the distribution of precipitating clouds in extratropical cyclones. Superimposed on the clouds produced by the frontal air motions are still smaller mesoscale *rainbands* and convection, within which the precipitation-producing cloud microphysical processes are most concentrated. In Secs. 11.4 and 11.5, we examine this superimposed structure by reviewing the observed structure of clouds and precipitation in extratropical cyclones. Sections 11.4.1 and 11.4.2 concentrate on the more general aspects of the observed pattern of clouds and precipitation in a developing frontal cyclone. Then Secs. 11.4.3–11.4.6 describe in more detail the clouds and rainbands that occur within specific parts of this pattern. Finally, Sec. 11.5 examines the clouds of polar lows.

11.1 Structure and Dynamics of a Baroclinic Wave

11.1.1 Idealized Horizontal and Vertical Structure

A baroclinic wave can develop in the atmosphere when the wind at a given pressure level is blowing parallel to the temperature (or potential temperature) isotherms and there is a nonzero temperature gradient across the flow. If certain other conditions are met, a small initial wave perturbation superimposed on the flow will amplify, and the situation is referred to as a state of *baroclinic instability*.[283] To obtain a qualitative understanding of the baroclinic wave-growth process, consider a region of strong north–south temperature gradient in the Northern Hemisphere. Suppose that a weak wave-like perturbation (e.g., Fig. 11.1) appears in a geostrophic flow that is otherwise uniform and purely zonal (i.e., west–east) at some particular pressure level. Let the wave perturbation be of sufficiently large scale that its wind field is also geostrophic. For simplicity, assume that the level under consideration for this wave-like perturbation is moving at exactly the same velocity as the zonal flow. Then, if we were to view the air motion in a coordinate system moving with the wave, the zonal flow would vanish, and the only horizontal motion remaining would be the meridional (i.e., northward and southward) velocity components associated with the wave-like perturbation. As shown in Fig. 11.1, the meridional motions distort the isotherms, which are assumed here to have been initially perfectly straight, with an east–west orientation and cold air lying to the north. The northerly wind in the vicinity of point A advects cold air southward, while the southerly wind at B carries warm air northward. These motions tend to produce a trough in the temperature isotherms that lags the trough in the geopotential field by a quarter wavelength. In the

---

[283] See Holton (1992) for a more detailed treatment of baroclinic instability.

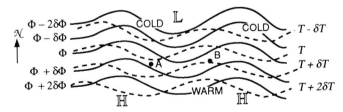

Figure 11.1 Distribution of geopotential height (Φ) and temperature (T) on a constant-pressure surface in a developing baroclinic wave in the Northern Hemisphere. The pressure surface is located near the level where the speed of the wave is the same as the speed of the mean zonal flow. (From Wallace and Hobbs, 1977.)

absence of other influences, the horizontal temperature advection by the perturbation flow will continue to distort the isotherms from their original east–west configuration, thus causing the east–west temperature contrasts in the wave to amplify further.

In a growing wave, in which the temperature perturbations are becoming larger, the kinetic energy associated with the wave disturbance increases. The mechanism by which the kinetic energy increases is a thermally direct circulation, in which the warmer air at B rises and the cooler air at A sinks, thus lowering the center of gravity of the fluid. The air at B is then on an upward-sloping poleward trajectory, while the air at A is on a downward-sloping equatorward trajectory. Implications for cloud formation are then clear: clouds are favored east of the trough of the baroclinic wave, while they are suppressed on the west side of the trough.

The range of slopes of parcel trajectories is limited. For example, the rise of air that we have postulated to occur at B cannot be too great, or else the effect of warm advection will be completely counteracted by the decrease of temperature, which must occur as the air rises. The allowable flow configurations can be determined theoretically, and the ones for which the disturbance grows most rapidly can be calculated.[284] All of these details are not necessary for our present discussion. We will confine ourselves to some general features of the developing wave that will enhance our appreciation of the vertical motions, which determine the regions of the wave where clouds may be favored or suppressed.

To gain a qualitative picture of the nature of the vertical motions, we may consider some aspects of the vertical structure of the idealized baroclinic wave that are implied by the out-of-phase relationship of the temperature and geopotential height contours. The location of the temperature trough to the west of the trough in the geopotential field (Fig. 11.1) indicates that the trough of the baroclinic wave tilts toward the west with height, as shown in Fig. 11.2. The tilt of the trough toward the cold air is required in order to maintain hydrostatic balance throughout the disturbance.[285] The ridge in the height field (or high in the pressure field) tilts westward for the same reason. Since the wave disturbance is in geostrophic balance, the winds between the ridges and troughs alternate between northerly and southerly as indicated in Fig. 11.2. Since the disturbance is both geostrophically and hydrostatically balanced, the thermal wind relation (Sec. 2.2.5) applies. Recall, however, that the northerly geostrophic winds advect colder air into the plane of the cross section, and the southerly geostrophic winds bring in warmer air. In the absence of other effects, these advective effects would destroy the thermal wind balance. The upward motion of the warm air and the downward motion of the cold air are needed to restore the balance. As the warm air ascends, its temperature is lowered as a result of the expansion. The temperature of the cold air increases as it sinks. These temperature changes thus mitigate the effect of the advection by lessening the west–east thermal contrast across the

---

[284] See Hoskins (1990) or Holton (1992).
[285] See Wallace and Hobbs (1977, Fig. 2.3c).

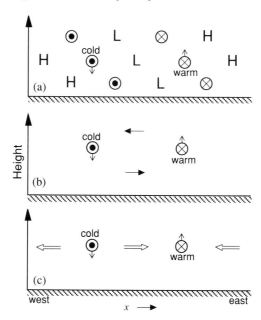

Figure 11.2   Longitude-height cross section through a growing two-dimensional baroclinic disturbance. Features indicated include poleward ($\otimes$) and equatorward (circled dot) meridional geostrophic air motions, zonal and vertical components of ageostrophic circulation [$(u_a, \bar{w})$ thin arrows], regions of warm and cold air, and Q-vectors (double arrows). To determine the Q-vectors, it is assumed that the geostrophic wind does not vary in the y-direction, that the zonal component of the base-state geostrophic current u_g does not vary in the x-direction, and that b_y (not shown) is a negative constant. Then the Q-vector reduces to $\mathbf{Q} = -v_{gx} b_y \mathbf{i}$, which changes direction depending on the local sign of v_{gx}. (Adapted from Hoskins, 1990. Reproduced with permission from the American Meteorological Society.)

disturbance. By continuity, the vertical motions must be compensated by horizontal ageostrophic motions. If we assume the idealized disturbance depicted in Fig. 11.2 to be two-dimensional, with no variability in the meridional direction, then the vertical motions must be compensated by ageostrophic motions in the zonal direction, as indicated in Fig. 11.2b. These motions also act to reduce the effect of the geostrophic temperature advection. The dynamical role of the ageostrophic circulation (u_a, w) is in fact to keep the disturbance in thermal wind balance. This fact can be demonstrated more formally, extended to three dimensions, and made quantitative by the following mathematical arguments.

11.1.2   Dynamics Governing Large-Scale Vertical Air Motion

Consider the flow to be Boussinesq, inviscid, dry adiabatic, geostrophic, hydrostatic, and at constant Coriolis parameter f. Then the governing equations are the dry adiabatic thermodynamic equation (2.11), the inviscid quasi-geostrophic equation of motion with constant f (2.23), the hydrostatic equation (2.38), and the Boussinesq continuity equation (2.55).

In this chapter we will view these relationships in a reference frame in which vertical variations are expressed in terms of the *pseudoheight* $\mathfrak{z}$,[286] defined as

$$\mathfrak{z} \equiv \left[1 - (p/\hat{p})^{\kappa}\right](1/\kappa)H_s \tag{11.1}$$

where $\hat{}$ indicates a constant reference state, taken to be representative of conditions near the earth's surface, $\kappa = R_d/c_p$, and $H_s = \hat{p}/(\hat{\rho}g)$. This vertical coordinate very nearly approximates the physical height z. It allows one to work with all the advantages of a pressure coordinate system, while maintaining the ability to visualize processes in a height framework. Taking $\partial(11.1)/\partial p$, noting from the hydrostatic relationship (2.38) that

$$\partial/\partial p = -\alpha g^{-1} \partial/\partial z \tag{11.2}$$

and making use of the definition of potential temperature (2.8), we find that the pseudoheight is related to the geopotential height z by

$$\theta \Delta \mathfrak{z} = \hat{\theta} \Delta z \tag{11.3}$$

We define a new thermodynamic variable b, given by

$$b \equiv g \frac{\theta}{\hat{\theta}} \tag{11.4}$$

This quantity is simply the potential temperature expressed as an equivalent buoyancy [as defined by (2.48)]. The use of this variable along with the pseudoheight $\mathfrak{z}$ gives a simplified form to the basic equations.

The horizontal equation of motion (2.23) for quasi-geostrophic flow is unaffected by the introduction of b and $\mathfrak{z}$. However, by combining (11.3) with (2.38) and making use of (11.4), we find that the hydrostatic relation may be rewritten in terms of b and $\mathfrak{z}$ as

$$b = \frac{\partial \Phi}{\partial \mathfrak{z}} \tag{11.5}$$

where Φ is the geopotential. By taking $\mathbf{k} \times \nabla$ (11.5) and making use of the definition of the geostrophic wind as expressed in (2.68), we obtain a convenient form of the thermal wind equation:

$$f \frac{\partial \mathbf{v}_g}{\partial \mathfrak{z}} = \mathbf{k} \times \nabla b \tag{11.6}$$

This form could have also been obtained by substituting from (2.3), (11.2), (11.3), and (11.6) in the thermal wind equation (2.41). The relations (11.3) and (11.4) furthermore imply that the Boussinesq continuity equation (2.55) in $\mathfrak{z}$-coordinates

[286] Following Hoskins and Bretherton (1972).

takes the form

$$u_x + v_y + \tilde{w}_{\mathfrak{z}} + \tilde{w}\frac{\partial \ln b}{\partial \mathfrak{z}} = 0 \tag{11.7}$$

where $\tilde{w} \equiv D\mathfrak{z}/Dt$. From the definition (11.4), it is evident that as long as the potential temperature does not deviate too much from the near-surface potential temperature $\hat{\theta}$, the Boussinesq continuity equation preserves its form in $\mathfrak{z}$-coordinates:

$$u_x + v_y + \tilde{w}_{\mathfrak{z}} \approx 0 \tag{11.8}$$

Under near-adiabatic and quasi-geostrophic conditions, the dry adiabatic form of the First Law of Thermodynamics (2.11) may be written with the aid of (11.3) in terms of b and $\mathfrak{z}$ as

$$\frac{D_g b}{Dt} + \tilde{N}^2 \tilde{w} \approx 0 \tag{11.9}$$

where $\tilde{N}^2$ is defined as in (2.98), except with $\mathfrak{z}$ replacing z. According to (2.25), the subscript g in D_g/Dt indicates that the only advection terms included in the total derivative are horizontal advection terms and that they are calculated with the geostrophic wind components u_g and v_g.

We may now use the basic equations written in terms of b and $\mathfrak{z}$ to see the role of the ageostrophic circulation in maintaining the thermal wind balance. First take $\mathbf{k} \times \nabla$ (11.9), to obtain

$$\frac{D_g}{Dt}(\mathbf{k} \times \nabla b) = \mathbf{k} \times \mathbf{Q} - \tilde{N}^2 \mathbf{k} \times \nabla \tilde{w} \tag{11.10}$$

where we have made use of the *Q-vector*, defined as

$$\mathbf{Q} \equiv \left(-u_{gx}b_x - v_{gx}b_y\right)\mathbf{i} + \left(-u_{gy}b_x - v_{gy}b_y\right)\mathbf{j} \equiv Q_1\mathbf{i} + Q_2\mathbf{j} \tag{11.11}$$

Now, taking the $\mathfrak{z}$-derivative of the horizontal equation of motion (2.23), substituting from (11.6), and noting from (2.68) that the geostrophic wind is nondivergent (i.e., $u_{gx} = -v_{gy}$), we obtain

$$\frac{D_g}{Dt}\left(f\mathbf{v}_{g\mathfrak{z}}\right) = -\mathbf{k} \times \mathbf{Q} - f^2\mathbf{k} \times \mathbf{v}_{a\mathfrak{z}} \tag{11.12}$$

From the first terms on the right-hand sides of (11.10) and (11.12), it is evident that the action of the geostrophic wind upon the temperature field, as expressed quantitatively by the Q-vector, tends to destroy the thermal-wind balance by changing the two sides of the thermal wind equation (11.6) by equal but opposite amounts. From the second terms on the right-hand sides of (11.10) and (11.12), it is evident that the role of the three-dimensional ageostrophic motion ($\mathbf{v}_a, \tilde{w}$) is to maintain the thermal-wind balance. This result expresses more quantitatively the conclusion drawn qualitatively from the idealized two-dimensional disturbance sketched in Fig. 11.2b and extends the conclusion to three dimensions. The role of the ageostrophic circulation in restoring the thermal wind balance at the same rate that it is being destroyed by the action of the large-scale wind on the thermal field can be seen most clearly by subtracting (11.12) from (11.10), which leads to

$$\tilde{N}^2 \nabla_H \tilde{w} - f^2 \mathbf{v}_{a_{\hat{3}}} = 2\mathbf{Q} \tag{11.13}$$

where ∇_H is the horizontal gradient operator. Taking $\nabla_H \cdot$ (11.13) and making use of the continuity equation (11.8), we obtain the *omega equation:*[287]

$$\tilde{N}^2 \nabla_H^2 \tilde{w} + f^2 \tilde{w}_{\hat{3}\hat{3}} = 2\nabla_H \cdot \mathbf{Q} \tag{11.14}$$

The solution of this equation for a given $\mathbf{Q}$ (i.e., for a given temperature and geostrophic wind field) determines the vertical velocity field that is required to maintain thermal wind balance. The omega equation is easily generalized to include additional sources of vertical motions arising from friction, diabatic heating, and the variation of f with latitude.

Since the left-hand side of (11.14) behaves like a three-dimensional Laplacian of $\tilde{w}$, we conclude that at a given pressure level (constant $\hat{3}$), convergence of $\mathbf{Q}$ is associated with upward air motion and divergence of $\mathbf{Q}$ with downward motion. This relation is readily applied to the schematic disturbance in Fig. 11.2. The Q-vectors associated with the middle-level geostrophic north–south flow and temperature field are shown as double arrows. They converge in the warm-air region and diverge in the cold region. Therefore, the vertical motion, which we previously postulated to be upward in the warm region and downward in the cold region, is seen via (11.14) to be necessary to keep the disturbance in thermal wind balance. Hence, a wave disturbance such as that shown in Fig. 11.1 must have warm air rising and cold air sinking if it is to remain in geostrophic and hydrostatic balance.

11.1.3   Application of the Omega Equation to a Real Baroclinic Wave

The omega equation is a *diagnostic relation,* meaning that it contains no time derivatives and is readily applied to a real weather map showing the distribution of temperature and geostrophic wind. A 700-mb chart showing a developing baroclinic wave is shown in Fig. 11.3. The geostrophic wind, indicated by the geopotential height contours, and the temperature isotherms are shown in Fig. 11.3a. The temperature trough lags the trough in the height field, as in the idealized example in Fig. 11.1. A developing frontal cyclone at lower levels was associated with the wave, as indicated by the superimposed surface frontal positions. The Q-vectors computed from the 700-mb temperature and geostrophic wind fields are indicated in Fig. 11.3b. The divergence of $\mathbf{Q}$ is shown by dashed contours, which indicate large-scale subsidence of cold air on the southwest side of the trough in

---

[287] The name omega equation refers to the fact that this relation is often written in terms of the "vertical motion" in pressure coordinates, $\omega \equiv Dp/Dt$.

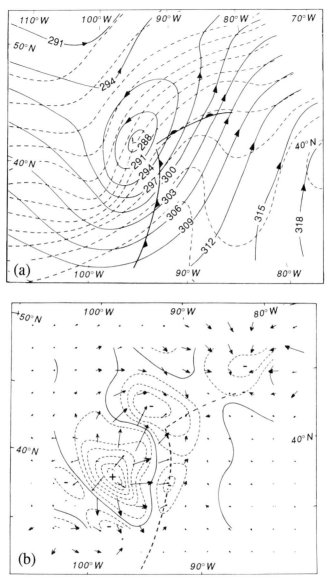

Figure 11.3   Example of a 700-mb chart showing a developing baroclinic wave. (a) Geopotential height contours (30-m interval) and temperature (2°C interval). Direction of geostrophic wind indicated by arrowheads. (b) Q-vectors computed from the temperature and geostrophic wind fields. Isopleths of $(2g/\hat{\theta})\nabla \cdot \mathbf{Q}$ for every 1×10^{-16} m^{-1}s^{-3}, with the zero isopleth being solid. Sea-level frontal positions are indicated by standard symbols in (a) and by heavy dashed lines in (b). (From Hoskins and Pedder, 1980. Reprinted with permission from the Royal Meteorological Society.)

the height field, centered well to the rear of the surface cold front. General ascent of the warm air is indicated on the east and northeast sides of the trough. The largest and most intense region of ascent is ahead of the surface warm front. However, the region of ascent extends southward over the surface cold front, so that altogether the region of upward motion (convergent **Q**) forms a pattern that resembles a giant comma. In view of this pattern of general ascent in the baroclinic wave, it is not surprising that cloud systems associated with baroclinic waves (not only comma cloud systems per se, but also the larger frontal cyclones and smaller polar lows) are typically in the form of a large comma shape (e.g., Fig. 1.29 and 1.30).

11.1.4 Low-Level Cyclone Development

The region of ascent at 700 mb in Fig. 11.3 is also seen to encompass an area lying over the low-pressure center at low levels (where the cold front and warm front meet). Upward motion at this level implies horizontal velocity convergence at lower levels, since the vertical motion vanishes at the lower boundary. Taking $\mathbf{k} \cdot \nabla \times$ (2.23) and applying the mass continuity relation (11.8), we obtain the geostrophic vorticity equation

$$\frac{D_g \zeta_g}{Dt} = f \frac{\partial \tilde{w}}{\partial ӟ} \tag{11.15}$$

which is the form taken by (2.59) under quasi-geostrophic conditions at constant f. At the center of the low-level cyclone the right-hand side is positive because of the convergence. Therefore, since there is no vorticity advection at the center of a vortex, the geostrophic vorticity in the low center must increase. According to (2.69), the geostrophic vorticity is proportional to the Laplacian of the geopotential. The increase in vorticity then implies strengthening of the low-level depression in the geopotential field. Since the strongest ascent implied by the Q-vectors in Fig. 11.3b is actually northwest of the surface low center, there is an indication that the low-level cyclone is in the process of shifting its position to be more in line with the upper-level trough.

The simultaneous evolution of the low-level cyclone and upper-level wave is pictured schematically in Fig. 11.4. The connection between the lower- and upper-level flow is found in the hydrostatic relation (11.5). Since the large-scale air motion is in hydrostatic balance, the geostrophic flows at upper and lower levels are related through the thermal field. Integration of (11.5) with respect to $ӟ$ shows that the geopotential height difference (or *thickness*) between any two pressure levels is directly proportional to the mean potential temperature of the intervening layer of air.[288] Figure 11.4 shows the geopotential height contours, geostrophic flow direction, and thickness pattern for the 1000- and 500-mb levels in an idealized developing baroclinic wave. As noted in discussing Fig. 11.1, continued thermal advection in the incipient baroclinic wave leads to further distortion of the

---

[288] The mathematical expression of this relation is called the *hypsometric equation*. See Chapter 2 of Wallace and Hobbs (1977).

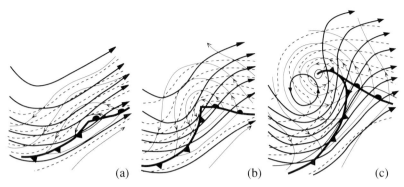

Figure 11.4   Idealized model of a midlatitude baroclinic wave disturbance with fronts (shown by standard frontal symbols) in three stages of development. Geopotential height contours at 1000 mb (thin solid) and 500 mb (heavy solid) are shown together with fronts at 1000 mb. Geastrophic flow direction indicated by arrowheads. Contours of the 1000–500-mb thickness (dashed). (From Palmén and Newton, 1969.)

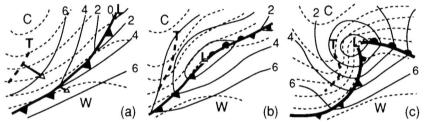

Figure 11.5   Schematic model showing stages in surface cyclone development that occurs when an upper-level baroclinic wave (whose trough is indicated by the heavy dashed line marked with a T) moves over a pre-existing low-level front. Thick arrows in (a) show velocities of trough and front. New cyclone appears in (b). Further development is shown in (c). Isobars are labeled in departure in millibars from a sea-level pressure of 1000 mb. Lighter dashed lines are isotherms. C and W indicate cold and warm areas, respectively. (Adapted from Petterssen, 1956. Reproduced with permission from McGraw-Hill, Inc.)

isotherms in time. The time sequence in Fig. 11.4 illustrates this distortion. The wave in the thickness field increases in amplitude as the storm ages. As anticipated in the discussion of Fig. 11.3, the low-level cyclone moves back toward a position more in line with the upper-level trough. By the time of Fig. 11.4c, the intensification of the upper-level trough has led to the formation of a closed cyclonic circulation, and the low-level cyclone is located almost under it. Eventually, the upper and lower parts of the disturbance become aligned. The thickness field, which according to (11.5) is obtained by graphical subtraction of the upper- and lower-level geopotential fields, becomes similarly aligned. Thermal advection then ceases, and the wave and frontal cyclone can no longer intensify.

Symbols are used in Fig. 11.4 to indicate the positions of *fronts* at the 1000-mb level. All of the fronts are found at the warm edges of *frontal zones*, which are zones of strong thermal gradient, identifiable in the figure by close packing of the thickness contours. The *cold front* marks the warm edge of the frontal zone in the cold polar air moving equatorward and eastward on the west side of the cyclone. The *warm front* marks the warm edge of the receding frontal zone on the east side of the cyclone; it indicates the advancing edge of the tropical air flowing poleward at the 1000-mb level. Since the cold front advances more rapidly than the warm front recedes (cf. Fig. 11.4a and b), the warm sector progressively narrows, and warm-sector air is lifted off the surface by the time of Fig. 11.4c. At this stage, the developing cyclone is said to be *occluded*. At 1000 mb an *occluded front* is drawn along the line of maximum temperature extending from the point where the cold and warm fronts meet to the center of the low-pressure system.

Figure 11.4 suggests that the low-level cyclone and fronts appear and develop simultaneously as a continuous process in the context of a baroclinic wave development. In many real cases, a baroclinic wave that is well defined at upper or middle levels, but weak in the lower atmosphere, moves over a *pre-existing* low-level frontal zone. This process is indicated schematically in Fig. 11.5. When the area of upward motion ahead of the trough of the upper-level wave (Figs. 11.2 and 11.3) appears over the pre-existing front, cyclogenesis and frontogenesis at low levels are hastened. The surface cyclone is excited at a location along the pre-existing front, as shown in Fig. 11.5. The surface front intensifies and becomes distorted with cold and warm fronts extending out of the low center and moving and developing in a manner not unlike that shown in Fig. 11.4.

11.2   Circulation at a Front

While the general vertical motion pattern in a developing baroclinic wave (Fig. 11.3) goes a long way toward explaining the cloud patterns associated with extratropical cyclones, it does not adequately account for the details of the circulations producing the clouds. The patterns of temperature and geostrophic wind in a baroclinic wave are frontogenetical, which means that the large-scale winds tend to concentrate the temperature gradients and form fronts. The vertical circulation required to maintain thermal-wind balance[289] is thus locally concentrated and intensified in the vicinity of frontal zones. In this section, we investigate the dynamics of frontogenesis to gain a better appreciation of the cloud-forming circulations in extratropical cyclones. To establish some basic features of frontogenesis, we will first review the dynamics of dry frontogenesis (Secs. 11.2.1 and 11.2.2). Then we will examine moist frontogenesis, the case most relevant to cloud dynamics, as an extension of the dry theory (Sec. 11.2.3). Finally, we will inspect some example calculations of the vertical circulation at a front (Sec. 11.2.4).

---

[289] Actually, the circulation at strong fronts is probably not exactly in thermal-wind balance, and this imbalance may be important in the frontal circulation. However, the balanced flow at a front is a good first approximation and is sufficient for our purposes in this chapter.

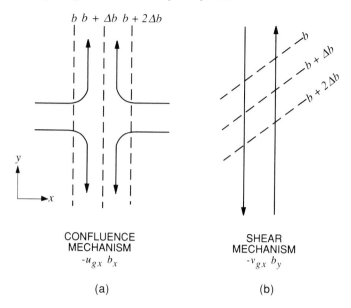

Figure 11.6

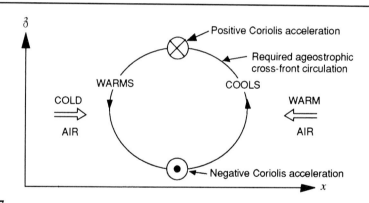

Figure 11.7

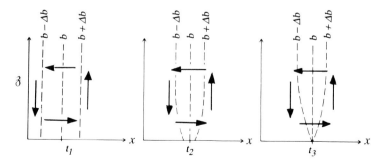

Figure 11.8

11.2.1   Quasi-Geostrophic Frontogenesis

To keep the discussion as simple as possible, we will consider fronts from a two-dimensional perspective, examining the circulation in a vertical plane oriented perpendicular to a front oriented parallel to the y-axis.[290] In the case of the quasi-geostrophic motions considered in Sec. 11.1, this circulation is expressed by the x-component of (11.13):

$$\tilde{N}^2 \tilde{w}_x - f^2 u_{a_{\tilde{3}}} = 2Q_1 \qquad (11.16)$$

The Q-vector component Q_1 is thus the forcing for the ageostrophic circulation $(u_a, \tilde{w})$ at the front, in the absence of friction and diabatic heating. From the y-component of (11.10):

$$\frac{D_g}{Dt}\left(b_x\right) = Q_1 - \tilde{N}^2 \tilde{w}_x \qquad (11.17)$$

It can be seen that Q_1 contributes, along with the horizontal gradient of $\tilde{w}$, to the rate at which the x-gradient of b is becoming concentrated within a fluid parcel. If the horizontal gradient of b is increasing, the parcel is said to be undergoing *frontogenesis*. If the gradient is decreasing, the situation is referred to as *frontolysis*. The two terms contributing to Q_1, as defined in (11.11), are $-u_{gx}b_x$, which is called the *confluence mechanism*, and $-v_{gx}b_y$, which is referred to as the *shear mechanism*. These mechanisms are illustrated in Fig. 11.6. An important difference in the two mechanisms is that the confluence mechanism can be active when there is no gradient of b in the y-direction, and the shear mechanism can be active when there is no gradient of b in the x-direction. Some theoretical analyses of frontogenesis are based entirely on either one or the other mechanism. In reality, both mechanisms are active, and the winds and isotherms in the vicinity of a front in a developing cyclone (e.g., the low-level geostrophic wind and thickness lines in Fig. 11.4) are composed of a combination of the two mechanisms.

[290] The discussion in this subsection is extracted mainly from the excellent review of Hoskins (1982). Further very useful reviews are given by Bluestein (1986) and Keyser (1986).

Figure 11.6   Confluence and shear mechanisms of frontogenesis. Streamlines of the horizontal geostrophic wind ($\mathbf{v} = u_g \mathbf{i} + v_g \mathbf{j}$) are superimposed on the field of b. (Adapted from Hoskins, 1982.)

Figure 11.7   Quasi-geostrophic frontogenesis forced by the deformation mechanism. Double arrows show geostrophic flow in x-direction. The geostrophic flow is in thermal wind balance. Temperature field is indicated by locations of warm and cold air. Flow is into the page at upper levels ($\otimes$) and out of the page at lower levels (circled dot). The ageostrophic flow required to maintain thermal wind balance is indicated by the streamline. The upper-level component of the flow normal to the page is being accelerated by the action of the Coriolis force on the ageostrophic circulation, while the lower-level component is negatively accelerated. (Adapted from Hoskins, 1982. Reprinted with permission from Annual Reviews, Inc.)

Figure 11.8   Idealization of the evolution of the b field at a sequence of times (t_1, t_2, and t_3) in the case of quasi-geostrophic frontogenesis. Arrows show the ageostrophic cross-front circulation. (Adapted from Bluestein, 1986. Reproduced with permission from the American Meteorological Society.)

The relations (11.16) and (11.17) are useful in formulating the theory of *quasi-geostrophic frontogenesis*, which is an oversimplification of real atmospheric frontogenesis but nonetheless a valuable foundation on which to build more realistic theories, to be considered below. Henceforth, it will be assumed that the geostrophic wind components u_g and v_g are invariant in the front-parallel y-direction and that the front-parallel flow has no ageostrophic component. Since geostrophic motion is horizontally nondivergent [as is evident from (2.68)], the continuity equation (11.8) becomes

$$u_{ax} + \tilde{w}_{\tilde{\delta}} \approx 0 \tag{11.18}$$

Mass continuity is then satisfied by a stream function Ψ of the form

$$\left(u_a, \tilde{w}\right) = \left(-\Psi_{\tilde{\delta}}, \Psi_x\right) \tag{11.19}$$

in which case we can rewrite (11.16) as

$$\tilde{N}^2 \Psi_{xx} + f^2 \Psi_{\tilde{\delta}\tilde{\delta}} = 2Q_1 \tag{11.20}$$

Solution of this elliptic equation for Ψ allows the ageostrophic circulation at a front characterized by a given Q_1 to be obtained.

The formulation of (11.16) and (11.20) provides insight into the vertical circulation at a front under hydrostatically and geostrophically balanced conditions. For example, Fig. 11.7 represents a case where frontogenesis is being forced by the confluence mechanism (Fig. 11.6a). Convergent geostrophic flow in the x-direction is advecting warm and cold air toward each other. The required ageostrophic circulation given by (11.20) is indicated qualitatively by the streamline. The physical sense of this solution is seen as follows. In the illustrated confluence case, $b_x > 0$, $u_{gx} < 0$, and $b_y = 0$. Hence, according to the definition (11.11), $Q_1 > 0$. It follows from (11.17) that the geostrophic wind is acting to *strengthen* the horizontal temperature gradient. At the same time, the positive value of Q_1 in the y-component of (11.12) acts to *weaken* the vertical shear of the along-front geostrophic current. Thus, the Q-vector component Q_1 acts to destroy the along-front part of the thermal-wind balance (11.6). As always in quasi-geostrophic flow, the role of the ageostrophic circulation is to compensate this destruction in order to maintain the thermal-wind balance. The two terms on the left-hand side of (11.16) must together exactly counteract the tendency of Q_1 to destroy thermal-wind balance. In the illustrated solution in Fig. 11.7, the ageostrophic circulation $(u_a, \tilde{w})$ has the characteristics

$$u_{a_{\tilde{\delta}}} < 0 \text{ and } \tilde{w}_x > 0, \quad \text{except on lower and upper boundaries.} \tag{11.21}$$

This effect of $\tilde{w}_x > 0$ is to counteract the effect of Q_1 in (11.17) through differential adiabatic temperature changes. The adiabatic cooling and warming associated with the upward and downward components, respectively, of the ageostrophic circulation act to weaken the buoyancy gradient at middle levels. Consequently, the horizontal thermal gradient at midlevels does not become as strong as at lower and upper levels. At those levels $\tilde{w}$ (and, hence, $\tilde{w}_x$) are zero. Therefore, the thermal wind balance is maintained at lower and upper boundaries (e.g., the ground and the tropopause) entirely by the horizontal component of the ageostrophic circulation, which everywhere in the example of Fig. 11.7 has vertical shear such that

$$-f^2 u_{a_{\hat{3}}} > 0 \tag{11.22}$$

which counteracts the effect of Q_1 in the along-front component of (11.12). The Coriolis acceleration of the horizontal component of the ageostrophic circulation thus acts to strengthen the vertical shear of the along-line geostrophic wind.

At lower and upper boundaries, (11.22) is the only effect mitigating the destruction of thermal wind balance. Adiabatic temperature changes are therefore absent, and it is possible to build up very concentrated gradients of potential temperature at the lower and upper boundaries. Given enough time, the thermal gradient on the boundaries may collapse to a near discontinuity. Such a collapse of the low-level field of b in a quasi-geostrophic front of the type we have been considering is indicated in Fig. 11.8.

11.2.2 Semigeostrophic Frontogenesis

As indicated in Fig. 11.8, quasi-geostrophic frontogenesis produces a front that is vertically oriented. However, real frontal zones are observed to have sloping leading edges. Another unrealistic feature, evident from Fig. 11.8, is that regions of static instability ($\partial \bar{\theta}/\partial z < 0$) can be formed at low levels. It turns out that there are further unrealistic features, among which is that the frontogenesis proceeds much too slowly.[291]

The major shortcomings of the quasi-geostrophic theory of frontogenesis are traced to the fact that it does not take into account that a front is highly nonisotropic, with different characteristic length and velocity scales in the along-front and cross-front directions. To take these differences in scale into account we make use of the *semigeostrophic* theory summarized in Sec. 2.2.2. The equation of motion for the along-front component of the wind is (2.30). Recall that the total derivative operator in the semigeostrophic system. (D_A/Dt) includes advection by the ageostrophic component of motion ($u_a, \bar{w}$) as well as by the geostrophic flow (u_g, v_g). Thus, the First Law of Thermodynamics (2.11) in the semigeostrophic system is

$$\frac{D_A b}{Dt} = 0 \tag{11.23}$$

The form of the semigeostrophic momentum and thermodynamic equations (2.30) and (11.23) can be simplified by a coordinate transformation in which a new cross-front coordinate is defined as

$$X \equiv x + \frac{v_g}{f} \tag{11.24}$$

---

[291] See Bluestein (1986) for further discussion.

From (2.110), it is seen that X is related to the absolute momentum M. Specifically,

$$X = f^{-1}M_g \tag{11.25}$$

where M_g is the value of M computed with the geostrophic wind v_g. According to (2.30) and the coordinate transformation (11.24), the wind in the new coordinate system, obtained by applying the total derivative operator D/Dt [defined by (2.2)] to (11.24), is the geostrophic wind:

$$\frac{DX}{Dt} = u_g \tag{11.26}$$

Using the transformation from x to X, we can recover the simple geostrophic form of the basic equations. Consider an arbitrary quantity

$$\mathscr{A} = \mathscr{A}(X, y, \mathfrak{z}, t) \tag{11.27}$$

Taking $D(11.27)/Dt$, applying the chain rule of differentiation, substituting from (11.26), and recalling that the wind component in the y-direction is geostrophic yields

$$\frac{D\mathscr{A}}{Dt} = \frac{\mathscr{D}\mathscr{A}}{\mathscr{D}\tau} + \tilde{w}\frac{\partial\mathscr{A}}{\partial Z} \tag{11.28}$$

where

$$\frac{\mathscr{D}}{\mathscr{D}\tau} \equiv \frac{\partial}{\partial\tau} + u_g\frac{\partial}{\partial X} + v_g\frac{\partial}{\partial y} \tag{11.29}$$

and the notation $\partial/\partial\tau$ and $\partial/\partial Z$ is used to indicate that X (not x) is being held constant in the differentiation; that is,

$$\frac{\partial}{\partial Z} \equiv \left(\frac{\partial}{\partial\mathfrak{z}}\right)\bigg|_{X,y,t}, \quad \frac{\partial}{\partial\tau} \equiv \left(\frac{\partial}{\partial t}\right)\bigg|_{X,y,\mathfrak{z}} \tag{11.30}$$

Applying (11.28) to $\mathscr{A} = v_g$ and b, and noting from (2.30) and (11.23) that $Dv_g/Dt = D_A v_g/Dt = -fu_a$ and $Db/Dt = D_A b/Dt = 0$, we obtain

$$\frac{\mathscr{D}v_g}{\mathscr{D}\tau} + fU_a = 0 \tag{11.31}$$

and

$$\frac{\mathscr{D}b}{\mathscr{D}\tau} + \tilde{w}\frac{\partial b}{\partial Z} = 0 \tag{11.32}$$

where

$$U_a \equiv \frac{\tilde{w}}{f}\frac{\partial v_g}{\partial Z} + u_a \tag{11.33}$$

By comparing (11.31) and (11.32) to the y-component of (2.23) and (11.9), the similarity in form to the quasi-geostrophic equations can be seen.

Before proceeding, it is useful to note some further properties of the geostrophic coordinate transformation. By the chain rule of differentiation,

$$\frac{\partial \mathcal{A}}{\partial x} = \frac{\partial \mathcal{A}}{\partial X} \frac{\partial X}{\partial x}$$

(11.34)

and

$$\frac{\partial \mathcal{A}}{\partial \mathfrak{z}} = \frac{\partial \mathcal{A}}{\partial X} \frac{\partial X}{\partial \mathfrak{z}} + \frac{\partial \mathcal{A}}{\partial Z}$$

(11.35)

Taking $\partial(11.24)/\partial x$, we obtain

$$\frac{\partial X}{\partial x} = \frac{\zeta_{ag}}{f}$$

(11.36)

where ζ_{ag} is the vertical component of the geostrophic absolute vorticity, as defined by (2.65). Since the geostrophic flow is invariant in y, according to our assumptions,

$$\zeta_{ag} = f + \frac{\partial v_g}{\partial x}$$

(11.37)

Taking $\partial(11.24)/\partial \mathfrak{z}$, we obtain

$$\frac{\partial X}{\partial \mathfrak{z}} = \frac{1}{f} \frac{\partial v_g}{\partial \mathfrak{z}}$$

(11.38)

With substitution from (11.36)–(11.38), (11.34) and (11.35) become

$$\frac{\partial \mathcal{A}}{\partial x} = \frac{\zeta_{ag}}{f} \frac{\partial \mathcal{A}}{\partial X}$$

(11.39)

and

$$\frac{\partial \mathcal{A}}{\partial \mathfrak{z}} = \frac{\partial \mathcal{A}}{\partial X} \frac{1}{f} \frac{\partial v_g}{\partial \mathfrak{z}} + \frac{\partial \mathcal{A}}{\partial Z}$$

(11.40)

In the special case where $\mathcal{A} = v_g$, (11.39) becomes

$$\frac{\partial v_g}{\partial x} = \frac{\zeta_{ag}}{f} \frac{\partial v_g}{\partial X}$$

(11.41)

and, with the aid of (11.41), (11.40) becomes

$$\frac{\partial v_g}{\partial \mathfrak{z}} = \frac{\zeta_{ag}}{f} \frac{\partial v_g}{\partial Z}$$

(11.42)

The identities (11.34)–(11.42) are useful in converting several of the basic equations from x to X space. First, it can be seen that the thermal-wind relation retains its form in X space. If $\mathcal{A} = b$, then from (11.39) we have

$$\frac{\partial b}{\partial x} = \frac{\zeta_{ag}}{f} \frac{\partial b}{\partial X} \tag{11.43}$$

From (11.42) and (11.43) and the thermal-wind equation in x-space (11.6), we obtain

$$fv_{gZ} = b_X \tag{11.44}$$

which is analogous to the along-front part (i.e., the y-component) of (11.6).

It follows from the thermal-wind balance (11.44) that there exists a function $\tilde{\Phi}$ that satisfies relations of the form

$$fv_g = \tilde{\Phi}_X \tag{11.45}$$

and

$$b = \tilde{\Phi}_Z \tag{11.46}$$

which are analogs to the geostrophic and hydrostatic relationships in x-space [(2.68) and (11.5)]. It can be seen, moreover, that the function $\tilde{\Phi}$ is

$$\tilde{\Phi} = \Phi + \frac{v_g^2}{2} \tag{11.47}$$

where Φ is the geopotential.[292] Thus, if $\tilde{\Phi}$ is used instead of Φ, the forms of the geostrophic and hydrostatic relationships are preserved in X-space.

The identities (11.39)–(11.42) can be further used to show that the form of the mass-continuity equation is also preserved in the X-coordinate system, if we define new velocity components. If we use the ageostrophic component U_a, as defined in (11.33), and define a new vertical-velocity component

$$W \equiv \tilde{w}\frac{f}{\zeta_{ag}} \tag{11.48}$$

then the continuity equation (11.18) can be rewritten, with the aid of (11.39) and (11.40), as

$$\frac{\partial U_a}{\partial X} + \frac{\partial W}{\partial Z} \approx 0 \tag{11.49}$$

A particularly useful quantity in the context of the X-coordinate system is

$$\mathcal{P}_g \equiv \frac{1}{f}\boldsymbol{\omega}_{ag} \cdot \nabla b \tag{11.50}$$

---

[292] To show that (11.47) satisfies (11.45) and (11.46), one must simply differentiate $\tilde{\Phi}$ and make use of (2.68), (11.5), (11.39), and (11.40).

where $\boldsymbol{\omega}_{ag}$ is the geostrophic absolute vorticity, defined in x, y, and z coordinates by (2.65). It is defined in pseudoheight coordinates as

$$\boldsymbol{\omega}_{ag} \equiv -\frac{\partial v_g}{\partial \mathfrak{z}}\mathbf{i} + \frac{\partial u_g}{\partial \mathfrak{z}}\mathbf{j} + \left(\zeta_g + f\right)\mathbf{k} \tag{11.51}$$

The quantity $\mathcal{P}_g$ is similar to the geostrophic Ertel's potential vorticity given by (2.64), and for the remainder of the chapter we will refer to $\mathcal{P}_g$ simply as potential vorticity. In the case of two-dimensional[293] flow ($\partial/\partial y = 0$), it can be shown with substitution of (11.39) and (11.40) into (11.50) that

$$\mathcal{P}_g = \frac{\zeta_{ag}}{f}\frac{\partial b}{\partial Z} \tag{11.52}$$

Substituting this form of $\mathcal{P}_g$ into (11.32), we obtain

$$\frac{\mathfrak{D}b}{\mathfrak{D}\tau} + W\mathcal{P}_g = 0 \tag{11.53}$$

From (11.52) and (2.98), it is apparent that the potential vorticity has the form of a buoyancy frequency in X-coordinates. Comparing (11.53) to the quasi-geostrophic form of the thermodynamic equation (11.9), we see that the potential vorticity indeed plays the role of buoyancy frequency in the thermodynamic equation in the X system. Additionally, the potential vorticity $\mathcal{P}_g$ has the important property that it is conserved following parcels of air:

$$\frac{D\mathcal{P}_g}{Dt} = 0 \tag{11.54}$$

This conservation property is obtained for the two-dimensional, semigeostrophic, adiabatic, inviscid case considered here by taking $D(11.52)/Dt$ and making use of (11.41) to obtain

$$\frac{D\mathcal{P}_g}{Dt} = \frac{1}{f}b_Z\frac{D\zeta_{ag}}{Dt} + \frac{\zeta_{ag}}{f}\frac{Db_Z}{Dt} = \frac{\zeta_{ag}^2}{f^3}b_Z\frac{Dv_{gX}}{Dt} + \frac{\zeta_{ag}}{f}\frac{Db_Z}{Dt} \tag{11.55}$$

It can be shown that the two terms on the right-hand side of (11.55) cancel by substituting from the equation of motion (11.31) and thermodynamic equation (11.53) and making use of mass continuity (11.49), thermal-wind balance for v_g [as expressed by (11.44)] and u_g [as given by the x-component of (11.6)], the assumption of no y-variation ($\partial/\partial y = 0$), and the nondivergence of the geostrophic wind [seen from (2.68)]. The usefulness of (11.54) will be seen below, as we explore the nature of the ageostrophic circulation in X-space.

[293] Note that by making an additional coordinate transformation, $Y = y - (u_g/f)$, we could have shown (11.52) to be valid also for three-dimensional flow.

By analogy to the quasi-geostrophic case, we can perform further operations on (11.31) and (11.53) to obtain

$$\frac{\mathcal{D}}{\mathcal{D}\tau}\left(b_X\right) = Q_1' - \left(\mathcal{P}_g W\right)_X \tag{11.56}$$

and

$$\frac{\mathcal{D}}{\mathcal{D}\tau}\left(fv_{gZ}\right) = -Q_1' - f^2 U_{aZ} \tag{11.57}$$

where Q_1' is defined the same as Q_1 in (11.11), but with X replacing x. Equations (11.56) and (11.57) are analogous to the y-components of (11.10) and (11.12), and subtraction of (11.57) from (11.56) leads to

$$\left(\mathcal{P}_g W\right)_X - f^2 U_{aZ} = 2Q_1' \tag{11.58}$$

Since U_a and W obey mass continuity according to (11.49), we can employ a new stream function Ψ', such that

$$\left(U_a, W\right) = \left(-\Psi_Z', \Psi_X'\right) \tag{11.59}$$

Substituting this stream function into (11.58), we obtain

$$\left(\mathcal{P}_g \Psi_X'\right)_X + f^2 \Psi_{ZZ}' = 2Q_1' \tag{11.60}$$

This relation[294] is similar in form to the quasi-geostrophic ageostrophic stream-function equation (11.20) in physical space, except that x has been replaced by X and the potential vorticity $\mathcal{P}_g$ has replaced the buoyancy frequency $\tilde{N}^2$. As long as $\mathcal{P}_g$ is positive, (11.60) remains an elliptic partial differential equation, meaning that it has a unique solution everywhere within the X–Z domain, if values on the boundaries of the domain are given.[295] Since $\mathcal{P}_g > 0$ is also the condition for the flow to be symmetrically stable [Sec. 2.9.1, Eq. (2.150)], it becomes evident that (11.60) has such a solution as long as the flow is symmetrically stable. If $\mathcal{P}_g < 0$ (i.e., the flow is symmetrically unstable), (11.60) is no longer elliptic.

From the similarity of (11.20) and (11.60), it is inferred that the semigeostrophic frontogenesis forced by the confluence mechanism in X-space would have the circulation shown in Fig. 11.9a, which is similar to that for the case of geostrophic frontogenesis illustrated in Fig. 11.7. Recalling from (11.30) that $(\partial x/\partial Z)$ is defined as $(\partial x/\partial \mathfrak{z})$ at constant X, we note from the definition (11.24) that

$$\frac{\partial x}{\partial Z} = \left.\frac{\partial x}{\partial \mathfrak{z}}\right|_X = -\frac{1}{f}\left.\frac{\partial v_g}{\partial \mathfrak{z}}\right|_X \tag{11.61}$$

Thus, the stronger the shear along a surface of constant X, the more the surface tilts. Hence, the circulation in Fig. 11.9a becomes skewed as shown in Fig. 11.9b

---

[294] Sometimes called the ''Sawyer–Eliassen'' equation in honor of its developers, J. F. Sawyer (1956) and A. Eliassen (1959, 1962).

[295] See Hildebrand (1976, p. 417).

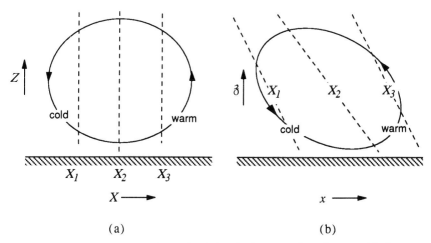

Figure 11.9 Ageostrophic cross-front circulation in the case of semigeostrophic frontogenesis forced by the deformation mechanism: (a) in X-space; (b) in physical (x) space.

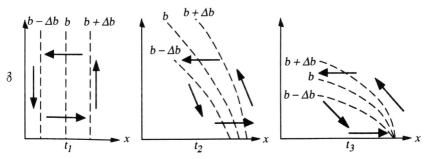

Figure 11.10 Idealization of the evolution of the b field at a sequence of times (t_1, t_2, and t_3) in the case of semigeostrophic frontogenesis. Arrows show the ageostrophic cross-front circulation. (Adapted from Bluestein, 1986. Reproduced with permission from the American Meteorological Society.)

when transformed back to physical space. Compared to the geostrophic case in Fig. 11.7, this tilt is a major improvement, lending the frontal circulation a more realistic character. The improved structure is also seen in the thermal field, as indicated schematically in Fig. 11.10. The tilting structure of the front with the tight gradient at low levels more closely resembles a real front. Also, there are no regions of buoyant instability in the semigeostrophic calculation. The tilt of the front appears in the semigeostrophic case since the ageostrophic circulation advects the thermal field, whereas in the geostrophic case b is advected only by the geostrophic wind.

The importance of the potential vorticity in calculating the evolution of the flow at a front is seen by first rearranging the expression for the potential vorticity $\mathcal{P}_g$ (11.52), which after substitution from (11.41), (11.45), and (11.46) becomes

$$1 = \mathcal{P}_g^{-1}\tilde{\Phi}_{ZZ} + f^{-2}\tilde{\Phi}_{XX} \tag{11.62}$$

It is now evident that knowledge of $\mathcal{P}_g$ (given appropriate boundary conditions) is sufficient to determine $\hat{\Phi}$, which, in turn, determines v_g and b from (11.45) and (11.46). From v_g and b, Q'_1 is known. Then (11.60) gives the stream function Ψ' and hence the transverse circulation (U_a, W). *The time dependence of the fronto-genesis then resides entirely in the potential vorticity equation* (11.54). Calculations show that the time scale of development of the front, indicated qualitatively in Fig. 11.10, is shorter and much more realistic than in the case of geostrophic frontogenesis (Fig. 11.8). Thus, the major deficiencies of the quasi-geostrophic frontogenesis (lack of frontal tilt, static instability, and long time scale) are removed when the advection by ageostrophic motion is included.

11.2.3   Moist Frontogenesis

For cloud dynamics, we are most interested in frontal air motions when clouds are forming and dissipating in concert with the motions. The foregoing discussion of dry frontogenesis forms a valuable background for a discussion of moist fronto-genesis. The ageostrophic circulation of the dry case remains the basic air-motion pattern. However, latent heat release modifies the basic ageostrophic circulation, and this modification must be taken into account to obtain a still more realistic representation of the vertical circulation at a front where clouds and precipitation are active.

To include the latent heat release associated with clouds and precipitation, we modify (11.53) to include a diabatic heating rate:

$$\frac{\mathcal{D}b}{\mathcal{D}\tau} + W\mathcal{P}_g = \frac{g}{\hat{\theta}}\dot{\mathcal{H}} \tag{11.63}$$

where $\dot{\mathcal{H}}$ is defined in (2.10). We will continue to regard the flow as inviscid and two-dimensional. By arguments similar to those leading to (11.60), we obtain a new version of the two-dimensional stream function equation:

$$\underbrace{\left(\mathcal{P}_g\Psi'_X\right)_X + f^2\Psi'_{ZZ}}_{\text{I}} = \underbrace{2Q'_1 + \frac{g}{\hat{\theta}}\frac{\partial\dot{\mathcal{H}}}{\partial X}}_{\text{II}} \tag{11.64}$$

Thus, the heating gradient acts along with the geostrophic Q-vector component Q'_1 to weaken or sharpen the thermal gradient, which must be compensated by the ageostrophic circulation, according to the left-hand side of (11.64).

Since the motions are not adiabatic, the potential vorticity [as defined by (11.50)] is not conserved. The rate of change of $\mathcal{P}_g$ found by substituting from (11.63) rather than (11.53) on the right-hand side of (11.55) is

$$\frac{D\mathcal{P}_g}{Dt} = \frac{\zeta_{ag}g}{f\hat{\theta}}\frac{\partial\dot{\mathcal{H}}}{\partial Z} \tag{11.65}$$

Thus, the potential vorticity of a parcel changes in proportion to the vertical gradient of heating along surfaces of constant X [or, equivalently, according to

(11.25), along a surface of constant geostrophic absolute momentum M_g]. If friction were included in (11.31), a second source would appear on the right-hand side of (11.65) that would involve the X-gradient of frictional forces. Thus, potential vorticity could be generated by either heating or friction. By neglecting the friction and considering only the heating associated with the latent heat of vaporization (from the liquid phase only), the equations retain much the same form as in the dry case. If we represent the condensation (or evaporation) rate as in (2.13) and assume that it occurs only under saturated conditions and that it is dominated by vertical advection of vapor, then we may write

$$\dot{\mathcal{H}} = -\frac{L}{c_p \Pi} \frac{Dq_v}{Dt} \approx -\frac{L}{c_p \Pi} \tilde{w} \frac{\partial q_{vs}}{\partial Z} \tag{11.66}$$

where q_{vs} is the saturation vapor pressure. If we further define a geostrophic equivalent potential vorticity similar to (2.71) as

$$\mathcal{P}_{eg} \equiv \frac{\zeta_{ag}}{f} \frac{\partial}{\partial Z}\left(g \frac{\theta_e}{\hat{\theta}_e}\right) \tag{11.67}$$

which is similar to the quantity P_{eg} defined in Sec. 2.5, then terms I and II in the stream function equation (11.64) can be combined so that (11.60) and (11.64) may be written jointly in the form

$$\left(\mathcal{P}_m \Psi'_X\right)_X + f^2 \Psi'_{ZZ} = 2Q'_1 \tag{11.68}$$

where

$$\mathcal{P}_m = \begin{cases} \mathcal{P}_{eg} & \text{if saturated} \\ \mathcal{P}_g & \text{if unsaturated} \end{cases} \tag{11.69}$$

Equation (11.68) is identical in form to (11.60), with $\mathcal{P}_m$ replacing $\mathcal{P}_g$. It is elliptic in saturated regions as long as the flow is conditionally symmetrically stable (i.e., $\mathcal{P}_{eg} > 0$).

11.2.4 Some Simple Theoretical Examples

We will now examine some results obtained by using the two-dimensional semigeostrophic equations to calculate the vertical circulation at a front where cloud processes are active. We first examine calculations[296] in which, to avoid time integration, $\mathcal{P}_m$ is set equal to a constant, but with different values in different regions, such that

$$\mathcal{P}_m = \begin{cases} \mathcal{P}_1, & X > \hat{L} \\ \mathcal{P}_2, & X < \hat{L} \end{cases} \tag{11.70}$$

The assumed configuration is shown in Fig. 11.11. The value of $\mathcal{P}_2$ is taken to be typical of dry frontogenesis. In some calculations, $\mathcal{P}_1$ is assumed equal to the

[296] Performed by Emanuel (1985).

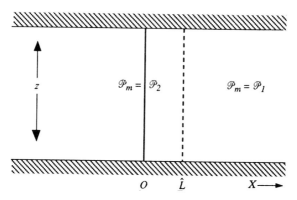

Figure 11.11 Configuration of potential vorticity $\mathscr{P}_m$ used to calculate idealized two-dimensional circulation at a front. (Adapted from Emanuel, 1985. Reproduced with permission from the American Meteorological Society.)

value assumed for $\mathscr{P}_2$, and the calculations reduce to the case of dry frontogenesis. In other calculations, $\mathscr{P}_1$ is assigned a value characteristic of cloudy conditions. $\mathscr{P}_1$ is assumed to be a small positive value, which assumes that the frontal cloud zone is characterized by nearly neutral conditional symmetric instability (Sec. 2.9.1). It remains unspecified how the neutral condition is achieved. However, there is empirical evidence from sounding and aircraft data in frontal zones to support this assumption.[297] According to this parameterization of a frontal-cloud zone, $\mathscr{P}_1$ can be regarded as $\mathscr{P}_{eg}$ in (11.69), while $\mathscr{P}_2$ is $\mathscr{P}_g$.

The calculations with $\mathscr{P}_m$ given by (11.70) have been made with the stream function equation (11.68) for two cases of confluence forcing [defined by term $-u_{gx}b_x$ in (11.11) and illustrated in Fig. 11.6]. In the first case, the thermal gradient is prescribed such that the vertical shear of v_g, implied by thermal-wind balance, is a constant. The transverse stream functions for the dry and moist versions of this case appear in Fig. 11.12 and 11.13, respectively. The results have been transformed to physical (x) space. The effect of the clouds clearly is to concentrate the lifting into a more narrow zone than in the case of dry semigeostrophic frontogenesis. In a second moist case, whose results are in Fig. 11.14, the vertical shear of v_g is specified to decrease with distance X from L. The effect of the horizontal variation of the shear is to give the front a curving shape in the vertical, with a steeper slope nearer the surface than in the midtroposphere.

Calculations with (11.68) can be made less restrictive by allowing the frontal circulation to develop in time by letting $\mathscr{P}_m$ evolve as a result of the condensational heating.[298] The evolution of the field of potential vorticity $\mathscr{P}_g$ is implied by (11.65) and (11.66). To represent frontal clouds, $\mathscr{P}_{eg}$ is again set equal to a small positive number, representing that the atmosphere has adjusted to a state of near conditionally symmetric neutrality. This condition is assumed to apply whenever

[297] See Emanuel (1985).
[298] This procedure was suggested by Thorpe and Emanuel (1985).

the air is saturated, which is presumed to be the case whenever the vertical air motion is upward. Thus,

$$\mathcal{P}_m = \begin{cases} \mathcal{P}_{eg} = \hat{\varepsilon}\,\mathcal{P}_g, & w > 0 \\ \mathcal{P}_g, & w \le 0 \end{cases} \tag{11.71}$$

where

$$\hat{\varepsilon} = \begin{cases} \text{small positive value} \;\sim 0.1 \text{ for "moist" conditions} \\ 1 \text{ for "dry" conditions} \end{cases} \tag{11.72}$$

The ageostrophic circulation is then derived as a function of time from (11.68) as each new field of $\mathcal{P}_m$ is obtained.

Equations (11.71) and (11.72) can be used in conjunction with (11.68) to estimate the frontogenetic circulation in a somewhat more realistic context. The calculations illustrated in Figs. 11.11–11.14 considered frontogenesis in the context of highly idealized circulations characterized by prescribed Q-vector confluence forcing. As we have seen in Sec. 11.1, atmospheric fronts are integral parts of baroclinic waves, and the fronts and wave develop in concert. The ageostrophic circulation in the context of a developing baroclinic wave has been investigated by letting the initial fields of potential vorticity and other quantities be those of a small-amplitude dry-baroclinic normal mode.[299] These calculations differ from those above both in this respect and in that the basic state is represented by a horizontally uniform westerly geostrophic current (i.e., u_g depends on height only). In this case, frontogenesis is forced by the shear term rather than the confluence term in Q_1 [i.e., in (11.11), $b_y = -\partial u_g/\partial \mathfrak{z}$ and $\partial u_g/\partial x = 0$].

The structure of the wave after 2 days is depicted in Fig. 11.15. The stream function in Fig. 11.15a shows again that the moisture produces an updraft that is narrow and intense compared to the dry descending part of the circulation. The updraft is particularly narrow and intense near the lower boundary, where the strongest front forms (Fig. 11.15b). As the gradient of heating by condensation develops, potential vorticity is generated in the vicinity of the front (Fig. 11.15c). The strongest generation is at low levels, where the most moisture is condensed. Another effect of the moisture is to concentrate and strengthen the low-level southerly jet just ahead of the frontal zone (Fig. 11.15d), and the absolute vertical vorticity in the low-level frontal zone (Fig. 11.15e).

11.3   Horizontal Patterns of Frontal Zones in Developing Cyclones

Since the vertical circulation in a baroclinic wave is concentrated in the active zones of frontogenesis, it is reasonable to expect that the horizontal distribution of clouds around a frontal cyclone should to some extent reflect the spatial pattern of

---

[299] By Emanuel *et al.* (1987).

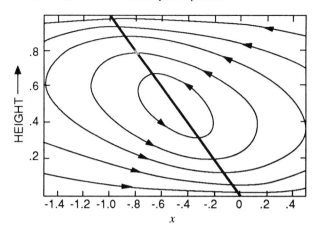

Figure 11.12

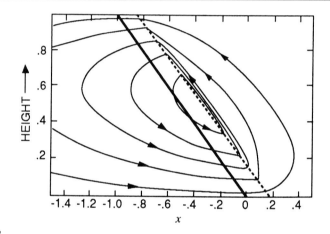

Figure 11.13

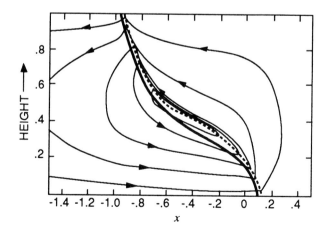

Figure 11.14

frontal zones in the wave. A first approximation to the horizontal pattern of frontal zones can be obtained if semigeostrophic calculations like those described in Secs. 11.2.2–11.2.4 are extended to the fully three-dimensional case by using the geostrophic momentum approximation (2.32) to examine the development of a slightly perturbed idealized baroclinic wave. Results of such a calculation are shown in Fig. 11.16. Between the times represented in Fig. 11.16a and b, the horizontal temperature gradient increases markedly. There are two regions of the developing cyclone where the frontogenesis is especially vigorous: along the incipient cold front, south of the cyclone center, and along the developing warm front, north to northeast of the cyclone. These regions of frontogenesis are depicted quantitatively in Fig. 11.17b by mapping the function $\mathfrak{D}(\nabla\theta)^2/\mathfrak{D}\tau$, which is a sensitive indicator of how rapidly the temperature gradient is increasing or decreasing. As time goes on, the warm-frontal zone is swept around the poleward side of the low center by the cyclonic winds, and, by the time of Fig. 11.16c, the occluded front can be identified with the narrow potential temperature maximum curling into the surface cyclone from the peak of the warm sector.

Another type of model used to simulate the development of frontal cyclones is the mesoscale model of the type mentioned in Sec. 9.2.3.6. These models use the full primitive equations and thus are not limited by the semigeostrophic assumption. They simulate the storm development in great detail, closely matching the structure of specific observed cases. An example of an oceanic storm is shown in Fig. 11.18. It displays the tendency seen in the semigeostrophic calculations to develop stronger frontal zones along the warm-front side of the occlusion, north to northeast of the surface low, and along the cold front, southeast of the low. The weaker temperature gradient located between the strong cold front to the south and the warm-frontal zone on the north side of the oceanic cyclone probably

---

Figure 11.12 Ageostrophic stream function obtained by solving the semigeostrophic Sawyer–Eliassen equation for a case of deformation forcing under dry conditions. Vertical shear of the along-front wind is assumed to be constant everywhere. Results have been transformed to physical (x) space. Height and horizontal distance are expressed in nondimensional units. Solid line denotes the $X = 0$ surface (see Fig. 11.11). (From Emanuel, 1985. Reprinted with permission from the American Meteorological Society.)

Figure 11.13 Ageostrophic stream function obtained by solving the semigeostrophic Sawyer–Eliassen equation for a case of deformation forcing under moist conditions. Vertical shear of the along-front wind is assumed to be constant everywhere. Results have been transformed to physical (x) space. Height and horizontal distance are expressed in nondimensional units. Solid line denotes the $X = 0$ surface. Dashed line denotes the $X = \hat{L}$ surface (see Fig. 11.11). (From Emanuel, 1985. Reprinted with permission from the American Meteorological Society.)

Figure 11.14 Ageostrophic stream function obtained by solving the semigeostrophic Sawyer–Eliassen equation for a case of deformation forcing under moist conditions. Vertical shear of the along-front wind is specified to decrease with distance X from $\hat{L}$ (see Fig. 11.11). Results have been transformed to physical (x) space. Height and horizontal distance are expressed in nondimensional units. Solid line denotes the $X = 0$ surface. Dashed line denotes the $X = \hat{L}$ surface. (From Emanuel, 1985. Reprinted with permission from the American Meteorological Society.)

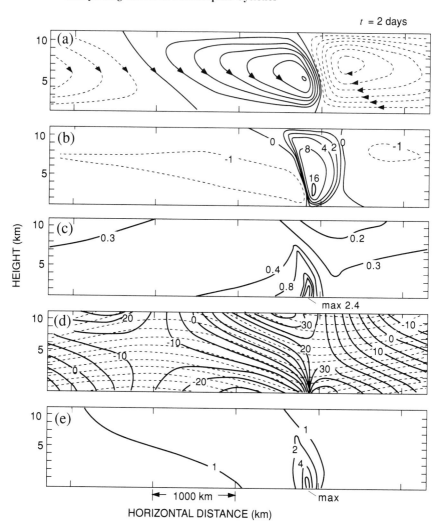

Figure 11.15 Two-dimensional, semigeostrophic, shear-forced, moist ($\hat{\varepsilon} = 0.1$) frontogenesis calculation after two days for an initial condition consisting of a small-amplitude dry-baroclinic normal mode. (a) Ageostrophic stream function (contour interval 4000 m^2 s^{-1}). (b) Vertical air velocity (cm s^{-1}). (c) Potential vorticity after 2 days (10^{-6} m^2 K s^{-1} kg^{-1}). (d) South–north flow (solid contours, m s^{-1}) and potential temperature (dashed, 4 K intervals). (e) Normalized geostrophic absolute vertical vorticity (ζ_{ag}/f). (From Emanuel *et al.*, 1987. Reprinted with permission from the American Meteorological Society.)

reflects the effect of the ocean surface on the surface temperature pattern outside the zone of strongest cold frontogenesis. Storms over land do not always exhibit such a weakening of the temperature gradient in this region. For example, a simulated overland storm, illustrated in Fig. 11.19, exhibits strong thermal gradients all the way into the center of the cyclone.

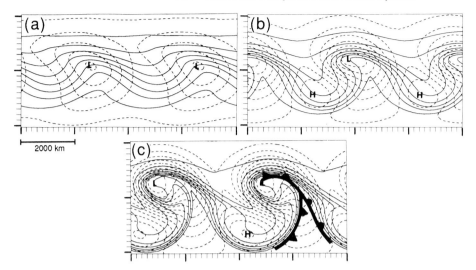

Figure 11.16   Semigeostrophic calculation of the development of the surface cyclone and thermal field in a slightly perturbed baroclinic wave. Isobars (dashed lines) are shown at intervals of 3 mb and potential temperature contours (solid lines) are given at intervals of 2 K with colder air at the top of the diagram. Time interval between (a) and (c) is 96 h. Frontal symbols have been added to indicate how the model fields would be represented on a standard surface weather chart. (Adapted from Schär, 1989.)

11.4   Clouds and Precipitation in a Frontal Cyclone

11.4.1   Satellite-Observed Cloud Patterns

From the theoretical treatment of the large-scale baroclinic wave (Sec. 11.1) and its frontal zones (Secs. 11.2 and 11.3), we have obtained an idea of the distribution of upward air motion in the wave. On the large scale, air generally rises within a comma-shaped zone to the east of the trough in midlevels (Fig. 11.3b). Within this broad zone of ascent, concentrated frontal zones form at low levels. They extend outward from the surface low center, as seen in Figs. 11.16–11.19. We have seen in Sec. 11.2 that the upward air motion in the wave is focused on the regions of frontogenesis, especially when condensation and cloud formation occur (see Secs. 11.2.3 and 11.2.4).

Since upward air motion is generally required for cloud formation and precipitation, the cloud and precipitation pattern in a baroclinic wave should be controlled to a large extent by the upward air motion pattern of the wave. However, the relationship between cloud and upward air motion is not a simple one-to-one correspondence. Horizontal motions are also important in determining the cloud and precipitation pattern. Parcels of air may traverse long horizontal distances while rising before they reach condensation. Also, once hydrometeors are formed, they may be advected large distances from the regions of upward motion where they formed and grew. It is therefore not surprising that when weather

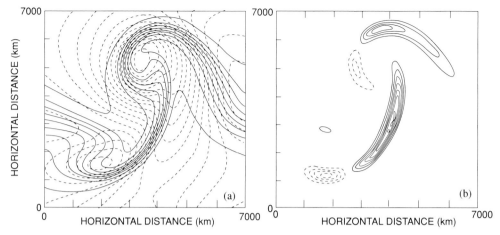

Figure 11.17

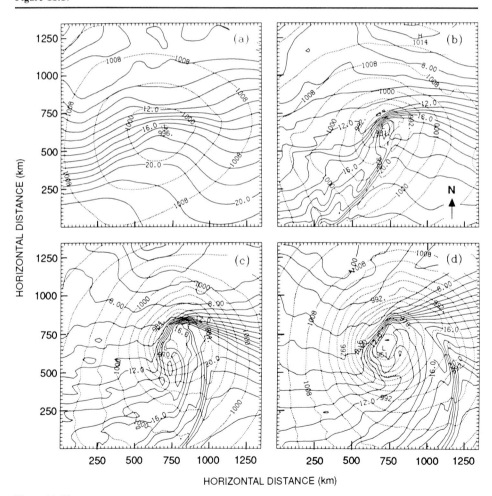

Figure 11.18

satellite data became available in the 1960s,[300] it quickly became evident that cloud patterns associated with developing cyclones exhibit a multiplicity of structures. Attempts have been made to catalog the variety of cloud images seen in satellite pictures under different combinations of upper-level waves and lower-level cyclones and fronts.[301] All of these variations are too numerous to examine in the present text. We will look only at the most obvious and general aspects of the satellite-observed cloud structure.

The basic upper-level cloud pattern associated with a baroclinic wave is in the shape of a comma, which reflects the comma-shaped pattern of the zone of upward motion on the eastern side of the trough of the wave (Fig. 11.3b). A good example is the polar-low comma cloud seen trailing the large frontal cloud system in Fig. 1.29. It is associated with a short-wavelength upper-level trough overtaking the frontal system. The frontal cloud system is associated with another, longer-wavelength baroclinic wave, and it too exhibits a comma configuration when viewed from a satellite perspective. The comma in this case is larger, and its head is formed by the broad cloud pattern centered near the northern head of the large frontal cloud band, at the location of the surface low-pressure center. The cold-frontal band forms the tail of the larger comma. This frontal-cloud comma is produced as a combined effect of the general comma-shaped region of upward motion associated with the baroclinic wave, the concentration of this upward motion into narrow frontogenetical zones, and the advection of the cloud material by the cyclonically circulating winds of the frontal low developing at low levels in the context of the baroclinic wave (as illustrated in Fig. 11.4). Although details vary from case to case, the general comma pattern is seen in nearly all cloud-producing baroclinic waves.

A significant limitation of the cloud patterns viewed in conventional infrared and visible satellite imagery is that one sees only the tops of the clouds. This view is often dominated by a large sheet of cirriform cloud, as in the idealized example

---

[300] The first weather satellite, TIROS I, was launched 1 April 1960.

[301] See *The Use of Satellite Pictures in Weather Analysis and Forecasting* (Anderson *et al.*, 1973), *Application of Satellite Data in Weather Analysis and Forecasting* (Anderson *et al.*, 1974), and *Satellite Imagery Interpretation for Forecasters*, Vols. 1–3 (National Weather Association, 1986a–c).

Figure 11.17 Semigeostrophic calculation showing in geostrophic space a surface cyclone and thermal field in an intermediate stage of development. (a) Isobars at intervals of 3.2 mb (dashed) and potential temperature isotherms at intervals of 2.1 K (solid). Colder air is located at the top of the figure. (b) Frontogenetic function $\mathfrak{D}(\nabla\theta)^2/\mathfrak{D}\tau$ [10.6(K/1000 km)^2 h^{-1}], with positive values indicated by solid contours and negative values by dashed contours. (From Schär and Wernli, 1993. Reprinted with permission from the Royal Meteorological Society.)

Figure 11.18 Mesoscale primitive-equation model calculation of the development of an extratropical cyclone. Four stages of development are shown: (a) incipient cyclone and (b) 12-h, (c) 18-h, and (d) 24-h forecast. Sea-level temperature (°C, solid lines) and pressure (mb, dashed lines). Spacing of tick marks is at 25-km intervals. (From Shapiro and Keyser, 1990. Reprinted with permission from the American Meteorological Society.)

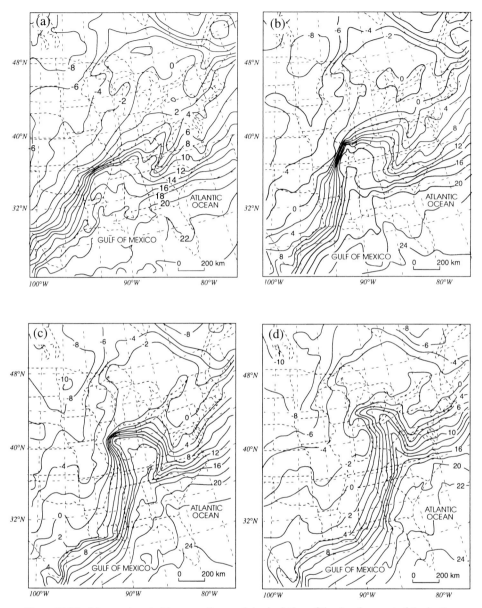

Figure 11.19   Mesoscale primitive-equation model calculation of the development of the isotherms in °C in an extratropical cyclone over land. (Courtesy of C. F. Mass.)

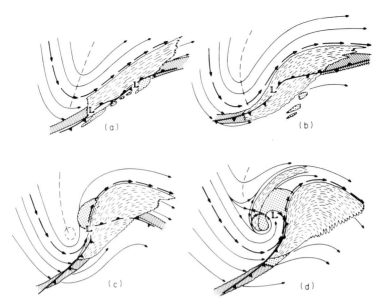

Figure 11.20 Schematic representing the clouds associated with an upper-level trough moving over a pre-existing low-level front and triggering cyclone development. Four stages of development are shown. Dashed shading indicates tops of cirriform cloud decks, crosshatching frontal cloud bands, and dots low- or middle-level cloud decks. Thin solid contours are streamlines of upper-tropospheric wind. Dashed line indicates the upper-level trough position. Heavy arrows show jet stream position. Frontal symbols indicate positions of fronts at the earth's surface. (Adapted from lecture notes of R. Weldon by Wallace and Hobbs, 1977.)

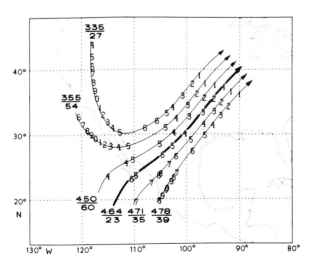

Figure 11.21 Trajectories entering the zone of upward motion to the west of the trough of a baroclinic wave. All trajectories terminate at 400 mb. The pressure level and the saturation pressure deficit are plotted in mb at the upstream end of each trajectory. The tens digit of the pressure level is plotted along the trajectory to indicate vertical motion. The northernmost trajectory along which air parcels reach saturation is drawn with a heavy line. (From Durran and Weber, 1988. Reprinted with permission from the American Meteorological Society.)

shown in Fig. 11.20. This schematic represents the clouds associated with the type of cyclone development illustrated in Fig. 11.5, wherein an upper-level trough moves over a pre-existing low-level front.

In Fig. 11.20a, the nearly undisturbed pre-existing front is characterized by a two-layer band of clouds lying parallel to the surface frontal position. The lower-level cloud band is produced by the upward branch of circulation associated with the surface front (Sec. 11.2). The upper-level band of cirriform cloud obscures the lower-level frontal cloud in the region east of the trough. The position of the jet stream (zone of strongest wind speed) in the trough is indicated by heavy arrows in Fig. 11.20a. The cirriform cloud layer tends to lie south of the jet. It has been suggested that the reason for the cirriform cloud in this location is that the trajectories entering the zone of upward motion to the east of the baroclinic wave trough (Fig. 11.3b) in the jet stream region are confluent (Fig. 11.21). All the indicated trajectories enter and pass through the zone of upward motion east of the trough. However, only the air parcels moving south of the heavy line marking the northern boundary of the cirriform cloud are initially moist enough to reach saturation. The confluent trajectories thus form a band of cirriform cloud south of this line. It is not clear from satellite data whether the upper-level cloud formed in this way remains a separate cloud layer or becomes attached to the lower frontal clouds.

As the cloud pattern indicated in Fig. 11.20 evolves, the cirriform cloud widens, apparently as the flow in the upper-level wave becomes more diffluent east of the intensifying trough. By the time of Fig. 11.20c, the surface low has moved from a position east of the jet stream to a location under the jet, as the baroclinic wave develops and the upper- and lower-level flow fields move toward a more aligned position (Fig. 11.4). At this phase in the development of the storm, the lower- or middle-level cloud top emerges from under the cirriform cloud on the poleward side of the jet stream. The lower cloud is beginning to evolve from being a purely linear frontal cloud band into the more comma-shaped pattern associated with the central region of the low-level frontal cyclone. This structure becomes more exaggerated as the wave and frontal cyclone continue to intensify and the low-level cyclone, now west of the upper-level jet stream, becomes more nearly aligned with the upper-level trough (Fig. 11.20d).

11.4.2 Distribution of Precipitation within the Cloud Pattern

Since routine satellite data are limited to a view of the cloud top, it is necessary to resort to other means to investigate the internal structure of the cloud system. For this purpose, radar is particularly useful since it penetrates the cloud to sense the three-dimensional structure and intensity of the precipitation forming within and falling from the cloud system (Chapter 4). Radar measurements can be supplemented by research aircraft data showing aspects of the microphysical structure of the clouds not possible to glean from radar or satellite data. Figure 11.22 summarizes typical features of the horizontal pattern of precipitation in a frontal cyclone, as have been determined from aircraft and radar studies. While the broad

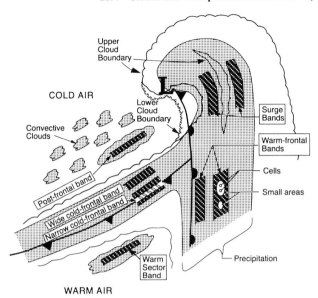

Figure 11.22 Idealization of the cloud and precipitation pattern associated with a mature extratropical cyclone. (Adapted from Matejka *et al.*, 1980 and Houze, 1981. Reprinted with permission from the Royal Meteorological Society; © American Geophysical Union.)

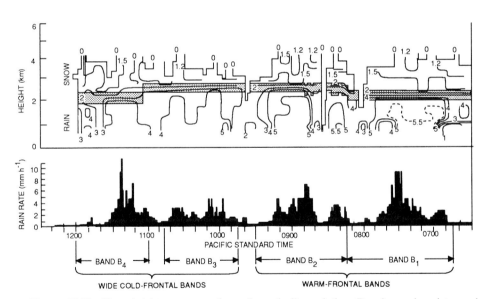

Figure 11.23 Time–height cross section of vertically pointing Doppler radar data and high-resolution rain gauge trace obtained during the passage of an extratropical cyclonic storm over Seattle, Washington. Radar data show 10-min average precipitation fall speeds (m s^{-1}). The melting layer, characterized by a large gradient of fall speed, is shaded for emphasis. Contours labeled with zeros outline the region of precipitation detected by the radar. (Adapted from Houze *et al.*, 1976. Reproduced with permission from the American Meteorological Society.)

outline of the frontal cloud in Fig. 11.22 follows the general outline of the large-scale vertical motion field (indicated by the Q-vector convergence in Fig. 11.3b), the overall pattern of precipitation (indicated by the light stippling in Fig. 11.22) is somewhat more focused and mostly confined to the regions of vertical motion associated with the surface cyclone and the major zones of active low-level frontogenesis (Secs. 11.2 and 11.3), namely the cold-frontal and warm-frontal zones. As we have seen in previous discussions, these frontal zones merge and extend into the region of the low center along the occluded front (Figs. 11.4, 11.16, 11.18, 11.19), and the precipitation pattern in Fig. 11.22 reflects this structure where the cold- and warm-frontal precipitation regions join and extend in toward the low center. Embedded within this general frontal precipitation pattern are various smaller-scale features, the elongated ones being the rainbands. Within the rainbands are still smaller mesoscale regions of enhanced precipitation and convective cells.

The categories of rainbands indicated in Fig. 11.22 include *warm-frontal*, *narrow cold-frontal*, *wide cold-frontal*, and *surge* rainbands, which occur within the envelope of the general frontal precipitation pattern, and *warm-sector* and *postfrontal* rainbands, which fall outside this region. The background precipitation within the main envelope is primarily stratiform (i.e., consists of the type of precipitation described in Chapter 6). The warm-frontal, wide cold-frontal, and surge rainbands (discussed in Secs. 11.4.3–11.4.6) are enhancements of the basic stratiform precipitation. An example illustrating this point is shown in Fig. 11.23, which is a time cross section of vertically pointing Doppler radar data obtained during the passage of a cyclonic storm similar to the one represented schematically in Fig. 11.22. Although four rainbands were embedded in the precipitation (two warm-frontal bands followed by two wide cold-frontal bands), a basic stratiform structure, with a well-defined melting band evident in the vertical gradient of precipitation fall speed, extended continuously across the whole storm. The melting layer rose slightly as the warm-front portion of the storm passed and lowered as the cold-frontal region went by. The basic vertical layering remained intact as the rainbands passed over, indicating that the rainbands were enhancements of the basic stratiform structure associated with the frontal system as a whole.

Although most of the rainbands superimposed on the main envelope of the frontal precipitation pattern in Fig. 11.22 are enhancements of the basic stratiform precipitation process that dominates the frontal system, one of these rainband types is distinctly nonstratiform. It is the *narrow cold-frontal* rainband. As we will see in Sec. 11.4.3, this type of band is a line of intense (sometimes forced rather than free) convection associated with the density-current action of the low-level leading edge of the cold front.

The warm-sector and postfrontal rainbands indicated in Fig. 11.22 are not discussed as special topics below since they consist of clouds and precipitation of types that are discussed elsewhere in the book. They often consist of lines of convective showers or thunderstorms. As such, they are governed by the same principles as other lines of convection already discussed in Chapters 8 and 9. In some situations, postfrontal bands may be associated with a comma cloud, which forms in the cold-air mass in connection with an approaching short-wave trough; the comma-cloud phenomenon was introduced in Sec. 1.4.3 and will be discussed further in Sec. 11.5.

We now turn (in Secs. 11.4.3–11.4.6) to a more detailed examination of the rainbands and other features embedded in the general cloud and precipitation pattern illustrated in Fig. 11.22. This discussion draws heavily from the results of field studies in which the radar-echo patterns, air motions, thermodynamics, air-craft-sampled cloud microphysics, and satellite imagery of frontal cloud systems have been obtained simultaneously. Also contributing to this discussion are the results of modeling studies, which have helped diagnose aspects of the observed phenomena that are beyond our present capability to observe directly.

11.4.3   Narrow Cold-Frontal Rainbands

The largest horizontal component of air motion in cold-frontal clouds is in the along-front direction. The rising air generally is part of the poleward-moving warm air current of the cyclone. It ascends when it is incorporated into the circulation associated with the strong frontogenesis at the cold front. Along a segment of the cold front located fairly far from the storm center a vertical section has the structure shown in Fig. 11.24. In the cross section AB, the cross-front and vertical components of the circulation are generally consistent with the theoretical picture of frontogenesis, discussed in Sec. 11.2. The warm air current generally has a high moisture content and is therefore characterized by high θ_e. Although this current is flowing in a predominantly along-line direction, it develops a rearward and

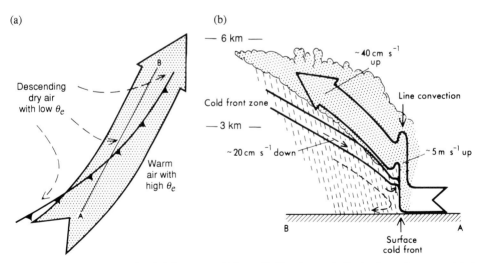

Figure 11.24   Schematic of airflow at a cold front. Bold trajectory indicates flow of warm moist air into and through the cloudy zone. (a) Horizontal projection. (b) Vertical cross section along AB. Dashed lines indicate precipitation. Dashed trajectories indicte flow of dry, θ_e air. (Adapted from Browning, 1986. Reproduced with permission from the American Meteorological Society.)

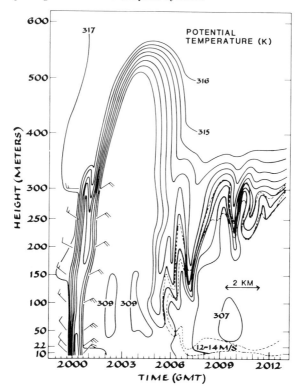

Figure 11.25

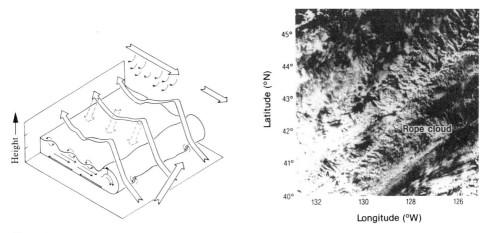

Figure 11.26   Figure 11.27

upward-sloping component, which is the rising branch of the ageostrophic circulation (u_a, w) associated with frontogenesis. The colder, drier, low-θ_e flow subsiding on the cold side of the front constitutes the downward branch of the circulation.

Along the line AB in Fig. 11.24, the leading nose of the front at low levels is characterized by strong, concentrated convergence and abrupt upward motion. It is also found that the thermal gradient behind this leading nose is nearly discontinuous. In Sec. 11.2 (Fig. 11.8 and Fig. 11.10), we saw that the temperature gradient in a frontogenetic zone can collapse to a near discontinuity at the lower boundary of the fluid, where the temperature changes by horizontal geostrophic and ageostrophic advection cannot be offset by adiabatic temperature changes produced by vertical air motions. This collapse is manifested as a strong cold front. Detailed observations of this sharp surface front show it to have the characteristics of a density current, similar to a thunderstorm gust front (Sec. 8.9). This density-current structure is found even in nonprecipitating fronts, as shown by the example in Fig. 11.25. The kinematic structure of the zone, as observed by Doppler radar in precipitating cases, is illustrated schematically in Fig. 11.26. The large head at the leading edge of the cold air mass and the turbulent wavy structure on the top boundary of the spreading cold air (indicated in both Figs. 11.25 and 11.26) resemble quite closely the thunderstorm gust front structure illustrated in Fig. 8.39. When the air lifted over the density current is sufficiently moist, a *rope cloud* forms (Fig. 11.27). If this narrow (~1–5 km wide) cloud precipitates, it produces a narrow cold-frontal rainband, as depicted in Fig. 11.22. The documented cases of this type of rainband indicate that it can be produced by the forced ascent of stable or only slightly unstable air. Although the forced ascent can be strong (of the order of a few meters per second), the precipitating cloud remains limited in vertical and horizontal extent to the zone of lifting forced by the density current. Were the air forced up over the density current particularly unstable, then a mesoscale convective system (Chapter 9) of deep thunderstorms (most likely in the form of a squall line, either with or without trailing stratiform precipitation), would be triggered and quickly obscure the narrow lifting zone at the frontal density current.

Distortions in the form of small vortices sometimes occur at intervals of about 15 km along the windshift line of the density current (Fig. 11.26). In the case

---

Figure 11.25 Density-current structure at the leading edge of a nonprecipitating cold front that passed over an instrumented tower near Boulder, Colorado. Isotherms (solid lines) of potential temperature are labeled in K. Dashed lines enclose region of a 12–14 m s^{-1} wind surge. Wind flags (full barb, 5 m s^{-1}) show tower winds preceding and following the passage of the front. (From Shapiro *et al.*, 1985. Reprinted with permission from the American Meteorological Society.)

Figure 11.26 Schematic of the density-current structure at the leading edge of a precipitating cold front. Air motions determined from Doppler radar data indicated by various arrows. (From Carbone, 1982. Reprinted with permission from the American Meteorological Society.)

Figure 11.27 Visible satellite image showing a rope cloud formed by air lifted over the density current at the leading edge of a cold front over the Pacific Ocean. The front was found by aircraft to be located between A and A′ (southwest corner). (From Shapiro and Keyser, 1990. Reprinted with permission from the American Meteorological Society.)

represented in Fig. 11.26, a weak tornado occurred in association with one of the vortices along the narrow cold-frontal rainband.[302] The kinematic structure of the vortex was generally similar to a tornadic thunderstorm (probably of the nonsupercell type discussed in Sec. 8.7). However, the thermodynamic structure indicated by soundings and Doppler radar retrieval analysis revealed no potential instability of the air forced to rise at the front. This tornadic storm thus differs in one important way from the tornadic thunderstorms produced in thermodynamically unstable environments (Chapter 8). In both cases, the density current plays the role of generating horizontal vorticity via the horizontal buoyancy gradient across its leading edge. However, the source of the density current is entirely different in the two cases. In the thunderstorms discussed in Chapter 8, buoyant instability is required for the formation of downdrafts, which spread out and form a gust front. In the case of the narrow cold-frontal band, the density current is apparently the result of strong low-level frontogenesis producing a nearly discontinuous thermal gradient at the earth's surface. While this density current may sporadically propagate independently, its generally forward progress is determined primarily by larger-scale dynamics and only secondarily by convective downdrafts.

The occurrence of a tornado along a narrow cold-frontal rainband is rare. However, the perturbed structure of the narrow band, from which the tornado occasionally arises, is quite common. Figure 11.28 shows the radar echo structure of four different narrow cold-frontal rainbands observed along the Washington coast in a period of less than 1 month. Each narrow cold-frontal rainband exhibits small-scale radar echoes oriented oblique to the mean position of the cold front. Doppler radar observations reveal that these small elongated cells are related systematically to the distortions of the windshift line (Fig. 11.29). The distortion of the windshift lines results in maximum low-level convergence along the portion of the line where the elongated precipitation core is located, while strong cyclonic shear occurs in the gap region. The tornado described above evidently formed at the point of maximum cyclonic shear.

It has been suggested that the distortion of the windshift line along the cold front is a manifestation of some sort of instability that arises in connection with the horizontal wind shear across the front.[303] It might also be associated with bulges and clefts of the type that are observed in density currents in laboratory tank experiments.[304] Whatever the origin of the distortions, their effect on the clouds and precipitation once they have formed can be seen from a diagram like that in Fig. 11.30. In Fig. 11.30a, the wind along an initially undisturbed front has

---

[302] This tornado was discovered by Carbone (1983) and described by him in some detail. He analyzed the vorticity budget and made comparisons to tornadic thunderstorms.

[303] Suggested by Matejka (1980), Carbone (1982, 1983), and Hobbs and Persson (1982). See these papers for more detailed discussion of this hypothesis. A similar suggestion was made by Wakimoto and Wilson (1989) to explain the vorticity centers long the boundary layer convergence lines over which nonsupercell tornadoes form (Sec. 8.7).

[304] The bulges and clefts in laboratory density currents are discussed and illustrated by Simpson (1987, pp. 150–153). The analogy between them and the distortions in the windshift line along a surface cold front was suggested by Hobbs and Persson (1982) as a possible alternative to an instability associated with the horizontal shear across the line.

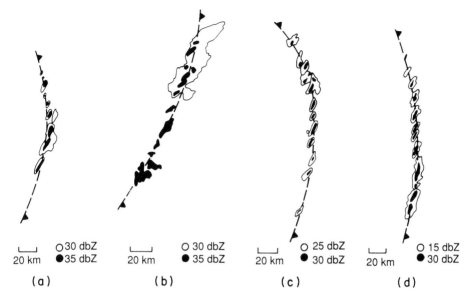

Figure 11.28   Low-level radar reflectivity patterns in narrow cold-frontal rainbands approaching the coast of Washington State on (a) 14 November 1976, (b) 17 November 1976, (c) 21 November 1976, and (b) 8 December 1976. Dashed line with frontal symbols attached indicates the position of the cold front, which in each case was moving with a component of motion from left to right (i.e., west to east) in the picture. Radar echoes from precipitation other than that associated with the narrow cold-frontal rainbands have been deleted. (From Hobbs and Biswas, 1979. Reprinted with permission from the Royal Meteorological Society.)

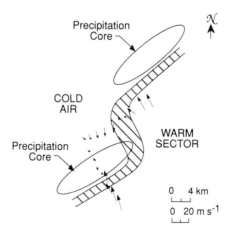

Figure 11.29   Schematic of the relative airflow (shown by arrows), at about 50-m altitude, across two precipitation cores and the gap region between them in a narrow cold-frontal rainband. The airflow was observed by Doppler radar. The surface windshift line observed by Doppler radar is indicated by the crosshatched region. The length of each arrow is proportional to the magnitude of relative wind velocity at the origin of the arrow. (From Hobbs and Persson, 1982. Reprinted with permission from the American Meteorological Society.)

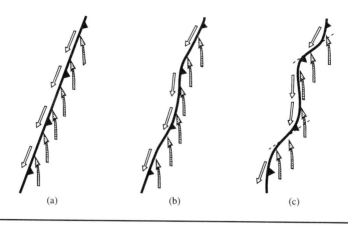

Figure 11.30

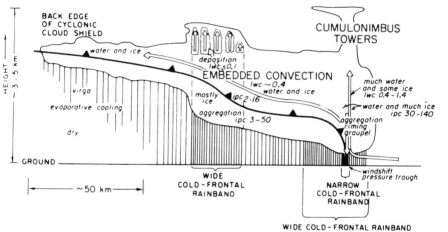

Figure 11.31

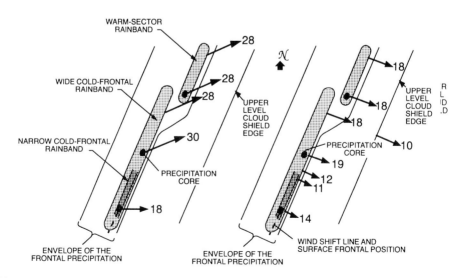

Figure 11.32

no variation in the along-front direction. In Fig. 11.30b, a weak wavelike perturbation appears along the front. In Fig. 11.30c, the perturbation amplifies, producing small-scale lines of enhanced horizontal convergence (dashed lines) between which are zones of suppressed convergence. The precipitation cells occur at the lines of enhanced convergence.

Details of the cloud structure and kinematics in the narrow cold-frontal rainband have been further revealed by instrumented aircraft used in conjunction with radar and other special observations[305] (e.g., Fig. 11.31). The updraft is ~1–5 km wide. A downdraft of similar width is observed just to the rear of the updraft. The up–down sequence at this location is consistent with the Doppler radar results summarized in Fig. 11.26. The updrafts are ~1–5 m s^{-1} and produce sufficient liquid water that riming and graupel formation are the principal ice particle growth mechanisms in this zone. The concentrations of ice particles are ~100 ℓ^{-1}, which indicate that some form of ice enhancement (Sec. 3.2.6) occurs in this zone.

11.4.4   Wide Cold-Frontal Rainbands

The example in Fig. 11.31 also shows a wide cold-frontal rainband superimposed on the frontal circulation. As noted in the discussion of Fig. 11.23, the wide cold-frontal rainband is a region of enhanced stratiform precipitation. Quite unlike the

---

[305] Details of aircraft penetrations of rope clouds and narrow cold-frontal band zones can be found in Matejka *et al.* (1980), Hobbs and Persson (1982), Bond and Fleagle (1985), and Shapiro and Keyser (1990).

Figure 11.30   Hypothesized flow of air at low levels in a frame of reference that is moving with a cold front that develops a wavelike distortion. Shaded arrows represent the horizontal flow of air near the surface in the low-level jet ahead of the cold front. This air undergoes convergence at the cold front, where it ascends in a narrow updraft. Open arrows represent the horizontal flow of air immediately behind the cold front. (a) Initially uniform frontal wind shift line. (b) A perturbation is introduced on the front. (c) The perturbation amplifies, producing small lines of enhanced horizontal convergence (dashed lines) between which are zones of suppressed convergence. (Adapted from Matejka, 1980 by Hobbs and Persson, 1982. Reprinted with permission from the American Meteorological Society.)

Figure 11.31   Cloud structure, air motion, and precipitation mechanisms at a cold front as revealed by instrumented aircraft, radar, and other observations of frontal cyclones passing over Washington State. Vertical hatching below cloud bases represents precipitation: the density of the hatching corresponds qualitatively to the precipitation rate. Open arrows depict airflow relative to the front: a strong convective updraft and downdraft above the surface front and pressure trough, and broader ascent over the cold front aloft. Ice particle concentrations (ipc) are given in numbers per liter; cloud liquid water contents (lwc) are in g m^{-3}. The motion of the rainband is from left to right. Horizontal and vertical scales are approximate but typical of aircraft and radar observations in specific cases. (From Matejka *et al.*, 1980. Reproduced with permission from the Royal Meteorological Society.)

Figure 11.32   Schematic of the clouds and precipitation associated with a cold front moving over Washington State. (a) Arrows indicate velocities (m s^{-1}) of the cloud shield, the envelope of the frontal precipitation, the wide and narrow cold-frontal rainbands, a warm-sector rainband, and the precipitation cores (heavily shaded areas). (b) Velocities normal to the cold front. (From Hobbs *et al.*, 1980. Reproduced with permission from the American Meteorological Society.)

narrow cold-frontal band, its most active region of upward motion is aloft, above the front, and is only very indirectly connected with processes near the earth's surface. The wide cold-frontal band rather appears to emanate from a layer aloft, which is characterized by enhanced mean ascent.

In addition to enhanced mean ascent, the wide cold-frontal rainband contains shallow convective generating cells (Sec. 6.2). The precipitation particles generated in these cells are thought to augment the general stratiform precipitation rate at lower levels by the feeder–seeder mechanism, as they fall through the lower layers of frontal cloud (Fig. 6.7). Figure 11.31 indicates almost no liquid water present in the lower levels of the cloud of the wide cold-frontal rainband. An ice particle concentration $\sim$3–50 ℓ^{-1} is large enough to indicate that some form of ice enhancement (Sec. 3.2.6) was occurring here, as well as in the narrow band.[306] The particles were growing by aggregation just above the melting layer, as is usual for stratiform precipitation (Secs. 6.1–6.3).

The wide cold-frontal band moves with the winds in the layer of enhanced vertical air motion and generating cells. This motion is independent of the narrow cold-frontal band, which moves with the surface cold front. The wide cold-frontal band appears to be generated in a position similar to that shown in Fig. 11.31. However, it can move faster than the front, sometimes ending up ahead of the position of the front at the surface. As it moves forward in this way, the wide cold-frontal rainband sometimes temporarily straddles the narrow cold-frontal band (Fig. 11.32). The formation and movement of the wide cold-frontal band thus suggest that while the frontal system provides a favorable environment for its formation, its dynamics are somewhat separable from the frontal dynamics. In this respect, the wide cold-frontal bands differ strikingly from the narrow cold-frontal rainband, which is anchored to the dynamics of the cold front at low levels.

The time and space scales of the wide cold-frontal rainband are smaller than those of the frontal system as a whole. Multiple wide cold-frontal rainbands can occur within the space and time domain of a single cold-frontal system (e.g., Figs. 11.22 and 11.23). Their essential dynamics are superimposed on the larger frontogenetical circulation. As suggested by Fig. 11.31, the enhanced mean ascent appears to be associated with a local, probably transient steepening of the frontal slope.[307] Whether the steepening of the front in this location is a cause or effect of the rainband is not clear. That it could be the latter is suggested by the fact that evaporation and melting of the precipitation particles falling from an established wide cold-frontal rainband cloud into the unsaturated air below the front could introduce the horizontal gradient of heating into (11.64) that would locally enhance the frontogenetic circulation.[308]

There is evidence that some wide cold-frontal rainbands are a manifestation of conditional symmetric instability (Sec. 2.9.1).[309] This possibility is illustrated by

---

[306] See Matejka *et al.* (1980) for details of the data obtained on the aircraft penetration.

[307] The steepening of the slope of the front in the vicinity of wide cold-frontal rainbands has been examined more recently by Locatelli *et al.* (1992).

[308] Locatelli et al. (1992), however, take the position that it is cause rather than effect.

[309] The possibility that rainbands in frontal systems could be manifestations of moist symmetric instability was suggested by Hoskins (1974) and Bennetts and Hoskins (1979).

Fig. 11.33, which shows schematically the results of a mesoscale numerical-model simulation of a cold-frontal system. Both narrow and multiple wide cold-frontal rainbands formed in the simulation. A narrow cold-frontal band was forced by boundary-layer convergence at the leading edge of the front, which is consistent with the observations described above. A region of negative geostrophic equivalent potential vorticity P_{eg} (defined at the end of Sec. 2.5) was present in the

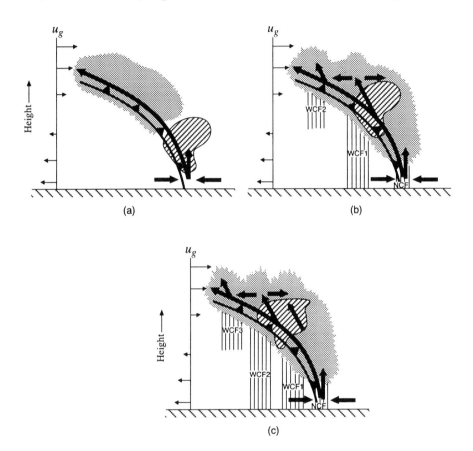

Figure 11.33   Schematic of the processes leading to the formation of cold-frontal rainbands in a mesoscale numerical model. The crosshatching shows the region of negative equivalent potential vorticity. Shading shows the region containing cloud water. Vertical hatching below cloud base represents precipitation. The geostrophic wind in the plane of the cross section (u_g) is indicated on the left-hand side of the figure. Broad arrows show ageostrophic air motions. All winds shown are relative to the motion of the baroclinic wave within which the frontal system developed. (a) A region of negative equivalent potential vorticity, which forms in the warm air, is advected up along the frontal surface by the ageostrophic motions. (b) When the region of conditional symmetric instability becomes saturated the instability is realized, resulting in a wide cold-frontal rainband (WCF1). A second rainband (WCF2) is forced by convergence behind WCF1. Convergence in the planetary boundary layer produces a narrow cold-frontal rainband (NCF). (c) WCF1 moves toward the warm air; WCF2 moves into the region of conditional symmetric instability and intensifies. A third band (WCF3) is forced by convergence behind WCF2. (Adapted from Knight and Hobbs, 1988. Reproduced with permission from the American Meteorological Society.)

boundary layer. As noted in 2.9.1, the condition $P_{eg} < 0$ is the criterion for potential symmetric instability. The region of negative P_{eg} in the boundary layer of the model case was partly present in the initial conditions and partly created in the boundary layer in connection with (parameterized) turbulence. The first wide cold-frontal band (WCF1) forms when the region of negative P_{eg} is lifted by the frontal circulation to saturation. Once the region is saturated, the potential symmetric instability can be released. The enhanced vertical motion arising from the release of the instability intensifies the precipitation, thus creating the rainband in the precipitation pattern. The second wide cold-frontal band (WCF2) arises differently from WCF1. It is forced by convergence behind WCF1 at a somewhat higher altitude. WCF1 moves toward the warm air faster than the front. WCF2 moves into the region of negative P_{eg}, where it intensifies, presumably as conditional symmetric instability is released. WCF3 is forced by convection behind and follows a life cycle similar to WCF2.

From the foregoing discussions, it appears that the enhancement of the stratiform precipitation in the wide cold-frontal rainbands is effected microphysically by generating cells aloft and the feeder–seeder process (Fig. 11.31) and dynamically by the enhancement of the mean upward air motion by the release of potential symmetric instability (Fig.11.33). Other factors such as frontal slope may also be important.

11.4.5   Warm-Frontal Rainbands

The warm-frontal rainbands indicated in Fig. 11.22 are similar in dimension to the wide cold-frontal rainbands. They differ in that they occur in the forward part of the cloud shield of the developing cyclone, where warm advection dominates, rather than in the trailing part where cold advection dominates. They tend to be parallel to the isotherms in that part of the storm, just as the wide cold-frontal bands tend to be parallel to the isotherms in their part of the storm. In this subsection, we will examine some actual examples illustrating the kinematics and microphysics of warm-frontal rainbands.

The 850-mb isotherms and geopotential height pattern for the first example are shown in Fig. 11.34. The rainbands were observed in the vicinity of station MDW during a 4-h period preceding the map time. The locations of the axes of three rainbands (labeled A1, A2, and A3) 3 h before the map time are shown by the solid lines in Fig. 11.35. The dashed lines labeled B1–B5 show the positions of five lines of generating cells (Sec. 6.2) situated between 5 and 6 km altitude. The generating cells, located above the rainbands, were of the type described in Sec. 6.2. They moved over the lower-level rainbands and seeded them with ice particles in the manner illustrated in the example of stratiform rain shown in Fig. 6.8. That example was taken from a warm-frontal rainband, and the precipitation mechanisms were illustrated in Figs. 6.8–6.10, where it was seen that a mesoscale region of enhanced updraft at lower levels was seeded from above by the generating cells. These mechanisms typify the precipitation processes in warm-frontal rainbands.

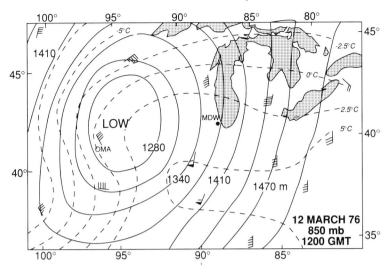

Figure 11.34   Conditions at 850 mb for a case in which warm-frontal rainbands were observed in the vicinity of Chicago, Illinois (station MDW) during a 4-h period preceding the map time. Isotherms (dashed) and geopotential height (solid lines) are analyzed. Full wind barb is for 5 m s⁻¹ and flag is for 25 m s⁻¹. (Adapted from Heymsfield, 1979. Reproduced with permission from the American Meteorological Society.)

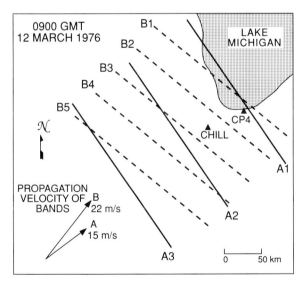

Figure 11.35   Locations of the axes of three rainbands (labeled A1–A3) and five bands of generating cells (labeled B1–B5) in the vicinity of Chicago, Illinois. CP4 and CHILL are Doppler radar sites. (Adapted from Heymsfield, 1979. Reproduced with permission from the American Meteorological Society.)

The rainbands A1–A3 were observed simultaneously by two Doppler radars. Fields of variables obtained by dual-Doppler synthesis (Sec. 4.4.6) from the data of these two radars are displayed in Fig. 11.36. The wind direction field (Fig. 11.36a) shows a sloping layer of veering wind, which according to the thermal wind relation (11.6) implies warm advection. The warm advection in this part of the baroclinic disturbance was concentrated in this layer, which sloped upward in the sense of a warm front. The dual-Doppler synthesized wind field in this zone

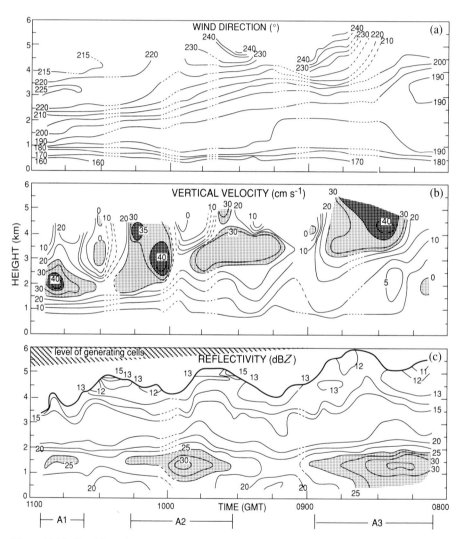

Figure 11.36   Dual-Doppler radar observations of rainbands A1–A3, whose positions are shown in Fig. 11.35. Time–height cross sections centered over station MDW (see Fig. 11.34) show (a) wind direction, (b) vertical velocity, (c) radar reflectivity factor. Dashed contours indicate regions of missing observations. (Adapted from Heymsfield, 1979. Reproduced with permission from the American Meteorological Society.)

was found to be frontogenetic.[310] A sloping zone of upward motion was found within the layer of warm advection, as is expected from the large-scale omega equation (11.14) and illustrated by Fig. 11.3b. In addition, the vertical velocity is enhanced in three mesoscale zones, where the maximum upward velocity exceeds 40 cm s^{-1} and the region of ascent extends upward out of the concentrated layer of warm advection. These three regions of enhanced ascent seem to correspond to the three rainbands A1–A3, each rainband exhibiting its peak reflectivity somewhat downstream (to the right) of the corresponding region of enhanced upward motion (cf. Fig. 11.36b and c).

The relation between the vertical motion and the radar reflectivity fields in Fig. 11.36 is subject to some uncertainty. The upward motion in Fig. 11.36b was calculated by substituting the divergence of the Doppler-synthesized wind field into the anelastic continuity equation (2.54). As discussed in Sec. 4.4.6, this procedure is especially prone to error in cases such as this one, in which the echo top is not at a level where an upper-boundary condition on the vertical velocity can be established. Hence, the vertical velocity field in Fig. 11.36b is only a rough estimate. Also, the time scale of the fallout of the ice particles growing in the upward motion zones, the wind velocity relative to the rainbands, and the lifetimes of the rainbands must all be commensurate for there to be a one-to-one correspondence between the rainbands and the three areas of enhanced upward motion. Variability in a direction perpendicular to the cross sections in Fig. 11.36 is another source of uncertainty. Nonetheless, the mesoscale regions of enhanced vertical motion are similar in size to the rainbands, and each rainband appears to have been associated with such a zone of vertical motion.

The final example of a warm-frontal rainband is the feature A positioned in the forward portion of the cloud shield of the frontal cyclone as shown in Fig. 11.37. A schematic of the clouds, precipitation, and thermal field deduced from rawinsonde and aircraft data is shown in Fig. 11.38. An intrusion of dry air aloft above the main cloud deck created a layer of potential instability ($\partial \bar{\theta}_e / \partial z < 0$; see Sec. 2.9.1). It is suggested that the generating cells were triggered in this layer as a result of the vertical motion in the layer of warm advection below.[311] Aircraft measurements below the layer of cells[312] indicate that the seeded region was glaciated. In contrast, the region of cloud adjacent to the seeded zone exhibited measurable amounts of liquid water. The ice particles were present in concentrations of 4–16 ℓ^{-1}, similar to the concentrations seen in wide cold-frontal rainbands (Fig. 11.31). Also similar to the wide cold-frontal bands, the particles were growing by aggregation in the region just above the melting layer.

The dynamical mechanism leading to the enhanced vertical air motions in warm-frontal rainbands (Fig. 6.9, Fig. 11.36b) has not been clearly identified. They may be produced by symmetric instability, as appears to be the case for some wide cold-frontal rainbands (Sec. 11.4.3). However, this has not yet been

---

[310] See Heymsfield (1979) for details.
[311] This mechanism was postulated by Kreitzberg and Brown (1970).
[312] See Matejka *et al.* (1980) for details.

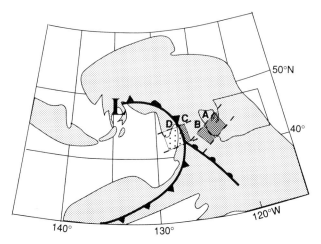

Figure 11.37 Configuration of mesoscale rainbands, fronts, surface low (L), and satellite-observed cloud shield on 7 January 1975. A and B are warm-frontal rainbands. C is a surge rainband, while D is a field of convective showers trailing the surge rainband. The positions of features A–D relative to the frontal structure, their widths, and their spacing are those observed as they passed across a special data-collection network in Washington State (outlined). (From Matejka, 1980.)

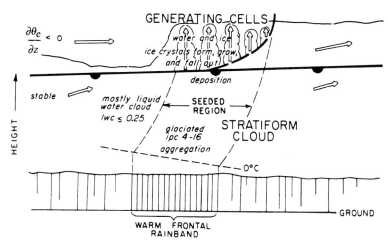

Figure 11.38 Schematic of the clouds, precipitation, and thermal field of a warm-frontal rainband deduced from rawinsonde and aircraft data obtained in frontal systems passing over Washington State. The structure of the clouds and the predominant mechanisms of precipitation growth are indicated. Vertical hatching below cloud base represents precipitation; the density of the hatching corresponds qualitatively to the precipitation rate. The heavy, broken line branching out from the front is the leading edge of an intrusion of dry air that was conditionally unstable ($\partial\theta_e/\partial z < 0$) and in which generating cells formed. Ice particle concentrations (ipc) are given in numbers per liter; cloud liquid water contents (lwc) are in g m^{-3}. The motion of the rainband in the figure is from left to right. Wide arrows indicate air motion. (Adapted from Matejka *et al.*, 1980. Reprinted with permission from the Royal Meteorological Society.)

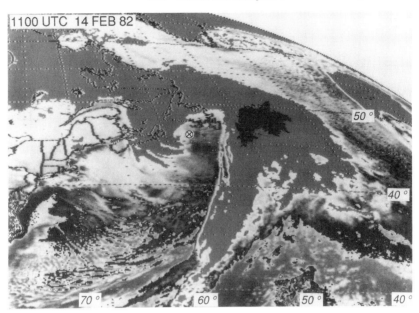

Figure 11.39 Infrared satellite image of an occluded cyclone at 1100 GMT on 14 February 1982. Symbol ⊗ indicates surface low position. (From Kuo *et al.*, 1992. Reprinted with permission from the American Meteorological Society.)

demonstrated by a modeling study comparable to that in Fig. 11.33. It has also been suggested that gravity waves ducted in the layer of warm advection might play a role in producing warm-frontal rainbands.[313]

11.4.6 Clouds and Precipitation Associated with the Occlusion

The clouds and precipitation found in the region of the occlusion (i.e., the region poleward of the point of merger of cold- and warm-frontal zones) is complicated by the three-dimensional circulation of the intertwined currents of air of differing thermal characteristics and moisture contents flowing into the center of the cyclone. In Fig. 11.22, this zone is characterized by the head of the comma pattern seen in the high, cirriform cloud top. In a well-developed cyclone, this high cloud has an indentation protruding eastward exposing low- to midlevel cloud tops over the surface occlusion. As we will see, this indentation in the upper cloud is due primarily to an intrusion of subsided dry air at upper levels. The precipitation in this region is in a broad area mostly ahead of the surface occlusion. Wide surge rainbands, which are rather similar to warm-frontal rainbands, occur ahead of the dry-air intrusion.

To illustrate the production of clouds and precipitation in the occluded portion of the cyclone, we refer to an example shown in Fig. 11.39. The upper-level cloud

[313] See Lindzen and Tung (1976) for the exposition of this hypothesis.

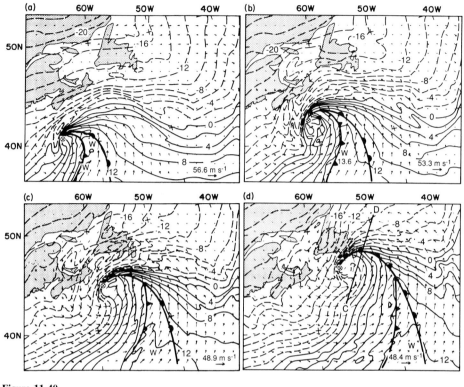

Figure 11.40

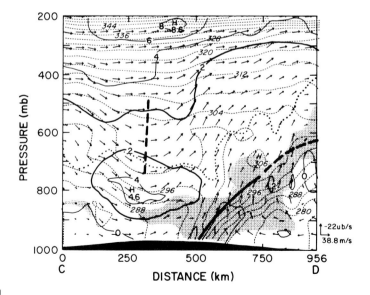

Figure 11.41

shield exhibits the characteristic indentation, revealing lower cloud tops over the surface occlusion. This case has been simulated accurately with a mesoscale numerical model in a manner similar to the two storms illustrated in Figs. 11.18 and 11.19. Like the case in Fig. 11.18, the storm illustrated here occurred over the Atlantic Ocean. The evolution of the low-level temperature pattern in Fig. 11.40 is similar to that in Fig. 11.18, except that by the time of the occluded stage (Fig. 11.40d) there is not an intensified zone of strong gradient of temperature just to the rear of the cold front to the southeast of the surface low center. The occlusion itself is, however, rather similar to the previous case.

A vertical cross section along line CD in Fig. 11.40d is shown in Fig. 11.41. The position of the occluded front (i.e., the warm ridge connecting the warm sector with the surface low center, as seen in Fig. 11.40d) is shown by a heavy solid line. It has the slope and general appearance of a warm-frontal zone and in fact transforms to a warm front at higher levels (heavy dashed extension of the heavy solid line). It is not intersected by a cold front aloft.[314] A cold front in the vicinity of the trough aloft is indicated by a heavy dashed line 250 km upwind of the occlusion; however, it is unconnected to the occlusion. The dominance of the warm-frontal structure in the region of the occlusion is consistent with the horizontal temperature pattern in Fig. 11.40, where the strong temperature gradient is on the warm side of the occlusion. The cross section cuts through the indentation in the upper cloud pattern seen in Fig. 11.39 and confirms that the clouds are confined to low levels. They are found in the general region of upward motion east of the trough of the upper-level baroclinic wave. They are concentrated roughly in a sloping upward zone straddling the occluded front. The air above the cloud layer has been advected from the west side of the upper-level trough after having subsided from high levels. Its potential vorticity, which is conserved (Sec. 2.5), indicates it is of stratospheric origin (the regions of potential vorticity units >2 in Fig. 11.41 are considered to have originated in the stratosphere). Although this air is rising on the east side of the trough, it is so dry that it cannot reach saturation and form cloud. Hence the clouds in this region are confined to low and middle levels.

---

[314] See Sec. 3 of Kuo *et al.* (1992) for a discussion of the criteria used in locating occlusions, warm fronts, and cold fronts in this analysis.

Figure 11.40 Mesoscale primitive-equation model calculation of the development of the extratropical cyclone shown in Fig. 11.39. Predicted 900-mb temperature (°C) and wind field valid at (a) 0000 GMT, (b) 0430 GMT, (c) 0900 GMT, (d) 1330 GMT on 14 February 1982. Line CD shows the location of the cross section in Fig. 11.41. (From Kuo *et al.*, 1992. Reprinted with permission from the American Meteorological Society.)

Figure 11.41 Vertical cross section along line CD in Fig. 11.40d. Equivalent potential temperature (dashed isopleths, in K), potential vorticity (thin solid isopleths in "potential vorticity units" defined as 1×10^{-6} K kg^{-1} m s^{-1}), 40% isohume (dotted line), and cloud area (stippling). Cloud boundary is defined by the 100% isohume. Air motion relative to the storm is indicated by arrows according to the attached scale. The potential vorticity contour for two potential vorticity units is emphasized because air with potential vorticity exceeding this value is probably of stratospheric origin. Heavy dashed line represents an upper cold front. (From Kuo *et al.*, 1992. Reprinted with permission from the American Meteorological Society.)

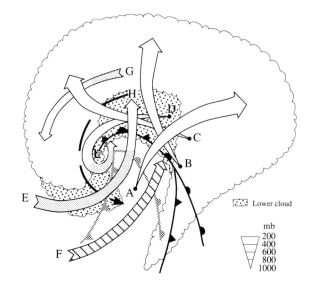

Figure 11.42

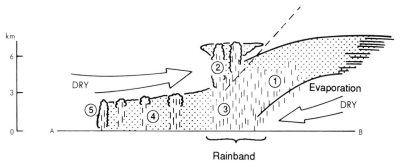

Figure 11.43

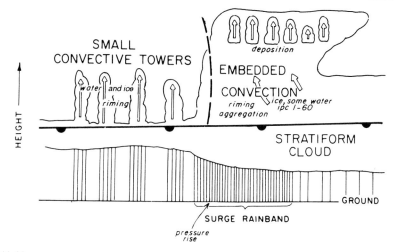

Figure 11.44

Figure 11.42 is a schematic based on model-computed air trajectories for the case illustrated in Figs. 11.39–11.41. It provides insight into the development of the clouds in the central region of the cyclone during the time that the occlusion is forming. The two sets of frontal symbols indicate the temperature pattern at the beginning and end of the period represented. The clouds shown are those present at the end of the period. The trajectories cover the time period bracketed by the two sets of fronts. Trajectory A shows that the upper cloud shield on the east side of the storm is produced by condensation in air rising from the boundary layer of the warm sector of the cyclone. It is only this portion of the cloud cover in the occluded cyclone that arises from the warm sector. Trajectories B and C show that the northern and northwestern parts of the cloud shield are produced by lifting of air originally on the cold side of the warm front. Although these air parcels are on the poleward side of the front, they are within the zone of warm advection, where the large-scale vertical air motion associated with the baroclinic wave is upward (Sec. 11.1). Trajectories D, G, and H also originate on the cold side of the warm front. D and G account for the cloud in the southward protruding hook of cloud on the west side of the surface low, while H is simply a surface-level parcel of cold air that encircles the cyclone center. Trajectory E traces the dry air of upper tropospheric to stratospheric air that has subsided on the west side of the upper-level trough and begun to rise on the east side but is unable to reach saturation; it is the flow represented in the vertical cross section in Fig. 11.41 and it accounts for the indentation in the upper cloud deck. As parcels C and D rise

Figure 11.42 Schematic diagram of the airflow in an occluded cyclone. Trajectories shown by arrows are based on model-output from the simulation of the storm illustrated in Figs. 11.39–11.41. Frontal pattern at the beginning time light frontal symbols of the trajectories is an open wave with the warm sector (i.e., the wedge of warm air between the cold and warm fronts) extending into the surface low-pressure center. Frontal pattern at the ending time heavy frontal symbols of the trajectories is fully occluded, with the warm sector located far from the low center. Width of arrows is proportional to altitude as indicated by the pressure scale in lower right. Subsiding trajectories are shaded; rising trajectories are open. Level trajectory is hatched. Cloud shield boundaries are shown by scalloped lines. Middle to upper clouds unshaded. Lower clouds shaded. (Adapted from Kuo *et al.*, 1992. Reproduced with permission from the American Meteorological Society.)

Figure 11.43 Schematic of the airflow in an occluded extratropical cyclone. Numbers represent precipitation type as follows: (1) warm-frontal precipitation, (2) convective precipitation-generating cells associated with dry air intrusion aloft, (3) precipitation from the upper cold-frontal convection descending through an area of warm advection, (4) shallow moist zone characterized by warm advection and scattered outbreaks of mainly light rain and drizzle, and (5) precipitation at the surface cold front. Dashed line represents leading edge of dry air. (Adapted from Browning and Monk, 1982. Reprinted with permission from the Royal Meteorological Society.)

Figure 11.44 Schematic vertical cross section across a surge rainband. The heavy, broken line branching out from the warm front is the leading edge of an intrusion of dry air aloft that is conditionally unstable and in which small convective towers form. The structure of the clouds and the predominant mechanisms of precipitation growth are indicated. Vertical hatching below cloud base represents precipitation; the density of the hatching corresponds qualitatively to the precipitation rate. Ice particle concentrations (ipc) are given in numbers per liter. The motion of the rainband and dry air intrusion aloft in the figure is from left to right. Wide arrows indicate air motion. (Adapted from Matejka *et al.*, 1980. Reprinted with permission from the Royal Meteorological Society.)

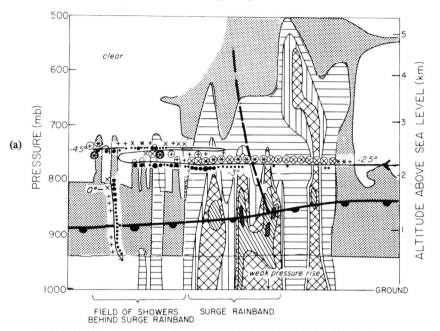

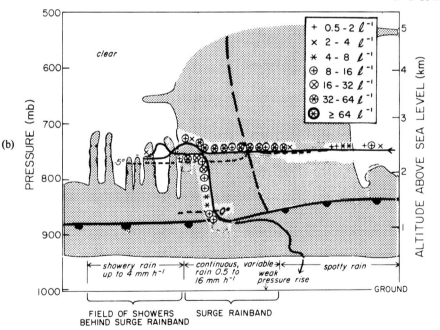

Figure 11.45

from the boundary layer to upper levels, they pass under trajectory E in the center of the region of low- to middle-level cloud just ahead of the surface occlusion. The lower cloud deck is evidently produced by condensation in parcels C and D on their way westward and upward. Since they are not initially so dry, condensation can occur at low levels and produce the lower-level frontal cloud, which is exposed in satellite view by the clear indentation left by the higher-level trajectory E. Trajectory F, south of E, also is characterized by dry air from the upper troposphere or stratosphere. It differs from E only in that it follows a nearly constant-level path during the time period shown.

As indicated in Fig. 11.22, the precipitation pattern ahead of the indentation in the upper-level cloud shield sometimes exhibits a substructure. The surge rainband tends to form just ahead of the indentation, and sometimes there is a multiple structure with more than one surge band and a break in the upper-level cloud in between.[315] Schematic vertical cross sections across surge rainbands are shown in Fig. 11.43 and Fig. 11.44. These cross sections, derived from entirely different studies, show the surge rainband just ahead of the dry intrusion, with a tendency for small convective showers to be embedded in the lower-level cloud pattern behind the rainband. The rainband itself is marked by a deepening of the cloud, with convective cells aloft producing snow to seed the lower cloud. In this respect, the precipitation structure is similar to that in the wide cold-frontal and warm-frontal rainbands. The precipitation is of the stratiform type discussed in Sec. 6.2, in which generating cells produce ice falling into the cloud layer below (Figs. 6.7–6.10). Also similar to the wide cold-frontal and warm-frontal rainbands are the concentrations of ice particles in the range of $1-60 \ \ell^{-1}$ and the growth of the ice particles by aggregation just above the melting layer. Although there were no multiple Doppler radar measurements to indicate whether or not enhanced mesoscale vertical air motion occurred in the lower parts of the surge rainband cloud measurements illustrated in Fig. 11.44, the presence of moderately rimed ice particles across a substantial portion of the cloud would be consistent with

[315] The name "surge rainband" was inspired by the fact that it occurs in conjunction with the surge of dry air producing the indentation in the upper-level cloud.

Figure 11.45 Vertical cross sections through a mesoscale surge rainband observed in Washington State on 10 January 1976 and the field of the convective showers following it. The leading edge of dry air is indicated by a heavy dashed line. (a) The path of the University of Washington B-23 research aircraft is shown by the continuous arrowed line. Small, medium, and large dots on the flight path denote cloud liquid water contents from 0.10 to 0.20, 0.25 to 0.50, and 0.55 to 0.95 g m^{-3}, respectively. Regions of radar reflectivity along the path of the aircraft are shown by hatching for contours of 12, 22, 27, 32, and 37 dBZ. (b) The path of the National Center for Atmospheric Research Sabreliner aircraft is shown by the continuous, arrowed line. The outlines of the clouds, shown in (a) and determined from observations made on board the B-23, have been duplicated in (b) to serve as a reference. Symbols indicating ice particle concentrations on the flight paths are defined in the key. (From Matejka *et al.*, 1980. Reprinted with permission from the Royal Meteorological Society.)

such enhanced upward motion. If it were present, the similarity to the warm-frontal and wide cold-frontal rainbands would be even more evident.

The schematic model in Fig. 11.44 was based on three cases.[316] An example of the type of data from which it was obtained is in Fig. 11.45. In this case, a surge rainband occurred just ahead of the clear, dry intrusion at middle to upper levels. It was observed nearly simultaneously by two research aircraft, a single Doppler radar, and serial rawinsondes. Figure 11.45 is a composite of those data.[317] The reflectivity field from the radar shows a cellular structure, which appears to have been produced by the fallstreaks from convective cells aloft, since reports from the aircraft did not indicate particularly strong vertical motions at their flight levels. The temperature, liquid water content, and ice particle concentration along the flight paths are shown. Ice particles were present in concentrations of 7–60 ℓ^{-1}, and liquid water contents of 0.1–0.5 g m^{-3} were observed continuously across about the rear two-thirds of the band. The leading portion was glaciated. The ice particles included needles, assemblages of needles, and aggregates as large as 3 to 4 mm in diameter. The cell ahead of the band was a transient feature that was not included in the band because it did not exhibit mesoscale continuity. At the position of the dashed line, the radar measured a wind shift and the aircraft encountered a temperature drop of ~0.5°C. The latter was likely the result of evaporation as the intruding dry air evaporated cloud and precipitating particles. A slight pressure rise occurred as the band passed over surface stations.

11.5   Clouds in Polar Lows

11.5.1   Comma-Cloud Systems

In Sec. 1.3.3, two examples of polar lows were shown. The first was a "comma cloud," which forms in connection with a short-wavelength baroclinic wave located a short distance to the rear of a major frontal system (Fig. 1.29).[318] Several important processes are involved in the development of a comma-cloud disturbance. As noted in Sec. 11.1, the region of upward air motion in a baroclinic wave, in accordance with the distribution of the geostrophic Q-vector (Fig. 11.3), is typically found in a comma-shaped region to the west of the trough. Cloud formation is favored in this region of ascent. The clouds in the short wave are found primarily in the envelope of this vertical-motion region. However, comma-cloud systems form mainly over oceans during the winter half of the year, and additional factors influence the cloud development. The potential instability of the air is high

---

[316] See Matejka *et al.* (1980) for details.

[317] In earlier studies, Kreitzberg (1963) and Kreitzberg and Brown (1970) inferred the existence of surge-type rainbands and deduced some qualitative aspects of their structure from experiments employing special rawinsonde data. The more recent studies generally confirm their findings and add detail on the microphysical and precipitation structure.

[318] Although the major frontal cloud system also exhibits a much larger comma shape in satellite pictures, the name "comma cloud" is commonly used to refer to the smaller comma associated with the short wave trailing the larger frontal cloud system.

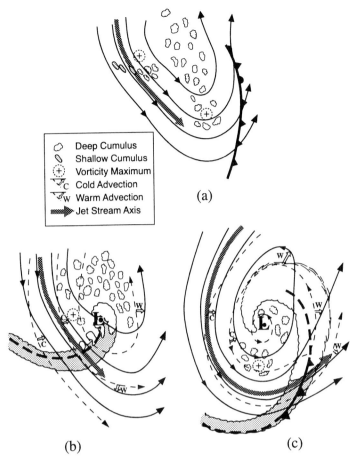

Deep Cumulus
Shallow Cumulus
Vorticity Maximum
Cold Advection
Warm Advection
Jet Stream Axis

(a)

(b)   (c)

Figure 11.46 Schematic diagram of comma-cloud development: (a) incipient stage, (b) intensifying stage, (c) mature stage. Solid lines are 500-mb height contours. Light dashed lines are surface isobars. Arrowheads on contours indicate direction of geostrophic wind. Surface front is shown in (a). Heavy broken line marks the surface trough in (b) and (c). Frontal symbol in (c) indicates that the trough may assume frontal characteristics in some cases. Further explanation at lower left of (a). (From Reed and Blier, 1986. Reprinted with permission from the American Meteorological Society.)

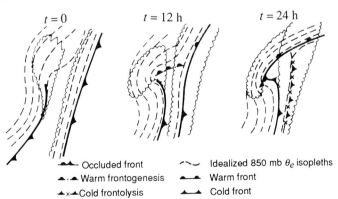

$t = 0$ $t = 12$ h $t = 24$ h

Occluded front   Idealized 850 mb θ_e isopleths
Warm frontogenesis   Warm front
Cold frontolysis   Cold front

Figure 11.47 Horizontal map showing empirical model of the instant occlusion process in which a comma-cloud polar low amalgamates with a pre-existing low-level frontal system. Dashed lines are Isotherms with cold air to the left. (Adapted from McGinnigle *et al.*, 1988. Produced with permission of the Controller of Her Brittanic Majesty's Stationery Office.)

and favors convective rather than stratiform clouds, at least during the initial phases of cloud formation in the wave. The oceanic regions where the comma clouds form are also characterized by appreciable surface fluxes of heat and moisture. These fluxes influence the vertical motion field directly through the diabatic heating term, which would have been included in the omega equation (11.14) had we used the nonadiabatic version of the thermodynamic equation (2.9) instead of (2.11) in the derivation of the equation.

The development of a comma-cloud system is depicted schematically in Fig. 11.46. The incipient comma cloud is associated with the western maximum of vertical vorticity west of the main front shown in Fig. 11.46a. This vorticity maximum is associated with a short-wavelength trough on the west side of the larger-scale trough directly associated with the front. The cold air mass flowing over the ocean behind the main frontal system to the east is potentially unstable, and the clouds in the air mass tend to be convective as indicated in Fig. 11.22). A large group of convective clouds is associated with the larger-scale trough. A smaller group of convective clouds is indicated to have concentrated into a small comma-shaped area corresponding to the region of ascent on the east side of the small-scale trough. Since the air is potentially unstable, the ascent associated with the wave encourages the formation of the convective clouds rather than the formation of a continuous area of stable upward motion and stratiform cloud. Eventually, the cumulonimbus clouds grow larger such that their anvils expand. Some may group together to form mesoscale convective systems (Chapter 9). By the time of the intensifying stage of the comma-cloud system (Fig. 11.46b), the expanded anvils of individual or grouped cumulonimbus have begun to exhibit small comma shapes, as they are distorted by the wind field, which is characterized by strong positive vertical vorticity. Also, the effect of the developing baroclinic wave has begun to be felt at the ocean surface, where a low-pressure system has begun to form below the region of upward motion at upper levels. This cyclogenetic behavior is characteristic of a baroclinic wave, as we have seen in Sec. 11.1 (Fig. 11.3). After this time, the cloud system evolves into a mature stage (Fig. 11.46c), in which the upper-level cloud shield evolves into a more continuous sheet of cirriform cloud. To some extent, this large comma-shaped cloud shield is the merger of the earlier cumulonimbus and mesoscale convective systems. However, the baroclinic wave and its frontogenetic wind field undoubtedly promote cloud organization in response to the vertical circulations associated with the wave and its associated frontogenesis (Sec. 11.2). The symbology used in Fig. 11.46c indicates that in its mature phase the comma-cloud system sometimes exhibits frontal characteristics.

The short wave with which the comma cloud is associated tends to overtake the larger frontal cloud system typically located to its east. When this happens, there is a dynamical interaction. We saw in Sec. 11.1 that when an upper-level baroclinic wave moves over a pre-existing front, surface cyclogenesis is favored (Fig. 11.5). Thus, when the wave with which the comma-cloud system is associated catches up with the front to the east, that frontal system begins to be disturbed, as in the stage of frontal low development (Fig. 11.4a and Fig. 11.5b). At the same time, however, the cyclogenesis and frontogenesis associated with the comma-cloud system itself continue. The two dynamical systems, along with their associated cloud patterns, merge and immediately form a cloud system that has all the appearances of a fully occluded frontal cyclone. In effect, there is a jump from Fig. 11.4a to Fig. 11.4c in terms of the stage of development of the surface cyclonic system. One empirical model of the amalgamation of the comma cloud and the original, larger-scale frontal system is presented in Fig. 11.47. Several variations on this theme of interacting systems have been observed. Sometimes the cyclone on the pre-existing front dominates and absorbs the comma cloud. In other cases the two systems retain separate identities throughout the development. In still other cases, the comma cloud is the dominant partner in the development.

11.5.2 Small Hurricane-like Vortex

The comma cloud considered in Sec. 11.5.1 is but one type of polar low. It is not possible to describe here all the types of polar lows that can occur. However one type of some interest is the second example of a polar low mentioned in Sec. 1.3.3. It is generally a smaller-scale feature than the comma-cloud system, and it exhibits a spiral shape and an eye, similar to that of a hurricane (Fig. 1.30).[319] This type of storm is sometimes found near large areas of pack ice at high latitudes, where, as a result of the continuous outgoing long-wave radiation during the arctic night, air temperatures can become extremely low (as low as −40°C) before the air streams out over neighboring ocean water. The air coming off the ice is quickly modified when it makes contact with the warmer ocean surface (recall the cloud streets in Fig. 5.21). However, as illustrated schematically in Fig. 11.48, this air remains markedly colder than air farther out to sea, which has had a longer

---

[319] For a more detailed summary of this case, see Businger (1991).

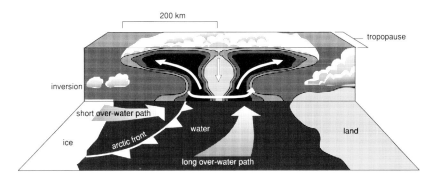

Figure 11.48 Conceptual model of a polar low characterized by a small hurricane-like vortex. Contrast between cold air sweeping off the arctic ice and air warmed by a long passage over open water comprises a temperature discontinuity called an arctic front. A cloud-free eye, in which air is sinking, is surrounded by a symmetric wall of convective clouds. The convection is fed by cyclonic converging flow of high winds at the surface. Below the tropopause, the outflow of air and moisture from the storm creates a broad, anvil-shaped deck of stratiform cloud. (From Businger, 1991.)

trajectory over the water. One result is the formation of an *arctic front* at the leading edge of the air that originated recently over the ice.

In this situation, there are several factors that can affect the dynamical evolution of the polar low. In particular: (i) A feature like the arctic front is always susceptible to cyclogenetic stimulation by the passage of an upper-level trough (as in Fig. 11.5). (ii) As we saw in Chapter 10, the tropical cyclone draws its energy almost wholly from the ocean surface as boundary-layer mixing is stimulated by strong winds over the warmer water. In the polar low, there is also a considerable disequilibrium at the sea surface, in this case because of the large temperature contrast between air and ocean. Thus, hurricane-like dynamics are feasible. (iii) The polar low forms in an environment where the vertical thermal stratification exhibits considerable buoyant instability (Sec. 2.9.1). As a result, some of the energy of the storm may be drawn from convective available potential energy (CAPE, Sec. 8.4) stored in the vertical temperature gradient. Any or all of these energy sources may be tapped in a given case of polar-low development. Given the multiplicity of viable energy sources, it is understandable that a variety of observed polar-low structures exist. The cases exhibiting hurricane-like cloud structure may be those that depend to a significant extent on (ii). However, this topic remains one of active research.

Chapter 12 | Orographic Clouds

> "...towering up the darkening mountain's side...
> It mantles round the mid-way height..."[320]

In previous chapters, we have seen a variety of dynamical features that govern the air motions in clouds. These include turbulence and entrainment in layer clouds, buoyancy, pressure perturbation, entrainment, and vorticity in cumulus and cumulonimbus, mesoscale circulations in complexes of thunderstorms, and secondary circulations associated with the winds and thermal patterns of hurricanes, baroclinic waves, and fronts. There remains one important source of air motions in clouds that we have not yet discussed: the flow of air over hills and mountains. In this chapter, we consider this subject by examining the dynamics of cloud-producing air motions induced by wind blowing over terrain.

It is evident that when a fluid on the earth is flowing over an uneven, solid lower boundary, the vertical velocity of the fluid at the interface will be upward or downward, depending on the horizontal direction of the fluid flow relative to the slope of the bottom topography. Since the fluid is a continuous medium, the vertical motion at the bottom will be felt through some depth extending above the lower boundary. Clouds can form if the air forced over the terrain is sufficiently moist. Moreover, since restoring forces exist in the fluid, the vertical motion produced there by the lower boundary can excite waves. Thus, the vertical motion produced in a fluid by flow over terrain can include alternating regions of upward and downward motion, which may extend above, downstream, or upstream of the hill. Clouds can form in the upward-motion areas of the waves. Nonprecipitating clouds that form in moist layers in direct response to the wave motions induced by flow over topography are referred to as wave clouds. These clouds often take the form of lenticular clouds (Figs. 1.18–1.22) and rotor clouds (Figs. 1.23 and 1.24), which are visually spectacular tracers of the atmospheric wave motions in mountainous regions. Precipitating clouds can also be formed or modified by the flow over orography. Stable nimbostratus clouds can be formed or enhanced by upslope motions and dried out by downslope motions, while the formation of cumulonimbus clouds may be triggered in several ways by flow over terrain.

In this chapter, we will first consider clouds that form where air in the boundary layer is forced to flow upslope (Sec. 12.1). Then we will consider the more com-

---

[320] Goethe's reference to a cloud enveloping the top of a mountain.

plex subject of clouds that form in association with waves excited by flow over varying topography. In Sec. 12.2, we will examine waves and clouds that form in response to flow over two-dimensional mountain ridges. In Sec. 12.3, we will extend the discussion to three dimensions by considering cloud formation produced by flow over isolated mountain peaks. Finally, in Sec. 12.4, we will examine the effect of flow over topography on precipitating clouds.

12.1 Shallow Clouds in Upslope Flow

A simple boundary condition applies in all of our considerations of flow over topography. Since the earth's surface is fixed, the component of air motion normal to the surface must vanish at the ground. The vertical wind component at the surface w_o is then

$$w_o = \left(\mathbf{v}_H\right)_o \cdot \nabla \hat{h} \tag{12.1}$$

where $(\mathbf{v}_H)_o$ is the horizontal wind at the surface and $\hat{h}$ is the height of the terrain. It follows that wherever a shallow layer of air is flowing horizontally toward rising terrain, cloud will form near the surface. Air can be directed up a slope for a host of reasons, ranging from purely local effects over a small hill to widespread synoptic-scale flow over gently sloping terrain (such as when low-level easterly flow prevails over the central United States and low clouds cover the entire Great Plains, which slope gradually upward toward the Rocky Mountains). The clouds that form in upslope flow are often in the form of fog or stratus confined to low levels (Fig. 12.1). However, they may be deep enough to produce drizzle or other light precipitation.

12.2 Wave Clouds Produced by Long Ridges[321]

In certain wind and thermodynamic stratifications, the boundary condition (12.1) is felt through a deep layer. As the surface air is forced to move up and down over the topography, restoring forces in the atmosphere come into play and a variety of wave motions can occur. Associated with the waves are substantial vertical motions, which can lead to clouds if the layers of air affected are sufficiently moist. To gain an appreciation of the dynamics of these clouds, we will examine the physics of the various waves excited by wind blowing over irregular terrain. To simplify matters, we restrict the discussion of this section to air flowing over uniform, infinitely long ridges. This case is physically distinct from the flow over three-dimensional hills or mountain peaks. In the case of a ridge, air has no opportunity to flow around the barrier; it must either flow over the ridge or be blocked. In contrast, for an isolated peak in the terrain, air may flow around the

---

[321] Much of the material in this section is based on review articles by Durran (1986b, 1990) and on Chapter 9 of Holton (1992).

Figure 12.1 Upslope fog. Cascade Mountains, Washington. (Photo by Steven Businger.)

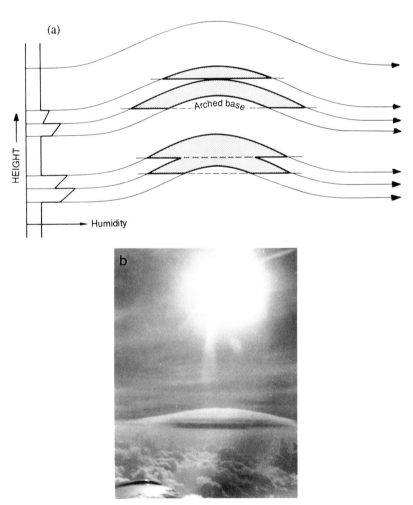

Figure 12.2 Layered structure of wave clouds. (a) Streamlines show airflow. On the left is an imagined profile of relative humidity, each layer with its own condensation level. The corresponding wave-cloud shapes downwind are outlined and shaded. Arched base occurs if the layer of air lifted is dry enough. (b) Photograph of a lenticular cloud downwind of Mt. Rainier, Washington. (Diagram adapted from Scorer, 1972; photo by Arthur L. Rangno.)

obstacle, and thus more possibilities for flow patterns arise. The three-dimensional case will be considered in Sec. 12.3, where we will treat it as an extension of the two-dimensional case.

To simplify the mathematics in both this section and Sec. 12.3, we will consider the air to be unsaturated. The thermodynamic equation thus reduces to the conservation of potential temperature (2.11). Since our primary objective here is to describe the dynamics of clouds, it may appear contradictory to ignore the effects of condensation; however, mathematical strategies exist for taking the condensation into account quantitatively. For example, one can switch to conservation of equivalent potential temperature (2.18) once condensation begins. For our purposes, however, the resulting modifications of the mathematics are unnecessarily complex, since it turns out that the air motions in mountain waves are not qualitatively altered by the occurrence of condensation in the flow. The patterns of cloud formation in the flow over topography can be readily inferred from the vertical air motion calculated ignoring condensation. Wherever a layer of air is subjected to upward motion as a result of the topographically induced wave motions and is sufficiently moist, a cloud will form. This process is illustrated in Fig. 12.2, which indicates how wave perturbations of the flow in the x–z plane can lead to lenticular cloud forms in layers of elevated relative humidity. Note how discontinuities in the humidity layering can lead to an arched cloud base and to stacks of lenticular clouds.

12.2.1   Flow over Sinusoidal Terrain

We first examine the dry-adiabatic flow over a series of sinusoidally shaped two-dimensional ridges of infinite length parallel to the y-axis. Although this is a rather idealized case, it is nonetheless useful. It illustrates the essential physics of the problem, and it is easily extended to the case of a ridge of arbitrary shape, since the latter can be considered a superposition of sinusoidal profiles of different wavelengths.

We have seen in previous chapters that vertical motions in clouds are often revealed through a consideration of the generation and redistribution of horizontal vorticity. This approach is again useful in considering air motions generated by flow over terrain. For two-dimensional Boussinesq flow in the x–z plane, vorticity about the y-axis (ξ) is generated by horizontal gradients of buoyancy, according to (2.61). As the paths of fluid parcels are tipped upward and downward by the surface topography, horizontal gradients of vertical velocity are created in the fluid. According to (2.11), the alternating regions of upward and downward motion, in turn, introduce alternating regions of adiabatic cooling and warming. The alternating cool and warm regions thus generated constitute horizontal gradients of buoyancy, which then generate horizontal vorticity ξ according to (2.61). If the flow over the terrain adjusts to steady state, (2.61) reduces to

$$u\xi_x + w\xi_z = -B_x \tag{12.2}$$

The variables in this equation may be written, as discussed in Sec. 2.6, in terms of perturbations (indicated by a prime) from a mean state (indicated by an overbar). If the perturbations of the flow are of small amplitude, they are described by linearizing (as discussed in Sec. 2.6.2) about a state of mean motion $[\bar{u}(z), \bar{w} = 0]$, where $\bar{u}(z)$ is the mean-state flow in the x-direction (i.e., normal to the ridges in the topography). First linearizing the right-hand side of (12.2) we obtain

$$u\xi_x + w\xi_z = \frac{w}{\bar{u}}\bar{B}_z \tag{12.3}$$

where we have substituted for B'_x from the steady-state version of (2.106). From (12.3), it is evident that the steady-state velocity field must be arranged so that the advection of vorticity just balances the buoyancy generation constituted by the adiabatic cooling and warming associated with the perturbations of vertical motion. Since the fluid is moving horizontally as well as vertically, the temperature perturbations produced by vertical motion w are spread over a horizontal distance by the horizontal velocity component $\bar{u}$, thus determining a horizontal gradient of B, the effect of which must be balanced by the advection.

When the advection terms on the left-hand side of (12.3) are linearized, after making use of the definition $\xi \equiv u_z - w_x$ and invoking the mass continuity equation (2.55), we obtain

$$w_{zz} + w_{xx} + \ell^2 w = 0 \tag{12.4}$$

where ℓ^2 is the *Scorer parameter*, defined as

$$\ell^2 \equiv \frac{\bar{B}_z}{\bar{u}^2} - \frac{\bar{u}_{zz}}{\bar{u}} \tag{12.5}$$

Equation (12.4) is simply the steady-state form of (2.107), whose solutions are internal gravity waves. The patterns of vorticity generated by the flow over topography thus take the form of gravity waves. The general solution of (12.4) when ℓ^2 is a constant is

$$w = \mathrm{Re}\left[\hat{w}_1\,e^{i(kx+mz)} + \hat{w}_2\,e^{i(kx-mz)}\right] \tag{12.6}$$

where $\hat{w}_1$ and $\hat{w}_2$ are constants and

$$m^2 = \ell^2 - k^2 \tag{12.7}$$

Thus, the vertical structure of the waves depends on the relative magnitude of the Scorer parameter and the horizontal wave number. If $k > \ell$, then the solution (12.6) decays or amplifies exponentially with height. If $k < \ell$, then the wave varies sinusoidally in z with wave number m.

The values of the basic-state wind profile $\bar{u}(z)$ and static stability $\bar{B}_z$ are considered to be those measured upstream of the mountain ridge. If these values are given, then so is ℓ^2 [according to (12.5)]. The field of w is obtained by solving (12.4) under the constraint of appropriate boundary conditions at the earth's surface and as $z \to \infty$. The horizontal velocity field follows readily from $w(x, z)$ by mass continuity since the two-dimensional form of the Boussinesq continuity equation (2.55) is simply

$$u_x + w_z = 0 \tag{12.8}$$

and the upstream horizontal velocity $\bar{u}(z)$ is given.

The lower-boundary condition imposed in solving (12.4) is simply the linearized form of (12.1). If the sinusoidal surface topography is represented mathematically by

$$\hat{h} = h_a \cos kx \tag{12.9}$$

where h_a is a constant and k is an arbitrary wave number, then (12.1) implies that

$$w_o = -\bar{u}_o h_a k \sin kx \tag{12.10}$$

This lower-boundary condition is called *free slip*. The appropriate upper-boundary condition depends on whether the wave solution varies exponentially with height ($k > \ell$) or sinusoidally ($k < \ell$). In either case, the fundamental factor is that the mountains at the lower boundary are the energy source for the disturbance. If $k > \ell$, solutions that amplify exponentially as z increases are regarded as physically unreasonable and set to zero. If $k < \ell$, any waves that transport energy downward are set to zero, while those that transport energy upward are retained; this assumption is called the *radiation boundary condition*. It can be shown that the waves retained by the radiation boundary condition are those whose phase lines tilt upstream as z increases.[322]

Application of these lower- and upper-boundary conditions[323] leads to

$$w(x, z) = \begin{cases} -\bar{u} h_a k e^{-\hat{\mu} z} \sin kx, & k > \ell \\ -\bar{u} h_a k \sin(kx + mz), & k < \ell \end{cases} \tag{12.11}$$

where $\hat{\mu}^2 \equiv -m^2$, and $\hat{\mu}$ is thus a positive number, emphasizing that in the case of $k > \ell$ the waves decay exponentially with height. These waves are referred to as *evanescent* (i.e., fading away). In the case of $k < \ell$ (i.e., for wider mountains), the waves propagate vertically without loss of amplitude. These two types of wave structure are illustrated in Fig. 12.3 for a case of constant $\bar{u}$ and $\bar{B}_z$. It is clear from these examples how the flow stays in phase with the mountains but dies out with height in the narrow-ridge case ($k > \ell$, Fig. 12.3a), while in the wide-ridge case ($k < \ell$, Fig. 12.3b), the waves retain their amplitude with height but have ridge and trough lines sloping upstream.

12.2.2   Flow over a Ridge of Arbitrary Shape

Although the infinite series of sinusoidal ridges considered in the previous subsection are useful to illustrate the two basic types of wave structure that can arise

---

[322] See Durran (1986b).
[323] *Ibid.*

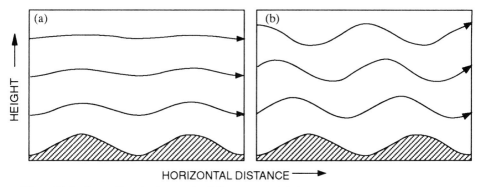

Figure 12.3   Streamlines in the steady airflow over an infinite series of sinusoidal ridges when (a) the wave number of the topography exceeds the Scorer parameter (narrow ridges) or (b) the wave number of the topography is less than the Scorer parameter (wide ridges). Shading indicates bottom topography. (From Durran, 1986b. Reprinted with permission from the American Meteorological Society.)

from airflow across the terrain, they are very special in that they excite only k, the horizontal wave number of the terrain itself. When the topography has any other shape, a spectrum of waves is activated. To see this, consider a single mountain ridge of arbitrary shape $\hat{h}(x)$. This shape can be represented by a Fourier series

$$\hat{h}(x) = \sum_{s=1}^{\infty} \mathrm{Re}\left[h_s e^{ik_s x}\right] \tag{12.12}$$

That is, the terrain profile is the superposition of sinusoidal profiles like (12.9). The solution of (12.4) for each such profile (of wave number k_s) is of the form (12.6). The solution for the arbitrary profile $\hat{h}(x)$ is the sum of all of these solutions for individual wave numbers. After application of the free slip and radiation boundary conditions, the total solution takes the form

$$w(x,z) = \sum_{s=1}^{\infty} \mathrm{Re}\left[i\bar{u}_o k_s h_s e^{i\left(k_s x + m_s z\right)}\right] \tag{12.13}$$

The individual Fourier modes contributing to this expression each behave as the total solution for periodic sinusoidal topography. The individual modes will thus be vertically propagating or vertically decaying, depending on whether m_s is real or imaginary (i.e., on whether $k_s > \ell$ or $k_s < \ell$). If the ridge is narrow, wave numbers $k_s > \ell$ dominate the solution and the resulting disturbance is primarily evanescent. If the ridge is wide, wave numbers $k_s < \ell$ dominate, and the disturbance propagates vertically.

A bell-shaped ridge of the form

$$\hat{h}(x) = \frac{h_a a_H^2}{a_H^2 + x^2} \tag{12.14}$$

illustrates the behavior of the solutions summed in (12.13). Since $\hat{h} = h_a$ at $x = 0$ and $\hat{h} = h_a/2$ at $x = \pm a_H$, a_H^{-1} is a scale characteristic of the dominant wave numbers forced by the mountain. The solution for a narrow mountain ($a_H^{-1} \gg \ell$) is illustrated in Fig. 12.4a. The disturbance, dominated by exponentially decaying solutions, is symmetric with respect to the crest of the ridge and decays strongly with height. An example of a wide-mountain solution ($a_H^{-1} \ll \ell$) is shown in Fig. 12.4c. If it is assumed that this case is equivalent to taking $k \ll \ell$, then the solution is hydrostatic and the dominant terms in (12.13) are of the form

$$i\bar{u}_o k_s h_s e^{i(k_s x + \ell z)} \tag{12.15}$$

Thus, the dependence of vertical wavelength on horizontal wavelength disappears, and the mountain profile is reproduced at every altitude that is an integral multiple of $2\pi/\ell$. It can be shown that the horizontal component of the group velocity (Sec. 2.7.2) of a stationary two-dimensional hydrostatic wave is zero. Hence, the energy propagation (which is in the direction of the group velocity) is purely vertical. Thus, for the mountain ridge that is wide enough to excite hydrostatic waves (but not large enough for Coriolis force to be important), the disturbance occurs directly over the ridge, and at any given height there is only one wave crest in the flow.

The intermediate case in which $a_H^{-1} = \ell$ is illustrated in Fig. 12.4b. In this case, the solution (12.13) is dominated by vertically propagating *nonhydrostatic* waves ($k < \ell$ but not $\ll \ell$). The phase lines of these waves still slope upstream, and energy is transported upward. However, unlike hydrostatic waves, the nonhydrostatic waves also have a horizontal component of group velocity (and hence energy propagation) in the downstream direction. As a result, additional wave crests appear aloft downstream of the mountain ridge.

12.2.3 Clouds Associated with Vertically Propagating Waves

We can now see how certain types of wave clouds are produced. If a stratum within the air flowing over the ridge undergoes strong upward displacement as part of a vertically propagating wave (either hydrostatic or nonhydrostatic), clouds will appear. For illustration, possible locations of clouds associated with vertically propagating waves are shaded in Fig. 12.4b and c. Lower-level clouds form upstream of and over the ridge. Upper-level clouds (usually cirrus or cirrostratus) may be found downstream of the ridge. These theoretical cases indicate that the horizontal scale of clouds associated with vertically propagating waves is ~10–50 km. An example of clouds associated with vertically propagating wave motion like that in Fig. 12.4c can be seen in Fig. 12.5. The lower-level cloud over the Continental Divide in the background and the upper-level cirriform cloud in the foreground are associated with the vertically propagating waves' upward air motion, which is found upstream of the crest of the ridge at low levels and downstream of the crest at upper levels.

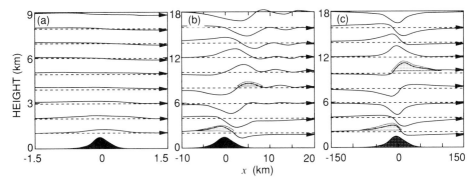

Figure 12.4 Streamlines in the steady airflow over an isolated bell-shaped ridge. (a) Narrow ridge. (b) Width of ridge comparable to the Scorer parameter. (c) Wide ridge. Lighter shading indicates possible locations of clouds. Darker shading indicates bottom topography. (From Durran, 1986b Reprinted with permission from the American Meteorological Society.)

Figure 12.5 Looking upwind at clouds associated with vertically propagating waves. In the center background, a lower-level wave cloud is evident directly over the Continental Divide, which is the main orographic barrier. The mountains in the foreground, which appear larger in the photo, are actually smaller foothills. The cirriform cloud deck at higher levels is produced by the upward air motion associated with the vertically propagating wave induced by the Divide. These clouds are like those indicated schematically in Fig. 12.4c. Boulder, Colorado. (Photo by Dale R. Durran.)

12.2.4 Clouds Associated with Lee Waves

So far we have considered only situations in which ℓ^2 is constant with respect to height. When ℓ^2 decreases suddenly with increasing height, a different type of mountain wave occurs. This type of disturbance is usually called a *lee wave*, but it is also known as a *resonance wave*, *trapped wave*, or *trapped lee wave*. According to (12.5), a decrease of ℓ^2 can be brought on by a decrease of $\bar{B}_z$, an increase of $\bar{u}$, or a change of the curvature of the wind profile $\bar{u}_{zz}$. The occurrence of lee waves can be illustrated by considering a two-level stratification of ℓ^2, in which ℓ_L^2 and ℓ_U^2 represent the Scorer parameters for the lower and upper layers, respectively. Then two equations of the form (12.4) can be applied, one with each value of ℓ^2. Solutions to the upper-layer equation decay exponentially with height, while solutions to the lower-layer equation are sinusoidal in z (i.e., propagate vertically). The solutions are required to match smoothly at the interface, which is taken to be $z = 0$. When the radiation boundary condition is applied, the solution of the upper-layer equation takes the form

$$w_u = \hat{A}e^{-\mu_u z}f(x) \tag{12.16}$$

where $f(x)$ represents the variability in x, and

$$\mu_u \equiv \sqrt{k^2 - \ell_U^2} \tag{12.17}$$

The solution in the lower layer is of the form

$$w_l = \left(\hat{B}\sin m_l z + \hat{C}\cos m_l z\right)f(x) \tag{12.18}$$

where $\hat{B}$ and $\hat{C}$ are constants and

$$m_l \equiv \sqrt{\ell_L^2 - k^2} \tag{12.19}$$

Both (12.16) and (12.18) are special cases of the general solution (12.6). In order for the solutions (12.16) and (12.18) to match smoothly at the interface, we must have both $w_u = w_l$ and $w_{uz} = w_{lz}$ at $z = 0$. These conditions applied to (12.16) and (12.18) imply that

$$\hat{A} = \hat{C} \quad \text{and} \quad -\mu_u\hat{A} = m_l\hat{B} \tag{12.20}$$

It follows from (12.18) and (12.20) that

$$w_l = \hat{A}\left(-\frac{\mu_u}{m_l}\sin m_l z + \cos m_l z\right)f(x) \tag{12.21}$$

If we consider a location far enough to the lee of the ridge, where the ground is level, then $w_l = 0$ at the ground. If we assign to ground level the height $z = -z_o < 0$, then for w_l to be zero, (12.21) implies that

$$\cot(m_l z_o) = -\frac{\mu_u}{m_l} \tag{12.22}$$

By graphing each side of this equation as a function of k^2, one finds that the graphs cross at one or more points only if the condition

$$\ell_L^2 - \ell_U^2 > \frac{\pi^2}{4z_o^2} \tag{12.23}$$

is satisfied. That is, a horizontal wavelength exists that satisfies the equation only if the depth of the lower layer exceeds a certain threshold before waves can be trapped in it. Furthermore, from the definitions (12.17) and (12.19), it is evident that the right-hand side of (12.22) exists only if

$$\ell_L > k > \ell_U \tag{12.24}$$

Thus, the condition (12.23) can be satisfied only for a wave number in this range, which is consistent with the fact that waves propagate vertically in the lower layer and decay exponentially in the upper layer.

Figure 12.6 shows the flow over the two-dimensional bell-shaped ridge described by (12.14) for a case in which the stratification of the undisturbed flow supports trapped waves. The interface level is at 500 mb (about 5 km altitude). Note that although the profiles of temperature and wind are continuous in z, the value of ℓ^2 is discontinuous, having different constant values above and below the interface. From Fig. 12.6a, it is evident that the trapped waves are indeed confined to the lower layer. It is also apparent that they have no tilt. The latter result is surprising since solutions in the lower layer are of the form of vertically propagating waves, whose phase lines tilt. The explanation lies in the fact that vertically propagating waves cannot continue to propagate upward when they reach the upper layer. Instead they are reflected as downward-propagating waves, which are reflected back up when they reach the ground. As the waves of various wave numbers undergo this multiple reflection process downstream, a superposition of upward- and downward-propagating waves is established. Since the upward- and downward-propagating waves have opposite tilt, their superposition results in no tilt at all. The trapped waves in the figure can be described as having the form

$$w(x,z) = \hat{\beta}\sin\hat{\alpha}z\cos kx \qquad (12.25)$$

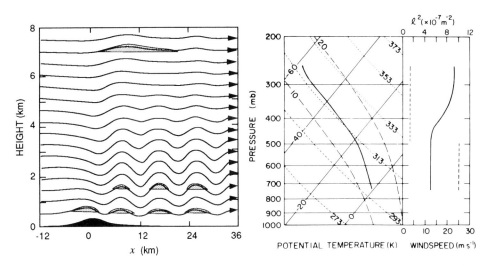

Figure 12.6   (a) Streamlines of steady flow over an isolated two-dimensional bell-shaped ridge for a case in which the stratification of the undisturbed flow supports trapped waves. Lighter shading indicates possible locations of clouds. Darker shading indicates bottom topography. (b) The vertical distribution of temperature and wind speed (solid lines) in the undisturbed flow. This temperature and wind layering implies a discontinuous two-layer structure for the Scorer parameter (dashed line on right panel). (From Durran, 1986b. Reprinted with permission from the American Meteorological Society.)

where $\hat{\beta}$ and $\hat{\alpha}$ are constants. From trigonometric identities, it can be seen that this expression is the sum of equal-amplitude upstream- and downstream-tilting waves

$$w(x,z) = \frac{\hat{\beta}}{2}\sin(\hat{\alpha}z + kx) + \frac{\hat{\beta}}{2}\sin(\hat{\alpha}z - kx) \tag{12.26}$$

Figure 12.6a indicates that the horizontal scale of the trapped lee waves is ~10 km. For typical values of stability and wind, wavelengths are in the range of 5–25 km. Thus, in general, lee waves are of shorter wavelength than the pure vertically propagating waves. Possible locations of clouds are shaded in Fig. 12.6a. The clouds associated with the lee waves are ~3–5 km in horizontal scale. They can often be distinguished from the clouds associated with vertically propagating waves by their repetition downstream of the ridge and by their smaller horizontal scale. In Fig. 12.7, a satellite photograph illustrates an example of numerous lee waves occurring in a regime of strong northwesterly winds aloft over the mountainous western United States.

It is not uncommon for lee wave clouds and clouds associated with a vertically propagating wave to be present in the same situation. Figure 12.6a shows this possibility schematically, where, in addition to the lee-wave clouds downstream of the mountain ridge, a cap cloud is indicated over the crest of the mountain ridge at low levels and a downstream cirrus cloud is indicated aloft. These latter two cloud types are qualitatively similar to those in Fig. 12.4. The weak vertically propagating wave structure appears in addition to the lee waves because the mountain ridge produces some forcing at wave numbers $k < \ell_U$, thus generating waves that can propagate through the upper layer. The larger horizontal scale of the clouds associated with the vertically propagating wave in the upper layer accounts for the greater width of the upper-level cloud compared to the lee-wave clouds below. An actual example of clouds similar to those indicated in Fig. 12.6a is shown in Fig. 12.8. The cap cloud over the crest of the main ridge is seen in the background. Ahead of it are two rows of lee-wave clouds at lower levels, and the wider sheet of cirriform cloud in the foreground aloft is generated by the vertically propagating wave motion.

12.2.5   Nonlinear Effects: Large-Amplitude Waves, Blocking, the Hydraulic Jump, and Rotor Clouds

The preceding analysis relies on linearized equations to describe the dynamics of wave-induced clouds. Under certain atmospheric conditions, nonlinear processes and large-amplitude waves become important in the flow over a mountain ridge. These processes can be studied with nonhydrostatic numerical models—the same type used to study convective clouds (Sec. 7.5.3). Figure 12.9 is an example of the type of model airflow that can develop when large-amplitude waves are present. Shading indicates locations where clouds could appear if layers of air were moist enough. Four types of clouds are indicated, three of which are the same features that appear in the linear case (cf. Fig. 12.6a). The low-level cloud over the crest

Figure 12.7 Visible satellite imagery showing wave clouds over the southwestern United States, 1715 GMT, 2 May 1984.

Figure 12.8 Looking upwind at wave clouds. In the background, lower right, a cap cloud is seen over the Continental Divide, which is the main orographic barrier. The mountains in the foreground, which appear larger in the photo, are actually smaller foothills. In the middle and foreground are two rows of lee-wave clouds induced by the Divide. The cirriform cloud deck at higher levels is produced by the upward air motion associated with a vertically propagating wave. Boulder, Colorado. (Photo by Dale R. Durran.)

and the upper-level cloud just downwind of the ridge are associated with the vertically propagating wave. In this case, the low-level cloud is usually referred to as the *Föhn* wall (Figs. 1.23–1.25). The large-amplitude wave structure is characterized by strong downslope winds (or *Föhn*), and the cloud over the ridge extends just over the crest such that it appears as a wall to an observer on the lee side of the ridge. Lee-wave clouds are again associated with trapped waves downstream of the ridge. These are shown in the shading pattern as the two small clouds

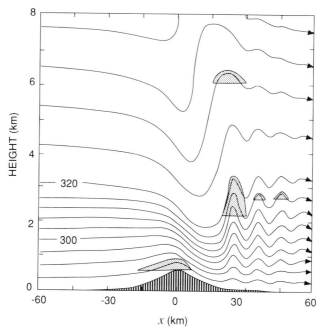

Figure 12.9 Numerical-model results illustrating two-dimensional adiabatic flow that can develop when large-amplitude waves are present. The atmosphere was specified to consist of a less stable layer aloft and a more stable layer below to trap waves in the lower layer. Contours of potential temperature (equivalent to streamlines) are labeled in K. Shading indicates locations where clouds could appear if layers of air were moist enough. (Adapted from Durran, 1986a. Reproduced with permission from the American Meteorological Society.)

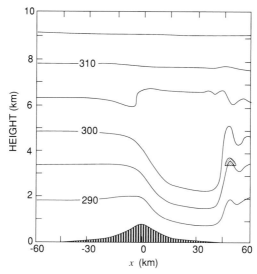

Figure 12.10 Numerical-model results illustrating two-dimensional adiabatic flow that can develop when large-amplitude waves are present. In this case the mean state has a critical layer and wave breaking occurs. Contours of potential temperature (equivalent to streamlines) are labeled in K. Shading indicates location where a cloud could appear if the layer of air were moist enough. (From Durran and Klemp, 1987 as reprinted in Durran, 1990. Reprinted with permission from the American Meteorological Society.)

farthest downwind. The new feature that appears as a result of nonlinear effects, is the rotor cloud. It appears at the sudden vertical jump at the downstream endpoint of the very strong downslope winds, which form immediately to the lee of the ridge and accelerate down the mountain.

Figure 12.10 shows another example of strong downslope winds and possible rotor cloud formation associated with a strong jump in the flow downstream of the ridge. The strong downslope winds, strong upward jump of the air motion, and rotor cloud in these examples are manifestations of large-amplitude wave motion that can occur in three kinds of situations:[324]

1. *Wave breaking.* This case occurs when downstream of the ridge, the θ surfaces overturn, producing a layer of low stability and nearly stagnant flow. This structure, sometimes referred to as a "local critical layer,"[325] is seen at 3.6 km altitude downstream of the mountain in Fig. 12.10. Vertically propagating waves cannot be transmitted through the layer of near-uniform θ centered at about 4 km. The waves excited by the mountain barrier are thus reflected, and wave energy is trapped between this layer and the surface on the lee side of the barrier. If the depth of the cavity between the self-induced critical layer and the mountain slope is suitable, the reflections at the critical layer produce a resonant wave that amplifies in time and produces the strong downslope surface winds and downstream jump.

2. *Capping by a mean-state critical layer.* A local critical layer need not be induced by wave breaking. If the air impinging on the mountain ridge contains a critical layer in its mean-state wind stratification, the necessary reflections, resonance, and amplification may occur on the lee side to produce the downslope winds and downstream jump, even if the mountain is otherwise too small to produce breaking waves. The example in Fig. 12.10 is a case in which both a critical layer is present in the mean state and breaking waves occur.

3. *Scorer parameter layering.* This situation arises when the mountain is too small to force breaking waves but large enough to produce large-amplitude waves in an atmosphere with constant $\bar{u}$ and a two-layer structure in $\bar{B}_z$. Figure 12.9 is an example of this case, in which vertically propagating waves are partially reflected. When the less stable (low $\bar{B}_z$) layer aloft and the more stable (high $\bar{B}_z$) lower-level layer are suitably tuned, superposition of the reflecting waves produces the strong downslope surface winds and downstream jump.

The strong downslope winds and downstream jump, where the rotor cloud occurs, have the characteristics of a special type of flow referred to as a *hydraulic jump*, which occurs in a variety of geophysical situations and is characterized by a sudden change in the depth and velocity of a layer of fluid. For example, a tidal flow up a river is sometimes characterized by such a jump.[326] A hydraulic jump

---

[324] See Durran (1990).

[325] In general, a *critical layer* is defined as one in which the horizontal phase speed of a gravity wave is equal to the speed of the mean flow.

[326] The river Severn in England is famous for its tidal hydraulic jump, which surfers have been known to ride for several kilometers upstream. See Lighthill (1978) or Simpson (1987).

can also occur downstream of the flow over an obstacle—as in the case of the flow over a rock in a stream, where a turbulent jump in the fluid depth is often seen just downstream of the rock. The latter case is analogous to the atmospheric flow that produces a rotor cloud.

To investigate this type of flow quantitatively, let us consider the steady flow of a homogeneous fluid over an infinitely long ridge. Such a flow is governed by the two-dimensional steady-state shallow-water momentum and continuity equations, which may be written as

$$u \frac{\partial u}{\partial x} = -g \frac{\partial \left(H + \hat{h} \right)}{\partial x} \tag{12.27}$$

and

$$\frac{\partial u H}{\partial x} = 0 \tag{12.28}$$

respectively, where x is perpendicular to the ridge, u is the velocity in the x-direction, H is the vertical thickness of the fluid, and $\hat{h}$ is the height of the obstacle. The pressure gradient acceleration on the right-hand side of (12.27) is equivalent to that on the right-hand side of (2.99), in the case where the term $\delta \rho$ in (2.99) is assumed to be $\approx \rho_1$ (which is the case if the upper layer is air and the lower layer is water).

Flow described by (12.27) and (12.28) may be classified according to the speed of the flow u in relation to $\sqrt{gH}$, which is approximately the phase speed of linear shallow-water gravity waves [$\sqrt{g\bar{h}}$ according to (2.104)]. To make this classification, it is useful to define a *Froude number* (Fr), such that

$$\mathrm{Fr}^2 \equiv \frac{u^2}{gH} \tag{12.29}$$

Then (12.27) may be written as

$$\tfrac{1}{2} u^2 + g \left(H + \hat{h} \right) = \text{constant} > 0 \tag{12.30}$$

or

$$\left(\tfrac{1}{2} \mathrm{Fr}^2 + 1 \right) H = \hat{h}_c - \hat{h} \tag{12.31}$$

where $\hat{h}_c$ is a positive constant. Since the left-hand side must be positive, the condition

$$\hat{h} < \hat{h}_c \tag{12.32}$$

must be satisfied. If $\hat{h} > h_c$, no solution of (12.27) exists and the flow is *blocked*, because it is not energetic enough to surmount the ridge. The smaller the Fr, the more likely the flow will be blocked.

Since (12.27) cannot be solved under blocking conditions, blocked flows must be either nonsteady or three-dimensional. In Sec. 12.3, we will examine three-dimensional flows at low Froude number around isolated mountain peaks and see that when the flow cannot go over the mountain, it turns laterally and goes around. Here we are concerned with two-dimensional steady-state flows for which (12.27) has solutions (i.e., cases for which (12.32) is satisfied and, hence, the flow can get over a purely two-dimensional barrier). To examine these cases, it is useful to combine (12.27), (12.28), and (12.29) to obtain

$$\left(1 - \text{Fr}^{-2}\right)\frac{\partial\left(H + \hat{h}\right)}{\partial x} = \frac{\partial \hat{h}}{\partial x} \tag{12.33}$$

This equation states that the free surface of the fluid can either rise *or fall* as the fluid encounters rising bottom topography. Figure 12.11 indicates various possibilities. In the case of Fr > 1 (Fig. 12.11a), called *supercritical flow*, the fluid thickens and [according to (12.28)] slows down as it approaches the top of the obstacle and reaches its minimum speed at the crest. In the case of Fr < 1 (Fig. 12.11b), called *subcritical flow*, the fluid thins and accelerates as it approaches the top of the obstacle.

Physical insight into supercritical and subcritical flows is obtained by noting that in (12.27) there is a three-way balance among nonlinear advection uu_x, the pressure gradient acceleration associated with the depth of the fluid, and the pressure gradient acceleration induced by the ridge displacing the fluid vertically. The nonlinear advection can be related to the Froude number, since the continuity equation (12.28) implies that

$$\text{Fr}^2 = -\frac{uu_x}{gH_x} \tag{12.34}$$

Thus, Fr^2 is the ratio of the nonlinear advection to the pressure gradient acceleration associated with the variation of fluid depth. Also, the minus sign shows that these two effects are always in opposition.

In the supercritical case (Fr > 1), nonlinear advection dominates the pressure gradient produced by a change in fluid thickness [according to (12.34)], and the height of the ridge must produce horizontal variations in pressure in the same sense as changes associated with fluid depth [according to (12.33)]. Hence, the fluid thickens as it passes over the windward side of the ridge (accumulating potential energy and losing kinetic energy) and thins on the lee side. In the subcritical case (Fr < 1), nonlinear advection is dominated by the effect of the fluid-thickness pressure gradient. The only way for this to occur is for the variation in the height of the ridge to produce horizontal variations in pressure that are in the opposite sense of the changes associated with fluid depth. That is, H_x and $\hat{h}_x$ must be of opposite sign, so that the fluid thins (loses potential energy and gains kinetic energy) as it passes over the windward side of the ridge and thickens on the lee side.

Figure 12.11c shows the flow regime that is characterized by a hydraulic jump on the lee side of the ridge. This case is subcritical on the windward side of the ridge to such a degree that the speed of the flow attains a supercritical value just at

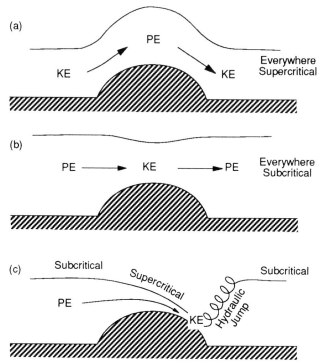

Figure 12.11 Behavior of shallow water flowing over an obstacle: (a) everywhere supercritical flow; (b) everywhere subcritical flow; (c) hydraulic jump. KE and PE refer to kinetic and potential energy, respectively. (Adapted from Durran, 1986a. Reproduced with permission from the American Meteorological Society.)

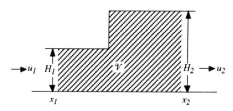

Figure 12.12 A volume of fluid containing a hydraulic jump.

the crest of the barrier. Thus, the flow continues to increase in speed and become still shallower as it runs down the lee slope. In the atmospheric analog to this shallow-water case, severe downslope winds are sometimes a manifestation of the lee-side supercritical flow. Downstream, the strong flow coming down the slope reverts suddenly back to ambient conditions. This sudden transition takes the form of a hydraulic jump.

An important characteristic of the hydraulic jump is that energy is dissipated at the jump. This fact can be inferred by considering a volume of incompressible fluid $\mathcal{V}$ containing a hydraulic jump, as idealized in Fig. 12.12. The volume $\mathcal{V}$ is bounded by x_1 and x_2 and y and $y + \Delta y$ in a coordinate system that moves with the discontinuity in fluid height. The flow is assumed to be homogeneous (i.e., of constant density) and steady state in this moving coordinate system.

We first consider the mass continuity in the fluid containing the jump by averaging (12.28) to obtain

$$\tilde{Q} \equiv \bar{u}H = \text{constant} \tag{12.35}$$

Next , we consider the conservation of momentum in the mass of fluid contained in $\mathcal{V}$. The equation of motion governing the mean fluid motion is obtained by applying the averaging procedure described in Sec. 2.6 to (2.1), ignoring Coriolis and molecular frictional forces and recalling that the density of an incompressible fluid is constant. The result is

$$\frac{D\bar{\mathbf{v}}}{Dt} = -\nabla \frac{\bar{p}}{\rho} - g\mathbf{k} + \bar{\mathcal{F}} \tag{12.36}$$

Integrating the x-component of (12.36) over the mass of fluid in $\mathcal{V}$ in the moving coordinate system, with substitution from (12.35), and recalling that the pressure at a given level is the weight of the fluid above leads to

$$\tilde{Q}^2 = \frac{g}{2} H_1 H_2 \left(H_1 + H_2 \right) + Q_F^2 \tag{12.37}$$

where

$$Q_F^2 = \frac{H_1 H_2}{\Delta y \left(H_1 - H_2 \right)} \iiint_{\mathcal{V}} \mathcal{F}_u \, d\mathcal{V} \tag{12.38}$$

and $\mathcal{F}_u$ is the x-component of $\bar{\mathcal{F}}$. The quantity Q_F^2 in (12.37) can be seen to arise from the drag of turbulence or other small-scale motion. The first term in (12.37) arises from the combination of the horizontal pressure gradient acceleration and the horizontal advection of u.

Now we may examine the energy of the mean flow in the volume $\mathcal{V}$ containing the hydraulic jump. Under the assumed steady-state, two-dimensional conditions, with uniformity in the y-direction, the kinetic energy equation obtained by taking $\bar{\mathbf{v}} \cdot (12.36)$ is

$$\tilde{\mathcal{D}} = \nabla \cdot \left(\frac{\bar{u}^2}{2} + \frac{\bar{w}^2}{2} + gz + \frac{\bar{p}}{\rho} \right) \bar{\mathbf{v}} \tag{12.39}$$

where

$$\tilde{\mathcal{D}} \equiv \bar{\mathbf{v}} \cdot \bar{\mathcal{F}} \tag{12.40}$$

A negative value of $\tilde{\mathcal{D}}$ indicates that energy is being dissipated locally. Integration of (12.39) over the mass of fluid in the volume $\mathcal{V}$ bounded by x_1 and x_2 and y and

$y + \Delta y$ leads to

$$\iiint\limits_{\mathcal{V}} \tilde{\mathscr{D}}\rho \, d\mathcal{V} = \iint\limits_{S} \rho v_n \left(\frac{\bar{u}^2}{2} + \frac{\bar{w}^2}{2} + gz + \frac{\bar{p}}{\rho} \right) dS \tag{12.41}$$

where v_n is the outward relative velocity component normal to the surface S surrounding the volume $\mathcal{V}$. If the term on the right is negative, there is a net convergence of energy into the fluid volume, and this convergence must be matched by net dissipation in the volume in order to maintain the steady state. Carrying out the integration on the right-hand side of (12.41), with substitution from (12.35) and (12.37) and again letting the pressure be given by the weight of the fluid above, leads to

$$\iiint\limits_{\mathcal{V}} \tilde{\mathscr{D}}\rho \, d\mathcal{V} = \frac{\rho \Delta y \tilde{Q} g (H_1 - H_2)^3}{4 H_1 H_2} - \frac{\rho \Delta y \tilde{Q} Q_F^2 \left(H_2^2 - H_1^2 \right)}{2 H_1^2 H_2^2} \tag{12.42}$$

In the case represented in Fig. 12.12, $\tilde{Q} > 0$ and $H_1 - H_2 < 0$. The right-hand side of (12.42) is therefore negative, indicating a net loss of energy in the volume containing the jump. Moreover, (12.42) is still obtained even if the volume $\mathcal{V}$ is made arbitrarily small by moving x_1 and x_2 as close as we like to the jump in fluid height. Thus, it is clear that all the energy loss occurs *at* the jump. This result implies that dissipative forces must be active in the vicinity of the jump in order to conserve energy. Thus, (12.42) is a clear indication that some energy dissipation must occur in the vicinity of the jump. The loss of energy at the jump is effected by turbulence and/or downstream-propagating waves, which are allowed if the flow to the right of the jump is subcritical. It is for this reason that rotor clouds located at a hydraulic jump can be quite turbulent and are known by pilots to be a sign of dangerous zones for light aircraft. Examples of spectacular rotor clouds were shown in Figs. 1.23 and 1.24.

The primary dynamical characteristics of rotor clouds that can be explained by the foregoing theory are the strong upward air motion that occurs at the hydraulic jump and the strong turbulence or gravity wave motion required to accomplish the dissipation of energy at the jump. There is some belief, especially from sail plane pilots, that the air in the rotor clouds overturns around a horizontal axis normal to the mean flow direction. It is from this aspect that the name rotor cloud is derived. However, neither careful observation of rotor clouds nor model simulations have yet confirmed whether or not such rotation is a predominant feature of the in-cloud air motions.

12.3 Clouds Associated with Flow over Isolated Peaks

We have seen that small-amplitude wind perturbations produced by flow over a long ridge take the form of vertically propagating and trapped lee waves. Somewhat similar responses occur in flow over an isolated mountain peak. However, parcel and wave motions are not restricted to only the x- and z-directions. Components of motion in the y-direction can appear as the air flows around as well as over the three-dimensional peak.

To illustrate this fact, we again consider the vorticity about the y-axis generated as a basic unidirectional horizontal flow encounters and passes over the mountain. We neglect Coriolis effects and use the steady-state form of the three-dimensional Boussinesq y-vorticity equation (2.58)

$$u\xi_x + v\xi_y + w\xi_z = -B_x + \xi v_y + \left(\zeta v_z + \eta v_x\right) \tag{12.43}$$

which states that the advection of ξ is balanced by baroclinic generation $(-B_x)$, stretching ξv_y, and tilting $(\zeta v_z + \eta v_x)$. We will concern ourselves first with small-amplitude motions by linearizing this equation about a purely horizontal, unidirectional mean wind $\bar{u}(z)$ and mean lapse rate $\bar{B}_z$. To illustrate the essential physics retained under linear conditions, we first linearize the right-hand side of (12.43). The result is

$$u\xi_x + v\xi_y + w\xi_z = \frac{w}{\bar{u}}\bar{B}_z + \bar{u}_z v_y \tag{12.44}$$

This relation differs from the two-dimensional form (12.3) in that the stretching of vorticity contained in the basic-state shear by perturbations lateral to the basic-state current $(\bar{u}_z v_y)$ now appears on the right-hand side, along with the baroclinic generation of ξ associated with air moving up and down adiabatically and the advection of ξ by the y-component of motion $(v\xi_y)$ on the left-hand side. If we now linearize the left-hand side of the equation and substitute from the perturbation form of the three-dimensional Boussinesq continuity equation (2.55), we obtain, after some rearrangement,

$$-\bar{u}v_{yz} - \bar{u}w_{zz} - \bar{u}w_{xx} = w\bar{u}\ell^2 + \bar{u}_z v_y \tag{12.45}$$

The only differences between this result and the two-dimensional case [cf. Eq. (12.4)] are the first and last terms, which involve the perturbation of the flow lateral to the basic current.

Since the variable v is now involved, a further physical relationship is required to close the system of equations. For this, we turn to the vertical vorticity equation (2.59), which under the present steady-state linearized conditions with no Coriolis effects reduces to

$$\bar{u}\zeta'_x = \bar{u}_z w_y \tag{12.46}$$

which says that the horizontal advection of vertical vorticity by the basic-state flow is just offset by the tilting of the base-state shear by y-variability of w. This relation may be rewritten as

$$v_{xx} = u_{yx} + \frac{\bar{u}_z}{\bar{u}}w_y \tag{12.47}$$

Taking the horizontal Laplacian $\nabla_H^2 \equiv \partial^2/\partial x^2 + \partial^2/\partial y^2$ of (12.45), making use of the mass continuity equation (2.55), and substituting from (12.47) leads to

$$\left(\nabla^2 + \ell^2\right)w_{xx} + \frac{\bar{B}_z}{\bar{u}^2}w_{yy} = 0 \tag{12.48}$$

where ∇ is the three-dimensional Laplacian. As in the two-dimensional case (12.4), this vorticity equation has solutions of the forms of evanescent, vertically propagating and trapped waves. However, the presence of the perturbation motions lateral to the basic current leads to the waves being distorted into interesting configurations, which can affect the forms of clouds that develop in the waves.

The vertically propagating modes can be obtained by further simplifying the problem such that $\bar{u}$ does not vary with height. In this case, the Scorer parameter given by (12.5) reduces to

$$\ell^2 = \frac{\bar{B}_z}{\bar{u}^2} \tag{12.49}$$

and (12.48) becomes

$$\frac{\partial^2}{\partial x^2}\nabla^2 w + \ell^2\nabla_H^2 w = 0 \tag{12.50}$$

which has solutions of the form

$$w = \hat{w}e^{i(kx+jy+mz)} \tag{12.51}$$

with the following relation among k, j, and m:

$$m^2 = \frac{k^2 + j^2}{k^2}\left(\ell^2 - k^2\right) \tag{12.52}$$

which is similar to (12.7), except for the factor $(k^2 + j^2)/k^2$. Thus, m^2 depends on j as well as k.

From (12.52) it is apparent that, as in the two-dimensional case, vertically propagating or evanescent waves occur depending on whether $\ell^2 > k^2$ or $\ell^2 < k^2$, respectively. As in the two-dimensional case, we can proceed by assuming some specific form for the surface topography and disallowing downward energy propagation at the top boundary. Figure 12.13 shows an example of the vertical motion pattern associated with steady, constant $\bar{B}_z$, and constant $\bar{u}$ flow over a wide ($\ell > k$) mountain. Alternating boomerang-shaped regions of upward and downward motion occur downwind of the peak. This pattern of vertical motion occurs even if the mountain is wide enough that the flow response is hydrostatic. That is, in the three-dimensional hydrostatic case, the wave disturbance is not confined to the region directly over the mountain. In this vertical motion pattern, a boomerang- or horseshoe-shaped cloud can be observed in the lee of the mountain. The example

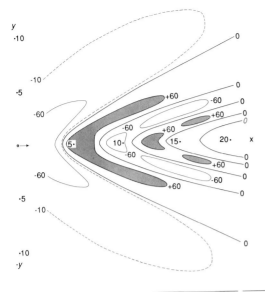

Figure 12.13 Vertical velocity field downwind of a symmetrical obstacle. Areas of upward motion are shaded. Nondimensional units are such that 100 units is 45.7 cm s^{-1}. (From Wurtele, 1953.)

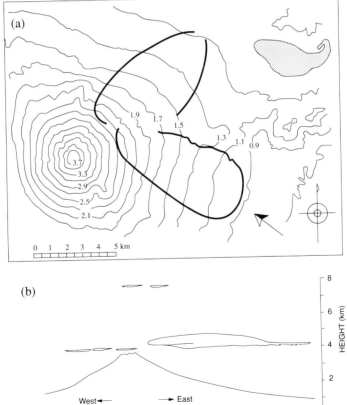

Figure 12.14 Plan view (a) and vertical cross section (b) of the horseshoe-shaped cloud in the lee of Mt. Fuji seen in Fig. 1.19. Wind was from the west to west–southwest at mountain top (3710 m) level. In (a), lighter contours are the terrain height contours labeled in km; darker lines are the cloud outline. (Adapted from Abe, 1932.)

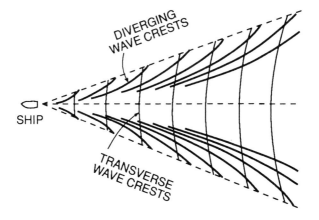

Figure 12.15

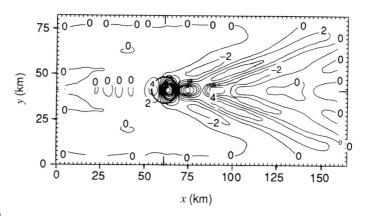

Figure 12.16

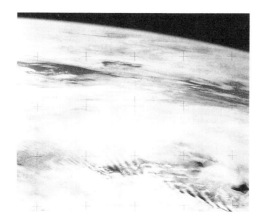

Figure 12.17

shown photographically in Fig. 1.19 was studied and mapped photogrammetrically as shown in Fig. 12.14.[327]

The vertical motion field in Fig. 12.13 is not the only form that mountain waves, and hence wave clouds, can take downstream of an isolated peak. So far we have only considered the simplified case of constant $\bar{u}$ and $\bar{B}_z$ and hence constant ℓ^2. The wave trapping that occurs when $\bar{u}$ varies with height while $\bar{B}_z$ is constant has been studied[328] by examining the case where $\bar{u}$ varies exponentially with height, such that

$$\bar{u} = U_o e^{z/\tilde{L}} \tag{12.53}$$

where U_o is the mean wind at the ground and $\tilde{L}$ is the wind scale height. It is found that even a slight wind shear is associated with a considerable amount of trapping.[329] The trapped waves that develop in the flow past a mountain peak are analogous to waves produced by a ship. These ship waves appear as distinct modes of the solution to (12.48), when the full expression (12.5) is used for ℓ^2, and $\bar{u}_{zz}$ is determined from (12.53). These modes are of two distinct types, as illustrated in Fig. 12.15. Inside a wedge-shaped zone downwind of the peak, there are *transverse* waves, which lie more or less perpendicular to the basic-state flow, and *diverging* waves, with crests that meet the incoming flow at a small angle. Outside the wedge, only diverging modes exist.[330]

The transverse waves are analogous to the trapped lee waves that occur in two-dimensional flow. They consist of superpositions of waves that have attempted to

---

[327] The detailed photogrammetric study by Abe (1932) of a horseshoe-shaped cloud in the lee of Mt. Fuji in Japan helped motivate Wurtele's (1957) study of the three-dimensional linear dynamics of the flow over isolated three-dimensional mountain peaks. Wurtele's solution of (12.50), shown in Fig. 12.12, was meant to explain Abe's photographs.

[328] By Sharman and Wurtele (1983).

[329] This exponential representation of the wind actually has the property that $\ell^2 \to -\tilde{L}^{-2}$ as $z \to \infty$, so that every wavelength is ultimately trapped.

[330] For further discussion of these wave types, see Smith (1979) and Sharman and Wurtele (1983).

Figure 12.15 Horizontal configuration of transverse and diverging phase lines for deep-water "ship waves." (From Sharman and Wurtele, 1983. Research sponsored by NASA Dryden FRF, L. J. Ehernberger, monitor. Reprinted with permission from the American Meteorological Society.)

Figure 12.16 Numerical-model simulation of highly sheared flow over a 300–500-m-high obstacle. Vertical velocity at the 5-km level is shown (in 2 cm s^{-1} contour intervals) after the flow has reached a steady state. Each tick mark on the frame enclosing the pattern indicates the location of a horizontal grid point (2.5 km). The obstacle is centered at the long tick marks. The contour representing the obstacle shape is shown as the heavy circle; the contour shown is the one for which the obstacle height is 0.1 the maximum height. (From Sharman and Wurtele, 1983. Research sponsored by NASA Dryden FRF, L. J. Ehernberger, monitor. Reprinted with permission from the American Meteorological Society.)

Figure 12.17 Example of clouds associated with diverging wave modes in the lee of Bouvet Island. The photo was taken by *Skylab* astronauts (photo reference number: SL4-137-3632). The exact scale of the picture is not known. Numerical simulations of this type of wave structure suggest that the crests are ~10–40 km apart.

propagate upstream but have been advected to the lee. The diverging waves have not attempted to propagate upstream but instead have propagated laterally away from the mountain while being advected to the lee by the basic current. Numerical model experiments for flow over a 300–500-m-high obstacle have been performed with the full set of time-dependent equations for a range of wind shear, indicated by the ratio $\tilde{R} \equiv N\tilde{L}/U_o$, where N is given by (2.98). Since N is constant, the range of $\tilde{R}$ is determined entirely by the shear, as measured by the wind scale height $\tilde{L}$. We will examine results, after the flow reaches a steady state, for three values of $\tilde{R}$.

Results at low $\tilde{R}$ (high shear) are shown in Fig. 12.16. Here only diverging modes exist. Strong motions are organized in bands forming a pattern similar to that of the idealized diverging wave crests in Fig. 12.15. The strong vertical motions are concentrated on the edges of the wedge, with relatively little disturbance directly downstream of the mountain in the center of the wedge. Figure 12.17 shows an example of clouds associated with diverging wave modes.

Figure 12.18 shows results at moderate $\tilde{R}$. In this case, both types of waves occur in the wedge-shaped region downwind of the mountain. Only diverging waves exist outside the wedge. Response at this particular $\tilde{R}$ includes a strong transverse pattern across the wedge. These waves turn into diverging structure outside the wedge. The effect of trapping in the case of the transverse waves is evident by the series of waves downstream, which resemble lee waves downwind of a two-dimensional ridge, and by the fact that the wave amplitude rapidly dies out with increasing altitude (cf. Fig. 12.18a and b). Figure 12.19 shows an example of clouds associated with predominantly transverse waves.

The case of infinite $\tilde{R}$ reduces to the no-shear ($\bar{u}$ = constant) case examined previously. Figure 12.20 illustrates this case, in which there is no trapping, and the horseshoe-shaped vertical velocity pattern is again seen downwind. This pattern is not readily distinguishable from that of moderate $\tilde{R}$ (cf. Fig. 12.18). If horseshoe-shaped clouds are observed to the lee of a mountain peak, they are probably associated with trapped waves, since the calculations show that even small shear produces significant trapping.

So far, we have considered only small-amplitude (linear) perturbations produced in flow over a three-dimensional mountain peak. In the two-dimensional case, we saw that in the case of shallow-water flow over an obstacle of finite height, the flow is blocked whenever Fr^2 [defined by (12.29)] is sufficiently small. A similar tendency is found in continuously stratified fluid if

$$\frac{\overset{\cdot\cdot}{}}{N^2 \hat{h}_m^2} \ll 1 \tag{12.54}$$

where $\hat{h}_m$ is the maximum height of the mountain. The condition (12.54) is met if the fluid is either moving slowly, extremely buoyantly stable, or both. If the obstacle is a three-dimensional peak, rather than a two-dimensional ridge, then instead of being blocked, the flow turns laterally and finds a way around the mountain. It can be shown[331] that in the case of (12.54) the flow reduces to a purely

---

[331] See Durran (1990) for further discussion and references.

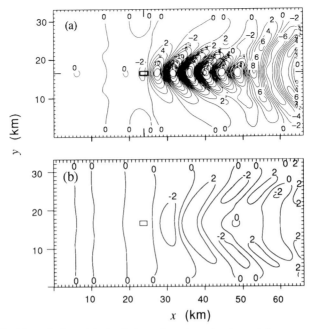

Figure 12.18 Numerical-model simulation of moderately sheared flow over a 300–500-m-high obstacle. Vertical velocity at the 5-km level (a) and 10-km level (b) is shown (in 2 cm s^{-1} contour intervals) after the flow has reached a steady state. Each tick mark on the frame enclosing the pattern indicates the location of a horizontal grid point (1 km). The obstacle is centered at the long tick marks and is represented by the rectangle at the leading edge of the wave pattern. (From Sharman and Wurtele, 1983. Research sponsored by NASA Dryden FRF, L. J. Ehernberger, monitor. Reprinted with permission from the American Meteorological Society.)

Figure 12.19 Example of clouds associated with predominantly transverse waves in the vicinity of the Aleutian Island chain. Time, date, and scale of the photograph are not known.

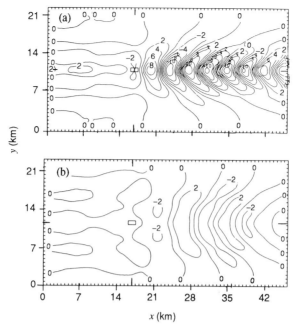

Figure 12.20 Numerical-model simulation of no-shear flow over a 300–500-m-high obstacle. Vertical velocity at the 4-km level (a) and 8-km level (b) is shown (in 2 cm s^{-1} contour intervals) after the flow has reached a steady state. Each tick mark on the frame enclosing the pattern indicates the location of a horizontal grid point (0.7 km). The obstacle is centered at the long tick marks and is represented by the rectangle at the leading edge of the wave pattern. (From Sharman and Wurtele, 1983. Research sponsored by NASA Dryden FRF, L. J. Ehernberger, monitor. Reprinted with permission from the American Meteorological Society.)

horizontal flow around the obstacle, while at the other extreme

$$\frac{\bar{u}^2}{N^2 \hat{h}_m^2} \gg 1 \tag{12.55}$$

the equations of motion reduce to their linear form, and the solutions described above apply.

The range of flow configurations that can exist between the two extremes of (12.54) and (12.55) have been investigated by numerically integrating the full set of nonhydrostatic equations. Figures 12.21 and 12.22 are examples of such results. The flow shown in vertical and horizontal cross sections in Figs. 12.21a and 12.22a are the results for $\bar{u}/N\hat{h}_m = 2.22$. In this case, the flow is in qualitative agreement with linear theory. Vertically propagating wave structure is evident over the mountain. Directly over the mountain top a cap cloud like that shown in Fig. 1.18 could form if the layer of air next to the mountain was moist enough. The possible location of a cap cloud is shaded in Fig. 12.21a.

Figures 12.21b and 12.22b show results obtained when $\bar{u}/N\hat{h}_m$ is decreased by an order of magnitude to 0.22. The flow becomes nearly horizontal. On the lee side

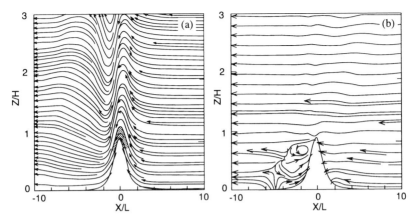

Figure 12.21 Vertical cross sections showing streamlines from a model simulation of the three-dimensional flow over a circular bell-shaped mountain when the ratio $\bar{u}/N\hat{h}_m$ is (a) 2.22; (b) 0.22. Shading shows possible location of cap cloud. Scales are normalized by specified vertical (H) and horizontal (L) scales. (From Durran, 1990 after Smolarkiewicz and Rotunno, 1989. Reprinted with permission from the American Meteorological Society.)

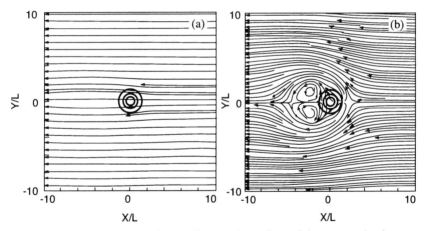

Figure 12.22 Horizontal maps of streamlines at the surface of the topography from a model simulation of the three-dimensional flow over a circular bell-shaped mountain when the ratio $\bar{u}/N\hat{h}_m$ is (a) 2.22; (b) 0.22. These maps correspond to the cross sections in Fig. 12.21. Concentric circles show the location of the mountain. Scales are normalized by a specified horizontal scale (L). (From Durran, 1990 after Smolarkiewicz and Rotunno, 1989. Reprinted with permission from the American Meteorological Society.)

of the mountain, counter-rotating eddies occur in the horizontal flow and an overturning eddy forms in the vertical plane. The upward branch of the vertical eddy rises up the side of the mountain. At the mountain top it merges with the main current blowing across the peak.

A somewhat similar eddy motion appears to be involved in the occurrence of the banner cloud, which is sometimes observed in the lee of sharp mountain peaks

(Fig. 1.26). A sketch illustrating the factors involved in formation of the banner cloud is shown in Fig. 12.23. As indicated, a very sharp-edged peak is usually involved in the formation of a banner cloud. When the flow encounters the peak, it is both forced up over the peak and split laterally into two branches. The separation of the flow in the vertical and horizontal leads to an increase of speed and corresponding reduction of the pressure [according to (8.42)] at the downwind edge of the mountain. A minimum of pressure perturbation is thus created in the lee of the top part of the peak. The pressure gradient force thus created generates [according to (7.1)] vertical motion up the lee slope. If the air drawn up the lee slope is sufficiently moist and reaches its condensation level below the top of the mountain, cloud forms. When the newly formed cloud reaches the height of the top of the mountain, it is incorporated into the main airstream over the mountain and is swept rapidly downwind in a plume resembling a wind sock or smoke from a chimney.[332]

12.4   Orographic Precipitation

So far we have discussed nonprecipitating clouds that form in shallow moist layers in flows perturbed as they flow over ridges or peaks in the surface topography. It is also possible for orographically induced flow to produce or influence precipitating clouds. Figure 12.24 summarizes the mechanisms of orographic control over precipitation. We will consider each of these mechanisms briefly in the following subsections.

12.4.1   Seeder–Feeder Mechanism over Small Hills

In Sec. 6.2, we saw how the stratiform precipitation process can be enhanced by the seeder–feeder mechanism, in which convective cells aloft can produce large precipitation particles, which, upon falling through a lower cloud layer, grow at the expense of the water content of the lower cloud. Stratus or small cumulus formed in the boundary-layer flow over small hills (Sec. 12.1) can be a particularly effective feeder cloud. As illustrated schematically in Fig. 12.24a, precipitation from another cloud layer aloft may be enhanced as it falls through the low-level feeder cloud. By itself, the low-level cloud might not precipitate. Precipitation particles from the upper cloud collect cloud particles from the low cloud, and the water thus collected is deposited on the ground.[333] Figures 12.25 and 12.26 illus-

---

[332] This plausible explanation of the banner cloud appeared in the extensive nineteenth-century treatment of the global system, *Die Erde als Ganzes* (The Earth as a Whole) *ihre Atmosphäre und Hydrosphäre*, by Hann (1896). An essentially similar explanation was offered by Douglas (1928). He, however, neglected to mention the possible role of horizontal flow separation. Unfortunately, a variety of specious explanations of the banner cloud have been suggested and have permeated the literature on the subject since these early studies.

[333] This mechanism was proposed by Bergeron (1950, 1968) to explain the enhancement of precipitation over low hills. As we have seen in Chapter 6, the idea not only was accepted for orographic enhancement but is now thought to be important in stratiform precipitation in general.

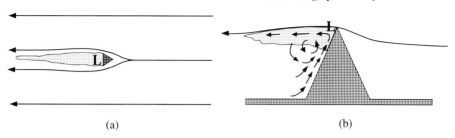

Figure 12.23 Schematic illustration of the formation of a banner cloud by flow past an isolated mountain peak. (a) Plan view (b) vertical cross section. Light stippling denotes the cloud.

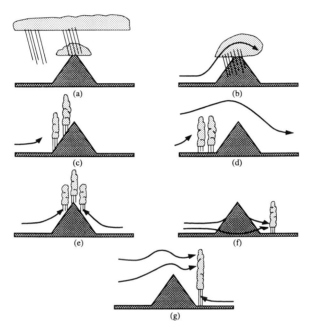

Figure 12.24 Mechanisms of orographic precipitation. (a) Seeder–feeder mechanism; (b) upslope condensation; (c) upslope triggering of convection; (d) upstream triggering of convection; (e) thermal triggering of convection; (f) lee-side triggering of convection; (g) lee-side enhancement of convection. Clouds are lightly stippled. Slanted lines below cloud base indicate precipitation.

trate an example of this process over low hills in South Wales. In this case, the upper-level clouds are moving, orographically triggered convective clouds. However, the process works equally well if the pre-existing clouds are of some other type (e.g., frontal).

12.4.2 Upslope Condensation

As we have seen, it is possible for stable ascent forced by flow over a mountain ridge or peak to be felt through a deep layer above the mountain. If the air forced over a mountain is sufficiently moist through a large portion of the lifted layer,

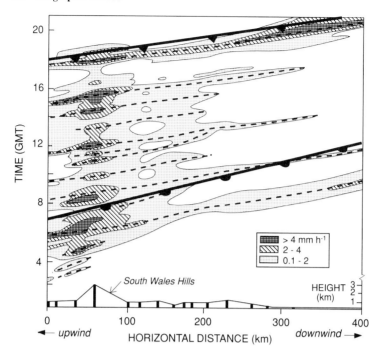

Figure 12.25   Time–distance diagram showing the rainfall rates (mm h^{-1}) and the movement of mesoscale precipitation areas over the hills of South Wales. The rainfall begins ahead of the cold front and continues in the warm sector. The rainfall rate over the hills is continuous but variable and closely associated with the passage of convective clouds aloft. (Adapted from Smith, 1979, after Browning *et al.*, 1974.)

condensation may occur through a deep layer [in contrast to shallow moist layers leading to boundary-layer clouds in upslope flow (Sec. 12.1) or wave clouds (Secs. 12.2 and 12.3)]. Figure 12.24b illustrates this process conceptually. Figure 12.27 illustrates the amount of cloud water that can be produced by flow over a mountain ridge. The streamlines, calculated with a nonlinear two-dimensional model (Fig. 12.27a), represent flow over the Cascade Mountain Range of Washington State under typical wintertime conditions. The isopleths of cloud liquid water content calculated by means of the bulk nonprecipitating warm-cloud water-continuity scheme [(3.63)–(3.64)] are shown in Fig. 12.27b. The maximum water content exceeds 1 g kg^{-1}, which is large enough to expect precipitation to develop [recall that the autoconversion threshold in (3.76) is often assumed to be 1 g kg^{-1}].

It is possible for precipitating clouds to occur in pure orographic flow of the type indicated in Fig. 12.27a. More often, though, the orographically generated precipitating cloud is superimposed on pre-existing clouds, especially those associated with a front passing over the mountain range, in which case the water field in Fig. 12.27b can be regarded as the potential enhancement of the frontal precipitation by the orographic upward motion. The downward motion on the lee side, of course, dries out the frontal clouds after they pass over the ridge.

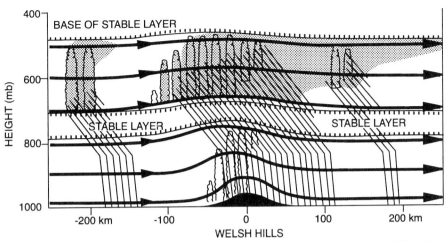

Figure 12.26 Schematic cross section of the rain clouds in a warm sector over hills of South Wales. The moving, orographically triggered convective clouds aloft produce precipitation which is locally enhanced by the low-level convective clouds over the hills (see Fig. 12.25). The slope of the hydrometeor trajectories changes abruptly at the freezing line as snow changes to rain. (Adapted from Smith, 1979, after Browning *et al.*, 1974.)

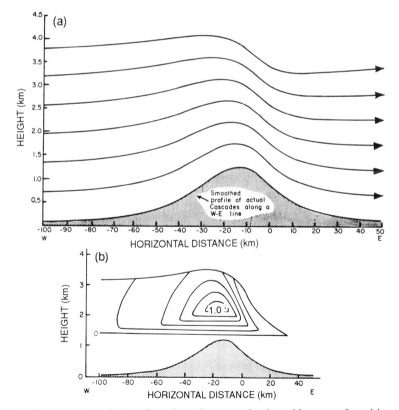

Figure 12.27 Model calculation of condensation occurring in stable ascent forced by a typical westerly flow over the Cascade Mountains of Washington State. (a) Streamlines of the flow. (b) Isopleths of adiabatic condensate at 0.2 g kg^{-1} intervals. (Panel a from Fraser *et al.*, 1973; panel b from Hobbs *et al.*, 1973. Reprinted with permission from the American Meteorological Society.)

The situation depicted in Fig. 12.27 is highly idealized in that real mountain ranges are not perfectly two-dimensional. We saw in Sec. 12.3 that when $\bar{u}/N\hat{h}_m$ is small, the flow tends to turn laterally and go horizontally around peaks rather than over them. It is indeed often the case that $\bar{u}/N\hat{h}_m$ is small, such that two-dimensional flow is blocked and three-dimensional flow around peaks and through gaps becomes significant, and vertical motions are correspondingly reduced.

12.4.3 Orographic Convection

When the air flowing over rugged terrain is potentially unstable, the lifting induced by the terrain can lead to the release of instability. In this case, the orographic clouds take the form of cumulus or cumulonimbus rather than fog, stratus, wave clouds, or stable precipitating clouds. Orographic cumulonimbi can be very important precipitation producers. Some of the rainiest areas of the world (e.g., the monsoon areas of India, Bangladesh, and Southeast Asia) are dominated by this type of precipitation.

Once formed and active, the orographic convective clouds are largely governed by the dynamics of convective clouds, as discussed in Chapters 7–9. As long as the clouds remain in the vicinity of the mountain, however, the cloud dynamics will be a complex interaction of convective and orographic dynamics. These complex interactions are only beginning to be understood. We will therefore not delve too deeply into this topic. We will concern ourselves only with the triggering and enhancement that occur during the earlier stages of the orographic convection. These processes are indicated in Fig. 12.24c–g, which indicate how various flows over and around topography can trigger or enhance convection upstream of the mountain, on the windward slope, directly over the peak, on the lee slope, or downwind of the mountain.

12.4.3.1 Upslope and Upstream Triggering

It is fairly obvious that any upslope motion can trigger convection if the air moving upslope is sufficiently moist and unstable. Figure 12.24c represents such upslope triggering. It is perhaps less obvious that topographically induced motions may lead to condensation and triggering of convection upstream of the mountain slope. However, as we have seen, flow over mountains can become complex above the surface layer, with lifting induced by the terrain sometimes being felt aloft some considerable distance upstream of the mountain. The idea of upstream triggering is indicated conceptually in Fig. 12.24d and further illustrated in Fig. 12.28, where the layer of air between the ground and the streamline is shown to be destabilized as the parcel upstream of the hill is lifted while going from A to B.

Two types of upstream lifting may occur. We have seen that vertically propagating waves described by linear theory, which are associated with mountains of any size, can tilt upstream (Figs. 12.3 and 12.4). The case shown in Fig. 12.26 illustrates how convective cells can be triggered upstream of small hills. Such a situation may be difficult to recognize in practice since as the clouds form and develop they are advected over the hills.

A second type of upstream lifting is associated with blocking (i.e., with nonlinear effects introduced by a large, finite mountain barrier). An intuitive feeling for how blocking by a two-dimensional obstacle can produce upstream lifting is provided by examining what happens in a controlled laboratory setting when a homogeneous layer of fluid flowing in a restricted channel at velocity $\bar{u}$ with low Fr is blocked by a barrier. Figure 12.29 idealizes this situation for two cases. In the case of complete blocking, the Fr is too low for the fluid to surmount the barrier; consequently, the fluid piles up against it. Mass continuity is satisfied by a hydraulic jump upstream (at x_j). Since mass continuity must be satisfied across the jump, then, according to (12.28), u decreases with increasing x across the jump. Thus, either the fluid parcels at x_j are rising as the jump travels out from the barrier, or the depth of the fluid to the right of x_j is increasing as mass piles up against the barrier. In either case, upward motion could be occurring at $x \geq x_j$. However, neither of these two situations is steady state.

A steady state can be achieved if the flow is partially blocked, as shown in Fig. 12.29b. In a laboratory setting, this situation is produced by towing the obstacle along the bottom of a tank. This case corresponds to the flow associated with severe downslope winds and rotor cloud formation on the lee side (Sec. 12.2.5). The partial blocking phenomenon is the windward-side portion of the flow. This case can be in a steady state since the blocking is incomplete and downstream and upstream conditions can now be matched. The upstream jump thus can become stationary in a frame moving with the laboratory obstacle, and a fixed region of upward motion can be established upstream of the obstacle. It is presumed that an analogous situation can occur in the atmosphere.

One area where upstream partial blocking is thought to play a role in triggering cumulonimbus is over the Arabian Sea west of the Western Ghats Mountain Range of southwestern India. Figure 12.30 contains the results of a nonlinear, two-dimensional model simulation (using a model of the type discussed in Sec. 7.5.3) of the atmospheric flow over the Western Ghats. The upstream flow was taken to be that characteristic of the Indian summer monsoon. The mountain profile is a smoothed representation of the Western Ghats, which slope up steeply from the ocean and level out into a plateau. The pressure perturbation field (Fig. 12.30a) indicates the piling up of mass against the windward slope. The horizontal wind field (Fig. 12.30b) shows a corresponding deceleration as the lower-tropospheric monsoon flow approaches the range. Finally, the vertical velocity field (Fig. 12.30c) shows how the lifting extends out over the ocean, where it could trigger convection upstream of the mountain range. It should be noted, however, that once convection is initiated, the flow can be altered by the convective heating. Thus, the situation portrayed in Fig. 12.30 may not be the final state of affairs.

12.4.3.2   Thermal and Lee-Side Triggering

Figure 12.24e and f indicate two additional ways in which mountains can trigger convection. Thermal forcing (Fig. 12.24e) occurs when daytime heating produces an elevated heat source and corresponding thermally direct circulation, with con-

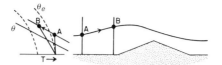

Figure 12.28

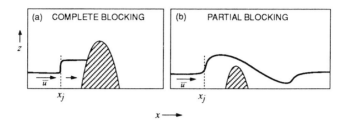

Figure 12.29

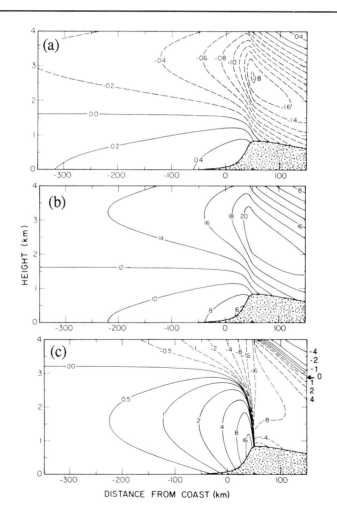

Figure 12.30

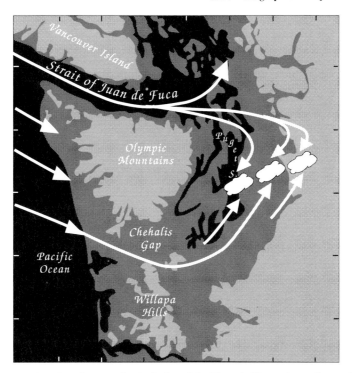

Figure 12.31 Triggering of convection in the lee of the Olympic Mountains under conditions of the Puget Sound Convergence Zone of Washington State. Streamlines represent surface winds. (Adapted from Mass, 1981. Reproduced with permission from the American Meteorological Society.)

vergence at the top of the mountain. It is well known that this type of circulation can trigger anything from small cumulus to incipient mesoscale convective systems, all of whose dynamics have been discussed in previous chapters.

The lee-side forcing indicated in Fig. 12.24f results from low-Froude-number flow around an isolated obstacle. One consequence of this diversion around the mountain is the formation of convection in the lee of the mountain. A good

---

Figure 12.28 Destabilization of a nearly saturated, nearly conditionally unstable air mass owing to lifting aloft upstream of the mountain. The air parcels aloft rise dry adiabatically, thus decreasing the stability of the air column against moist convection. (From Smith, 1979.)

Figure 12.29 Blocking of a homogeneous layer of fluid flowing in a restricted channel at a low Froude number. (a) Complete blocking. (b) Partial blocking. Stationary obstacle shown by hatching. Fluid enters from left at velocity $\bar{u}$. (Adapted from Simpson, 1987. Reproduced with permission from Ellis Horwood Ltd., Chichester.)

Figure 12.30 Nonlinear two-dimensional model simulation illustrating upstream partial blocking over the Arabian Sea west of the Western Ghats Mountain Range of southwestern India. Cross sections perpendicular to mountains of (a) pressure perturbation (mb); (b) horizontal wind speed (m s^{-1}); and (c) vertical velocity (cm s^{-1}). (From Grossman and Durran, 1984. Reprinted with permission from the American Meteorological Society.)

example of this phenomenon is the Puget Sound Convergence Zone, which forms and triggers convection in the lee of the Olympic Mountains of Washington State (Fig. 12.31).

12.4.3.3   Lee-Side Enhancement of Deep Convection

The sketch in Fig. 12.24g indicates how convection triggered on the windward slope over the crest of a ridge can be enhanced on the lee side. The enhancement results from combined effects of midlevel upward motion associated with a vertically propagating wave induced by flow over a mountain and low-level thermally induced upslope flow. This combination of effects is thought to be significant in the enhancement of deep convection that forms over the Rocky Mountains and subsequently develops into mesoscale convective complexes that move eastward across the central United States.[334]

---

[334] See Tripoli and Cotton (1989a) for a detailed conceptual model of this phenomenon.

References

Abe, M., 1932: The formation of cloud by the obstruction of Mount Fuji. *Geophys. Mag.*, **6**, 1–10.

Abe, M., 1941: Mountain clouds, their forms and connected air current. Part II. *Bull. Central Met. Obs. Japan*, VII, **3**, 93–145.

Acheson, D. J., 1990: *Elementary Fluid Dynamics*. Clarendon Press, Oxford, 397 pp.

Ackerman, T. P., K.-N. Liou, F. P. J. Valero, and L. Pfister, 1988: Heating rates in tropical anvils. *J. Atmos. Sci.*, **45**, 1606–1623.

Agee, E. M., 1982: An introduction to shallow convective systems. *Cloud Dynamics* (E. M. Agee and T. Asai, Eds.), D. Reidel Publishing, Dordrecht, 423 pp.

Agee, E. M., T. S. Chen, and K. E. Doswell, 1973: A review of mesoscale cellular convection. *Bull. Amer. Meteor. Soc.*, **10**, 1004–1012.

Anderson, R. K., J. P. Ashman, G. R. Farr, E. W. Ferguson, G. N. Isayeva, V. J. Oliver, F. C. Parmenter, T. P. Popova, R. W. Skidmore, A. H. Smith, and N.F. Veltishchev, 1973: *The Use of Satellite Pictures in Weather Analysis and Forecasting*. Technical Note No. 124, World Meteorological Organization, Geneva, 275 pp.

Anderson, R. K., *et al.*, 1974: Application of Meteorological Satellite Data in Analysis and Forecasting. ESSA Technical Report NESC 51, U.S. Dept. of Commerce, NTIS AD-786 137.

Anthes, R. A., 1982: *Tropical Cyclones: Their Evolution, Structure and Effects*. American Meteorological Society, Boston, 208 pp.

Árnason, G., and R. S. Greenfield, 1972: Micro- and macro-structures of numerically simulated convective clouds. *J. Atmos. Sci.*, **29**, 342–367.

Arya, S. P., 1988: *Introduction to Micrometeorology*. Academic Press, New York, 303 pp.

Asai, T., 1964: Cumulus convection in the atmosphere with vertical wind shear: Numerical experiment. *J. Met. Soc. Japan*, **42**, 245–259.

Asai, T., 1970a: Three-dimensional features of thermal convection in a plane couette flow. *J. Met. Soc. Japan*, **48**, 18–29.

Asai, T., 1970b: Stability of a plane parallel flow with variable vertical shear and unstable stratification. *J. Met. Soc. Japan*, **48**, 129–139.

Asai, T., 1972: Thermal instability of a shear flow turning direction with height. *J. Met. Soc. Japan*, **50**, 525–532.

Asai, T., and A. Kasahara, 1967: A theoretical study of the compensating downward motions associated with cumulus clouds. *J. Atmos. Sci.*, **24**, 487–496.

Asai, T., and I. Nakasuji, 1973: On the stability of the Ekman boundary-layer flow with thermally unstable stratification. *J. Met. Soc. Japan*, **51**, 29–42.

Atlas, D., Ed., 1990: *Radar in Meteorology*. American Meteorological Society, Boston, 806 pp.

Atlas, D., K. R. Hardy, R. Wexler, and R. J. Boucher, 1963: On the origin of hurricane spiral bands. *Technical Conference on Hurricanes and Tropical Meteorology*, Mexico, *Geofisica Internacional*, **3**, 123–132.

Atlas, D., D. Rosenfeld, and D. Wolff, 1990: Climatologically tuned reflectivity rain rate relations and links to area–time integrals. *J. Appl. Meteor.*, **29**, 1120–1135.

Atwater, M. A., 1972: Thermal effects of organization and industrialization in the boundary layer: A numerical study. *Boundary-Layer Met.*, **3**, 229–245.

Auer, A. H., Jr., 1972: Distribution of graupel and hail with size. *Mon. Wea. Rev.*, **100**, 325–328.

Austin, P. H., M. B. Baker, A. M. Blyth, and J. B. Jensen, 1985: Small-scale variability in warm continental cumulus clouds. *J. Atmos. Sci.*, **42**, 1123–1138.

Austin, P. M., 1987: Relation between measured radar reflectivity and surface rainfall. *Mon. Wea. Rev.*, **115**, 1053–1070.

Baker, M. B., R. G. Corbin, and J. Latham, 1980: The influence of entrainment on the evolution of cloud droplet spectra. I: A model in inhomogeneous mixing. *Quart. J. Roy. Met. Soc.*, **106**, 581–598.

Balsley, B. B., W.L. Ecklund, D. A. Carter, A. C. Riddle, and K. S. Gage, 1988: Average vertical motions in the tropical atmosphere observed by a radar wind profiler on Pohnpei (7°N latitude, 157°E longitude). *J. Atmos. Sci.*, **45**, 396–405.

Barnes, G. M., E. J. Zipser, D. Jorgensen, and F. D. Marks, Jr., 1983: Mesoscale and convective structure of a hurricane rainband. *J. Atmos. Sci*, **40**, 2125–2137.

Barton, I. J., 1983: Upper-level cloud climatology from an orbiting satellite. *J. Atmos. Sci.*, **40**, 435–447.

Batchelor, G. K., 1967: *An Introduction to Fluid Dynamics*. Cambridge University Press, Cambridge, 615 pp.

Battan, L. J., 1959: *Radar Meteorology*. University of Chicago Press, Chicago, 161 pp.

Battan, L. J., 1973: *Radar Observations of the Atmosphere*. University of Chicago Press, Chicago, 324 pp.

Baynton, H. W., R. J. Serafin, C. L. Frush, G. R. Gray, P. V. Hobbs, R. A. Houze, Jr., and J. D. Locatelli, 1977: Real-time wind measurement in extratropical cyclones by means of Doppler radar. *J. Appl. Tech.*, **16**, 1022–1028.

Beard, K.V., 1976: Terminal velocity and shape of cloud and precipitation drops aloft. *J. Atmos. Sci.*, **33**, 851–864.

Beard, K. V., 1985: Simple altitude adjustments to raindrop velocities for Doppler-radar analysis. *J. Atmos. Oceanic Tech.*, **2**, 468–471.

Beard, K. V., and H. T. Ochs, 1986: Charging mechanisms in clouds and thunderstorms. *The Earth's Electrical Environment*. National Academy Press, Washington D.C., 114–130.

Beard, K. V., and H. R. Pruppacher, 1969: A determination of the terminal velocity and drag of small water drops by means of a wind tunnel. *J. Atmos. Sci*, **26**, 1066–1072.

Bénard, M. H., 1901: Les tourbillons cellulaires dans une nappe liquide transportant de la chaleur par convection en régime permanent. *Annales de Chimie et de Physique*, **23**, 62–144.

Bennetts, D. A., and B. J. Hoskins, 1979: Conditional symmetric instability—a possible explanation for frontal rainbands. *Quart. J. Roy. Met. Soc.*, **105**, 945–962.

Bergeron, T., 1950: Über der mechanismus der ausgiebigan niederschläge. *Ber. Deut. Wetterd.*, **12**, 225–232.

Bergeron, T., 1954: The problem of tropical hurricanes. *Quart. J. Roy. Meteor. Soc.*, **80**, 131–164.

Bergeron, T., 1968: On the low-level redistribution of atmospheric water caused by orography. *Proceedings, International Cloud Physics Conference*, Toronto.

Berry, E. X., and R. L. Reinhardt, 1973: Modeling of condensation and collection within clouds. Final report, NSF grant GA-21350, July.

Betts, A. K., 1986: A new convective adjustment scheme. Part I: Observational and theoretical basis. *Quart. J. Roy. Met. Soc.*, **112**, 677–691.

Biggerstaff, M. I., and R. A. Houze, Jr., 1991a: Kinematic and precipitation structure of the 10–11 June 1985 squall line. *Mon. Wea. Rev.*, **119**, 3035–3065.

Biggerstaff, M. I., and R. A. Houze, Jr., 1991b: Midlevel vorticity structure of the 10–11 June 1985 squall line. *Mon. Wea. Rev.*, **119**, 3066–3079.

Biggerstaff, M. I., and R. A. Houze, Jr., 1993: Kinematics and microphysics of the transition zone of a midlatitude squall-line system. Accepted, *J. Atmos. Sci*.

Black, P. G., and F. D. Marks, Jr., 1987: Environmental interactions associated with hurricane supercells. *Proceedings, 17th International Conference on Tropical Meteorology*, Miami, FL, American Meteorological Society, Boston, 416–419.

Black, P. G., F. D. Marks, Jr., and R. A. Black, 1986: Supercell structure in tropical cyclones.

Proceedings, 23rd Conference on Radar Meteorology, Snowmass, CO, American Meteorological Society, Boston, JP255–JP259.

Bluestein, H. B., 1986: Fronts and jet streaks: A theoretical perspective. *Mesoscale Meteorology and Forecasting* (P. S. Ray, Ed.), American Meteorological Society, Boston, 173–215.

Bluestein, H. B., and W. P. Unruh, 1989: Observations of the wind field in tornadoes, funnel clouds, and wall clouds with a portable Doppler radar. *Bull. Amer. Meteor. Soc.*, **70**, 1514–1525.

Blyth, A. M., and J. Latham, 1985: An airborne study of vertical structure and microphysical variability within a small cumulus. *Quart. J. Roy. Met. Soc.*, **111**, 773–792.

Böhm, J. P., 1989: A general equation for the terminal fall speed of solid hydrometeors. *J. Atmos. Sci.*, **46**, 2419–2427.

Bond, N. A., and R. G. Fleagle, 1985: Structure of a cold front over the ocean. *Quart. J. Roy. Met. Soc.*, **111**, 739–760.

Borovikov, A. M., I. I. Gaivoronskii, E. G. Zak, V. V. Kostarev, I. P. Mazin, V. E. Minervin, A. K. Khrgian, and S. M. Shmeter, 1963: *Cloud Physics (Fizika Oblakov)*, U.S. Dept. of Commerce, Washington D.C., 392 pp.

Brandes, E. A., 1990: Evolution and structure of the 6–7 May 1985 mesoscale convective system and associated vortex. *Mon. Wea. Rev.*, **118**, 109–127.

Bringi, V. N., and A. Hendry, 1990: Technology of polarization diversity radars for meteorology. *Radar in Meteorology* (D. Atlas, Ed.), American Meteorological Society, Boston, 153–190.

Bringi, V. N., R. M. Rasmussen, and J. Vivekanandan, 1986a: Multiparameter radar measurements in Colorado convective storms. Part I: Graupel melting studies. *J. Atmos. Sci.*, **43**, 2545–2563.

Bringi, V. N., J. Vivekanandan and J. D. Tuttle, 1986b: Multiparameter radar measurements in Colorado convective storms. Part II: Hail detection studies. *J. Atmos. Sci.*, **43**, 2564–2577.

Brown, J. M., 1974: *Mesoscale motions induced by cumulus convection: A numerical study*. Ph.D. thesis, Department of Meteorology and Physical Oceanography, Massachusetts Institute of Technology, Cambridge, 207 pp.

Brown, J. M., 1979: Mesoscale unsaturated downdrafts driven by rainfall evaporation: A numerical study. *J. Atmos. Sci.*, **36**, 313–338.

Brown, R. A., 1980: Longitudinal instabilities and secondary flows in the planetary boundary layer: A review. *Rev. Geophys. Space. Phys.*, **18**, 683–697.

Brown, R. A., 1983: The flow in the planetary boundary layer. *Eolian Sediments and Processes* (M. E. Brookfield and T. S. Ahlbrandt, Eds.), Elsevier Publishers, Amsterdam, 291–310.

Brown, R., and W. T. Roach, 1976: The physics of radiation fog: II—A numerical study. *Quart. J. Roy. Met. Soc.*, **102**, 335–354.

Browning, K. A., 1964: Airflow and precipitation trajectories within severe local storms which travel to the right of the winds. *J. Atmos. Sci.*, **21**, 634–639.

Browning, K. A, 1986: Conceptual models of precipitating systems. *Wea. Forecasting*, **1**, 23–41.

Browning, K. A., and G. B. Foote, 1976: Airflow and hail growth in supercell storms and some implications for hail suppression. *Quart. J. Roy. Met. Soc.*, **102**, 499–534.

Browning, K. A., and G. A. Monk, 1982: A simple model for the synoptic analysis of cold fronts. *Quart. J. Roy. Met. Soc.*, **108**, 435–452.

Browning, K. A., and R. Wexler, 1968: A determination of kinematic properties of a wind field using Doppler radar. *J. Appl. Meteor.*, **7**, 105–113.

Browning, K. A., F. F. Hill, and C. W. Pardoe, 1974: Structure and mechanism of precipitation and the effect of orography in a wintertime warm sector. *Quart. J. Roy. Met. Soc.*, **100**, 309–330.

Browning, K. A., J. C. Fankhauser, J.-P. Chalon, P. J. Eccles, R. G. Strauch, F. H. Merrem, D. J. Musil, E. L. May, and W. R. Sand, 1976: Structure of an evolving hailstorm, Part V: Synthesis and implications for hail growth and hail suppression. *Mon. Wea. Rev.*, **104**, 603–610.

Burgers, J. M., 1948: A mathematical model illustrating the theory of turbulence. *Advan. Appl. Mech.*, **1**, 197–199.

Burgess, D., and P. S. Ray, 1986: Principles of radar. *Mesoscale Meteorology and Forecasting* (P. S. Ray, Ed.), American Meteorological Society, Boston, 85–117.

Burgess, D. W., V. T. Wood, and R. A. Brown, 1982: Mesocyclone evolution statistics. *Preprints,*

Twelfth Conference on Severe Local Storms, San Antonio, TX, American Meteorological Society, Boston, 422–424.

Businger, S., 1991: Arctic hurricanes. *American Scientist*, **79**, 18–33.

Businger, S., and P. V. Hobbs, 1987: Mesoscale structures of two comma cloud systems over the Pacific Ocean. *Mon. Wea. Rev.*, **115**, 1908–1928.

Businger, S., and R. J. Reed, 1989: Cyclogenesis in cold air masses. *Wea. Forecasting*, **4**, 133–156.

Byers, H. R., 1959: *General Meteorology*. McGraw-Hill Book Company, New York, 540 pp.

Byers, H. R., and R. R. Braham, Jr., 1949: *The Thunderstorm*. U.S. Government Printing Office, Washington, D.C., 287 pp.

Byrne, G. J., A. A. Few, and M. F. Stewart, 1989: Electric field measurements within a severe thunderstorm anvil. *J. Geophys. Res.*, **94**, 6297–6307.

Calheiros, R. V., and I. Zawadzki, 1987: Reflectivity rain-rate relationships for radar hydrology in Brazil. *J. Clim. Appl. Meteor.*, **26**, 118–132.

Carbone, R. E., 1982: A severe frontal rainband. Part I: Stormwide hydrodynamic structure. *J. Atmos. Sci.*, **39**, 258–279.

Carbone, R. E., 1983: A severe frontal rainband. Part II: Tornado parent vortex circulation. *J. Atmos. Sci.*, **40**, 2639–2654.

Carbone, R. E., J. W. Conway, N. A. Crook, and M. W. Moncrieff, 1990a: The generation and propagation of a nocturnal squall line. Part I: Observations and implications for mesoscale predictability. *Mon. Wea. Rev.*, **118**, 26–49.

Carbone, R. E., N. A. Crook, M. W. Moncrieff, and J. W. Conway, 1990b: The generation and propagation of a nocturnal squall line. Part II: Numerical simulations. *Mon. Wea. Rev.*, **118**, 50–65.

Charba, J., 1974: Application of gravity-current model to analysis of squall-line gust front. *Mon. Wea. Rev.*, **102**, 140–156.

Chauzy, S., P. Raizonville, D. Hauser, and F. Roux, 1980: Electrical and dynamical description of a frontal storm deduced from the LANDES 79 experiment. *J. Rech. Atmos.*, **14**, 457–467.

Chauzy, S., M. Chong, A. Delannoy, and S. Despiau, 1985: The June 22 tropical squall line observed during the COPT 81 experiment: Electrical signature associated with dynamical structure and precipitation. *J. Geophys. Res.*, **90**, 6091–6098.

Chen, S. S., 1990: *A numerical study of the genesis of extratropical convective mesovortices*. Ph.D. dissertation, Department of Meteorology, Pennsylvania State University, University Park, 178 pp.

Chen, S. S., and W. M. Frank, 1993: A numerical study of the genesis of extratropical convective mesovortices. Part I: Evolution and dynamics. *J. Atmos. Sci.* **50**, 2401–2426.

Cheng, C.-P., and R. A. Houze, Jr., 1979: Sensitivity of diagnosed convective fluxes to model assumptions. *J. Atmos. Sci.*, **37**, 774–783.

Chisholm, A. J., and J. H. Renick, 1972: Supercell and multicell Alberta hailstorms. *Preprints, International Cloud Physics Conference*, London, England, 1–8.

Chong, M., P. Amayenc, G. Scialom, and J. Testud, 1987: A tropical squall line observed during the COPT 81 experiment in west Africa. Part I: Kinematic structure inferred from dual-Doppler radar data. *Mon. Wea. Rev.*, **115**, 670–694.

Churchill, D. D., 1988: *Radiation, turbulence and microphysics in the stratiform regions of tropical cloud clusters*. Ph.D. dissertation, Dept. of Atmospheric Sciences, University of Washington, 121 pp.

Churchill, D. D., and R. A. Houze, Jr., 1987: Mesoscale organization and cloud microphysics in a Bay of Bengal depression. *J. Atmos. Sci.*, **44**, 1845–1867.

Churchill, D. D., and R. A. Houze, Jr., 1991: Effects of radiation and turbulence on the diabatic heating and water budget of the stratiform region of a tropical cloud cluster. *J. Atmos. Sci.*, **48**, 903–922.

Clark, T. L., 1973: Numerical modeling of the dynamics and microphysics of warm cumulus convection. *J. Atmos. Sci.*, **30**, 857–878.

Clark, T. L., 1979: Numerical simulations with a three-dimensional cloud model: Lateral boundary condition experiments and multicellular severe storm simulations. *J. Atmos. Sci.*, **36**, 2191–2215.

Colman, B., 1992: The operational consideration of non-supercell tornadoes, particularly those where local terrain is of foremost importance to tornadogenesis. *Proceedings for the Forest Workshop on Operational Meteorology*, Canadian AES/CMOS, Whistler, B.C., Canada, September.

Cotton, W. R., 1972: Numerical simulation of precipitation development in supercooled cumuli. Part I. *Mon. Wea. Rev.*, **100**, 757–763.

Davies-Jones, R. P., 1986: Tornado dynamics. *Thunderstorm Morphology and Dynamics* (E. Kessler, Ed.), University of Oklahoma Press, Norman, 197–236.

Davies-Jones, R. P., and E. Kessler, 1974: Tornadoes. *Weather and Climate Modification* (W. N. Hess, Ed.), John Wiley and Sons, New York, 552–595.

de Rudder, B., 1929: Luftkörperwechsel und atmosphärische Unstetigkeitsschichten als Krankheitsfaktoren. *Ergebn. inn. Med.* 36 S. 273.

Deardorff, J. W., 1972: Numerical investigation of neutral and unstable planetary boundary layers. *J. Atmos. Sci.*, **29**, 91–115.

Doneaud, A. A., P. L. Smith, A. S. Dennis, and S. Sengupta, 1981: A simple method for estimating convective rain volume over an area. *Water Resour. Res.*, **17**, 1676–1682.

Doneaud, A. A., S. I. Niscov, D. L. Priegnitz, and P. L. Smith, 1984: The area–time integral as an indicator for convective rain volumes. *J. Climate. Appl. Meteor.*, **23**, 555–561.

Douglas, C. K. M., 1928: Some alpine cloud forms. *Quart. J. Roy. Met. Soc.*, **54**, 175–178.

Doviak, R. J., and D. S. Zrnic, 1984: *Doppler Radar and Weather Observations.* Academic Press, Orlando, FL, 458 pp.

Droegemeier, K. K., and R. B. Wilhelmson, 1987: Numerical simulation of thunderstorm outflow dynamics. Part I: Outflow sensitivity experiments and turbulence dynamics. *J. Atmos. Sci.*, **44**, 1180–1210.

Durran, D. R., 1986a: Another look at downslope windstorms. Part I: The development of analogs to supercritical flow in an infinitely deep, continuously stratified field. *J. Atmos. Sci.*, **43**, 2527–2543.

Durran, D. R., 1986b: Mountain waves. *Mesoscale Meteorology and Forecasting* (P. S. Ray, Ed.), American Meteorological Society, Boston, 472–492.

Durran, D. R., 1989: Improving the anelastic approximation. *J. Atmos. Sci.*, **46**, 1453–1461.

Durran, D. R., 1990: Mountain waves and downslope winds. *Atmospheric Processes Over Complex Terrain* (W. Blumen, Ed.), American Meteorological Society, Boston, 59–81.

Durran, D. R., and J. B. Klemp, 1987: Another look at downslope winds. Part II: Nonlinear amplification beneath wave-overturning layers. *J. Atmos. Sci.*, **44**, 3402–3412.

Durran, D. R., and D. B. Weber, 1988: An investigation of the poleward edges of cirrus clouds associated with midlatitude jet streams. *Mon. Wea. Rev.*, **116**, 702–714.

Durst, C. S., and R. C. Sutcliffe, 1938: The importance of vertical motion in the development of tropical revolving storms. *Quart. J. Roy. Met. Soc.*, **64**, 75–84.

Ekman, V. W., 1902: Om jordrotationens inverkan pa vindströmmar i hafvet. *Nyt. Mag. f. Naturvid.*, **40**, 37–63.

Eliassen, A., 1948: The quasi-static equations of motion with pressure as an independent variable. *Geophys. Publ.*, **17**, 44 pp.

Eliassen, A., 1959: On the formation of fronts in the atmosphere. *The Atmosphere and the Sea in Motion* (B. Bolin, Ed.), Rockefeller Institute Press, New York, 277–287.

Eliassen, A., 1962: On the vertical circulation in frontal zones. *Geophys. Publ.*, **24**, 147–160.

Elmore, K. L., D. McCarthy, W. Frost, and H. P. Chang, 1986: A high-resolution spatial and temporal multiple Doppler analysis of a microburst and its application to aircraft flight simulation. *J. Clim. Appl. Meteor.*, **25**, 1398–1425.

Emanuel, K. A., 1985: Frontal circulations in the presence of small, moist symmetric stability. *J. Atmos. Sci.*, **42**, 1062–1071.

Emanuel, K. A., 1986a: An air–sea interaction theory for tropical cyclones. Part I: Steady-state maintenance. *J. Atmos. Sci.*, **43**, 585–604.

Emanuel, K. A., 1986b: Overview and definition of mesoscale meteorology. *Mesoscale Meteorology and Forecasting* (P.S. Ray, Ed.), American Meteorological Society, Boston, 1–17.

Emanuel, K. A., M. Fantini, and A. J. Thorpe, 1987: Baroclinic instability in an environment of small stability to slantwise moist convection. Part I: Two-dimensional models. *J. Atmos. Sci.*, **44**, 1559–1573.

Ferrier, B. S., and R. A. Houze, Jr., 1989: One-dimensional time-dependent modeling of GATE cumulonimbus convection. *J. Atmos. Sci.*, **46**, 330–352.

Fiedler, B. H., 1984: The mesoscale stability of entrainment into cloud-topped mixed layers. *J. Atmos. Sci.*, **41**, 92–101.

Fletcher, N. H., 1966: *The Physics of Rainclouds*. Cambridge University Press, Cambridge, 390 pp.

Foote, G. B., and P. S. DuToit, 1969: Terminal velocity of raindrops aloft. *J. Appl. Meteor.*, **8**, 249–253.

Forkel, R., W. G. Panhaus, R. Welch, and W. Zdunkowski, 1984: A one-dimensional numerical study to simulate the influence of soil moisture, pollution and vertical exchange on the evolution of radiation fog. *Contrib. Atmos. Phys.*, **57**, 72–91.

Fovell, R. G., 1990: Influence of the Coriolis force on a two-dimensional model storm. *Preprints, Fourth Conference on Mesoscale Processes*, Boulder, American Meteorological Society, Boston, 190–191.

Fovell, R. G., and Y. Ogura, 1988: Numerical simulation of a midlatitude squall line in two dimensions. *J. Atmos. Sci.*, **45**, 3846–3879.

Fovell, R. G., and Y. Ogura, 1989: Effect of vertical wind shear on numerically simulated multicell storm structure. *J. Atmos. Sci.*, **46**, 3144–3176.

Fovell, R. G., D. R. Durran, and J. R. Holton, 1992: Numerical simulations of convectively generated gravity waves in the atmosphere. *J. Atmos. Sci.*, **49**, 1427–1442.

Frank, W. M., 1977: The structure and energetics of the tropical cyclone. Part I: Storm structure. *Mon. Wea. Rev.*, **105**, 1119–1135.

Frank, W. M., 1983: The structure and energetics of the east Atlantic intertropical convergence zone. *J. Atmos. Sci.*, **40**, 1916–1929.

Frank, W. M., and C. Cohen, 1987: Simulation of tropical convective systems. Part I: A cumulus parameterization. *J. Atmos. Sci.*, **44**, 3787–3799.

Fraser, A. B., R. C. Easter and P. V. Hobbs, 1973: A theoretical study of the flow of air and fallout of solid precipitation over mountainous terrain. Part I: Airflow model. *J. Atmos. Sci.*, **30**, 801–812.

Fritsch, J. M., and C. F. Chappell, 1980: Numerical prediction of convectively driven mesoscale pressure systems. Part I: Convective parameterization. *J. Atmos. Sci.*, **37**, 1722–1733.

Fujita, T. T., 1981: Tornadoes and downbursts in the context of generalized planetary scales. *J. Atmos. Sci.*, **38**, 1511–1534.

Fujita, T. T., 1985: *The Downburst—Microburst and Macroburst*. Report of Projects NIMROD and JAWS. SMRP, University of Chicago, Chicago, 122 pp.

Fujita, T. T., 1986: *DFW Microburst on August 2, 1985*. University of Chicago Press, Chicago, 154 pp.

Fujita, T. T., and R. M. Wakimoto, 1983: Microbursts in JAWS depicted by Doppler radars, PAM, and aerial photographs. *Preprints, 21st Conference on Radar Meteorology*, Edmonton, Canada, American Meteorological Society, Boston, 638–645.

Fujita, T. T., D. L. Bradbury, and C. F. Van Thullenar, 1983: Palm Sunday tornadoes of April 11, 1965. *Mon. Wea. Rev.*, **98**, 29–69.

Gal-Chen, T., 1978: A method for the initialization of the anelastic equations: Implications for matching models with observations. *Mon. Wea. Rev.*, **106**, 587–606.

Gamache, J. F., and R. A. Houze, Jr., 1982: Mesoscale air motions associated with a tropical squall line. *Mon. Wea. Rev.*, **110**, 118–135.

Gamache, J. F., and R. A. Houze, Jr., 1985: Further analysis of the composite wind and thermodynamic structure of the 12 September GATE squall line. *Mon. Wea. Rev.*, **113**, 1241–1259.

Gill, A. E., 1982: *Atmosphere–Ocean Dynamics*. Academic Press, New York, 662 pp.

Goff, R. C., 1975: Thunderstorm outflow kinetics and dynamics. NOAA Tech. Memo, ERL NSSL-75, National Severe Storms Laboratory, Norman, OK, 63 pp.

Golden, J. H., 1974a: The life cycle of Florida Keys' waterspouts. I. *J. Appl. Meteor.*, **13**, 676–692.

Golden, J. H., 1974b: Scale-interaction implications for the waterspout life cycle. II. *J. Appl. Meteor.*, **13**, 693–709.

Golden, J. H., and D. Purcell, 1978a: Life cycle of the Union City, Oklahoma tornado and comparison with waterspouts. *Mon. Wea. Rev.*, **106**, 3–11.

Golden, J. H., and D. Purcell, 1978b: Airflow characteristics around the Union City tornado. *Mon. Wea. Rev.*, **106**, 22–28.

Gossard, E. E., 1990: Radar research on the atmospheric boundary layer. *Radar in Meteorology* (D. Atlas, Ed.), American Meteorological Society, Boston, 477–527.

Gray, W. M., 1979: Hurricanes: Their formation, structure, and likely role in the tropical circulation. *Meteorology over the Tropical Oceans* (D. B. Shaw, Ed.), Royal Meteorological Society, 155–218.

Grossman, R. L., and D. R. Durran, 1984: Interaction of low-level flow with the western Ghat Mountains and offshore convection in the summer monsoon. *Mon. Wea. Rev.*, **112**, 652–672.

Gultepe, I., and A. Heymsfield, 1988: Vertical velocities within a cirrus cloud from Doppler lidar and aircraft measurements during FIRE: Implications for particle growth. *Preprints, Tenth International Cloud Physics Conference*, Bad Homburg, Federal Republic of Germany, 476–478.

Hall, W. D., 1980: A detailed microphysical model within a two-dimensional dynamic framework: Model description and preliminary results. *J. Atmos. Sci.*, **37**, 2486–2507.

Hallet, J., and S. C. Mossop, 1974: Production of secondary ice particles during the riming process. *Nature*, **249**, 26–28.

Haltiner, G. J., and F. L. Martin, 1957: *Dynamical and Physical Meteorology*. McGraw-Hill Book Company, New York, 470 pp.

Hane, C. E., R. B. Wilhelmson, and T. Gal-Chen, 1981: Retrieval of thermodynamic variables within deep convective clouds: Experiments in three dimensions. *Mon. Wea. Rev.*, **109**, 564–576.

Hann, J. V., 1896: *Die Erde als Ganzes: Ihre Atmosphäre und Hydrosphäre*. F. Tempsky, Wien, 368 pp.

Harimaya, T., 1968: On the shape of cirrus uncinus clouds: A numerical computation—Studies of cirrus clouds: Part III. *J. Met. Soc. Japan*, **46**, 272–279.

Haurwitz, B., 1935: The height of tropical cyclones and the eye of the storm. *Mon. Wea. Rev.*, **63**, 45–49.

Hauser, D., F. Roux, and P. Amayenc, 1988: Comparison of two methods for the retrieval of thermodynamic and microphysical variables from Doppler-radar measurements: Application to the case of a tropical squall line. *J. Atmos. Sci.*, **45**, 1285–1303.

Hawkins, H. F., and S. M. Imbembo, 1976: The structure of a small, intense hurricane—Inez 1966. *Mon. Wea. Rev.*, **104**, 418–442.

Herman, G. F., and R. Goody, 1976: Formation and persistence of summertime arctic stratus clouds. *J. Atmos. Sci.*, **33**, 1537–1553.

Herzegh, P. H., and A. R. Jameson, 1992: Observing precipitation through dual-polarization radar measurements. *Bull. Amer. Meteor. Soc.*, **73**, 1365–1374.

Heymsfield, A. J., 1975a: Cirrus uncinus generating cells and the evolution of cirriform clouds. Part I: Aircraft observations and the growth of the ice phase. *J. Atmos. Sci.*, **32**, 799–808.

Heymsfield, A. J., 1975b: Cirrus uncinus generating cells and the evolution of cirriform clouds. Part II: The structure and circulations of the cirrus uncinus generating head. *J. Atmos. Sci.*, **32**, 809–819.

Heymsfield, A. J., 1977: Precipitation development in stratiform ice clouds: A microphysical and dynamical study. *J. Atmos. Sci.*, **34**, 367–381.

Heymsfield, A. J., and N. C. Knight, 1988: Hydrometeor development in cold clouds in FIRE. *Preprints, Tenth International Cloud Physics Conference*, Bad Homburg, Federal Republic of Germany, 479–481.

Heymsfield, A. J., A. R. Jameson, and H. W. Frank, 1980: Hail growth mechanisms in a Colorado storm. Part II: Hail formation processes. *J. Atmos. Sci.*, **37**, 1779–1813.

Heymsfield, G. M., 1979: Doppler-radar study of a warm frontal region. *J. Atmos. Sci.*, **36**, 2093–2107.

Hildebrand, F. B., 1976: *Advanced Calculus for Applications*. Prentice-Hall, Englewood Cliffs, NJ, 733 pp.

Hill, T. A., and T. W. Choularton, 1986: A model of the development of the droplet spectrum in a growing cumulus turret. *Quart. J. Roy. Met. Soc.*, **112**, 531–554.

Hobbs, P. V., 1973b: Ice in the atmosphere: A review of the present position. *Physics and Chemistry of Ice* (E. Whalley, S. J. Jones, and L. W. Gold, Eds.), Royal Society of Canada, Ottawa, 308–319.

Hobbs, P.V., 1974: *Ice Physics*. Oxford Press, Bristol, 837 pp.

Hobbs, P. V., 1981: The Seattle workshop on extratropical cyclones: A call for a national cyclone project. *Bull. Amer. Meteor. Soc.*, **62**, 244–254.

Hobbs, P. V., 1985: Holes in clouds: A case of scientific amnesia. *Weatherwise*, **38**, 254–258.

Hobbs, P. V., 1989: Research on clouds and precipitation past, present, and future. *Bull. Amer. Meteor. Soc.*, **70**, 282–285.

Hobbs, P. V., and K. R. Biswas, 1979: The cellular structure of narrow cold-frontal rainbands. *Quart. J. Roy. Met. Soc.*, **105**, 723–727.

Hobbs, P. V., and O. G. Persson, 1982: The mesoscale and microscale structure and organization of clouds and precipitation in midlatitude cyclones. Part V: The substructure of narrow cold-frontal rainbands. *J. Atmos. Sci.*, **39**, 280–295.

Hobbs, P. V., and A. L. Rangno, 1985: Ice particle concentrations in clouds. *J. Atmos. Sci.*, **42**, 2523–2549.

Hobbs, P. V., and A. L. Rangno, 1990: Rapid development of high ice particle concentrations in small polar maritime cumuliform clouds. *J. Atmos. Sci.*, **47**, 2710–2722.

Hobbs, P. V., R. C. Easter, and A. B. Fraser, 1973: A theoretical study of the flow of air and fallout of solid precipitation over mountainous terrain. Part II: Microphysics. *J. Atmos. Sci.*, **30**, 813–823.

Hobbs, P. V., T. J. Matejka, P. H. Herzegh, J. D. Locatelli, and R. A. Houze, Jr., 1980: The mesoscale and microscale structure and organization of clouds and precipitation in midlatitude cyclones. I: A case study of a cold front. *J. Atmos. Sci.*, **37**, 568–596.

Holton, J. R., 1992: *An Introduction to Dynamic Meteorology*, 3rd ed. Academic Press, New York, 511 pp.

Hoskins, B. J., 1974: The role of potential vorticity in symmetric stability and instability. *Quart. J. Roy. Met. Soc.*, **100**, 480–482.

Hoskins, B. J., 1975: The geostrophic momentum approximation and the semi-geostrophic equations. *J. Atmos. Sci.*, **32**, 233–242.

Hoskins, B. J., 1982: The mathematical theory of frontogenesis. *Annu. Rev. Fluid Mech.*, **14**, 131–151.

Hoskins, B. J., 1990: Theory of extratropical cyclones. *Extratropical Cyclones: The Erik Palmén Memorial Volume* (C. W. Newton and E. O. Holopainen, Eds.), American Meteorological Society, Boston, 64–80.

Hoskins, B. J., and F. P. Bretherton, 1972: Atmospheric frontogenesis models: Mathematical formulation and solution. *J. Atmos. Sci.*, **29**, 11–37.

Hoskins, B. J., and M. A. Pedder, 1980: The diagnosis of middle latitude synoptic development. *Quart. J. Roy. Met. Soc.*, **106**, 707–719.

Houghton, H. G., 1950: A preliminary quantitative analysis of precipitation mechanisms. *J. Meteor.*, **7**, 363–369.

Houghton, H. G., 1968: On precipitation mechanisms and their artificial modification. *J. Appl. Meteor.*, **7**, 851–859.

Houze, R. A., Jr., 1977: Structure and dynamics of a tropical squall-line system. *Mon. Wea. Rev.*, **105**, 1540–1567.

Houze, R. A., Jr., 1981: Structures of atmospheric precipitation systems—A global survey. *Radio Science*, **16**, 671–689.

Houze, R. A., Jr., 1989: Observed structure of mesoscale convective systems and implications for large-scale heating. *Quart. J. Roy. Met. Soc.*, **115**, 425–461.

Houze, R. A., Jr., and C.-P. Cheng, 1977: Radar characteristics of tropical convection observed during GATE: Mean properties and trends over the summer season. *Mon. Wea. Rev.*, **105**, 964–980.

Houze, R. A., Jr., and D. D. Churchill, 1987: Mesoscale organization and cloud microphysics in a Bay of Bengal depression. *J. Atmos. Sci.*, **44**, 1845–1867.

Houze, R. A., Jr., and P. V. Hobbs, 1982: Organization and structure of precipitating cloud systems. *Adv. Geophys.*, **24**, 225–315.

Houze, R. A., Jr., J. D. Locatelli, and P. V. Hobbs, 1976: Dynamics and cloud microphysics of the rainbands in an occluded frontal system. *J. Atmos. Sci*, **33**, 1921–1936.

Houze, R. A., Jr., P. V. Hobbs, P. H. Herzegh, and D. B. Parsons, 1979: Size distributions of precipitation particles in frontal clouds. *J. Atmos. Sci.*, **36**, 156–162.

Houze, R. A., Jr., S. A. Rutledge, T. J. Matejka, and P. V. Hobbs, 1981: The mesoscale and microscale structure and organization of clouds and precipitation in midlatitude cyclones. III: Air motions and precipitation growth in a warm-frontal rainband. *J. Atmos. Sci.*, **38**, 639–649.

Houze, R. A., Jr., S. A. Rutledge, M. I. Biggerstaff, and B. F. Smull, 1989: Interpretation of Doppler weather-radar displays in midlatitude mesoscale convective systems. *Bull. Amer. Meteor. Soc.*, **70**, 608–619.

Houze, R. A., Jr., B. F. Smull, and P. Dodge, 1990: Mesoscale organization of springtime rainstorms in Oklahoma. *Mon. Wea. Rev.*, **117**, 613–654.

Huschke, R. E., 1959: *Glossary of Meteorology*. American Meteorological Society, Boston, 638 pp.

Huschke, R., 1969: Arctic cloud statistics from "air-calibrated" surface weather observations. RAND Corp., RM-6173-PR.

International Commission for the Study of Clouds, 1932: *International Atlas of Clouds and Study of the Sky*, Vol. I, *General Atlas*. Paris, 106 pp.

International Commission for the Study of Clouds, 1932: *International Atlas of Clouds and Study of the Sky*, Vol. II, *Atlas of Tropical Clouds*. Paris, 27 pp.

Itoh, Y., and S. Ohta, 1967: *Cloud Atlas: An Artist's View of Living Cloud*. Chijin Shokan Co. Ltd., Tokyo, 71 pp.

Jameson, A. R., and D. B. Johnson, 1990: Cloud microphysics and radar. *Radar in Meteorology* (D. Atlas, Ed.), American Meteorological Society, Boston, 323–340.

Janowiak, J. E., and P. A. Arkin, 1991: Rainfall variations in the tropics during 1986–1989, as estimated from observations of cloud-top temperature. *J. Geophys. Res. Oceans*, **96**, supplement, 3359–3373.

Jensen, J. B., 1985: *Turbulent mixing, droplet spectral evolution and dynamics of warm cumulus clouds*. Ph.D. dissertation, Department of Civil Engineering, University of Washington, 185 pp.

Johnson, R. H., and P. J. Hamilton, 1988: The relationship of surface pressure features to the precipitation and airflow structure of an intense midlatitude squall line. *Mon. Wea. Rev.*, **116**, 1444–1472.

Johnson, R. H., W. A. Gallus, Jr., and M. D. Vescio, 1990: Near-tropopause vertical motion within the trailing-stratiform region of a midlatitude squall line. *J. Atmos. Sci.*, **47**, 2200–2210.

Jorgensen, D. P., 1984: Mesoscale and convective-scale characteristics of mature hurricanes. Part II: Inner-core structure of Hurricane Allen (1980). *J. Atmos. Sci*, **41**, 1287–1311.

Jorgensen, D. P., E. J. Zipser, and M. LeMone, 1985: Vertical motions in intense hurricanes. *J. Atmos. Sci.*, **42**, 839–856.

Joss, J., and C. G. Collier, 1991: An electronically scanned antenna for weather radar. *Preprints, 25th Conference on Radar Meteorology*, Paris, American Meteorological Society, Boston, 748–751.

Joss, J., and A. Waldvogel, 1990: Precipitation measurement and hydrology. *Radar in Meteorology* (D. Atlas, Ed.), American Meteorological Society, Boston, 577–597.

Kelley, K. W., 1988: *The Home Planet*. Addison-Wesley, Reading, MA, 160 pp.

Kennedy, P. J. (Ed.), 1982: ALPEX Aircraft Atlas. National Center for Atmospheric Research, Boulder, CO, 226 pp.

Kessinger, C. J., M. Hjelmfelt, and J. Wilson, 1983: Low-level microburst wind structure using Doppler radar and PAM data. Preprints, *21st Radar Meteorology Conf.*, Edmonton, American Meteorological Society, Boston, 609–615.

Kessinger, C. J., P. S. Ray, and C. E. Hane, 1987: The Oklahoma squall line of 19 May 1977. Part I: A multiple-Doppler analysis of convective and stratiform structure. *J. Atmos. Sci.*, **44**, 2840–2864.

Kessinger, C. J., D. B. Parsons, and J. W. Wilson, 1988: Observations of a storm containing misocyclones, downbursts, and horizontal vortex circulations. *Mon. Wea. Rev.*, **116**, 1959–1982.

Kessler, E., 1969: On the distribution and continuity of water substance in atmospheric circulations. *Meteor. Monogr.*, **10**, No. 32, 84 pp.

Kessler, E., 1974: Model of precipitation and vertical air currents. *Tellus*, **26**, 519–542.

Keyser, D., 1986: Atmospheric fronts: An observational perspective. *Mesoscale Meteorology and Forecasting* (P. S. Ray, Ed.), American Meteorological Society, Boston, 216–258.

Klemp, J. B., 1987: Dynamics of tornadic thunderstorms. *Annu. Rev. Fluid. Mech.*, **19**, 369–402.

Klemp, J. B., and R. B. Wilhelmson, 1978a: The simulation of three-dimensional convective storm dynamics. *J. Atmos. Sci.*, **35**, 1070–1096.

Klemp, J. B., and R. B. Wilhelmson, 1978b: Simulations of right- and left-moving thunderstorms produced through storm splitting. *J. Atmos. Sci.*, **35**, 1097–1110.

Knight, D. J., and P. V. Hobbs, 1988: The mesoscale and microscale structure and organization of clouds and precipitation in midlatitude cyclones. Part XV: A numerical modeling study of frontogenesis and cold-frontal rainbands. *J. Atmos. Sci.*, **45**, 915–930.

Koch, S. E., 1984: The role of an apparent mesoscale frontogenetical circulation in squall line initiation. *Mon. Wea. Rev.*, **112**, 2090–2111.

Krehbiel, P. R., 1986: The electrical structure of thunderstorms. *The Earth's Electrical Environment*. National Research Council, Washington, DC, 263 pp.

Kreitzberg, C. W., 1963: *The structure of occlusions as determined from serial ascents and vertically directed radar*. Ph.D. thesis, Department of Atmospheric Sciences, University of Washington, Seattle, 164 pp.

Kreitzberg, C. W., and H. A. Brown, 1970: Mesoscale weather systems within an occlusion. *J. Appl. Meteor.*, **9**, 417–432.

Kreitzberg, C. W., and D. J. Perkey, 1977: Release of potential instability. Part II: The mechanism of convective/mesoscale interactions. *J. Atmos. Sci.*, **34**, 1569–1595.

Kuettner, J. P., 1947: Der Segelflug in Aufwindstrassen. *Schweizer Aero Revue*, **24**, 480.

Kuettner, J. P., 1959: The band structure of the atmosphere. *Tellus*, **11**, 267–294.

Kuettner, J. P., 1971: Cloud bands in the earth's atmosphere: Observations and theory. *Tellus*, **23**, 404–425.

Kuo, Y.-H., R. J. Reed, and S. Low-Nam, 1992: Thermal structure and airflow of a model simulation of an occluded marine cyclone. *Mon. Wea. Rev.*, **120**, 2280–2247.

Lafore, J.-P., and M. W. Moncrieff, 1989: A numerical investigation of the organization and interaction of the convective and stratiform regions of tropical squall lines. *J. Atmos. Sci.*, **46**, 521–544.

Lamb, H., Sir, 1932: *Hydrodynamics*. Dover Publications, New York, 738 pp.

Leary, C. A., and R. A. Houze, Jr., 1979a: The structure and evolution of convection in a tropical cloud cluster. *J. Atmos. Sci.*, **36**, 437–457.

Leary, C. A., and R. A. Houze, Jr., 1979b: Melting and evaporation of hydrometeors in precipitation from the anvil clouds of deep tropical convection. *J. Atmos. Sci.*, **36**, 669–679.

Leary, C. A., and E. N. Rappaport, 1987: The life cycle and internal structure of a mesoscale convective complex. *Mon. Wea. Rev.*, **115**, 1503–1527.

Lemon, L. R., and C. A. Doswell, III, 1979: Severe thunderstorm evolution and mesoscyclone structure as related to tornadogenesis. *Mon. Wea. Rev.*, **107**, 1184–1197.

LeMone, M. A., 1973: The structure and dynamics of horizontal roll vortices in the planetary boundary layer. *J. Atmos. Sci.*, **30**, 1077–1091.

LeMone, M. A., G. M. Barnes, and E. J. Zipser, 1984: Momentum flux by lines of cumulonimbus over the tropical oceans. *J. Atmos. Sci.*, **41**, 1914–1932.

Levine, J., 1959: Spherical vortex theory of bubble-like motion in cumulus clouds. *J. Meteor.*, **16**, 653–662.

Lewellen, W. S., 1976: Theoretical models of the tornado vortex. *Proceedings of the Symposium on Tornadoes: Assessment of Knowledge and Implications for Man*. Institute for Disaster Research, Texas Tech University, Lubbock, 107–143.

Lhermitte, R. M., 1970: Dual-Doppler radar observation of convective storm circulation. *Preprints, 14th Radar Meteorology Conference*, Tucson, American Meteorological Society, Boston, 139–144.

Lighthill, M. J., Sir, 1978: *Waves in Fluids*. Cambridge University Press, Cambridge, 504 pp.

Lilly, D. K., 1968: Models of cloud-topped mixed layers under strong conversion. *Quart. J. Roy. Met. Soc.*, **94**, 292–309.

Lilly, D. K., 1979: The dynamical structure and evolution of thunderstorms and squall lines. *Annu. Rev. Earth Planet. Sci.*, **7**, 117–171.

Lilly, D. K., 1986: Instabilities. *Mesoscale Meteorology and Forecasting* (P. S. Ray, Ed.), American Meteorological Society, Boston, 259–271.

Lilly, D. K., 1988: Cirrus outflow dynamics. *J. Atmos. Sci.*, **45**, 1594–1605.

Lin, Y.-L., R. D. Farley, and H. D. Orville, 1983: Bulk parameterization of the snow field in a cloud model. *J. Clim. Appl. Meteor.*, **22**, 1065–1092.

Lindzen, R. S., and K. K. Tung, 1976: Banded convective activity and ducted gravity waves. *Mon. Wea. Rev.*, **104**, 1602–1607.

Liou, K.-N., 1980: *An Introduction to Atmospheric Radiation*. Academic Press, New York, 404 pp.

Liou, K.-N., 1986: Influence of cirrus clouds on weather and climate processes. *Mon. Wea. Rev.*, **114**, 1167–1199.

Locatelli, J. D., and P. V. Hobbs, 1974: Fall speeds and masses of solid precipitation particles. *J. Geophys. Res.*, **79**, 2185–2197.

Locatelli, J. D., P. V. Hobbs, and K. R. Biswas, 1983: Precipitation from stratocumulus clouds affected by fallstreaks and artificial seeding. *J. Clim. Appl. Meteor.*, **22**, 1393–1403.

Locatelli, J. D., J. E. Martin, and P. V. Hobbs, 1992: The structure and propagation of a wide cold-frontal rainband and their relationship to frontal topography. *Preprints, Fifth Conference on Mesoscale Processes*, Atlanta, GA, American Meteorological Society, Boston, 192–196.

López, R. E., 1977: The lognormal distribution and cumulus cloud populations. *Mon. Wea. Rev.*, **105**, 865–872.

López, R. E., 1978: Internal structure and development processes of C-scale aggregates of cumulus clouds. *Mon. Wea. Rev.*, **106**, 1488–1494.

Lord, S. J., H. E. Willoughby, and J. M. Piotrowicz, 1984: Role of parameterized ice-phase microphysics in an axisymmetric nonhydrostatic tropical cyclone model. *J. Atmos. Sci.*, **41**, 2836–2848.

Ludlam, F. H., and R. S. Scorer, 1953: Convection in the atmosphere. *Quart. J. Roy. Met. Soc.*, **79**, 94–103.

Maddox, R. A., 1976: An evaluation of tornado proximity wind and stability data. *Mon. Wea. Rev.*, **104**, 133–142.

Maddox, R. A., 1980: A satellite-based study of midlatitude mesoscale convective complexes. *Preprints, Eighth Conference on Weather Forecasting and Analysis*, Denver, CO, American Meteorological Society, Boston, 329–338.

Maddox, R. A., 1981: The structure and life cycle of midlatitude mesoscale convective complexes. Colorado State University, Atmos. Sci. Paper No. 36.

Maddox, R. A., D. M. Rodgers, and K. M. Howard, 1982: Mesoscale convective complexes over the United States in 1981: Annual summary. *Mon. Wea. Rev.*, **110**, 1501–1514.

Magono, C., and C. Lee, 1966: Meteorological classification of natural snow crystals. *J. Fac. Sci. Hokkaido Univ.*, *Ser. VII*, **2**, 321–325.

Malkus, J. S., and R. S. Scorer, 1955: The erosion of cumulus towers. *J. Meteor.*, **12**, 43–57.

Mapes, B. E., and R. A. Houze, Jr., 1992: An integrated view of the 1987 Australian monsoon and its mesoscale convective systems. Part I: Horizontal structure. *Quart. J. Roy. Met. Soc.*, **118**, 927–963.

Mapes, B. E., and R. A. Houze, Jr., 1993a: An integrated view of the 1987 Australian monsoon and its mesoscale convective systems. Part II: Vertical structure. *Quart. J. Roy. Met. Soc.*

Mapes, B. E., and R. A. Houze, Jr., 1993b: Cloud clusters and superclusters over the oceanic warm pool. *Mon. Wea. Rev.*, **121**, 1398–1415.

Marks, F. D., Jr., and R. A. Houze, Jr., 1987: Inner-core structure of Hurricane Alicia from airborne Doppler-radar observations. *J. Atmos. Sci.*, **44**, 1296–1317.

Marshall, J. S., 1953: Frontal precipitation and lightning observed by radar. *Can. J. Phys.*, **31**, 194–203.

Marshall, J. S., and W. M. Palmer, 1948: The distribution of raindrops with size. *J. Meteor.*, **5**, 165–166.

Martin, D. W., and A. J. Schreiner, 1981: Characteristics of West African and East Atlantic cloud clusters: A survey of GATE. *Mon. Wea. Rev.*, **109**, 1671–1688.

Mason, B. J., 1971: *The Physics of Clouds*, 2nd ed. Clarendon Press, Oxford, 671 pp.

Mass, C. F., 1981: Topographically forced convergence in western Washington state. *Mon. Wea. Rev.*, **109**, 1335–1347.

Matejka, T. J., 1980: *Mesoscale organization of cloud processes in extratropical cyclones*. Ph.D. dissertation, Dept. of Atmospheric Sciences, University of Washington, Seattle, 361 pp.

Matejka, T. J., and R. C. Srivastava, 1991: An improved version of the extended velocity–azimuth display analysis of single Doppler-radar data. *J. Ocean. Atmos. Tech.*, **8**, 453–466.

Matejka, T. J., P. V. Hobbs, and R. A. Houze, Jr., 1978: Microphysical and dynamical structure of the mesoscale cloud features in extratropical cyclones. *Proceedings, Conference on Cloud Physics and Atmospheric Electricity*, Issaquah, WA, American Meteorological Society, Boston, 292–299.

Matejka, T. J., R. A. Houze, Jr., and P. V. Hobbs, 1980: Microphysics and dynamics of clouds associated with mesoscale rainbands in tropical cyclones. *Quart. J. Roy. Met. Soc.*, **106**, 29–56.

McGinnigle, J. B., M. V. Young, and M. J. Bader, 1988: The development of instant occlusions in the North Atlantic. *Meteor. Mag.*, **117**, 325–341.

Meneghini, R., J. Eckerman, and D. Atlas, 1983: Determination of rain rate from spaceborne radar using measurements of total attenuation. *IEEE Trans. Geosci. Remote Sensing*, **GE-21**, 34–43.

Moncrieff, M. W., and J. S. A. Green, 1972: The propagation and transfer properties of steady, convective overturning in shear. *Quart. J. Roy. Met. Soc.*, **98**, 336–352.

Moore, C. B., 1976: Theories of thunderstorm electrification: Reply. *Quart J. Roy. Met. Soc.*, **102**, 225–240.

Morton, B. R., 1970: The physics of fire whirls. *Fire Research Abstr. Rev.*, **12**, 1–19.

Morton, B. R., Sir G. Taylor, and J. S. Turner, 1956: Turbulent gravitational convection from maintained and instantaneous sources. *Proc. R. Soc. London Ser. A.*, **235**, 1–23.

Nakaya, U., and T. Terada, 1935: Simultaneous observations of the mass, falling velocity, and form of individual snow crystals. Hokkaido University, Ser. II, **1**, 191–201.

National Meteorological Service of China, 1984: *The Cloud Atlas of China*. Gordon and Breach Scientific Publishers, New York, 336 pp.

National Weather Association, 1986a: *Satellite Imagery Interpretation for Forecasters*. Vol. I: *General Interpretation Synoptic Analysis* (P. S. Parke, Ed.). National Weather Association, Temple Hills, ISSN 0271-1044.

National Weather Association, 1986b: *Satellite Imagery Interpretation for Forecasters*. Vol. II: *Precipitation Convection* (P. S. Parke, Ed.). National Weather Association, Temple Hills, ISSN 0271-1044.

National Weather Association, 1986c: *Satellite Imagery Interpretation for Forecasters*. Vol. III: *Tropical Weather, Fog and Stratus, Aerosols, Winds and Turbulence, Glossary* (P. S. Parke, Ed.). National Weather Association, Temple Hills, ISSN 0271-1044.

Nicholls, M. E., 1987: A comparison of the results of a two-dimensional numerical simulation of a tropical squall line with observations. *Mon. Wea. Rev.*, **115**, 3055–3077.

Nicholls, S., 1984: The dynamics of stratocumulus: Aircraft observations and comparisons with a mixed-layer model. *Quart. J. Roy. Met. Soc.*, **110**, 783–820.

Nicholls, S., and J. D. Turton, 1986: An observational study of the structure of stratiform cloud sheets. Part II: Entrainment. *Quart. J. Roy. Met. Soc.*, **112**, 461–480.

Ogura, Y., and N. A. Phillips, 1962: Scale analysis of deep and shallow convection in the atmosphere. *J. Atmos. Sci.*, **19**, 173–179.

Ogura, Y., and T. Takahashi, 1971: Numerical simulation of the life cycle of a thunderstorm cell. *Mon. Wea. Rev.*, **99**, 895–911.

Ogura, Y., and T. Takahashi, 1973: The development of warm rain in a cumulus cloud. *J. Atmos. Sci.*, **30**, 262–277.

Orlanski, I., 1975: A rational subdivision of scales for atmospheric processes. *Bull. Amer. Meteor. Soc.*, **56**, 527–530.

Page, R. M., 1962: *The Origins of Radar*. Doubleday and Company, New York, 196 pp.

Palmén, E., and C. W. Newton, 1969: *Atmospheric Circulation Systems: Their Structure and Physical Interpretation*. Academic Press, New York, 603 pp.

Paluch, I. R., 1979: The entrainment mechanism in Colorado cumuli. *J. Atmos. Sci.*, **36**, 2467–2478.

Panofsky, H. A., and J. A. Dutton, 1984: *Atmospheric Turbulence: Models and Methods for Engineering Applications*. John Wiley and Sons, New York, 397 pp.

Pellew, A., and R. V. Southwell, 1940: On maintained convective motion in a fluid heated from below. *Proc. Roy. Soc. London, Ser. A.*, **176**, 312–343.

Petterssen, S., 1956: *Weather Analysis and Forecasting*. Vol. I. McGraw-Hill Book Company, New York, 428 pp.

Probert-Jones, J. R., 1962: The radar equation in meteorology. *Quart. J. Roy. Met. Soc.*, **88**, 485–495.

Proctor, F. H., 1988: Numerical simulations of an isolated microburst. Part I: Dynamics and structure. *J. Atmos. Sci.*, **45**, 3137–3160.

Pruppacher, H. R., and J. D. Klett, 1978: *Microphysics of Clouds and Precipitation*. D. Reidel Publishers, Dordrecht, 714 pp.

Purdom, J. F. W., 1973: Meso-highs and satellite imagery. *Mon. Wea. Rev.*, **101**, 180–181.

Purdom, J. F. W., 1979: The development and evolution of deep convection. *Preprints, 11th Conference on Severe Local Storms*, Kansas City, American Meteorological Society, Boston, 143–150.

Purdom, J. F. W., and K. Marcus, 1982: Thunderstorm trigger mechanisms over the southeast United States. *Preprints, 12th Conference on Severe Local Storms*, San Antonio, American Meteorological Society, Boston, 487–488.

Raga, G., 1989: *Characteristics of cumulus bandclouds off the east coast of Hawaii*. Ph.D. dissertation, Dept. of Atmospheric Sciences, University of Washington, Seattle, 151 pp.

Randall, D. A., J. A. Coakley, Jr., C. W. Fairall, R. A. Kropfli, and D. H. Lenschow, 1984: Outlook for research on subtropical marine stratiform clouds. *Bull. Amer. Meteor. Soc.*, **65**, 1290–1301.

Rangno, A. L., and P. V. Hobbs, 1988: Criteria for significant concentrations of ice particles in cumulus clouds. *Atmos. Res.*, **22**, 1–13.

Rangno, A. L., and P. V. Hobbs, 1991: Ice particle concentrations and precipitation development in small polar maritime cumuliform clouds. *Quart. J. Roy. Meteor. Soc.*, **117**, 207–241.

Ray, P. S., 1990: Convective dynamics. *Radar in Meteorology* (D. Atlas, Ed.), American Meteorological Society, Boston, 348–390.

Rayleigh, Lord, 1916: Convection currents in a horizontal layer of fluid. *Phil. Mag.*, **32**, 531–546.

Raymond, D. J., and A. M. Blyth, 1986: A stochastic mixing model for nonprecipitating cumulus clouds. *J. Atmos. Sci.*, **43**, 2708–2718.

Redelsperger, J.-L., and J.-P. Lafore, 1988: A three-dimensional simulation of a tropical squall line: Convective organization and thermodynamic vertical transport. *J. Atmos. Sci.*, **45**, 1334–1356.

Reed, R. J., and W. Blier, 1986: A case study of comma cloud development in the eastern Pacific. *Mon. Wea. Rev.*, **114**, 1681–1695.

Rinehart, R. E., 1991: *Radar for Meteorologists*, 2nd ed. Available from the Department of Atmospheric Sciences, University of North Dakota, Grand Forks, ND 58202-8216.

Rodgers, D. M., K. W. Howard, and E. C. Johnston, 1983: Mesoscale convective complexes over the United States in 1982: Annual summary. *Mon. Wea. Rev.*, **111**, 2363–2369.

Rodgers, D. M., M. J. Magnano, and J. H. Arns, 1985: Mesoscale convective complexes over the United States in 1983: Annual summary. *Mon. Wea. Rev.*, **113**, 888–901.

Rodi, A. R., 1978: Small-scale variability of the cloud-droplet spectrum in cumulus clouds. *Proceedings, Conference on Cloud Physics and Atmospheric Electricity*, Issaquah, WA, American Meteorological Society, Boston, 88–91.

Rodi, A. R., 1981: *The study of the fine-scale structure of cumulus clouds*. Ph.D. thesis, University of Wyoming, Laramie, 328 pp.

Rogers, R. R., 1976: *A Short Course in Cloud Physics*. Pergamon Press, Oxford, 224 pp.

Rogers, R. R., and M. K. Yau, 1989: *A Short Course in Cloud Physics*, 3rd ed. Pergamon Press, Oxford, 293 pp.

Roll, H. U., 1965: *Physics of the Marine Atmosphere*. Academic Press, New York, 426 pp.

Rosenfeld, D., D. Atlas, and D. A. Short, 1990: The estimation of rainfall by area integrals. Part 2: The height–area rain threshold (HART) method. *J. Geophys. Res.*, **95**, 2161–2176.

Rosenhead, L., 1931: The formation of vortices from a surface of discontinuity. *Proc. Roy. Soc. London, Ser. A.*, **134**, 170–192.

Rott, N., 1958: On the viscous core of a line vortex. *Z. Angew. Math. Physik*, **96**, 543–553.

Röttger, J., and M. F. Larsen, 1990: UHF/VHF radar techniques for atmospheric research and wind profiler applications. *Radar In Meteorology* (D. Atlas, Ed.), American Meteorological Society, Boston, 235–281.

Rotunno, R., 1977: Numerical simulation of a laboratory vortex. *J. Atmos. Sci.*, **34**, 1942–1956.

Rotunno, R., 1979: A study in tornado-like vortex dynamics. *J. Atmos. Sci.*, **36**, 140–155.

Rotunno, R., 1981: On the evolution of thunderstorm rotation. *Mon. Wea. Rev.*, **109**, 171–180.

Rotunno, R., 1986: Tornadoes and tornadogenesis. *Mesoscale Meteorology and Forecasting* (P. S. Ray, Ed.), American Meteorological Society, Boston, 414–436.

Rotunno, R., and K. A. Emanuel, 1987: An air–sea interaction theory for tropical cyclones. Part II: An evolutionary study using a hydrostatic axisymmetric numerical model. *J. Atmos. Sci.*, **44**, 543–561.

Rotunno, R., and J. B. Klemp, 1982: The influence of shear-induced pressure gradient on thunderstorm motion. *Mon. Wea. Rev.*, **110**, 136–151.

Rotunno, R., J. B. Klemp, and M. L. Weisman, 1988: A theory for strong, long-lived squall lines. *J. Atmos. Sci.*, **45**, 463–485.

Roux, F., 1985: Retrieval of thermodynamic fields from multiple Doppler-radar data using the equations of motion and the thermodynamic equation. *Mon. Wea. Rev.*, **113**, 2142–2157.

Roux, F., 1988: The west African squall line observed on 23 June 1981 during COPT 81: Kinematics and thermodynamics of the convective region. *J. Atmos. Sci.*, **45**, 406–426.

Rust, W. D., W. L. Taylor, and D. R. MacGorman, 1981: Research on electrical properties of severe thunderstorms in the Great Plains. *Bull. Amer. Meteor. Soc.*, **62**, 1286–1293.

Rutledge, S. A., and P. V. Hobbs, 1983: The mesoscale and microscale structure and organization of clouds and precipitation in midlatitude cyclones. VIII: A model for the feeder–seeder process in warm frontal rainbands. *J. Atmos. Sci.*, **40**, 1185–1206.

Rutledge, S. A., and P. V. Hobbs, 1984: The mesoscale and microscale structure and organization of clouds and precipitation in midlatitude cyclones. XII: A diagnostic modeling study of precipitation development in narrow, cold-frontal rainbands. *J. Atmos. Sci.*, **41**, 2949–2972.

Rutledge, S. A., and R. A. Houze, Jr., 1987: A diagnostic modeling study of the trailing stratiform region of a midlatitude squall line. *J. Atmos. Sci.*, **44**, 2640–2656.

Rutledge, S. A., and D. R. MacGorman, 1988: Cloud-to-ground lightning activity in the 10-11 June 1985 convective system observed during the Oklahoma–Kansas PRE-STORM project. *Mon. Wea. Rev.*, **116**, 1393–1408.

Rutledge, S. A., C. Lu, and D. R. MacGorman, 1990: Positive cloud-to-ground lightning in mesoscale convective systems. *J. Atmos. Sci.*, **47**, 2085–2100.

Saucier, W. J., 1955: *Principles of Meteorological Analysis*. University of Chicago Press, Chicago, 438 pp.

Saunders, C., W. Keith, and R. Mitzeva, 1991: The effect of liquid water on thunderstorm charging. *J. Geophys. Res.*, **96**, 11,007–11,017.

Sawyer, J. S., 1956: The vertical circulation at meteorological fronts and its relation to frontogenesis. *Proc. Roy. Soc. London*, **A234**, 346–362.

Schaefer, V. J., and J. A. Day, 1981: *A Field Guide to the Atmosphere*. Houghton Mifflin, Boston, 359 pp.

Schär, C. J., 1989: *Dynamische Aspekte der aussertropischen Zyklogenese. Theorie und numerische Simulation in Limit der balancierten Strömungssysteme*. Ph.D. dissertation, nr. 8845, Eidgenössische Technische Hochschule, Zürich, 241 pp.

Schär, C. J., and H. Wernli, 1993: Structure and evolution of an isolated semigeostrophic cyclone. *Quart. J. Roy. Met. Soc.*, **119**, 57–90.

Schemm, C. E., and F. B. Lipps, 1976: Some results from a simplified three-dimensional numerical model of atmospheric turbulence. *J. Atmos. Sci.*, **33**, 1021–1041.

Schlesinger, R. E., 1975: A three-dimensional numerical model of an isolated deep convective cloud: Preliminary results. *J. Atmos. Sci.*, **32**, 934–957.

Schubert, W. H., and J. J. Hack, 1982: Inertial stability and tropical cyclone development. *Mon. Wea. Rev.*, **39**, 1687–1697.

Schuur, T. J., B. F. Smull, W. D. Rust, and T. C. Marshall, 1991; Electrical and kinematic structure of the stratiform precipitation region trailing an Oklahoma squall line. *J. Atmos. Sci.*, **48**, 825–842.

Scorer, R. S., 1957: Experiments on convection of isolated masses of buoyant fluid. *J. Fluid Mech.*, **2**, 583–594.

Scorer, R. S., 1958: *Natural Aerodynamics*. Pergamon Press, New York, 312 pp.

Scorer, R. S., 1972: *Clouds of the World: A Complete Colour Encyclopedia*. David and Charles Publishers, Newton Abbot, 176 pp.

Scorer, R. S., and A. Verkask, 1989: *Spacious Skies.* David and Charles Publishers, London, 192 pp.

Scott, B. C., and P. V. Hobbs, 1977: A theoretical study of the evolution of mixed-phase cumulus clouds. *J. Atmos. Sci.,* **34,** 812–826.

Scott, D. F. S. (Ed.), 1976: *Luke Howard (1772–1864): His Correspondence with Goethe and His Continental Journey of 1816.* William Sessions Limited, York, 99 pp.

Sellers, W.D., 1965: *Physical Climatology.* University of Chicago Press, Chicago, 272 pp.

Shapiro, L. J., and H. E. Willoughby, 1982: The response of balanced hurricanes to local sources of heat and momentum. *J. Atmos. Sci.,* **39,** 378–394.

Shapiro, M. A., and D. A. Keyser, 1990: *Extratropical Cyclones: The Erik Palmén Memorial Volume* (C. W. Newton and E. O. Holopainen, Eds.), American Meteorological Society, Boston, 167–191.

Shapiro, M. A., T. Hampel, D. Rotzoll, and F. Mosher, 1985: The frontal hydraulic head: A micro-alpha scale (~1 km) triggering mechanism for mesoconvective weather systems. *Mon. Wea. Rev.,* **113,** 1150–1165.

Sharman, R. D., and M. G. Wurtele, 1983: Ship waves and lee waves. *J. Atmos. Sci.,* **40,** 396–427.

Shy, S. S., 1990: *A chemically reacting, turbulent stratified interface.* Ph.D. dissertation, Dept. of Aeronautics and Astronautics, University of Washington, Seattle, 107 pp.

Siems, S. T., C. S. Bretherton, M. B. Baker, S. Shy, and R. E. Breidenthal, 1990: Buoyancy reversal and cloud top entrainment instability. *Quart. J. Roy. Met. Soc.,* **116,** 705–739.

Sievers, U., R. Forkel, and W. Zdunkowski, 1983: Transport equations for heat and moisture in the soil and their application to boundary-layer problems. *Beitr. Phys. Atmos.,* **56,** 58–83.

Silverman, B. A., 1970: An Eulerian model of warm fog modification. *Proceedings of the Second National Conference on Weather Modification,* Santa Barbara, CA, American Meteorological Society, Boston, 91–95.

Silverman, B. A., and M. Glass, 1973: A numerical simulation of warm cumulus clouds: Part I. Parameterized vs. non-parameterized microphysics. *J. Atmos. Sci.,* **30,** 1620–1637.

Simpson, J. E., 1969: A comparison between laboratory and atmospheric density currents. *Quart. J. Roy. Met. Soc.,* **95,** 758–765.

Simpson, J. E., 1987: *Gravity Currents in the Environment and the Laboratory.* John Wiley and Sons, New York, 244 pp.

Simpson, J. S., 1988: TRMM: A satellite mission to measure tropical rainfall. NASA, 94 pp.

Simpson, J. S., and G. van Helvoirt, 1980: GATE cloud–subcloud interactions examined using a three-dimensional cumulus model. *Contrib. Atmos. Phys.,* **53,** 106–134.

Simpson, J. S., and V. Wiggert, 1969: Models of a precipitating cumulus tower. *Mon. Wea. Rev.,* **97,** 471–489.

Simpson, J. S., and V. Wiggert, 1971: 1968 Florida cumulus seeding experiment: Numerical model results. *Mon. Wea. Rev.,* **99,** 87–118.

Simpson, J. S., N. E. Westcott, R. J. Clerman, and R. A. Pielke, 1980: On cumulus mergers. *Arch. Met. Geoph. Biokl.* Ser. A., **29,** 1–40.

Simpson, J. S., B. R. Morton, M. C. McCumber, and R. S. Penc, 1986: Observations and mechanisms of GATE waterspouts. *J. Atmos. Sci.,* **43,** 753–782.

Sinclair, P. C., and J. F. W. Purdom, 1982: Integration of research aircraft data and three-minute interval GOES data to study the genesis and development of deep convective storms. *Preprints, 12th Conference on Severe Local Storms,* San Antonio, American Meteorological Society, Boston, 269–271.

Sirmans, D., and B. Bumgarner, 1975: Numerical comparison of five mean frequency estimators. *J. Appl. Meteor.,* **14,** 991–1003.

Skolnik, M. I., 1980: *Introduction to Radar Systems.* McGraw-Hill Book Company, New York, 581 pp.

Smith, R. B., 1979: The influence of mountains on the atmosphere. *Adv. Geophys.,* **21,** 87–230.

Smolarkiewicz, P. K., and R. Rotunno, 1989: Low Froude number flow past three-dimensional obstacles. Part I: Baroclinically generated lee vortices. *J. Atmos. Sci.,* **46,** 1154–1164.

Smull, B. F., and R. A. Houze, Jr., 1987: Dual-Doppler radar analysis of a midlatitude squall line with a trailing region of stratiform rain. *J. Atmos. Sci.,* **44,** 2128–2148.

Soong, S.-T., 1974: Numerical simulation of warm rain development in an axisymmetric cloud model. *J. Atmos. Sci.,* **31,** 1232–1240.

Sorbjan, Z., 1989: *Structure of the Atmospheric Boundary Layer*. Prentice-Hall, Englewood Cliffs, NJ, 317 pp.

Srivastava, R. C., 1971: Size distribution of raindrops generated by their breakup and coalescence. *J. Atmos. Sci.*, **28**, 410–415.

Srivastava, R. C., 1985: A simple model of evaporatively driven downdraft application to microburst downdraft. *J. Atmos. Sci.*, **42**, 1004–1023.

Srivastava, R. C., 1987: A model of intense downdrafts driven by the melting and evaporation of precipitation. *J. Atmos Sci.*, **44**, 1752–1773.

Srivastava, R. C., T. J. Matejka, and T. J. Lorello, 1986: Doppler-radar study of the trailing anvil region associated with a squall line. *J. Atmos. Sci.*, **43**, 356–377.

Stage, S. A., and J. A. Businger, 1981a: A model for entrainment into the cloud-topped marine boundary layer. Part I: Model description and application to a cold-air outbreak episode. *J. Atmos. Sci.*, **38**, 2213–2229.

Stage, S. A., and J. A. Businger, 1981b: A model for entrainment into the cloud-topped marine boundary layer. Part II: Discussion of model behavior and comparison with other models. *J. Atmos. Sci.*, **38**, 2230–2242.

Starr, D. O'C., and S. K. Cox, 1985a: Cirrus clouds. Part I: A cirrus cloud model. *J. Atmos. Sci.*, **42**, 2663–2681.

Starr, D. O'C., and S. K. Cox, 1985b: Cirrus clouds. Part II: Numerical experiments on the formation and maintenance of cirrus. *J. Atmos. Sci.*, **42**, 2682–2694.

Stephens, G. L., 1984: The parameterization of radiation for numerical prediction and climate models. *Mon. Wea. Rev.*, **112**, 826–867.

Stommel, H., 1947: Entrainment of air into a cumulus cloud. Part I. *J. Appl. Meteor.*, **4**, 91–94.

Stull, R. B., 1988: *An Introduction to Boundary-Layer Meteorology*. Kluwer Academic Publishers, Dordrecht, 666 pp.

Sullivan, R. D., 1959: A two-cell vortex solution of the Navier–Stokes equations. *J. Aerosp. Sci.*, **26**, 767–768.

Sun, J., and F. Roux, 1988: Thermodynamic structure of the trailing-stratiform regions of two west African squall lines. *Ann. Geophysicae*, **6**, 659–670.

Sun, J., and F. Roux, 1989: Thermodynamics of a COPT 81 squall line retrieved from single-Doppler data. *Preprints, 24th Conference on Radar Meteorology*, Tallahassee, American Meteorological Society, Boston, 50–53.

Sun, J., and R. A. Houze, Jr., 1992: Validation of a thermodynamic retrieval technique by application to a simulated squall line with trailing-stratiform precipitation. *Mon. Wea. Rev.*, **120**, 1003–1018.

Süring, R., 1941: *Die Wolken*. Akademische verlagsgesellschaft Becker and Erler, Liepzig, 139 pp.

Sverdrup, H. U., M. W. Johnson, and R. H. Fleming, 1942: *The Oceans*. Prentice-Hall, Englewood Cliffs, NJ, 1087 pp.

Tag, P. M., D. B. Johnson, and E. E. Hindman II, 1970: Engineering fog-modification experiments by computer modeling. *Proceedings, Second National Conference on Weather Modification*, Santa Barbara, CA, 97–102.

Takahashi, T., 1978: Riming electrification as a charge generation mechanism in thunderstorms. *J. Atmos. Sci.* **35**, 1536–1548.

Takeda, T., 1971: Numerical simulation of a precipitating convective cloud: The formation of a long-lasting cloud. *J. Atmos. Sci.*, **28**, 350–376.

Tao, W.-K., and J. Simpson, 1989: Modeling study of a tropical squall-type convective line. *J. Atmos. Sci.*, **46**, 177–202.

Taylor, G., 1987: *A numerical investigation of sulfate production and deposition in midlatitude continental cumulus clouds*. Ph.D. dissertation, Dept. of Geophysics, University of Washington, 311 pp.

Thomson, J., 1881: On a changing tessellated structure in certain liquids. *Proc. Phil. Soc. Glasgow*, **XIiI**, 464–468.

Thorpe, A. J., and K. A. Emanuel, 1985: Frontogenesis in the presence of small stability to slantwise convection. *J. Atmos. Sci.*, **42**, 1809–1824.

Thorpe, A. J., and M. J. Miller, 1978: Numerical simulations showing the role of the downdraught in cumulonimbus motion and splitting. *Quart. J. Roy. Met. Soc.*, **104**, 873–893.

Thorpe, S. A., 1971: Experiments on the instability of stratified shear flows: Miscible fluids. *J. Fluid Mech.*, **46**, 299–319.

Tripoli, G. J., and W. R. Cotton, 1989a: Numerical study of an observed orogenic mesoscale convective system. Part I: Simulated genesis and comparison with observations. *Mon. Wea. Rev.*, **117**, 273–304.

Tripoli, G. J., and W. R. Cotton, 1989b: Numerical study of an observed orogenic mesoscale convective system. Part II: Analysis of governing dynamics. *Mon. Wea. Rev.*, **117**, 305–328.

Turner, J. S., 1962: The starting plume in neutral surroundings. *J. Fluid Mech.*, **13**, 356–368.

Turner, J. S., 1973: *Buoyancy Effects in Fluids*. Cambridge University Press, Cambridge, 368 pp.

Untersteiner, N., 1961: On the mass and heat budget of Arctic sea ice. *Arch. Met. Bioklim. Ser. A.*, **12**, 151–182.

van Delden, A. J., 1987: *On cumulus cloud patterns and the theory of shallow convection*. Ph.D. dissertation, Rijks Universitet, Utrecht, The Netherlands, 186 pp.

Velasco, I., and J. M. Fritsch, 1987: Mesoscale convective complexes in the Americas. *J. Geophys. Res.*, **92**, 9591–9613.

Wakimoto, R. M., 1982: The life cycle of the thunderstorm gust fronts as viewed with Doppler radar and rawinsonde data. *Mon. Wea. Rev.*, **110**, 1060–1082.

Wakimoto, R. M., and V. N. Bringi, 1988: Dual-polarization observations of microbursts associated with intense convection: The 20 July storm during the MIST Project. *Mon. Wea. Rev.*, **116**, 1521–1539.

Wakimoto, R. M., and J. W. Wilson, 1989: Non-supercell tornadoes. *Mon. Wea. Rev.*, **117**, 1113–1140.

Wallace, J. M., and P. V. Hobbs, 1977: *Atmospheric Science: An Introductory Survey*. Academic Press, New York, 467 pp.

Warner, C., J. Simpson, G. Van Helvoirt, D. W. Martin, and D. Suchman, 1980: Deep convection on Day 261 of GATE. *Mon. Wea. Rev.*, **108**, 169–194.

Warner, J., 1950: The microstructure of cumulus cloud. Part III. The nature of the updraft. *J. Atmos. Sci.*, **27**, 682–688.

Warner, J., 1969a: The microstructure of cumulus cloud. Part I: General features of the droplet spectrum. *J. Atmos. Sci.*, **26**, 1049–1059.

Warner, J., 1969b: The microstructure of cumulus cloud. Part II: The effect on droplet size distribution of the cloud nucleus spectrum and updraft velocity. *J. Atmos. Sci.*, **26**, 1272–1282.

Warner, J., 1970: On steady-state one-dimensional models of cumulus convection. *J. Atmos. Sci.*, **27**, 1035–1040.

Warren, S. G., C. J. Hahn, J. London, R. M. Chervin, and R. L. Jenne, 1988: *Global Distribution of Total Cloud Cover and Cloud Type Amounts Over the Ocean*. U.S. Dept. of Energy DOE/ER/60085-H1, NCAR Technical Notes NCAR/TN-317+STR, NTIS-PR-360.

Webster, P. J., and R. A. Houze, Jr., 1991: The Equatorial Mesoscale Experiment (EMEX): An overview. *Bull. Amer. Meteor. Soc.*, **72**, 1481–1505.

Weinstein, A. I., and P. B. MacCready, Jr., 1969: An isolated cumulus cloud modification project. *J. Appl. Meteor.*, **8**, 936–947.

Weisman, M. L., and J. B. Klemp, 1982: The dependence of numerically simulated convective storms on vertical wind shear and buoyancy. *Mon. Wea. Rev.*, **110**, 504–520.

Weisman, M. L., J. B. Klemp, and R. Rotunno, 1988: The structure and evolution of numerically simulated squall lines. *J. Atmos. Sci.*, **45**, 1990–2013.

Welch, R. M., M. G. Ravichandran, and S. K. Cox, 1986: Prediction of quasi-periodic oscillations in radiation fogs. Part I: Comparison of simple similarity approaches. *J. Atmos. Sci.* **43**, 633–651.

Wilhelmson, R., 1974: The life cycle of the thunderstorm in three dimensions. *J. Atmos. Sci.*, **31**, 1629–1651.

Wilhelmson, R., and Y. Ogura, 1972: The pressure perturbation and the numerical modeling of a cloud. *J. Atmos. Sci.*, **29**, 1295–1307.

Williams, E. R., 1988: The electrification of thunderstorms. *Sci. Amer.*, **269**, 88–99.

Williams, E. R., 1989: The tripole structure of thunderstorms. *J. Geophys. Res.*, **94**, 13151–13168.

Williams, E. R., M. E. Weber, and R. E. Orville, 1989: The relationship between lightning type and convective state of thunderclouds. *J. Geophys. Res.*, **94**, 13213–13220.

Williams, M., and R. A. Houze, Jr., 1987: Satellite-observed characteristics of winter monsoon cloud clusters. *Mon. Wea. Rev.*, **115**, 505–519.

Willoughby, H. E., 1988: The dynamics of the tropical hurricane core. *Aust. Meteor. Mag.*, **36**, 183–191.

Willoughby, H. E., J. A. Clos, and M. G. Shoreibah, 1982: Concentric eyes, secondary wind maxima, and the evolution of the hurricane vortex. *J. Atmos. Sci.*, **39**, 395–411.

Willoughby, H. E., H.-L. Lin, S. J. Lord, and J. M. Piotrowicz, 1984a: Hurricane structure and evolution as simulated by an axisymmetric nonhydrostatic numerical model. *J. Atmos. Sci.*, **41**, 1169–1186.

Willoughby, H. E., F. D. Marks, Jr., and R. J. Feinberg, 1984b: Stationary and propagating convective bands in asymmetric hurricanes. *J. Atmos. Sci.*, **41**, 3189–3211.

Wilson, J. W., R. D. Roberts, C. Kessinger, and J. McCarthy, 1984: Microburst wind structure and evaluation of Doppler radar for airport wind shear detection. *J. Clim. Appl. Meteor.*, **23**, 898–915.

Woodcock, A., 1942: Soaring over the open sea. *Scientific Monthly*, **55**, 226.

Woodward, B., 1959: The motion in and around thermals. *Quart. J. Roy. Met. Soc.*, **85**, 144–151.

World Meteorological Organization, 1956: *International Cloud Atlas*, Vol. I. Geneva, Switzerland, 155 pp.

World Meteorological Organization, 1956: *International Cloud Atlas*, Vol. II. Geneva, Switzerland, 224 pl., app.

World Meteorological Organization, 1969: *International Cloud Atlas, Abridged Atlas*. Geneva, Switzerland, 72 pl., app.

World Meteorological Organization, 1975: *Manual on the Observation of Clouds and Other Meteors*. Geneva, Switzerland, 155 pp.

World Meteorological Organization, 1987: *International Cloud Atlas*, Vol. II. Geneva, Switzerland, 196 pl., app.

Wurtele, M. G., 1953: The initial-value lee-wave problem for the isothermal atmosphere. Scientific Report No. 3, Sierra Wave Project, Contract No. AF 19 (122)-263, Air Force Cambridge Research Center, Cambridge, MA.

Wurtele, M. G., 1957: The three-dimensional lee wave. *Beitr. Phys. frei. Atmos.*, **29**, 242–252.

Yagi, T., 1969: On the relation between the shape of cirrus clouds and the static stability of the cloud level. Studies of cirrus clouds: Part IV. *J. Met. Soc. Japan*, **47**, 59–64.

Yagi, T., T. Hariyama, and C. Magono, 1968: On the shape and movement of cirrus uncinus clouds by the trigonometric method utilizing stereo photographs—Studies of cirrus clouds. Part I. *J. Met. Soc. Japan*, **46**, 266–271.

Zhang, D.-L., and J. M. Fritsch, 1987: Numerical simulation of the meso-β scale structure and evolution of the 1977 Johnstown flood. Part II: Inertially stable warm core vortex and the mesoscale convective complex. *J. Atmos. Sci.*, **44**, 2593–2612.

Zhang, D.-L., and J. M. Fritsch, 1988a: Numerical sensitivity experiments of varying model physics on the structure, evolution, and dynamics of two mesoscale convective systems. *J. Atmos. Sci.*, **45**, 261–293.

Zhang, D.-L., and J. M. Fritsch, 1988b: A numerical investigation of a convectively generated, inertially stable, extratropical warm-core mesovortex over land. Part I: Structure and evolution. *Mon. Wea. Rev.*, **116**, 2660–2687.

Zhang, D.-L., and K. Gao, 1989: Numerical simulation of an intense squall line during the 10-11 June 1985 PRE-STORM. Part II: Rear inflow, surface pressure perturbations, and stratiform precipitation. *Mon. Wea. Rev.*, **117**, 2067–2094.

Zipser, E. J., 1977: Mesoscale and convective-scale downdrafts as distinct components of squall-line circulation. *Mon. Wea. Rev.*, **105**, 1568–1589.

Index

International Geophysics Series

EDITED BY

RENATA DMOWSKA
Division of Applied Sciences
Harvard University
Cambridge, Massachusetts

JAMES R. HOLTON
Department of Atmospheric Sciences
University of Washington
Seattle, Washington

*Out of print.

ISBN 0-12-356880-3